Handbook of

Acoustics

Handbook of
Acoustics

Editors

M. Venkatesan
J.C. Kannan
P. Murugesan
V. Devabharathi

K S R Institute for Engineering and Technology
&
Acoustical Society of India

BLOOMSBURY
LONDON • NEW DELHI • NEW YORK • SYDNEY

First Published, 2013

BLOOMSBURY PUBLISHING INDIA PVT. LTD.
London New Delhi New York Sydney

ISBN: 978-93-82563-48-8

Published by Bloomsbury Publishing India Pvt. Ltd.
VISHRUT Building, DDA Complex, Building No. 3, Pocket C-6 & 7
Vasant Kunj, New Delhi 110 070

Laser Typeset by SHINU
A-5/56, Moti Nagar, New Delhi 110 015

Printed at Anvi Composers
1 DDA Market, Block A-1B, Pachim Vihar, New Delhi 110 063

Preface

I feel immense pleasure and pride to let you know that K S R Institute for Engineering and Technology has instigated to organize National Symposium on Acoustics (NSA 12) with an aspiration of providing wide knowledge and massive opening to the researchers and engineering students of our country.

Acoustical engineering is the branch of engineering which deals with sound and vibration. It is the application of acoustics, the science of sound and vibration, in technology. In general, Acoustical engineers are concerned with the handling and control of sound. The prime objective of acoustical engineering is to lessen unwanted sounds, which is referred to as noise control.

In addition to reducing unwanted sounds, acoustical engineers sometimes generate useful sounds or study sound waves to collect information. I strongly believe that this symposium will be an outstanding opportunity for the cheerful engineers to learn the latest developments in the field of Acoustics.

It is the right moment for every participant to realize the need for the hour and to make better use of the opportunity available at hand. This shall be the precise platform for the creative intellectuals to exhibit their skills and to merge the theories into action plan.

The task of this symposium has multiple roles to be played. An enthusiastic technical committee has reviewed 75 papers for presentation and publication in the proceedings of NSA 2012. Along with 15 invited talks are incorporated in the proceedings.

The following are the core areas under which the selected papers are categorized.

- Physical Acoustics
- Engineering and Applied Acoustics
- Speech, Hearing and Music
- Underwater Acoustics
- Architectural Acoustics
- Ultrasonics
- Medical and Bio Acoustics
- Quieter Technologies
- Vibration/Noise/Acoustic Emission Based Condition Monitoring
- Acoustic Sensors and Signal Processing
- Environmental Acoustics and Regulations
- Nano Technology in Acoustics

The financial assistance received from various Governments, Private organizations and industries are thankfully acknowledged. It is admiringly avowed appreciation to various committees for their diligent efforts and I deeply appreciate all for their efforts to make NSA 2012 a great success. Kudos on this outstanding accomplishment.

Dr. M. Venkatesan
Chairman—NSA 2012

Contents

Section–VII: Underwater Acoustics

Section–VIII: Ultrasonics

Section–X: Noise

Section–XI: Acoustic Emission Basedcondition Monitoring

Section–XII: Acoustics Sensor Signal Processing

Section–I

Invited Talks

Acoustic Techniques for Detection and Characterization of Defects, Deformation and Crack Growth and for Monitoring Structural Integrity of Components

T. Jayakumar, C.K. Mukhopadhyay and Anish Kumar

Metallurgy and Materials Group, Indira Gandhi Centre for Atomic Research, Kalpakkam, Tamil Nadu

E-mail: tjk@igcar.gov.in

ABSTRACT: *Non-destructive evaluation (NDE) techniques such as ultrasonic and acoustic emission are employed for detection and evaluation of defects, characterization of deformation and crack growth and for monitoring structural integrity of components. In ultrasonic testing (UT), high frequency sound waves are used for detection of defects/ microstructural changes in materials. On the other hand, acoustic emission technique (AET) is based on the generation of transient elastic waves by rapid release of energy from localized dynamic sources and is used for detecting dynamic changes occurring in materials. In this paper, an overview of the application of ultrasonic techniques for enhanced characterization and imaging of defects in structural materials and components will be discussed. The application of acoustic emission technique for detection and characterization of deformation and crack growth in materials and components, and for monitoring integrity of components will also be discussed.*

Ultrasonic technique has been applied for several first of its kind developments in the authors' laboratory in the field of enhanced characterization and imaging of defects in structural materials and components. Ultrasonic attenuation based C-scan imaging technique has been used for imaging of Type–IV cracking in ferritic steel weldments exposed to creep damage. Time-frequency based analysis of ultrasonic signals has been found to be useful for better understanding of the frequency dependent scattering in polycrystalline materials and for extracting the signals submerged in noise. Three dimensional imaging of defects and sampling phased array methodology have been carried out for characterization of defects. A new time of flight based methodology has been developed for precise measurement of depth of surface breaking cracks. Guided wave based ultrasonic techniques have been developed for detection of defects in hexcan sheath and hexcan welds of fuel subassemblies of Prototype Fast Breeder Reactor (PFBR), and for defect detection in thick stainless steel plates.

AET has been used for understanding deformation and crack growth in materials and components. Acoustic emission (AE) generated during tensile deformation of smooth and notched specimens of near Alpha titanium alloy has been useful for characterizing yielding in this material and revealing localized deformation at notch tip. AET has been used for understanding deformation and crack growth during hydrotesting of pressure vessels of stainless steel with artificial defects. Different parameters of the AE signals and temperature rise in the vicinity of the notch measured by infrared thermography have been found to have good correlation.

A new approach for determining initiation fracture toughness has been developed based on AE signals and has been validated on SA333 Gr.6 steel. The fracture toughness values determined by AET are in agreement with those obtained from conventional approaches of fracture toughness determination.

AET has also been used for the study of crack growth behavior in plain and fly ash concretes with different curing periods during uniaxial compression loading.

Acoustic emission technique has been used to monitor integrity of components which include monitoring cutting tools during machining operation and monitoring structural integrity of pressurized components during hydrotesting. The progress of tool wear during turning of a Al alloy metal matrix composite has been studied. It has been shown that peak amplitude distribution of AE signals, statistical analysis of the AE signals based on assumed β distribution, and uncertainties associated with the measurement of AE signals can be used in a complimentary manner for comprehensive evaluation of tool wear and thus integrity of the cutting tool during the machining operation.

AET has also been applied for monitoring structural integrity of several industrial components during hydrotesting. A few examples include mounded LPG storage vessel of petrochemical industry; and H_2S storage vessels and exchange unit towers of nuclear industry. AE monitoring during hydrotesting of the mounded LPG storage vessel has been carried out jointly by IGCAR, Kalpakkam and NML, Jamshedpur. This vessel showed the presence of cracks during NDT examinations and was subsequently repair welded. The AE generated during the hydrotests was due to structural noise and microyielding of the vessel material. The absence of AE during repressurising confirmed that there was no growing discontinuity in the repair welded region monitored by AE, thus confirming structural integrity of the welded region. The H_2S storage vessels and exchange unit towers showed defect indications by ultrasonic testing in a few critical regions. AE generated during pressurisation, pressure hold and repressurisation cycles of these vessels was studied. AE generated during pressurisation is attributed to microyielding of the material of the structures. The absence of AE during repressurising of the vessels and towers was used to confirm the absence of growing defects, thus confirming their structural integrity. Ultrasonic testing carried out subsequent to hydrotesting did not also indicate any detectable growth of the defects detected earlier.

Keywords: Ultrasonic Technique, Acoustic Emission Technique, Defects, Deformation, Crack Growth, Integrity Monitoring.

Analysis of Breakout Noise from the Coupled Acoustic-Structural HVAC Systems

B. Venkatesham

Department of Mechanical Engineering, Indian Institute of Technology Hyderabad, A.P.
E-mail: venkatesham@iith.ac.in

ABSTRACT: *Noise control in the heating, ventilation and air-conditioning (HVAC) systems is one of the critical design parameters in measuring the occupant comfort. The noise generated by air-handling units propagates through the ducts in the axial as well as transverse direction. Noise radiated in the transverse direction from the duct walls excited by the internal sound field is called the breakout noise. An analytical formulation has been discussed in order to predict the breakout noise from a cavity with one or four compliant walls (plenum) and a duct with all four compliant walls by incorporating three-dimensional effects along with the acoustical and structural wave coupling phenomena.*

The first step in the breakout noise prediction is to calculate the interior acoustic response and flexural vibration displacement of the compliant walls. Dynamic interaction between the internal acoustic subsystem and flexible structural subsystem has been expressed in terms of the modal characteristics of the uncoupled response of the acoustic and structural sub-systems. Solutions of the inhomogeneous wave equation are rearranged in terms of impedance and mobility, and the equations describing the complete system are expressed in terms of matrices, which result in a compact matrix formulation.

The second step is to calculate the sound power radiated from complaint wall. The wall vibration velocity is a linear combination of the uncoupled flexural modes of the structural subsystem. It is substituted into the Rayleigh integral and Kirchhoff–Helmholtz (KH) integral formulation to predict the sound pressure radiated by the vibrating compliant wall. Sound power radiation from flexible walls of the plenum and duct walls has been calculated using an equivalent plate model. Analytical results are corroborated with numerical models.

INTRODUCTION

The air handling systems find application in the HVAC systems, aircraft industry, the appliances, gas turbine power generation, etc. A basic HVAC duct setup is depicted in Figure 1 with arrows indicating flow path and noise propagation. It consists of equipment like motor, fan or blower, and also components like plenum and duct system, to provide cool and conditioned air. Incidentally, they produce unwanted noise. Generally, sources for the unwanted noise in the air handling systems are: i) airflow velocity, ii) turbulence due to narrow and complex shaped ducts, iii) fan noise and motor noise, and iv) the motor and fan noise transmission through ducts.

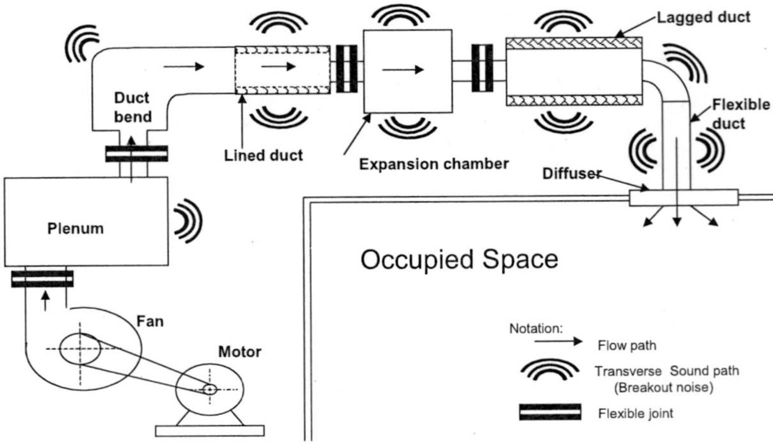

Fig. 1: Schematic of a Simple HVAC Setup

The HVAC systems consist of large air-handling units with thin flexible ducts, and plenums of different geometrical shapes and sizes connected in parallel and in series. The noise generated by air-handling units propagates through the ducts in the axial as well as transverse direction. Noise radiation in the transverse direction due to flexural vibration of the duct walls excited by the internal sound field is called the breakout noise. Generally, air-handling units generate most of the sound at low frequencies where the coupling phenomena are strong. The acoustic-structural coupling at low frequencies plays a critical role in understanding the breakout noise phenomena.So, any worthwhile analytical solution should be very accurate in the low frequency region [1].

Another source of thin rectangular duct wall excitation other than acoustic field is flow turbulence. The flow turbulence excitation is most likely to happen in the high-velocity ductwork and would dominate in the very low frequency range. The formulation proposed here will not consider this phenomenon, being below the frequency range of interest. Also, the flow speed is of magnitude less than 10 m/s.

The acoustic-structural coupling is the branch that investigates the interaction between the acoustic and the structural sub-systems. Dynamic interaction between acoustic space and flexible structure is an important problem in the field of acoustics for different practical applications like the HVAC duct breakout noise reduction, thin wall enclosures, aircraft skin panel vibration analysis, active noise control of panels, sound transmission measurement in building acoustics, etc. HVAC systems noise performance is characterized by the air-structural coupling. The response of the air-structural coupled system would not be essentially different from that of the uncoupled subsystems. The energy involved in the coupled interaction is much smaller than the energies in the subsystems. Mathematically, the coupled response can be expressed in terms of uncoupled responses of the constituent subsystems. Cummings [2] provided an excellent research overview of sound transmission through duct walls and discussed the basic phenomena of vibro-acoustic coupling and possible methods to solve this problem by using the common axial wave number approach and the finite element methods (FEM).A literature survey of the structural-acoustic coupling shows that a large amount of work has been done in calculating the free and forced response of different systems. Dowell [3] developed a comprehensive theoretical model for coupled responses in a structural-acoustic coupled system. Kim and Brennan [4]re-examined the formulation used by Dowell [3] and expressed the same in the impedance and mobility terms.

THEORETICAL FORMULATION

A cavity with one or four compliant walls (plenum) and a duct with all four compliant walls are considered to study breakout noise phenomena. General formulation has been discussed briefly in this article and further details are provided in references [1, 5–7].

Calculation of the Inside Pressure and Compliant Wall Vibration

Figure 2 shows the schematic diagram of a flexible structural system coupled with acoustical system. General equations of the inside system pressure p at location x and the compliant wall vibration velocity w at location z on the flexible wall for the uncoupled acoustic modes (N) and structural modes (M) are:

$$p(\mathbf{x},\omega)=\sum_{n=1}^{N}\psi_n(\mathbf{x})\;a_n(\omega)=\boldsymbol{\psi}^{\mathrm{T}}\mathbf{a} \quad \ldots (1)$$

$$w(\mathbf{z},\omega)=\sum_{m=1}^{M}\phi_m(\mathbf{z})\;b_m(\omega)=\boldsymbol{\varphi}^{\mathrm{T}}\mathbf{b} \quad \ldots (2)$$

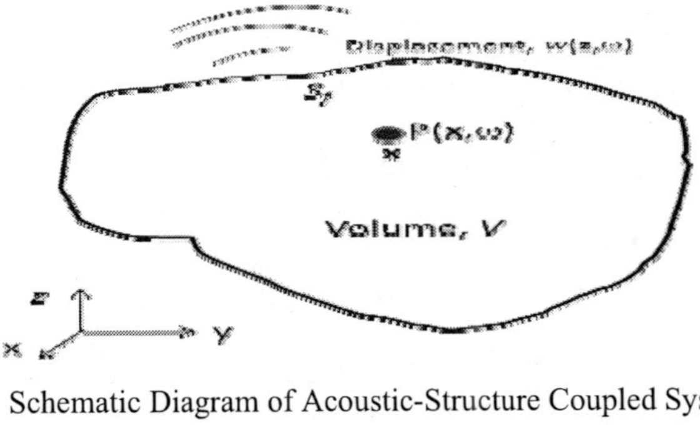

Fig. 2: Schematic Diagram of Acoustic-Structure Coupled System

where $\psi_n(\mathbf{x})$ is the uncoupled acoustic mode shape function, $a_n(\omega)$ is the complex amplitude of the nth acoustic pressure mode, $\phi_m(\mathbf{z})$ is the uncoupled vibration mode shape function, and $b_m(\omega)$ is the complex amplitude of the mth vibration velocity mode [4].

Complex amplitude of the nth acoustic mode under structural and acoustic excitation is given by,

$$a_n(\omega)=\frac{\rho_0 c_0^2}{V}A_n(\omega)\left(\int_V \psi_n(\mathbf{x})q_{vol}(\mathbf{x},\omega)\,dV+\int_{S_f}\psi_n(\mathbf{y})w(\mathbf{y},\omega)\,dS\right) \quad \text{...(3)}$$

Similarly, the complex vibration velocity amplitude of the mth mode can be expressed as,

$$b_m(\omega)=\frac{1}{\rho_s h S_f}B_m(\omega)\left(\int_{S_f}\phi_m(\mathbf{z})f(\mathbf{z},\omega)\,dS-\int_{S_f}\phi_m(\mathbf{z})p(\mathbf{z},\omega)\,dS\right) \quad \text{... (4)}$$

where, ρ_0 and c_0 denote the density and speed of sound in air, respectively. Function $q_{vol}(\mathbf{x},\omega)$ denotes the acoustic source strength density function (volume velocity for unit volume) in the system volume V and $w(\mathbf{z},\omega)$ denotes normal velocity of the surrounding flexible structure of surface area S_f. The firstintegral within brackets of Equation 3 represents the generalized acoustic source strength. Substitution of Equation (2) in second integral termof Equation (3) gives the coupling acoustic-structural mode shape coefficient $C_{n,m}$. It provides a relationship between the mode matching of acoustic and structural uncoupled modes. ρ_s and h denote density of the plate material and thickness of the plate, respectively. $f(\mathbf{z},\omega)$ denotes the force distribution on the surface of the plate. $p(\mathbf{z},\omega)$ denotes the cavity acoustic pressure distribution on the surface of the plate. The first integral term in Equation(4) represents the generalized modal force, and substituting Equation (1) in the second integral term of equation (4) for $p(\mathbf{z},\omega)$ gives the acoustic-structural coupling matrix coefficient.

Equations (3) and (4) can be expressed in the matrix form as:

$$a=Z_a(q+q_s),\ b=Y_s(g-g_a),\ \mathbf{g}_a=\mathbf{C}^T\mathbf{a} \quad \text{... (5–7)}$$

where, **a** isthe modal acoustic pressure vector, $\mathbf{Z_a}=\mathbf{A}\rho_0 c_0^2/V$ is an (N × N) diagonal matrix defined as the uncoupled acoustic modal impedance matrix: $\mathbf{q_s}=\mathbf{Cb};\mathbf{q}$ is the N-length modal source strength vector, and $\mathbf{q_s}$ is the modal source vector due to vibration of the structure. Matrix **A** is an (N × N) diagonal matrix. It contains acoustic mode resonance [3–4]. **C** is an (N × M) matrix defined as the coupling coefficient between the nth acoustic mode and the mth structural mode.

Combining Equations (5) and (6) yields the acoustic and structural modal amplitude vectors **a** and **b** in terms of the modal excitation vectors **q** and **g**:

$$a=\left(I+Z_aCY_sC^T\right)^{-1}Z_a\left(q+CY_sg\right) \quad \ldots (8)$$

$$b=\left(I+Y_sC^TZ_aC\right)^{-1}Y_s\left(g-C^TZ_aq\right) \quad \ldots (9)$$

Sound Power Radiation Calculation

Sound power radiation from the flexible wall in terms of the radiation impedance and complex amplitude of the vibration velocity modes is [5–8],

$$W_{rad}=\frac{1}{2}b^H\,\mathrm{Re}[Z]\,b \quad \ldots (10)$$

Radiation impedance, **Z** might be $\mathbf{Z}^{UBF}$ or $\mathbf{Z}^{BF}$ based on selected radiation model. The superscript H denotes the Hermitian of amatrix.

The radiation impedance of an unbaffled plate can be written as [6]

$$Z^{UBF}=j\omega S_t^T A^{-1} S_t \quad \ldots (11)$$

where $\mathbf{S}_t^T$ is transpose of the matrix, and

$$A=-\omega^2 M^A+K^A \quad \ldots (12)$$

$$S_{tm_1m_2m_1'm_2'}=\iint_{s_p}\phi_m(\vec{r})\phi_{m'}(\vec{r}')\,ds \quad \ldots (13)$$

$$M^A_{m_1m_2m_1'm_2'}=\frac{1}{4\pi\rho_0c_0^2}\iiint_{s_p\times s_p}\!\!\int\phi_m(\vec{r})\phi_{m'}(\vec{r}')\frac{e^{-jkR}}{R}ds\,ds' \quad \ldots (14)$$

$$K^A_{m_1m_2m_1'm_2'}=\frac{1}{4\pi\rho_0}\iiint_{s_p\times s_p}\!\!\int\left\{\frac{\partial\phi_m(\vec{r})}{\partial x}\frac{\partial\phi_{m'}(\vec{r}')}{\partial x'}+\frac{\partial\phi_m(\vec{r})}{\partial x}\frac{\partial\phi_{m'}(\vec{r}')}{\partial x'}\right\}\frac{e^{-jkR}}{R}ds\,ds' \quad \ldots (15)$$

Radiation impedance of a baffled plate can be written as [7]

$$\mathbf{Re}\left[\mathbf{Z}^{\mathbf{BF}}\right]=\int_0^{L_2}\int_0^{L_1}\int_0^{L_2}\int_0^{L_1}\phi_{mn}(x,y)\phi_{m'n'}(x',y')\frac{\sin kR}{R}dx'dy'dxdy \quad \ldots (16)$$

Transverse Transmission Loss

Transverse Transmission Loss is defined as [5–8]

$$TL_{tp} = 10\log(W_{\text{int}}/W_{rad}) \quad \ldots (17)$$

The radiated power (W_{rad}) is calculated by using Equation (10). Incidence power (W_{int}) to the acoustic subsystem can be calculated based on wave equation.

FREE VIBRATION ANALYSIS

Expressions of the uncoupled acoustic modes of a cavity and duct can be found in references [4–8]. The structural uncoupled modes for a simple supported plate are available in literature. Expressions for uncoupled structural modes of a duct are obtained by representing it as unfolded plate [9].

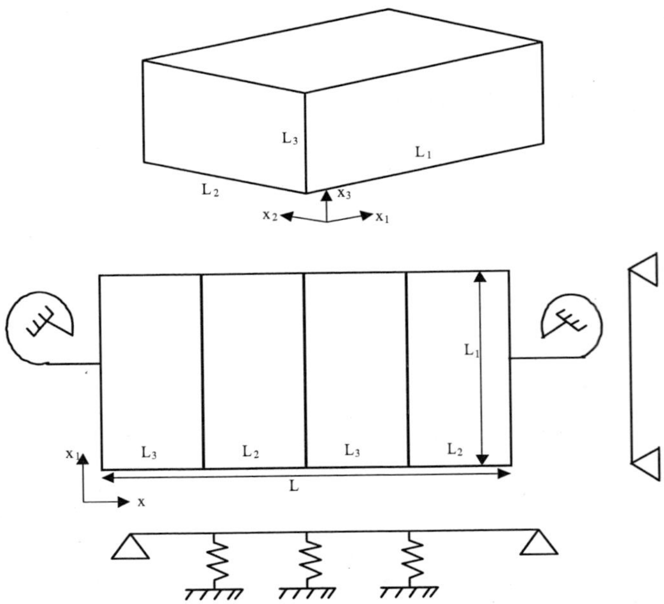

Fig. 3: Simply Supported Rectangular Duct and its Unfolded Plate Representation

Fig. shows an unfolded plate representation of the folded plate (duct) with four strips of the rectangular plate, made of the same isotropic material and of uniform thickness. The plate is considered to lie in the x-x1, plane, bounded by simple supported edges at $x_1 = 0$ and L_1. It is further assumed that thickness and flexural deflections of each plate are small compared with the wavelength of flexural vibrations, so that the thin plate theory is applicable.

NUMERICAL MODEL

The FEM-BEM based numerical model has been used here to validate the analytical method discussed in previous section in order to calculate the transverse transmission loss from different HVAC systems. The FEM-BEM analysis has been performed in the commercial package named SYSNOISE [10]. The procedure used inthe analysis of the coupled acoustic structural analysis by linking the 'FEM structure' and 'FEM fluid' models to simultaneously predict the inside pressure and compliant wall vibration displacement. The vibration displacement is then used as input for 'Indirect/Direct BEM' model to predict the radiated power from the flexible plate. Sound source input to the cavity is modeled by the equivalent constant velocity source in the SYSNOISE model. Weak coupling has been assumed in the calculation due to the presence of air medium inside and outside the cavity.

RESULTS AND DISCUSSION

Typical results based on the current approach have been discussed here. Consider a rectangular cavity and plenum of L_1 = 0.3 m, L_2 = 1.5 m, and L_3 = 0.4 m, and let the thickness of the flexible aluminum plate be 5 mm. Cavity with one wall flexible and four wall flexible (plenum), and duct with four wall flexible are considered to generate these results.

Free Vibration analysis of coupled system provides coupled natural frequency and decay time. The mechanical damping of the panel is represented in the modal decay times of the uncoupled panel modes. Fig. 4 shows the variation of the decay time of the cavity-controlled mode with respect to panel decay time. There is an optimum condition for cavity controlled mode decay time. The physical explanation of this phenomenon is that increasing panel damping will dissipate energy until it reaches the minimum decay time of the cavity-controlled mode (1, 0, 0). Beyond this, the increased damping behaves like

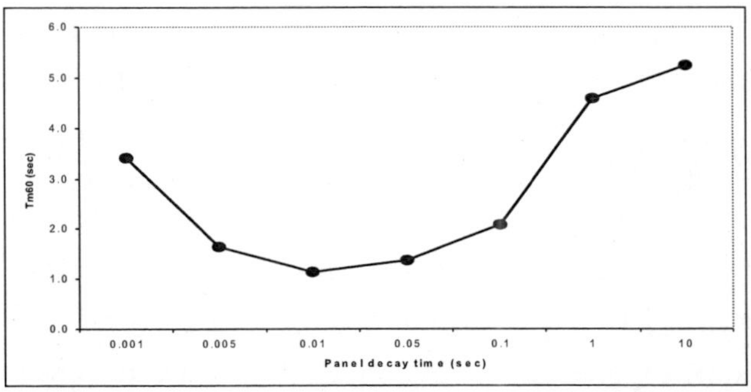

Fig. 4: Decay Time of Cavity Controlled Mode (1, 0, 0) as Function of Panel Modal Decay Time (h = 5 mm, cavity decay time of 5 sec.)

mass. It would restrict the panel motion, and effectively the acoustic energy would not transfer to the panel. Thus the modal characteristics would be similar to those of the uncoupled cavity subsystem. This physical understanding will be useful in designing the breakout noise control solutions.

Figures 5 show the effect of coupling on the inside cavity pressure. Equations (8) and (9) are modified for weak coupling by approximating $\left(I+Z_aCY_sC^T\right)\cong I$, and $\left(I+Y_sC^TZ_aC\right)\cong I$. Generally, the coupling is stronger for thin cavity walls compared to thick cavity walls.

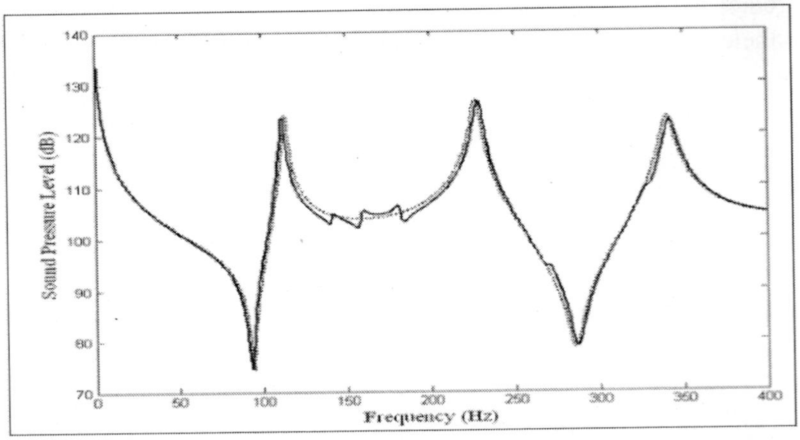

(- - -) uncoupled; (——) coupled.

Fig. 5: Effect of Coupling on the Sound Pressure Level Prediction Inside a Rectangular Cavity of Dimensions (0.3 m × 0.4 m × 1.5 m) with 5 mm Aluminum Flexible Cavity Wall, Due to Acoustic Excitation of the Structural-Acoustic System

Figure 6 shows the transverse transmission loss of a rectangular plenum with flexible walls. It is observed that analytical and computational results (FE-BE) are in good agreement. The "unbaffled equivalent unfolded plate" model is appropriate for analytical prediction of the breakout noise from a rectangular plenum with flexible walls. It is much faster than the numerical (FEM-BEM) model and needs little pre-computation effort.

Consider a rectangular duct of dimensions L1 = 1.5 m, L_2 = 0.3 m, and L_3 = 0.4 m, and the thickness of the flexible aluminum wall 5 mm. The acoustic and structural damping ratio loss factor = 0.01. Air properties are: sound speed = 340 m/sec and density = 1.2 kg/m^3. Three types of terminal impedances are investigated: (i) anechoic termination conditions $\left(Z_t \rightarrow \rho_0 c_0\right)$,(ii) rigid termination impedance condition $\left(Z_t \rightarrow \infty\right)$, and (iii) open end termination. Figure 8 shows the effect of the acoustic termination of the duct on breakout noise for three types of terminations.

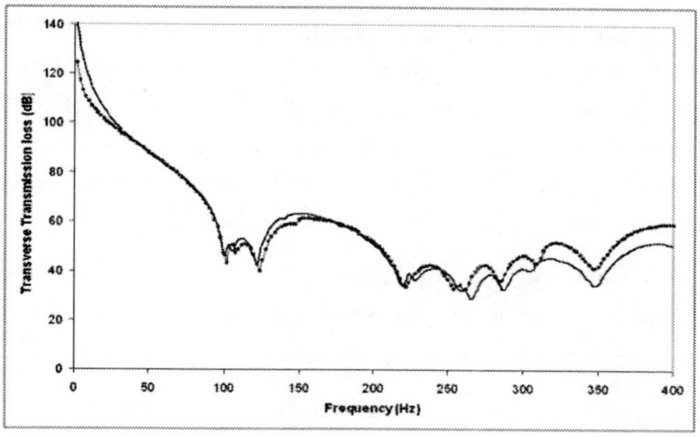

(—•—Sysnoise model; —— Analytical model).

Fig. 6: Numerical Corroboration of the Analytical Radiation Model for a Plenum

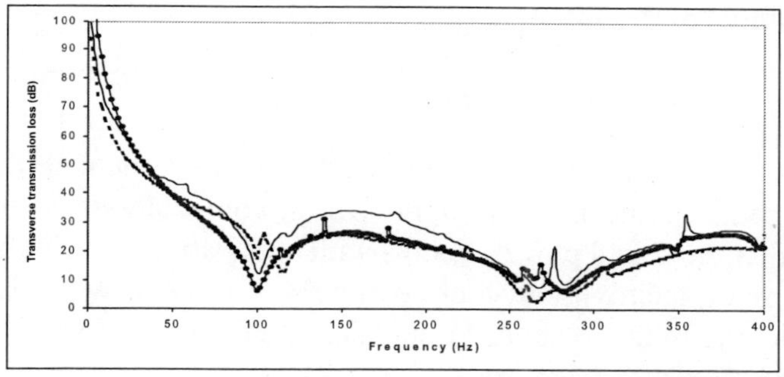

(—•— Open termination; —— anechoic termination; - - - - rigid termination).

Fig. 7: Effect of the Duct Termination Conditions on the Transverse Transmission Loss

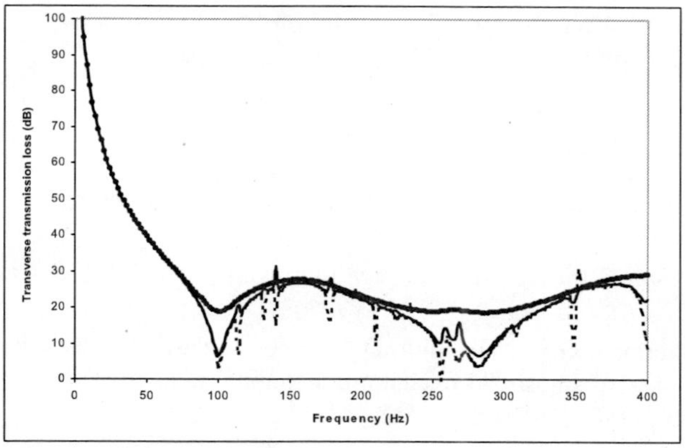

(—•—η= 0.1; ——η= 0.01; - - - - η= 0.001).

Fig. 8: Effect of Damping on the Transverse Transmission Loss in a Rectangular Duct with Open Terminal Impedance

It may be noted that the anechoic termination condition over predicts the transverse transmission loss for the rigid and the open-end termination conditions at low to medium frequencies but this frequency range is important for the breakout noise problems. Acoustic boundary conditions will affect the inside standing wave pressure pattern. It is important for the rigid termination condition that would create a strong standing wave pattern. It also increases the breakout noise from the HVAC ducts.

The effect of structural damping on transverse transmission loss is shown in Figure 8. The transverse transmission loss values are improved at the first transverse resonance (100 Hz) and also in the structural-acoustic coupling region at 250–280 Hz.

Acoustic-structural coupling is critical and limited to narrow frequency range. Most of the sound transmission occurs near the coupled natural frequencies, and dips in the transverse transmission loss relate to the transverse resonance frequencies and their magnitude controlled by damping. At very low frequencies, the transmission loss is controlled by stiffness of the structure.

SUMMARY

A comprehensive theoretical model has been developed to calculate the breakout noise, based on the impedance and mobility approach. A computationally efficient method is suggested to predict the sound power models with application to a cavity with one or four compliant walls and a duct with all four compliant walls. A new intuitive way has been provided to model the folded plate as "Equivalent plate model" in free vibration analysis and sound power radiation calculation.

It is worthwhile to extend the current approach to predict breakout noise from an acoustically lined rectangular duct and plenum chamber with compliant walls. Further studies are required to develop analytical models in understanding the effect of stiffeners and damping (anti-drumming) on the duct breakout noise.

REFERENCES

[1] Venkatesham, B., Breakout Noise from the Coupled Acoustic-Structural HVAC Systems, PhD Thesis, 2008.

[2] Cummings, A., Sound transmission through duct walls, *Journal of Sound and Vibration* 239(4) (2001) 731–765.

[3] Dowell, E.H., Gorman, G.F. and Smith, D.A., Acoustoelasticity: general theory, acoustic modes and forced response to sinusoidal excitation, including comparisons with experiment, Journal of Sound and Vibration, 52 (1977), 519–542.

[4] Kim, S.M. and Brennan, M.J., A compact matrix formulation using the impedance and mobility approach for the analysis of structural-acoustic systems, *Journal of Sound and Vibration,* 223(1) (1999) 97–113.

[5] Venkatesham, B., MayankTiwari and Munjal, M.L., Analytical prediction of the breakout noise from a rectangular cavity with one compliant wall, *Journal of the Acoustical Society of America*, 124(5), (2008), 2952–2962.

[6] Venkatesham, B., MayankTiwari and Munjal, M.L., Analytical prediction of the breakout noise from a reactive rectangular plenum with four flexible walls, *Journal of Acoustical Society of America*, 2010.

[7] Venkatesham, B., MayankTiwari and Munjal, M.L., Prediction of Breakout Noise from a Rectangular Duct with Compliant Walls, *International Journal of Acoustics and Vibration*, Vol. 16 (4), 180–190, 2011.

[8] Li, W.L. and Gilbeling, H.J., Determination of the mutual radiation resistance of a rectangular plate and their impact on the radiated sound power, *Journal of Sound and Vibration,* 229 (2000).

[9] Lee, H.P., Natural frequencies and modes of cylindrical polygonal ducts, Journal of Sound and Vibration, 164(1), (1993), 182–187.

[10] SYSNOISE Rev 5.6, User's Manual, LMS International, 2003.

Measurement of Acoustics in ISRO Launch Vehicle Missions

A.R. Krishnan

Avionics Entity, VSSC, ISRO PO, Thiruvananthapuram

INTRODUCTION

ISRO is in the process developing bigger launch vehicles to cater to future needs of heavier satellites in polar and GTO and re-entry missions. During the launch vehicle mission from lift off to satellite injection, the launch vehicle is subjected to acoustics and vibration fields. For a successful flight, it is important to ensure that the acoustic levels are within design limits of the various subsystems .The acoustic environment is severe during liftoff and atmospheric phase of the flight. At liftoff, the violent and turbulent mixing of the high temperature and high velocity exhaust of the jet with the surrounding air is the main cause of the acoustic noise. In the atmospheric flight the flight profile with the vehicle configuration determines the acoustic environment. The acoustic excitation induces broadband vibration responses of the vehicle structure and the get transmitted to the interior of the vehicle where sensitive assemblies are kept. The acoustic levels (SPL) normally encountered are of 120 to 170 dB during these regions.

During the development phase of the launch vehicle, large no of acoustic measurements are made to characterize the vehicle structure, deduce test levels for various subsystems and generate database for improvements. A typical of 10–15 measurements are made during the development phase. A typical acoustic measurement chain consists of an acoustic sensor, charge amplifier and digital signal processing unit. Acoustic sensor converts sound pressure level to electric charge. The

common types are condenser type and MEMS based transducers. The dynamic range of these sensors are in the range of 100–180 dB SPL and bandwidth more than 10 KHz. They are rugged and operate over a wide temperature range. The charge output from these sensors are converted to analog voltage using charge amplifiers and band limited to 8 KHz using higher order switched capacitor filters. The overall telemetry available in a Launch vehicle is 2–4 Mbps and each acoustic channel if transmitted as raw data would need 200 Kbps. Hence onboard signal processing is needed to reduced bit rate. The DSP based on board system is used to process these signal in real time and transmit the information at reduced bandwidth. The frequency analysis is done in the 1/3 rd octave band covering a frequency range of 31.5 Hz to 8 KHz. The details of the sensors, interface electronics and the DSP based processing system are explained below.

SENSORS

The condenser type and the MEMS type sensors are normally used in launch vehicles. Sensor 2510 from m/s Endevco was originally used in all measurements. Later it is replaced by indigenous MEMS acoustic sensor jointly developed by VSSC and CEERI, Pilani. The endevco sensor uses Piezite as crystal material with a nominal capacitance of 5000 pF. It is rugged and hermitically sealed construction and operates over a wide temperature range (up to 250 deg.C). The MEMS sensor is a bulk micro machined device, etched out of silicon bulk. A sense layer of silicon is deposited over the diaphragm. Appropriately patterned aluminum electrodes sandwich the ZnO layer. For strength the silicon chip is bonded to glass. The diaphragm converts the acoustic pressure into lateral stresses in silicon and Zinc oxide layers. ZnO layer converts lateral stresses in it into electric charge. A vent tunnel allows movement of air in and out of cavity under the diaphragm when any static ambient pressure changes (for low frequency acoustic pressure)while blocking the measurement band. The charge variation in the acoustic sensor is converted to voltage using a front end charge amplifier circuit and further amplified using a voltage amplifier. The picture, important specifications and the block diagram of acoustic measurement in launch vehicle are given below.

- Range 100 to 180 dB
  (2Pa to 20 KPa)
- Frequency Range
  31.5 Hz to 8 KHz
- Sensitivity
  150 to 200 uV/Pa

STRUCTURAL DESIGN

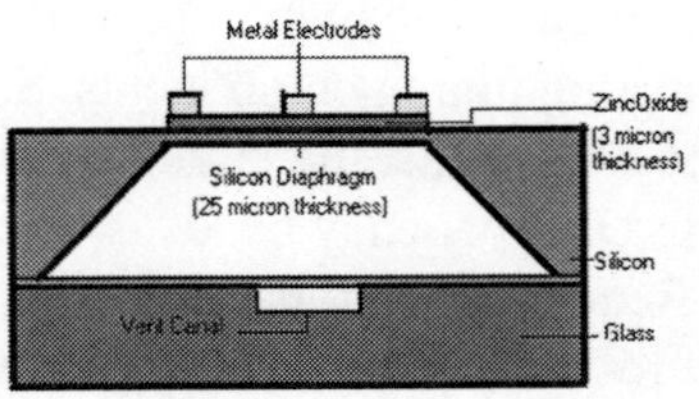

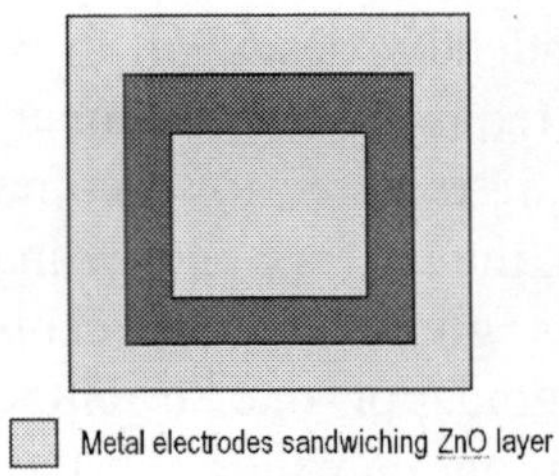

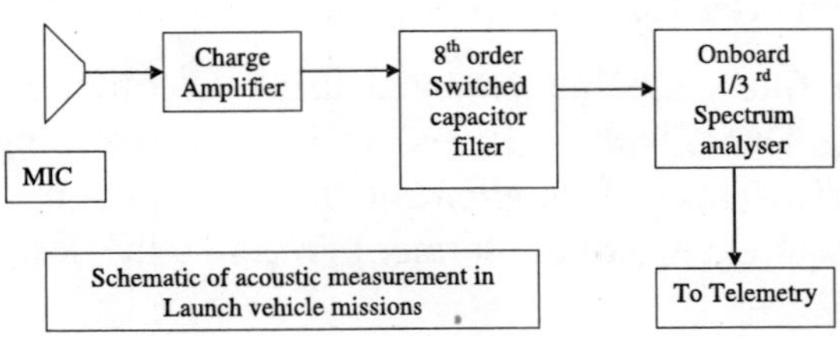

Schematic of acoustic measurement in Launch vehicle missions

The electrical function of the sensor can be illustrated by its Thevenin equivalent as shown in the Figure below.

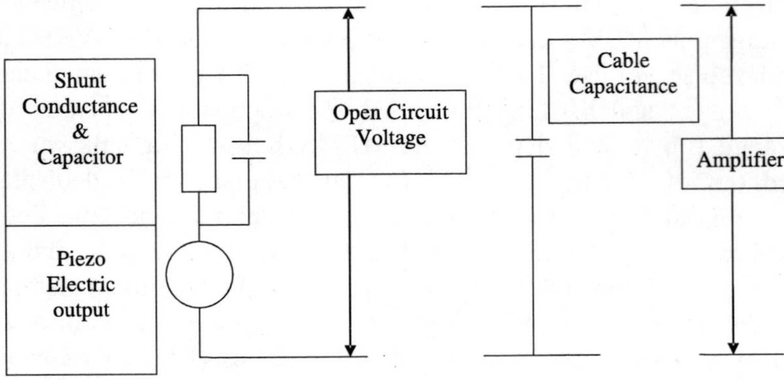

ONBOARD PROCESSING APPROACH

The general method for frequency analysis of any Signal is the bank of filter method. The measurement is done in the standard1/3rd octave bands depending on the application and required resolution. The two main types of frequency analysis used for signals are constant percentage bandwidth (filter bandwidth is constant percentage of centre frequency) and constant bandwidth (constant bandwidth independent of centre frequency of the filter). These techniques permit detailed analysis of the Spectrum Finally the rms value of each filter output is computed for finite period of time. For normal applications the rate is fixed as 250 milliseconds. For acoustics constant percentage bandwidth analysis is done·

The processing can be implemented both in analog or digital methods· In the analog approach, separate filters are employed for each frequency band. Even though each filter can be independently designed to have required bandwidth, this method has the disadvantage of requiring large number of analog tuning components. Since higher values of capacitances are required for low frequency analysis, the circuit becomes bulky. In the digital approach, the average energy in different frequency bands is calculated using DSP techniques. This gives accurate centre frequency and band width and there is no performance degradation due to temperature and ageing. The measurement parameters can be made programmable. Further there is repeatability of performance during production. This method is selected for processing of acoustics in launch vehicles.

The individual band filters are implemented using 6th order Elliptic Infinite Impulse Response (IIR) filters. This 6th order filter is obtained by cascading three bi-quad sections with appropriate coefficients. IIR implementation was preferred over FIR due to the lower computation requirements of the former in meeting the design specifications.

The acoustic signals are analyzed in nine 1/3rd octave frequency bands The centre frequencies are logarithmically spaced and range from 25 Hz to 6.3 KHz. The filters corresponding to highest octave (6.3 KHz, 5 KHz and 4 KHz) are computed for all input samples. The next lower octave (3.15 KHz, 2.5 KHz and 2 KHz) is computed for alternate samples only. This means the input samples are decimated by a factor of 2. This is because the input was originally sampled for twice the maximum frequency of this filter band. Before doing computation for this octave, the input samples are low pass filtered so as to remove any higher frequencies corresponding to higher octave and then decimated by a factor of 2.This process is repeated for all lower octaves. This will reduce the computation requirement very much. The output of each filter bank is squared and averaged for a time period of 250ms.Finally square root is taken before transmission and each byte represents the rms output in the corresponding frequency band. A typical output is shown below.

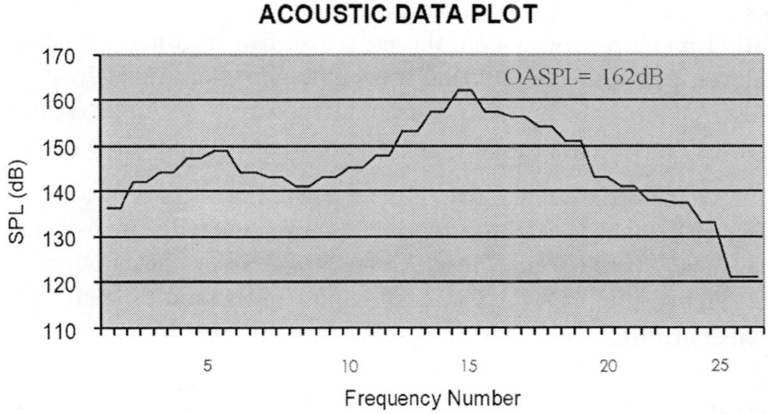

DESIGN CONSIDERATIONS

The following criteria were considered to meet the design objective:

1. Flexibility to choose different processing algorithm during development phase.
2. Programmability of important measurement parameters.
3. Sufficient dynamic range for measurement.
4. Elimination of analog frequency trimming components.
5 Hardware programmability of sampling clock generation.

The task identified were highly computation intensive and had to be carried out in real time to maintain the design flexibility, programmability and to get sufficient dynamic range, a floating point DSP processor (ADSP 21060) was selected. The processor is fast and powerful to process multiple acoustic and also vibration channels simultaneously.

The hardware is based on a floating point DSP processor ADSP21060 as shown below.

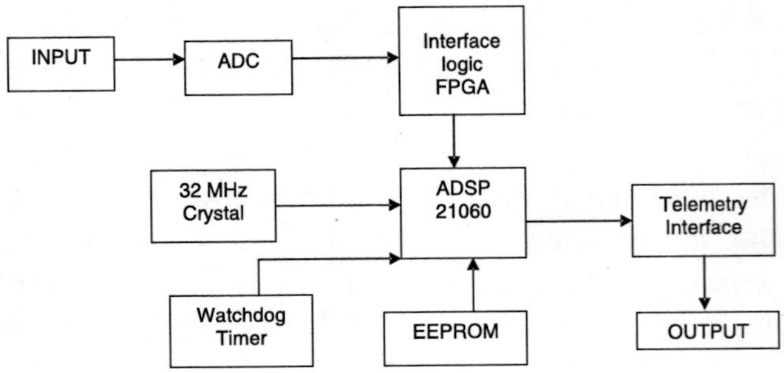

It can execute all instructions in single cycle and runs at 32 MHz on target hardware. The input channels are taken through a multiplexer and given to ADC. The ADC outputs are written to a FIFO memory from where the processor can read it periodically. Glue logic and control signal generation circuitries are implemented using a field programmable logic array (FPGA). Anti aliasing filters are realized using 8 th order switched capacitor filter. To increases the system reliability, hardware watch dog .timer is incorporated.

In order to achieve the programmability of the system, the outputs of the filters are scaled by suitable factors which could be programmed from outside. In the case of acoustic channel, for a given overall sound pressure level (OASPL), the maximum SPL (Sound Pressure level) that can be measured in each band can be programmed. An EEPROM is provided for this purpose.

CONCLUSION

The presentation covered the details of acoustic measurements in launch vehicles including purpose, details of sensor and electronics and on-board processing using DSP techniques. With the high speed floating point architecture of ADSP21060, a dynamic range of 48db and a telemetry bit rate reduction of 250:1 was achieved. This enabled more number of such measurements to be made in each launch vehicle missions during development phase.

Modeling and Recognition of Consonant-Vowel Units Using Statistical Learning Algorithms

N.K. Narayanan and T.M. Thasleema

Department of Information Technology, Kannur University, Kannur, Kerala

E-mail: nknarayanan@gmail.com; thasnitm1@hotmail.com

ABSTRACT: *This paper investigates the use of statistical learning algorithms for the classification of Malayalam Consonant-Vowel (CV) utterances in speaker independent, noisy environments. The techniques we adopted here are k-Nearest Neighborhood (k-NN) and Support Vector Machine (SVMs) based Decision Directed Acyclic Graph Support Vector Machine (DDAGSVM) classifiers. Three different feature extraction algorithms viz., Mel Frequency Cepstral Coefficients (MFCC), Normalized Wavelet Hybrid Feature (NWHF) and State Space Point Distribution (SSPD) parameter, are extracted to evaluate the performance of the above two classifiers. From the experimental results it is found that the DDAGSVM classifier is performing well compared with k-NN in additive noisy condition.*

INTRODUCTION

In the past few decades, significant progress has been made by researchers in the area of speech and language processing. In various traditional pattern recognition approaches, the statistical approach has been extensively studied and used in practice. More recently methods imported from statistical learning theory have been receiving increasing attention [1]. In the present work we are discussing two statistical pattern classifiers namely k-Nearest Neighborhood (k-NN) and Support Vector Machine (SVM) to classify Malayalam CV speech unit database.

The present research work is motivated by the knowledge that a little attempts were rendered for the automatic speech recognition of CV speech unit in Indian languages like Hindi, Tamil, Bengali, Marathi etc. But very less works have been found to be reported in the literature on Malayalam CV speech unit recognition, which is the principal language of South Indian state of Kerala. Very few research attempts were reported so far in the area of Malayalam vowel recognition. So more basic research works are essential in the area of Malayalam CV speech unit recognition.

Malayalam is one of the major languages from Dravidian language family. Other major languages include Kannada, Tamil, Telugu, Tulu and Konkani. Malayalam is the principal language of the South Indian state of Kerala and also of the Lakshadweep Islands off the west coast of India spoken by about 36 million people. Malayalam language now contains 51 V/CV units includes 15 long and short vowel sounds and the remaining 36 basic consonant sounds.

In the present work, all the experiments are carried out using 36 Malayalam CV speech unit database uttered by 96 different speakers. For the recognition experiments, database is divided into five different phonetic classes based on the manner of articulation of the consonants which are given in Table 1.

Table 1

Phonetic Classes	*Sounds*
Unspirated	/ka/, /ga/, /cha/, /ja/, /ta/, /da/, /tha/, /d_ha/, /pa/, /ba/
Aspirated	/kha/, /gha/, /chcha/, /jha/, /tta/, /dda/, /ththa/, /dha/, /pha/, /bha/
Nasals	/nga/, /na/, /nna/, /na/, /ma/
Approximants	/ya/, /zha/, /va/, /lha/, /la/
Fricatives	/sha/, /shsha/, /sa/, /ha/, /ra/, /rha/

FEATURE EXTRACTION

Feature extraction is the one of the essential components of the study to develop an ASR system. In this paper three different approaches to phonemic feature extraction is discussed which are explained below.

Mel Frequency Cepstral Coefficient

The MFCC method uses the bank of filters scaled according to the mel scale to smooth the spectrum and to perform similar to that executed by the human ear. The filters with mel scale spaced linearly at low frequencies up to 1 kHz and logarithmically at higher frequencies are used to capture the phonetical characteristics of the CV speech signals. Thus MFCCs are used to represent human speech perception models. MFCCs are computed as follows.

After preemphasize the signal is divided into a sequences of frames with 256 samples in each frame that can be analyzed independently and represented by a single feature vector. In order to reduce the signal discontinuities at edges of each frame a hamming window is applied. Hamming window mathematically represented as,

$$w_n = 0.54 - 0.46Cos\left(\frac{2\pi n}{N-1}\right), 0 \le n \le N-1 \qquad \ldots (1)$$

Compute the Fast Fourier Transform (FFT) of each frame and obtain its magnitude. Twenty four triangular-shaped filter banks with logarithmic spacing (called Mel banks) are applied to each frame. A filter bank, in its simplest form, is a set of band pass filters with different frequencies covering the interesting part of the spectrum (the fundamental frequency and the formants). The output of the filters in each frame can be used as features. The center frequencies of the filters can be chosen in several ways. Usually they are set according to some perceptually motivated scale. It was created by modeling human auditory system and has been showed that this model empirically improves recognition accuracy. The following formula is used to compute the mels for a given frequency F in Hz.

$$F_{mel} = 2595\log_{10}(1 + F_{Hz}/700) \qquad \ldots (2)$$

For calculation of mel-frequency spectrum, a bank of filters with triangular bandpass frequency response is used. Such filters compute the spectrum around each center frequency with increasing bandwidth. The log energy at the output of each filter is computed afterwards. The Discrete Cosine Transform (DCT) is taken to give the cepstral coefficients. The first 12 coefficients, excluding the 0th coefficient are the features that are known as MFCCs. The advantage of computing MFCCs by using filter energies is that they are more robust to noise and spectral estimation errors.

Normalized Wavelet Hybrid Feature

Certain ideas of wavelet theory appeared quite a long time ago [2]. Over the last decades wavelet analysis has turned to be a standard technique in the areas of geophysics, meteorology, audio signal processing and image compression [3][4][5]. Wavelet transform can be defined as the transformation of the signal under analysis into another representation which presents the signal in a more useful form [6].

The Discrete Wavelet Transform (DWT) has been treated as a Natural Wavelet Transform (NWT) for discrete time signals by different authors [7][8]. For computing the wavelet coefficients several discrete algorithms have been established [9]. Daubechies and some others had invented DWT peculiarly designed for analyzing finite set of observations over the set of scales using dyadic discretization [10][11]. As Daubechies mentioned in his work DWT can be interpreted as continuous wavelets with discrete scaling and translation factors. The WT is then evaluated at discrete scale and translation. More effective implementation of DWT is obtained using Multi Resolution Analysis (MRA) based subband coding technique. MRA is an effective tool for interpreting the information content of a signal. A MR representation produces a simple hierarchical model to develop the algorithms for the time-frequency analysis of the signal. At different resolution, the details of the speech signal generally characterize vocal characteristics of the speech signal. Each signal to be characterized is analyzed at a coarse resolution and then gradually increases the resolution. Thus MRA can be defined as a technique that permits us to analyze the signal in multiple frequency bands.

The main advantage of DWT over FT is the multi resolution analysis of signals with localization on both time and frequency. In the present work we utilize these characteristics of wavelet transform using two major wavelet decomposition techniques viz., Classical Wavelet Decomposition (CWD) and Wavelet Packet Decomposition (WPD) for Malayalam CV speech unit recognition.

Normalized Wavelet Hybrid Feature (NWHF) vector for the present work is generated using CWD and WPD method. The process for extracting NWHF feature vector is described below.

In the first step for the CWD, the sound signal is made to undergo recursively to decompose into k[th] level of resolutions; therefore the approximation coefficient matrix at this level is a sufficiently small representative of the original speech signal and carries enough information content to describe s characteristics coarsely.

Let A_k represents this approximation matrix at decomposition level k, which can be written as,

$$A_k = \begin{bmatrix} A_{1,1} & A_{1,2} & \cdots & A_{1,n} \\ A_{2,1} & A_{2,2} & \cdots & A_{2,n} \\ \cdots & \cdots & \cdots & \cdots \\ A_{k,1} & A_{k,2} & \cdots & A_{k,n} \end{bmatrix} \qquad \text{.... (3)}$$

Then, the first component of NWHF feature vector v_1 is,

$$v_1 = \bigcup_{i=1}^{k} \bigcup_{j=1}^{n} \{A_{i,j}\} \qquad \text{... (4)}$$

In the second step, for Wavelet Packet Decomposition, decompose each speech segment using k[th] level of resolutions for the best level of wavelet packet decomposition tree. The first coefficient matrix at the best level tree contains enough information to represent the given input consonant CV speech unit without loss of much speech features. Let *m* represent mean of one row vector in the coefficient matrix then the WPD feature vector v_2 is given by

$$v_2 = m_i,\ \mathrm{i} = 1, 2, \ldots, \mathrm{m} \qquad \text{... (5)}$$

(number of rows in the best level coefficient matrix)

In the third step we combined v_1 and v_2 to fusion CWD and WPD coefficients.

$$V = \bigcup_{i=1}^{2} \{v_i\} \qquad \text{... (6)}$$

Then the final feature vector F after z-score normalization is given as

$$F = \frac{V - \mu(V)}{\sigma(V)} \qquad \text{... (7)}$$

Thus an NWHF vector is extracted using the proposed algorithm.

The feature vector of size 20 is estimated by concatenating CWD and WPD features for hybrid feature vectors. The NWHF vector for different speaker shows the identity of the same sound so that an efficient feature vector can be formed using NWHF vectors. The NWHF obtained for different sounds seems to be distinguishable, giving discriminate feature vector for speech unit recognition.

State Space Point Distribution Parameter

In dynamical system approach, by embedding a signal into adequately high dimensional space, a topologically equivalent to the original state space structure of the system generating the signal is formed [12][13]. This embedding is known as Reconstructed State Space (RSS), is typically constructed by mapping time-lagged copies of the original signal onto axes of the new high dimensional space. The time evolution within the RSS traces out a trajectory pattern referred to as its attractor which is a representation of the dynamics of the underlying system [14]. Since the attractor of an RSS captures all the relevant information about the underlying system, it is an efficient choice for signal analysis, processing and classifications. The RSS approach proposed here has the advantage of extracting both linear and non-linear aspects of the entire system.

Takens' theorem states that under certain assumptions, state space of a dynamical system can be constructed through the use of time delayed versions of the original scalar measurements [15]. Thus a RSS can be considered as a powerful tool for signal processing domain in non-linear or even chaotic dynamical systems [16][17]. According to Takens embedding theorem, a RSS for a dynamical system can be produced for a measured state variable S_n, n = 1, 2, 3,….., N via method of delays by creating vectors given by,

$$\mathbf{S}_n = [s_n \; s_{n+\tau} \; s_{n+2\tau} \; \ldots\ldots\ldots \; s_{n+(d-1)\tau}] \qquad \text{... (8)}$$

where d is the embedding dimension and τ is the time delay value. The row vector $\mathbf{S}_n$ defines the position of a single point in the RSS. To completely define the dynamics of the system and to create a d dimensional RSS, corresponding trajectory matrix is given as

$$S_d = \begin{bmatrix} s_1 & s_{1+\tau} & \cdots & s_{1+(d-1)\tau} \\ s_2 & s_{2+\tau} & \cdots & s_{2+(d-1)\tau} \\ \cdots & \cdots & \cdots & \cdots \\ s_N & s_{N+\tau} & \cdots & s_{N+(d-1)\tau} \end{bmatrix} \quad \ldots (9)$$

A speech signal with amplitude values can be treated as a dynamical system with one dimensional time series data. Based on the above theory, this study investigates a method to model a RSS for speech signal through the use of time delayed versions of original scalar measurements. Thus a trajectory matrix $\mathbf{S}_2$ with embedding dimension d = 2 and τ = 1 can be constructed by considering the speech amplitude values $\mathbf{S}_n$ as one dimensional time series data. Thus $\mathbf{S}_2$ is given as,

$$S_2 = \begin{bmatrix} s_1 & s_2 \\ s_2 & s_3 \\ \cdots & \cdots \\ s_{N-1} & s_N \end{bmatrix} \quad \ldots (10)$$

The State Space Map (SSM) for the Malayalam consonant CV unit is constructed as follows. The normalized N samples values for each CV unit is the scalar time series S_n where n = 1, 2, 3…, N. For every consonant speech signal a trajectory matrix is formed with embedding dimension d = 2 and time delay τ = 1. Now the scatter plot SSM is generated by plotting the row values of the above constructed trajectory matrix by plotting S_n versus S_{n+1}. Figure 1 shows the SSM for the first consonant sound /ka/.

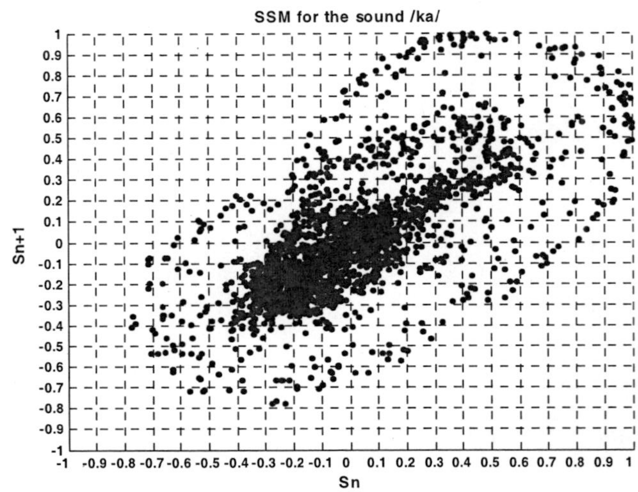

Fig. 1: Scatter Plot for the Sound /ka/with d = 2

In Automatic Speech Recognition (ASR), selection of distinctive features is certainly the most important factor for the high recognition performance. Present study uses non linear feature extraction technique called State Space Point Distribution (SSPD) from their SSM. For this purpose the SSM of the speech unit is divided into grids with 20 × 20 boxes. The box defined by co-ordinates (–1, 0.9), (–0.9, 1) is taken as box 1 and box just right side to it as taken as box 2 and so on in the x-direction with the last box being (0.9, 0.9), (1, 1) is taken as box 20. The process is repeated for all the rows and boxes are numbered consecutively for the 400 boxes. The SSPD for each pattern is calculated by estimating the number of points distributed in each of these 400 boxes. This can be mathematically represented as follows.

The reconstructed SSPD parameter for location 'i' in two dimensions can be defined as,

$$(SSPD)_i = \sum_{n=1}^{N} f([s_n, s_{n+1}], i) \qquad \ldots (11)$$

where $f([s_n, s_{n+1}]), i) = 1,$ if state space point defined by the row vector $[s_n, s_{n+1}]$ is in the location 'i'

0, otherwise

Using this information the SSPD plot is plotted by taking the box number along x-axis and the number of points in each box along y-axis. The SSPD plot for the first Malayalam CV sound /ka/ is given in Figure 2.

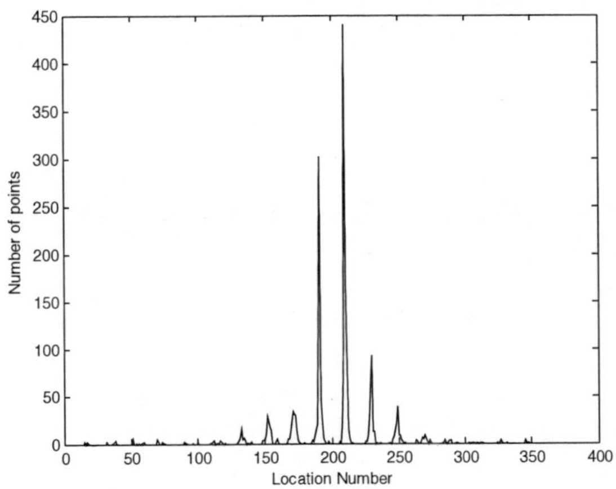

Fig. 2: SSPD Plot for the Sound /ka/

The SSM and the corresponding SSPD plot obtained for different speaker shows the identity of the sound so that an efficient feature vector can be formed using SSPD. The feature vector of size 20 is estimated by taking the average distribution of each row in the SSPD graph.

CLASSIFICATION

In the statistical learning approach, each pattern is represented in terms of a set of *d* features and can be viewed as a point in *d* dimensional feature space. The goal is to choose those features that allow pattern vectors belonging to different categories to occupy compact and disjoint regions in a *d*-dimensional feature space. The effectiveness of the representation space (feature set) is determined by how well patterns from different classes can be separated. Given a set of training patterns from each class, the objective is to establish decision boundaries in the feature space which separate patterns belonging to different classes. In the statistical decision theoretic approach, the decision boundaries are determined by the probability distributions of the patterns belonging to each class, which must either be specified or learned [18] [19].

k-Nearest Neighborhood

Pattern classification using distance function is an earliest concept in pattern recognition [20][21]. Here the proximity of an unknown pattern to a class serves as a measure of its classifications. k-NN is a well known non-parametric classifier, where a posteriori probability is estimated from the frequency of the nearest neighbors of the unknown pattern [22]. For classifying each incoming pattern k-NN requires an appropriate value of k. A newly introduced pattern is then classified to the group where the majority of k nearest neighbor belongs [23]. Hand proposed an effective trial and error approach for identifying the value of k that incur highest recognition accuracy [24]. Various pattern recognition studies with highest performance accuracy are also reported based on these classification techniques [25].

Consider the cases of m classes c_i, where i = 1, 2,…, m, and a set of N samples pattern y_i, i = 1, 2,….. N whose classification is priory known. Let $\boldsymbol{x}$ denote an arbitrary incoming pattern. The nearest neighbor classification approach classifies '$\boldsymbol{x}$' in the pattern class of its nearest neighbor in the set y_i

i.e. If $\|x - y_j\|^2 = \min\|x - y_i\|^2$, where $1 \le i \le N$ then $\boldsymbol{x}$ in c_j

This is 1-NN rule since it employs only one nearest neighbor to $\boldsymbol{x}$ for classification. This can be extended by considering k-Nearest Neighbors to $\boldsymbol{x}$ and using a majority-rule type classifier.

Support Vector Machines

SVM is a linear algorithm with some specific properties. The basic principle of SVM in pattern recognition application is to build an optimal separating hyperplane in such a way to separate two classes of pattern with maximal margin [26]. SVM accomplish this desirable property based on the idea of Structural Risk Minimization (SRM) from statistical learning theory which shows that the error rate of a learning machine on test data (i.e. generalization error report) is bounded by the sum of training error rate and the term that depending on the Vapnik-Chervonenkis (VC) dimension of the learning system [27][28]. By minimizing this upper bound high generalization performance can be obtained. For separable patterns SVM produces a value of 0 for first term and minimizes the second term. Furthermore, SVMs are quite different from other machine learning techniques in generalization of errors which are not related to the input dimensionality of the problem, but to the margin with which it separates data. This is the reason why SVMs can have good performance even in large number of input problems [29] [30].

SVMs are mainly used for binary classifications. For combining the binary classification into multiclass classification a relatively new learning architecture namely Decision Directed Acyclic Graph (DDAG) is used. For N class problem, the DDAG contains, one for each pair of classes. DDAGSVM works in a kernel induced feature space and uses two class maximal margin hyperplane at each decision node of the DDAG. The DDAGSVM is considerably faster to train and evaluate comparable to other algorithms.

SIMULATION EXPERIMENTS AND RESULTS

All the simulation experiments are carried out using Malayalam CV speech unit database, uttered by 96 different speakers. We used 8 kHz sampled speech signal which is low pass filtered to band limit to 4 kHz. We divide the dataset into training and test set which contains CV speech unit database, uttered by first 48 speakers for training and next 48 for testing. Thus training and test set contains total of 1728 samples each. The main objective of this article is to design a robust CV speech unit recognizer in additive noisy condition.

Table 2: Recognition Accuracies of 5 Different Phonetic Classes Using MFCC of Different SNR Values

Class	*Recognition Accuracy*											
	k-NN						*DDAGSVM*					
	0 dB	*3 dB*	*10 dB*	*20 dB*	*30 dB*	*Clean*	*0 dB*	*3 dB*	*10 dB*	*20 dB*	*30 dB*	*Clean*
Unaspirated	20.4	24.1	30.1	37.8	42.1	50.3	25.1	30.2	41.7	52.4	55.8	60.7
Aspirated	22.5	26.9	32.5	36.1	43.7	51.8	28.6	31.6	43.2	54.2	58.1	62.2
Nasals	26.9	30.2	37.3	39.2	45.9	54.8	30.7	33.8	45.9	57.1	61.5	64.9
Approximant	21.6	28.7	35.9	38.1	44.4	53.9	29.2	32.6	43.2	52.5	58.3	63.4
Fricatives	23.5	28.7	36.8	40.3	43.7	54.8	30.1	33.2	46.7	53.7	59.8	64.5

To perform this task each speech signal is corrupted by Additive White Gaussian Noise (AWGN) of different Signal to Noise Ratio (SNR) levels. The recognition accuracies obtained for Malayalam CV speech database for the five different phonetic classes at various SNR levels using MFCC, NWHF and SSPD parameters based on k-NN and DDAGSVM classifiers are tabulated in Table 2, Table 3 and Table 4 respectively.

Table 3: Recognition Accuracies of 5 Different Phonetic Classes Using NWHF of Different SNR Values

Class	*Recognition Accuracy*											
	k-NN						*DDAGSVM*					
	0 dB	*3 dB*	*10 dB*	*20 dB*	*30 dB*	*Clean*	*0 dB*	*3 dB*	*10 dB*	*20 dB*	*30 dB*	*Clean*
Unaspirated	25.7	28.9	45.5	56.2	62.4	62.6	45.8	52.2	61.2	72.7	75.2	76.8
Aspirated	29.1	36.4	47.9	59.6	63.9	63.3	48.3	55.3	63.7	74.6	78.8	79.9
Nasals	46.1	50.9	57.6	61.2	65.7	65.7	55.6	60.2	65.3	77.8	81.9	83.6
Approximant	41.7	48.3	55.2	60.6	64.8	64.9	50.3	55.7	63.8	72.8	78.5	81.9
Fricatives	43.3	49.1	56.5	60.9	63.1	63.5	53.6	59.4	66.9	78.6	82.1	84.2

Table 4: Recognition Accuracies of 5 Different Phonetic Classes using SSPD of Different SNR Values

Class	*Recognition Accuracy*											
	k-NN						*DDAGSVM*					
	0 dB	*3 dB*	*10 dB*	*20 dB*	*30 dB*	*Clean*	*0 dB*	*3 dB*	*10 dB*	*20 dB*	*30 dB*	*Clean*
Unaspirated	41.2	44.5	49.8	53.2	58.7	64.9	58.7	62.2	74.8	78.7	81.2	86.1
Aspirated	40.4	43.4	48.1	50.9	56.8	63.1	58.3	65.9	71.9	78.6	81.8	85.6
Nasals	48.1	51.2	53.1	59.3	63.6	70.2	65.4	70.8	75.3	81.4	88.5	94.2
Approximant	46.8	49.9	52.9	58.2	62.8	68.5	60.3	65.5	73.4	82.8	88.2	93.9
Fricatives	46.1	49.3	51.9	57.8	62.1	67.4	57.5	59.4	66.9	78.6	89.1	92.4

CONCLUSION

This paper describes two statistical learning approaches such as k-Nearest Neghborhod (k-NN) and Support Vector Machine (SVM) for the classification of Malayalam CV speech unit database in speaker independent environments. Three different modeling algorithms viz., MFCC, NWHF and SSPD parameters are discussed to evaluate the performance of these two classifiers. The recognition accuracies are calculated and compared using SVM and k-NN and it is observed that the time domain based SSPD parameters using DDAGSVM algorithms is performing well compared with NWHF and MFCC in speaker independent noisy environments. Classification of Malayalam consonants inherent with all the vowels are some of our future research direction.

REFERENCES

[1] Anil K. Jain, Robert P.W. Duin and Jianchang Mao, "Statistical Pattern Recognition: A Review", IEEE Trans. on Pattern Analysis and Machine Intelligence, Vol. 22(1), Jan 2000.

[2] A. Grossman, J. Morlet and P. Gaoupillaud, "Cycle octave and related transforms in seismic signal Analysis", Geoexploration, Vol. 23, pp. 85–102, 1984.

[3] C. torrence and G.P. Compo, "A practical guide to Wavelet Analysis", Bull. Amer. Meteorological. Soc., Vol. 79(1), pp. 61–78, 1998.

[4] Hongyu Liao, Mrinal Kr and Bruce F. Cockburn, "Efficient Architectures for 1-D and 2-D Lifting Based Wavelet Transform", IEEE Trans. on Signal Processing, Vol. 52(5), pp. 1315–1326, 2004.

[5] S. Mallat, "A wavelet Tour of Signal Processing, The Sparse Way", NewYork: Academic, 2009.

[6] K.P. Soman and K.I. Ramachandran, "Insight into Wavelets, from Theory to Practice", Prentice Hall of India, 2005.

[7] M. Vetterly and C. Herley, "Wavelets and Filter banks: Theory and Design", IEEE Trans. on Signal Processing, Vol. 40(9), pp. 2207–2232, 1992.

[8] M.J. Shensa, "Affine Wavelets: Wedding the Atrous and mallat Algorithms", IEEE Trans. on Signal Processing, Vol. 40, pp. 2464–2482, 1992.

[9] S. Mallat, "Multi frequency Channel Decomposition of Images and Wavelet Models", IEEE. Trans on Acoustics, Speech and Signal Processing, Vol. 37, pp. 2091–2110, 1989.

[10] I. Daubechies, "Ten Lectures on Wavelets", Philadelphia, PA:Soc. Appl. Math, 1992.

[11] Ronald W. Lindsay, Donald B. Percival and D. Andrew Rothrock, "The Discrete Wavelet Transform and the scale analysis of the surface properties of Sea Ice", IEEE Trans. on Geo Science and Remote Sensing, Vol. 34(3), pp. 771–787, 1996.

[12] E. Ott, *Chaos in Dynamical Systems*, Cambridge University Press, 1993.

[23] G.L. Baker and J. Gollub, *Chaotic Dynamics: An Introduction*, Cambridge University Press, 1996.

[14] Michael T. Jhonson, Rchard J. Povinalli, Andrew C. Lindgren, Jinjin Ye, Xiaolin Liu and Kevin Indrebo, "Time Domain Isolated Phoneme Classification using Reconstructed Phase Space", IEEE Trans. on Speech and Audio Processing, Vol. 13, No. 4, pp. 458–466, 2005.

[15] F. Takens, (1980), "Detecting Strange Attractors in Turbulence", *in Proc. Dynamical Systems and Turbulence*, Warwick, U.K., pp. 366–381.

[16] H. Kantz and T. Schreiber, (1997) *Non Linear Time Series Analysis*, Cambridge University Press.

[17] D.S. Broomhead and G.P. King, (1986). "Extracting qualitative Dynamics from experimental data", *Physica D*, pp. 217–236.

[18] Tou J.T. and Gonzalez R.C., "Pattern Recognition Principles", Addison-Wesley, London, 1974.

[19] Friedmen M. and Kandel A., "Introduction to Pattern Recognition: Statistical, Structural, Neural and Fuzzy Logic Approach", World Scientific, 1999.

[20] Cover T.M. and Hart P.E., "Nearest Neighbor Pattern Classification", IEEE trans. on Information Theory, Vol. 13 (1), pp. 21–27, 1967.

[21] Min-Chun Yu, "Multi-Criteria ABC analysis using artificial-intelligence based classification techniques", Elsevier- Expert Systems With Applications, Vol. 38, pp. 3416–3421, 2011.

[22] Hand D.J., "Discrimination and classification", NewYork, Wiley, 1981.

[23] Ray A.K and Chatterjee B, "Design of a Nearest Neighbor Classifier System for Bengali Character Recognition", Journal of Inst. Elec. Telecom. Eng, Vol. 30, pp. 226–229, 1984.

[24] Zhang B. and Srihari S.N., "Fast k-Nearest Neighbor using Cluster Based Trees", IEEE trans. on Pattern Analysis and Machine Intelligence, Vol. 26(4), pp. 525–528, 2004.

[25] Pernkopf F., "Bayesian Network Classifiers versus selective k –NN Classifier", Pattern Recognition, Vol. 38, pp. 1–10, 2005.

[26] Ying Tan and Jun Wang (2004). "A Support Vector Machine with a Hybrid Kernel and Minimal Vapnik-Chervonenkins Dimension", IEEE *Trans. On Knowledge and Data Engineering*, Vol. 10, No. 4, pp. 385–395 .

[27] Vladimir N Vapnik, (1999), "An Overview of Statistical Learning Theory", IEEE Trans. On Neural Networks, Vol. 10, No. 5, pp. 988–999 .

[28] Ravi Gupta, Ankush Mittal and Kuldip Singh, "A time Series based Feature Extraction Approach for Prediction of Protein Structured Class", *EURASIP Journal on Bioinformatics and System Biology*, 2008.

[29] E. Osuna, R. Freund, and F. Girosi (1997). "Training Support Vector Machines: An Application to Face Detection," Proc. IEEE *Conf.Computer Vision and Pattern Recognition*, pp. 17–19.

[30] M. Pontil and A. Verri (1998). "Support Vector Machines for 3D Object ecognition," *IEEE Trans. Pattern Analysis and Machine Intelligence*, Vol. 20, No. 6, pp. 637–646.

Evaluating Accurate Values of the Reverberation Chamber Sound Absorption Coefficient

Mahavir Singh and Dharam Pal Singh

Ultrasonics and Vibration (AUV) Section, Apex Level Standards and Industrial Metrology Division, CSIR-National Physical Laboratory, New Delhi

E-mail: mahavir@nplindia.org; mahavir.acoustics@gmail.com

ABSTRACT: *In order to check that the reverberation chamber measurement of absorption coefficient was consistent, a short run of repeatability measurements were carried out as per standard procedure. The standard ISO/IEC 17025:2005 on the competence of testing and calibration laboratories requires that these laboratories shall apply procedures for estimating the accurate values of their measurement results. One of the possibilities is to evaluate the accurate values, taking into account all components that contribute significant value to the final result. In case of the sound absorption coefficient value, carried out according to the standard EN ISO 354:2003, the overall value is first of all influenced by the reverberation times* T_1, T_2 *and the power attenuation coefficients* m_1 *and* m_2, *calculated according to the ISO 9613–1 standard and representing the climatic conditions in the reverberation room. In spite of very little difference between the values* m_1 *and* m_2 *representing the change of climatic conditions usually, it is the case in laboratory, exponential form of the coefficient's function causes that the accurate value of measurement results increase with frequency very fast. Particularly for the high frequencies, the accurate values of measurements are so important that the evaluation of the sound absorption coefficient is practically not possible.*

INTRODUCTION

However there is at present no wholly accepted calculations of repeatability and standard error in reverberation chamber absorption coefficient and the data quoted in the literature varies widely. Figure 1 shows a typical RT spectrum for the empty room, along with the IS 8225 minimum curve for the particular room volume of 257 m^3 and surface area of 244 m^2.

Because of lack of uncertainty evaluation based on inter-laboratory validation approach, it has to be carried out by the laboratory itself. Using the general methods specified in GUM[1], first of all we need to establish a relationship between the mesurand Y and other quantities $X_1, X_2, \ldots X_n$ (input values) through a function f, called the measurement equation

$$Y = f(X_1, X_2, \ldots \ldots . . X_n) \qquad \ldots (1)$$

Such equation must express not only the physical relation from which we can obtain the value of mesurand Y but also it should be accompanied by a quantitative statement of its uncertainty which arises from the uncertainties of the input values of directly measured quantities $x_1, x_2 \ldots, x_n$ being the estimates of $X_1, X_2, \ldots, X_n$.

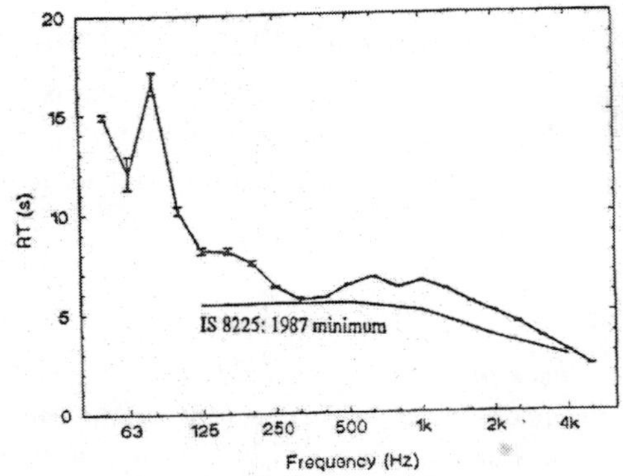

Fig. 1: Average Reverberation Time in Empty Chamber Plotted with IS 8225–1987 Minimum Curve. The Error Bars Represent ± One Standard Error

In general, uncertainty components (each of them-represented by estimated standard deviation, termed standard uncertainty and noted u(xi) or, shortly, ui) are categorized according to the method used to evaluate them. There are two methods of evaluating standard uncertainty:

- Type A is based on any valid statistical method, for example the standard deviation of the mean of a series of independent observations s(xi); in such case the standard uncertainty u(xi) = s(xi).
- Type B takes advantage of an outside source (for example, data provided in calibration) and/or of an assumed distribution.

The standard deviation of the estimated measurement result *y,* called combined standard uncertainty $u_c(y)$, is obtained by combining the individual standard uncertainties $u(x_i)$ using the usual statistic method of combining standard deviations "root-sum-squares" according to the formula:

$$u_c^2(y) = \sum_{i=1}^{n} \left(\frac{\partial f}{\partial x_i}\right)^2 u^2(x_i) \qquad \ldots (2)$$

and calculating the positive root square of the result.

Equation (2) is called the law of propagation of uncertainty and the partial derivatives $\delta f/\delta x_i$ is referred to sensitivity coefficients. The products of module $\delta f/\delta x_i$ and u (x_i) are usually presented in a table, as "uncertainty budget". This practice is very useful to identify the dominant terms that contribute significant uncertainty to the result.

The measure of uncertainty, that defines an interval in which the value of the quantity subject to measurement (the mesurand *Y*) can be confidently asserted to lie, is named expanded uncertainty *U* (it means that one can confidently believe that the value of *Y*, estimated by *y*, lies within the limits $y - U \leq Y \leq y + U$).

The expanded uncertainty is obtained by multiplying a combined standard uncertainty by a coverage factor *k* depending on the desired level of confidence and the type of statistical distribution.

$$U(Y) = ku_c\,(y) \qquad \ldots (3)$$

UNCERTAINTY ELEMENTS

Repeatability of Measured Absorption Coefficient

In order to check that the reverberation chamber measurement of absorption coefficient was consistent, a short run of repeatability measurements was made, as recommended by IS 8225. A sample of 12.5 mm thick gypsum board panel (test sample area of 12.8 m^2) was measured six times in the same configuration, being removed from the chamber and replaced for each measurement. The repeatability "r" of the system was the found from

$$r = t\sqrt{2}.\sigma_{n-1} \qquad \ldots (4)$$

where t is Student's factor for 95% probability and five degrees of freedom and σ_{n-1} is the standard deviation of the six measurements. The six absorption coefficients are plotted in Figure 2 along with *r* and a typical standard error curve from one of the measurements.

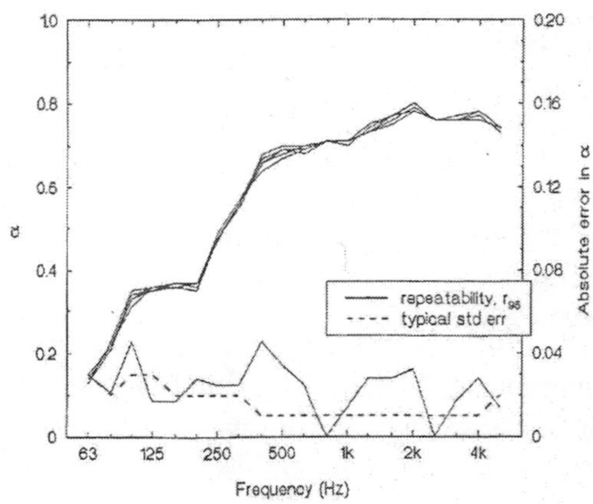

Fig. 2: Repeatability 95% of Six Gypsum Tiles Absorption Measurements and Typical Standard Error of One

The average α value of repeatability (r_{95}) is 0.11 in terms of reverberation chamber absorption coefficient and 0.02 in terms of absolute error. The average α value of typical standard error is 0.02 in terms of absolute error as shown in Figure 2.

Reverberation Time

The reverberation time is measured in *n* points for each frequency band. The mean value T_f calculated from *n* measurements is taken as the estimated measurement result. In this case, the standard uncertainty of T_f is equal to the experimental standard deviation $s\ (T_f)$ of the mean of a series of independent observations $T_{f,\ i=1,\dots n}$.

$$u(T_f) = s(T_f) = \frac{s(T_{f,i})}{\sqrt{n}}, s \qquad \dots (5)$$

Student's-t distribution with *n* freedom degrees is assumed.

Area of Sample

Assuming the rectangular distribution of estimated measurement result, with the boundary of ±0.005 m^2, the standard uncertainty of sample's area amounts to,

$$u(S) = \frac{0.005\text{m}^2}{\sqrt{3}} = 0.0029m^2 \qquad \dots (6)$$

Volume of Room

Assuming the rectangular distribution of estimated measurement result, with the boundary of resolution ±0.5 m^3, the standard uncertainty of room volume amounts to.

$$u(S) = \frac{0.5\text{m}^3}{\sqrt{3}} = 0.2887\ m^3 \qquad \dots (7)$$

Environment Factors

The measurement equations for the environmental factors like temperature, atmospheric pressure and relative humidity assume the form:

$$e = e_0 + \delta e_1 + \delta e_2 + \delta e_3 + \delta e_4 \qquad \dots (8)$$

where, e_0 the reading value of the environmental factor in a room,
δe_1 the dispersion of sensor indications,
δe_2 the resolution of sensor indications,
δe_3 the error of sensor indications,
δe_4 the uncertainty of indication error.

The standard uncertainty of dispersion δe_1 is calculated on the basis of the estimated standard deviation of *m* series of *n* measurements.

$$u(\partial e_1) = \frac{S_i(e_1)}{n} \quad \text{... (9)}$$

where, $S_i(e_1)$ the experimental standard deviation of ith series equal to

$$S_i(e_1) = \sqrt{\frac{\sum_{i=1}^{m} S_i(e_{1,i})}{m}} \quad \text{... (10)}$$

The standard uncertainty of resolution δe_2 is evaluated with the assumption of rectangular distribution with the boundary of resolution supposed, respectively, as b = ±0.05 °C, ± 0.05 kPa or ±0.05%.

$$u(\delta e_2) = \frac{b}{\sqrt{3}} \quad \text{... (11)}$$

Contribution in uncertainty of δe_3 and δe_4 are calculated from the expanded uncertainty *U* evaluated for normal distribution with confidence level 95% and stated in the calibration certificate of the device.

$$u(\delta e_3) = \frac{U\ (\delta e_3)}{k_{N,\ 95\%}} = \frac{U(\delta e_3)}{2} \quad \text{... (12)}$$

$$u(\delta e_4) = \frac{U\ (\delta e_4)}{k_{N,\ 95\%}} = \frac{U(\delta e_4)}{2} \quad \text{... (13)}$$

The equation of propagation law of uncertainty has the form:

$$u^2\ (e) = u^2(\delta e_1) + u^2(\delta e_2) + u^2(\delta e_3) + u^2(\delta e_4) \quad \text{... (14)}$$

UNCERTAINTY OF THE SOUND ABSORPTION COEFFICIENT MEASUREMENT

The sound absorption coefficient α_s of a plane absorber or a specified array of test objects is calculated using the formula,

$$\alpha_s = \frac{A_T}{S} \quad \text{... (15)}$$

where, A_T the equivalent sound absorption area of the test specimen, m^2
S the area covered by the test specimen, m^2.

The equivalent sound absorption area of the test specimen A_T, according to ISO 354:203 [3] is given as follows:

$$A_T = A_2 = A_1 = 55.3\ V \left[\frac{1}{C_2T_2} - \frac{1}{C_1T_1}\right] \text{-}\ 4V\ (m_2 - m_1) \quad \text{... (16)}$$

Where,

V	the volume of the empty reverberation room, m^3
T_1	the reverberation time of the empty reverberation room,
T_2	the reverberation time of the reverberation room after the test specimen has been introduced, s
c_1, c_2	the propagation speed of sound in air at the temperature t_1 and t_2 (respectively), m/s
m_1, m_2	the power attenuation coefficients, calculated according to ISO 9613–1[4] using the climatic conditions that have been present in the empty reverberation room and after the test specimen has been introduced, m^{-1}

Table 1: Sound Absorption with its Uncertainties and Sensitivity Coefficients

f, *Hz*	*αS*	$u_c(\alpha_S)$	$U(\alpha_S)$	*Sensitivity Coefficients of Environment Factors*					
				$\delta\alpha_S/\delta pa_1$	$\delta\alpha_S/\delta pa_2$	$\delta\alpha_S/\delta hr_1$	$\delta\alpha_S/\delta hr_2$	$\delta\alpha_S/\delta t_1$	$\delta\alpha_S/\delta t_2$
100	0.14	0.009	0.018	0.00000	0.00000	–0.00023	0.00018	0.00062	0.00085
125	0.31	0.014	0.029	0.00000	0.00000	–0.00035	0.00027	0.00061	0.00116
160	0.54	0.017	0.035	0.00000	0.00000	–0.00056	0.00045	0.00072	0.00181
200	0.64	0.020	0.041	0.00000	0.00000	–0.00082	0.00070	0.00072	0.00227
250	0.82	0.023	0.048	0.00000	0.00000	–0.00120	0.00109	0.00059	0.00301
315	0.92	0.025	0.052	0.00000	0.00000	–0.00177	0.00174	0.00020	0.00382
400	0.94	0.027	0.056	0.00000	0.00000	–0.00267	0.00280	–0.00064	0.00468
500	1.02	0.020	0.041	0.00000	0.00000	–0.00398	0.00437	–0.00182	0.00571
630	0.99	0.019	0.040	0.00001	–0.00001	–0.00617	0.00694	–0.00312	0.00680
800	1.00	0.024	0.049	0.00001	–0.00001	–0.00994	0.01119	–0.00448	0.00811
1000	0.96	0.019	0.039	0.00001	–0.00002	–0.01568	0.01748	–0.00568	0.00929
1250	0.98	0.026	0.053	0.00002	–0.00002	–0.02478	0.02729	–0.00692	0.01062
1600	0.98	0.023	0.047	0.00002	–0.00003	–0.04105	0.04464	–0.00852	0.01224
2000	0.96	0.030	0.060	0.00003	–0.00004	–0.06455	0.06960	–0.01033	0.01418
2500	0.99	0.045	0.093	0.00004	–0,00005	–0.10116	0.10837	–0.01285	0.01702
3150	1.03	0.068	0.139	0.00005	–0.00007	–0.16047	0.17109	–0.01665	0.02146
4000	1.01	0.106	0.217	0.00007	–0.00011	–0.25730	0.27334	–0.02279	0.02829
5000	1.02	0.163	0.332	0.00009	–0.00015	–0.39768	0.42128	–0.03153	0.03816
6300	0.98	0.253	0.516	0.00011	–0.00020	–0.61920	0.65403	–0.04515	0.05338
8000	0.95	0.394	0.804	0.00011	–0.00023	–0.96576	1.01642	–0.06626	0.07662

The value of m is calculated from the attenuation coefficient α dependent on temperature t, atmospheric pressure p_a and relative humidity h_r, according to the formula,

$$m = \frac{\alpha}{10\log(e)} \quad \ldots (17)$$

The detailed analyze of the uncertainty budget leads to a conclusion that the standard uncertainty of sound absorption depends very strongly on the sensitivity coefficients of humidity, particularly at the high frequencies (see Table 1).

In the previous version of standard ISO 354:1985 [2] a possible change of climatic conditions was not taken into consideration ($m_1 = m_2 = 0$). Table 2 presents the calculation results of the sound absorption coefficient α_s and its standard and expanded uncertainties carried out with respect to environment factors and without them. It can be observed that in spite of very little difference between the values m_1 and m_2 (representing the change of climatic conditions), the values of α_s calculated according to [3] in the frequency bands $f \geq 3150$ Hz are considerably greater. Also, the uncertainties of measurement results increase with frequency very fast. Particularly for the high frequencies, the values of uncertainties are so important that the evaluation of α_s is practically not possible.

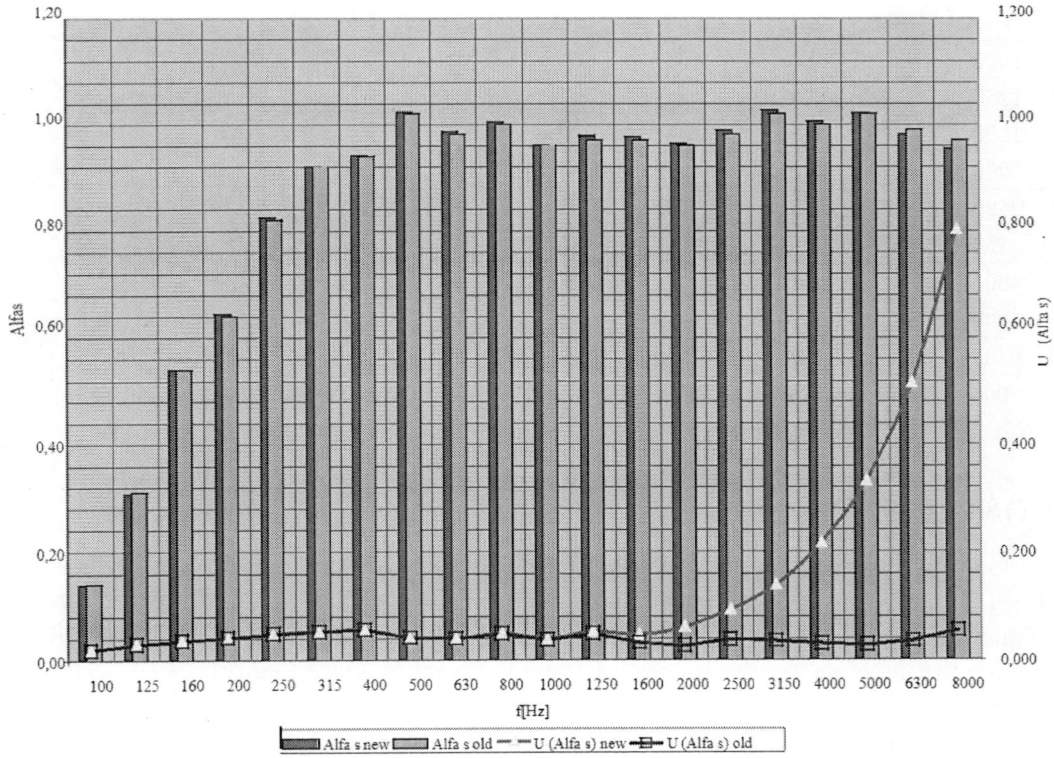

Fig. 3: Comparison the Calculation Results of the Sound Absorption Coefficient Expanded Uncertainty Carried Out According to Both Version of ISO 9613–1 Standard

Table 2: Set of Calculation Results of Sound Absorption Coefficient α_s and its Uncertainties Conducted According to Both Versions of Standard ISO 354 [3–4, 6]

f, Hz	*Acc. to ISO 354:2003 [3]*			*Acc. to ISO 354:1985 [2]*			m_2-m_1	*Difference* α_S
	α_S	$u_c(\alpha_S)$	$U(\alpha_S)$	α_S	$u_c(\alpha_S)$	$U(\alpha_S)$		
100	0.14	0.009	0.018	0.14	0.009	0.018	0.000	–0.001
125	0.31	0.014	0.029	0.31	0.014	0.029	0.000	–0.004
160	0.54	0.017	0.035	0.54	0.017	0.035	0.000	0.000
200	0.64	0.020	0.041	0.64	0.020	0.041	0.000	0.005
250	0.82	0.023	0.048	0.82	0.023	0.048	0.000	0.004
315	0.92	0.025	0.052	0.92	0.025	0.052	0.000	0.000
400	0.94	0.027	0.056	0.94	0.027	0.056	0.000	0.001
500	1.02	0.020	0.041	1.02	0.020	0.041	0.000	0.004
630	0.99	0.019	0.040	0.98	0.019	0.039	0.000	0.006
800	1.00	0.024	0.049	1.00	0.024	0.048	0.000	0.005
1000	0.96	0.019	0.039	0.96	0.018	0.037	0.000	0.000
1250	0.98	0.026	0.053	0.97	0.024	0.048	0.000	0.008
1600	0.98	0.023	0.047	0.97	0.015	0.031	0.000	0.005
2000	0.96	0.030	0.060	0.96	0.013	0.026	0.000	0.003
2500	0.99	0.045	0.093	0.98	0.018	0.038	0.000	0.008
3150	1.03	0.068	0.139	1.02	0.017	0.035	0.000	0.007
4000	1.01	0.106	0.217	1.00	0.014	0.029	0.000	0.006
5000	1.02	0.163	0.332	1.02	0.013	0.027	0.000	0.001
6300	0.98	0.253	0.516	0.99	0,017	0.035	0.000	–0.010
8000	0.95	0.394	0.804	0.97	0.026	0.054	0.000	–0.016

CONCLUSION

The standard error curve was computed from the variance of the measured RTs according to [3]. The evaluation of the uncertainty is very important because it testifies to the quality of measurements. Observing the uncertainty budget for each band of frequency allows finding the deciding factors. In case of sound absorption measurements, it could be very difficult to obtain accurate results due to the important uncertainty at high frequencies bands. The function circumscribing the power attenuation coefficient *m* calculated according to ISO 9613–1, thanks to the non-linear shape, influences the uncertainty of measurement results in spite of very small changes in environmental conditions in laboratory.

REFERENCES

[1] Guide to the Expression of Uncertainty in Measurement. International Organization for Standardization, Geneva, First Edition, 1993. Corrected and reprinted 1995.

[2] IS 8225–1987, Measurement of Sound Absorption in a Reverberation Room, Bureau of Indian Standard, New Delhi.

[3] Mahavir Singh, Omkar Sharma and V. Mohanan, Repeatability and standard error in reverberation chamber absorption coefficient, ICA2004 (Fr2.B1.4), pp. 3393–3396, 2004.

[4] EN ISO 354:1985, Acoustics-Measurement of sound absorption in a reverberation room.

[5] EN ISO 354:2003, Acoustics-Measurement of sound absorption in a reverberation room.

[6] ISO 9613–1:1993, Acoustics-Attenuation of sound during propagation outdoors—Part 1: Calculation of the absorption of sound by the atmosphere.

Ultrasonic Investigation in Organic Solvent Mixtures: Applicablity of Linear Free Energy Relationships

S.V. Ranganayakulu
Guru Nanak Institutions, Technical Campus, Ibrahimpatnam, Hyderabad
E-mail: nayakulu@rediffmail.com

ABSTRACT: *Ultrasonic Investigation studies were carried out in o-cresol, p-cresol an m-cresol in Ethyl acetate mixture. Ultrasonic velocity (v) is measured by using Ultrasonic Pulse Echo Overlap Technique, density (ρ) is measured by Specific Gravity bottle, Viscosity (η) is measured by Cannon Fenske Viscometer. Internal Pressure (π_i) and Free Volume (V_f) are calculated and is correlated with Linear Free Energy Relationships such as, Kosower Constant, Dielectric Constant and Dimroath constants. Linear plots of all these studies confirm the rationalization of data. The relative merits are discussed in this paper.*

INTRODUCTION

The sound velocity in liquid mixtures is known to be a quantity, which depends on concentration in a variety of manners according to the nature of each component liquid. Agarwal, V.C *et al.* [1] have been reported the ultrasonic velocity and absorption of critical mixture of water and phenol by an improved Pulse-Echo technique. Jaya Kumar, S *et al.* [2] measured the ultrasonic velocities (v) in the binary mixtures of o-Cresol, p-Cresol, o-Chloro phenol and p-Chloro phenol with Carbon tetra chloride at 303.15 K. Palaniappan, PLMR *et al.* [3] measured the ultrasonic velocity (v) in binary liquid mixtures with m-Cresol as the common component using Ultrasonic interferometer technique. Viswanatha Sarma, A *et al.* [4] computed the ultrasonic velocity (v) in two binary liquid mixture viz., m-Cresol-phenylhdrazine and m-Cresol-Quinoline using Nomoto and Van Dael ideal mixing relations at three temperatures. Muraliji, D *et al.* [5] evaluated internal pressure (π_i) and free volume (V_f) parameters of aqueous solution of phenols from the measured ultrasonic velocity (v), density (ρ) and viscosity (η). Muraliji, D *et al.* [6] reported the ultrasonic velocity (v), viscosity (η) and density (ρ) for binary mixture of Cresol in aqueous solutions at 304 K temperature. Satyanarayana, T *et al.* [7] investigated the ultrasonic studies of binary mixture of o-Chlorophenol, Cresols with N, N-diethyl acetamide at 308.15 K, 313.15 K, 318.15 K and 323.15 K temperatures over the whole composition range using ultrasonic Pulse Echo-Interferometer. Viswanatha Sarma, A *et al.* [8] have been reported the ultrasonic velocity (v) and absorption in binary liquid system of Phenols with N-N Dimethyl acetamide using Pulse Echo technique at a single frequency of 10 MHz at four different temperatures over the entire composition range. Vittal Prasad, T.E *et al.* [9] has been carried out activity coefficients (at 95.5 k Pa) over the entire composition range are obtained for the binary mixtures formed by m-Cresol with bromobenzene and o, m and p-xylenes.

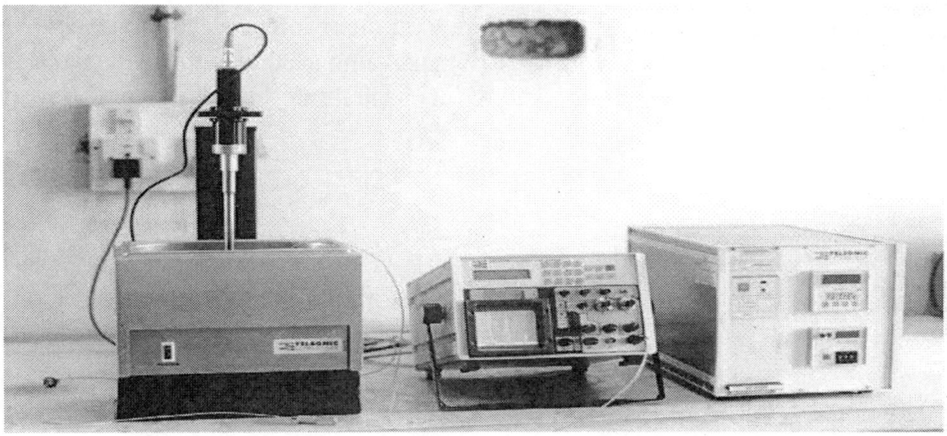

Fig. A: Ultrasonic Pulse Echo System and Temperature Control System

RESULTS AND DISCUSSION

It is noteworthy that, with increase of the Cresol component, there is an increase in internal pressure (π_i) and decrease in effective volume (V_f). This is quite in consonance with the usual pressure-volume interaction in binary mixtures. This is also confirmed by the linear plots obtained between effective volume (V_f) and internal pressure (π_i) with the density. Thus it is reasonably assumed that, use of Kosower's parameter [11] is sound. The abnormality at the extreme component volume percentage of Ethyl acetate (i.e., at 10% and 90% compositions) m-Cresol shows anomalous behaviour of higher ultrasonic velocity (v) indicating that, acidity it self is not the controlling factor at these compositions. It appears that intermolecular hydrogen bonding between the Cresol molecule and cresol—solvent molecules may be the factor causing the abnormality. Earlier this has been traced to transmission coefficients in organic reactions reported by Anantharaman, A *et al.* [12] and Armis, E.S.T [13]. Meta cresol has lower ultrasonic velocity (v) because of high acidity. Para cresol has higher ultrasonic velocity (v), because of the lower acidity. Hence the correlation was done using acidity constant of the three isomers and linearity is observed. This means that, acidity and electron releasing nature contribute to the variation of ultrasonic velocity (v) of the three isomers. The correlation of ultrasonic velocity (v) is compared with dielectric constant (∈) [John. A. Dean., 14] reveals that there is increase in ultrasonic velocity (v) with the increase in dielectric constant (∈). The more polar the medium, there is larger ultrasonic velocity (v). The variation of ultrasonic velocity (v) depends upon the (i) Polarity of the medium (or) Macro dielectric constant (∈), (ii) Acidity of the medium and (iii) Electron releasing nature of the substitute in the cresols. Ultrasonic velocity (v) increases with increasing dielectric constant (∈) (Figure 1) of the mixture of the two components. Similarly ultrasonic velocity (v) varies linearly in increasing proportion with increasing (Z) values (Figure 2). Hence, the order of increased is in consonance with the macro dielectric constant '∈', as well as 'Z' values. The Kosower parameter (Z) takes into account factors like cohesion of the liquid components of the mixture, acidity (or) basicity of the solvent mixtures and the

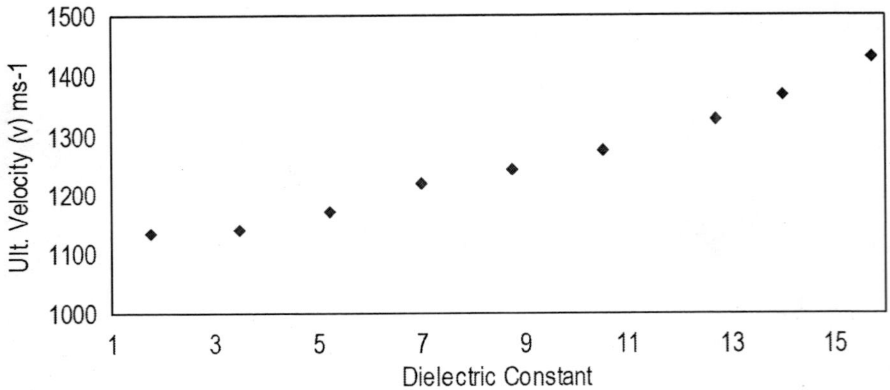

Fig. 1: Dielectric Constant (∈) Vs Ultrasonic Velocity (v) of Mixture of Ethyl Acetate and o- Cresol at 38°C Temperature

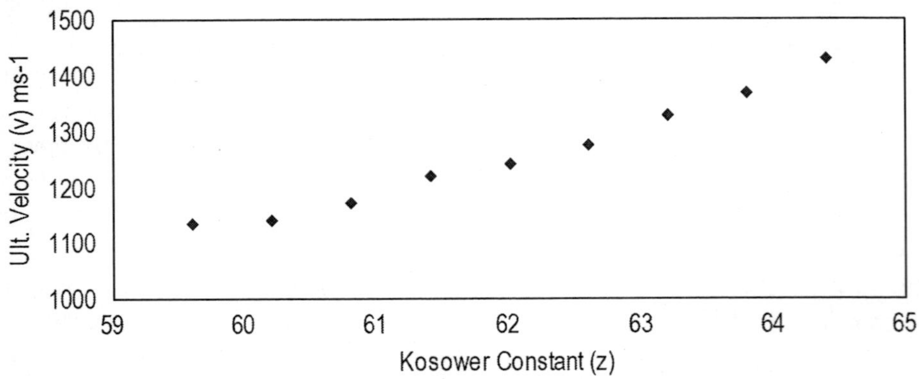

Fig. 2: Kosower Constant (Z) Vs Ultrasonic Velocity (v) of Mixture of Ethyl Acetate and o-Cresol at 38°C Temperature

internal dielectric constant ($\in$) of the solvents of the mixtures. Thus it is clear that, the ultrasonic velocity (v) increases with the polarity of the medium, which includes macroscopic and microscopic dielectric constants ($\in$) of the medium. Free volume (V_f) decreases with increase in dielectric constant ($\in$) as well as Kosower constant (Z) (Figure 3 and Figure 4) and internal pressure (π_i) increases with increase in dielectric

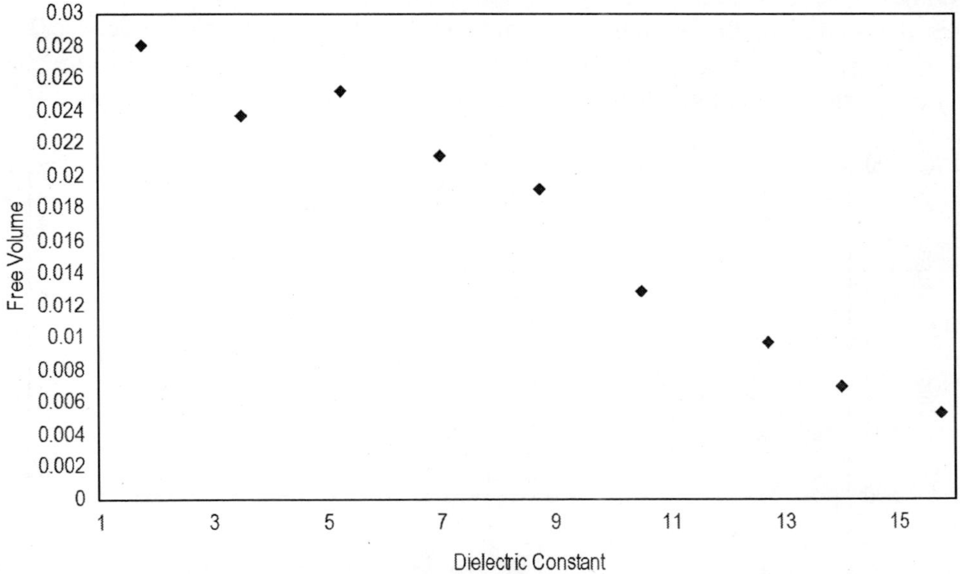

Fig. 3: Dielectric Constant ($\in$) Vs Free Volume (V_f) of Mixture of Ethyl Acetate and Cresol at 38°C Temperature

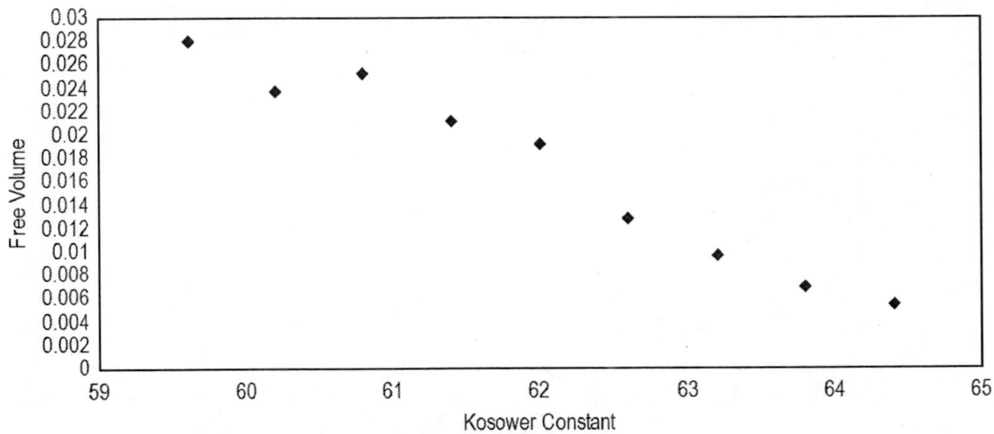

Fig. 4: Kosower Constant (Z) Vs Free Volume (V_f) of Mixture of Ethyl Acetate and o-Cresol at 38°C Temperature

constant ($\in$) and Kosower parameter (Z) (Figure 5 and Figure 6). The internal pressure (π_i) increased through marginally with increasing 'Z'. This is quite reasonable as effective free volume (V_f) and internal pressure (π_i) should be opposite in character. Hence use of Kosower parameter (Z) helps in understanding the internal pressure (π_i) variation and free volume (V_f) variation due to the corrections made. To sum up the observations are: ultrasonic velocity (v) is dependent on the polarity of the medium as well as the acidity of the medium. Both factors affect the ultrasonic velocity (v). High polarity of the medium leads to larger ultrasonic velocity (v) where as the increase in acidity of the solvent mixture, there will be decrease in ultrasonic velocity (v). So in solvents

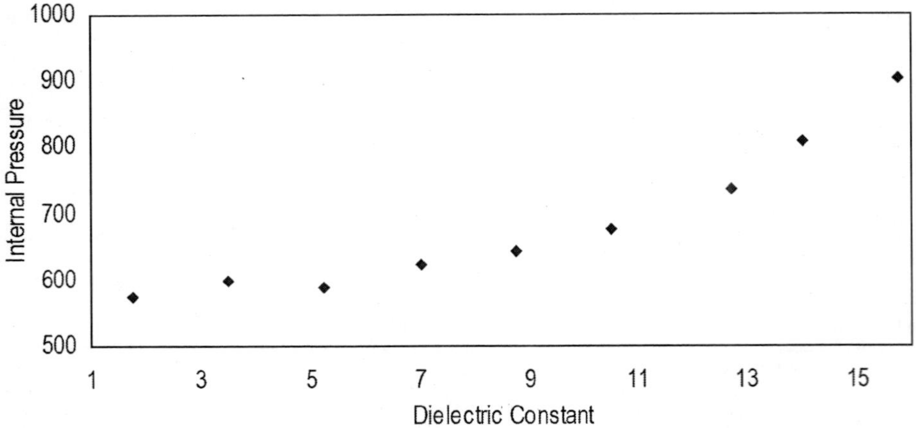

Fig. 5: Dielectric Constant ($\in$) Vs Internal Pressure (π_i) of Mixture of Ethyl Acetate and o-Cresol at 38°C Temperature

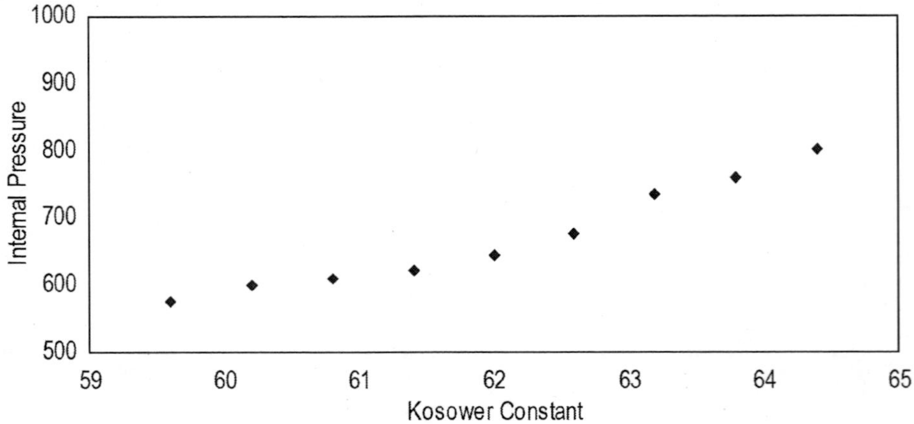

Fig. 6: Kosower Constant (Z) Vs Internal Pressure (π_i) of Mixture of Ethyl Acetate and o-Cresol at 38°C Temperature

which are polar as well as acidic cumulatively, there is increase and decrease in ultrasonic velocity (v) and as a result counter balancing each other. In the present case, the polarity of the medium outweighs the acidity of the medium leading to increase in ultrasonic velocity (v). In addition to the above studies, correlation has been done with ultrasonic velocity (v) and parameters like Dimroth parameter (E_T). The direction of ultrasonic velocity (v), internal pressure (π_i), free volume (V_f) is similar as observed with dielectric constant ($\in$) and Kosower's parameter (Z).Thus the parameters ultrasonic velocity (v), internal pressure (π_i) and effective volume (V_f) are controlled by factors like macro dielectric constant ($\in$) of the solvent mixture, acidity of the solvent mixture. Refined treatment of solvent effects like Kosower and Dimroth are extremely useful in rationalizing the measurements of ultrasonic velocity (v), free volume (V_f) and internal pressure (π_i) (Figure 7 and Figure 8).

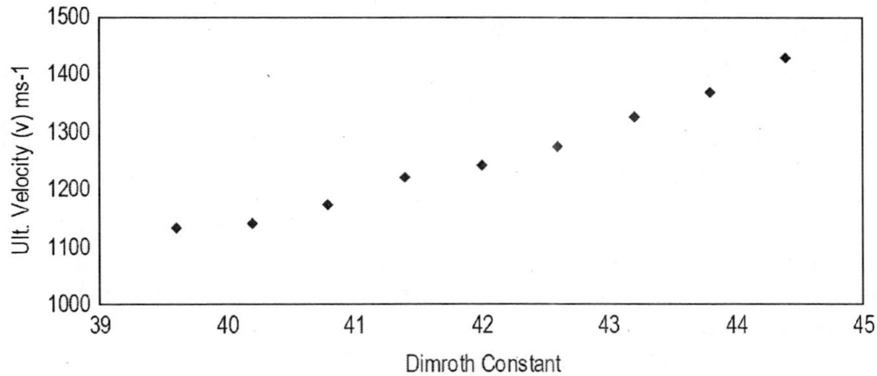

Fig. 7: Dimroth Constant (E_T) Vs Ultrasonic Velocity (v) of Mixture of Ethyl Acetate and o-Cresol at 38°C Temperature

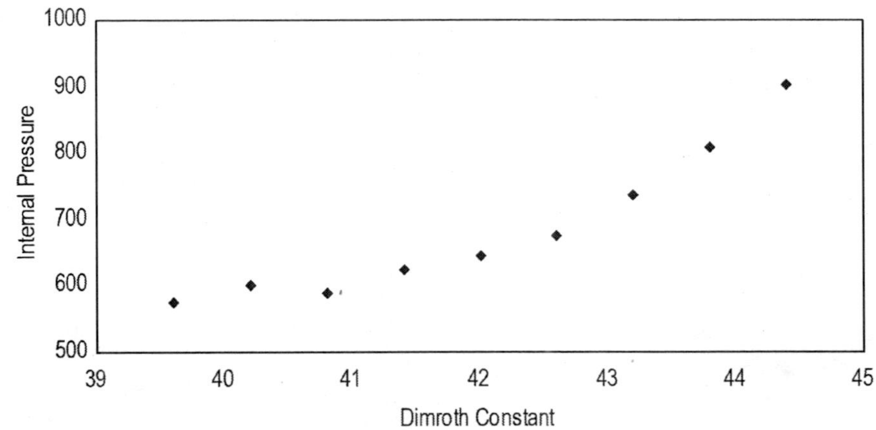

Fig. 8: Dimroth Constant (E_T) Vs Internal Pressure (π_i) of Mixture of Ethyl Acetate and o-Cresol at 38°C Temperature

CONCLUSIONS

Linear Energy Relationships such as Kosower constant, Dimroath constant and Dielectric Constant are correlated to Ultrasonic velocity, Internal Pressure, Free Volume parameters and found that they are in good agreement.

ACKNOWLEDGEMENTS

Author is thankful to Dr H S Saini, Managing Director and Sardar G S Kohli, Vice Chairman of GNI for valuable encouragement.

REFERENCES

[1] Agarwal, V.C. and Gupta, A.K., *J. Phys. D: Appl. Phys.*, 8 (1975) 1206.

[2] Jaya Kumar, S., Karunanidhi, N. and Kannappan, V., *Ind. J. Pure and Appl. Phy.*, 34 (1996). 761.

[3] Palaniappan, PLRM. Pichaimuthu, A and Kannappan, A.N., *Indian. J. Phys., 72B* (2) (1998). 175.

[4] Viswanatha Sarma, A and Viswanatha Sastry, J., *J. Acoust. Soc. Ind.*, 27 (1999) 309.

[5] Muraliji, D., Sekar, S, Dhanalakshmi, A and Ram Kumar, A.R., *J. Acoust. Soc. Ind.*, 29 (2001), 270.

[6] Muraliji, D., Sekar, S., Dhanalakshmi, A and Ramkumar, A.R., *J. Pure and Appl. Ultrasonics.*, 24 (2002) 71.

[7] Satyanarayana Rao, T., Veeraiah, N and Rambabu, C., *Ind. J. Pure and Appl. Phy.*, 40 (2002) 85.

[8] Viswanatha Sarma, A., Samatha, K and Murthy, J.S.R., *J. Acoust. Soc. Ind.*, 31(2003) 334.

[9] Vittal Prasad, T.E., Malleswara Reddy, N, Anil Kiran, N and Prasad, D.H.L., *Physics and Chemistry of Liquids.*, 44 (2006)

[10] Kosower, *J., Am. Chem. Soc.,* 80 (1958) 3253.

[11] Finar, I.L., Organic chemistry, *The Fundamental Principles,* Volume-I, 6th Edition (1995).

[12] Anantharaman, A and Krishnan, S.V., *Proc. Ind. Aca. Sci.*, 49 (1959) 174.

[13] Armis, E.S.J., *Analyt. Chem.*, 27 (1955) 1672.Dimroth, J., Liebigs, *Anal. Chem.*, 1(1969) 727.

Role of Ultrasonic and Spectroscopic Techniques in the Study of Molecular Interactions in Solutions

E. Jasmine Vasantha Rani

PG and Research Department of Physics, Seethalakshmi Ramaswami College, Trichy

ABSTRACT: *Interest in liquids is as old as science itself and it has always been realized that their properties are of significance in almost all sciences. In recent years ultrasonic studies in aqueous and non-aqueous electrolytic and non-electrolytic solutions have led to new insights in research. The study of liquid/solution using acoustical methods constitutes the mining field for scientist. Ultrasonic has found wide application in medical and biological fields.*

The velocity of propagation of ultrasonic waves in a liquid is sensitive to the thermodynamic fluctuation of pressure, volume and temperature. The most important interactions in chemical and biological systems are occurring in solutions link in the study of environmental conditions. Additionally liquids are utilized in a numerous industrial applications. It is for these reasons that the understanding of the properties of the molecular condition evokes keen interest in many fields of science and engineering. Ultrasonic technique is the simple and accurate method of analyzing the interactions between the molecules of liquids and solutions.

Sonic waves which couple fundamental thermodynamic properties constitute a valuable tool of the studies of large variety of molecular processes. Ultrasonic velocity is indispensable in evaluating the thermodynamic parameters such as internal pressure and free volume. Internal pressure is a qualitative measurable parameter of any condensed system which plays a dominant role in transport properties.In the solutions of electrolytes, the interaction between ion and solvent molecules is described in terms of ion solvation. Compressibility of the solution is assumed to be mainly due to the contribution of free solvent molecules. So the solvation effect of solutions is studied by computing compressibility using ultrasonic velocity. Apparent molal volume and apparent molal compressibility have been proven to be very useful in explaining the structural changes happening in solutions.

Spectroscopy has emerged as a powerful and an accurate tool in investigating the interaction between solute-solvent and solute-solute in solutions. Spectroscopic techniques are being widely used as research tools with unique capability and sensitivity. They provide information about the structure of solvent around the solute molecules. The structural changes occurring in the solution are envisaged through spectroscopy. These microscopic investigations support the acoustical, thermo dynamical and thermo chemical results obtained from so many research activities of liquid state physics.

Radiated Noise of Underwater Vehicles—Measurements and Mitigation

N.S. Naidu

Research Consultant, NSTL, Visakhapatnam

BRIEF SUMMARY: *Noise of Ships and Submarines is very important part in the field of Underwater Acoustics. The monitoring and reduction is very crucial and important in the habitability of the crew and passengers of the vessel. The performance of underwater detection systems such as Sonars depends on the noise generated by their own vessels and also of hostile vessels to be detected. Hostile Naval forces use the underwater noise signatures radiated by ships and submarines for detection and classification. It is essential to make periodic and regular radiated underwater noise measurements of Naval vessels to assess their vulnerability of detection and also to identify dominant noise sources if any, than expected, for application of retrofit noise control measures. This presentation describes the different components of total noise of ship or submarine, methodologies of conducting trials for collection of data, analysis and interpretation of results.*

INTRODUCTION

The subject of the underwater radiated noise by ships has been, in the past, studied and developed with particular attention to the silencing of warships and submarines for military applications. In fact, quietness is one of the most important aspect to deal with in order to improve and maintain stealth. A vessel with low underwater radiated noise remains undetected while accomplishing an operative mission and can represent the most effective countermeasure against enemy. The Underwater threat strongly depends on the acoustic performance of the platform and hence the war worthiness. It may be pertinent to note that all the technical data related to the underwater noise in military application are not at disposal in the open literature.

Recently a growing interest has been observed in the underwater radiated noise even for civilian ships like, oceanographic and fisheries research ships as well as merchant ships sailing in special sea protected environments. Radiated noise, in fact, can affect the ship performance causing a listening degradation of the own sensors. Similar aspects are today considered also for the most advanced fisheries research ships and for ships sailing in special sea protected areas, on the basis of the research centre studies on the fish reaction to noise. For fishery ships and for the sea protected areas the reduction of the ship emitted noise has the objective to reduce, as far as possible, to avoid any fish disturbance in their natural habitat.

The studies performed and the experience acquired for the naval constructions, in the field of the underwater noise reduction and control, represent now an important aid for the underwater noise policy in merchant ships. The noise reduction procedures can, in fact, be transferred, adapted and utilized to any kind of ship having particular requirements, in general expressed in terms of maximum emitted noise versus frequency.

Measurements of Radiated Underwater noise levels of sea going vessels during commissioning and even after commissioning are common in all civil as well as military sectors. The main purpose of these measurements are:

1. Making sure that the platform's Underwater Noise levels meet the requirements of the vessel to perform its function effectively and not more than designed specified values.
2. Obtaining the information required to identify main noise sources and then to develop the recommendations to improve the next in series platforms or to mitigate the signature by added noise control measures.

In case of military vessels, the final goal of Underwater Noise measurements is to improve the overall platform acoustic performance to reduce sonar detection and mine detection risks. To achieve these goals both 1/3 octave and narrow band noise levels are to be measured during platform tests in all operational modes of the vessel.

The measurements are normally carried out in the frequency range 5–100 Hz and 100 Hz –20 kHz.

UNDERWATER RADIATED NOISE SOURCES

Any underwater platform is generally a complicated elongated source of Underwater Noise radiation caused by various factors of physical nature. Usually the radiated no ise contributions can be divided in three classes:

1. **Propellers** are simple-looking devices but are extremely complex to design when a good performance in terms of noise is needed. Propeller operates in non-uniform wake field behind the platform. When it rotates, it is subjected to unsteady forces which can lead to discrete tonals at blade rate and its harmonics. The level of blade rate noise will depend on the level of non-uniformity in the wake, the number and type of blades and shaft speed. Skew -backed blades generally have lower noise levels than other types of blades. Propeller noise makes difficult to apply noise control measures other than good initial design; excluding cavitation noise, only the blade frequency tonal can be lowered with an optimized blade shape. It is reported that tip speed, no. of blades, blade thickness and width are some of the controlling parameters of propeller noise. It is shown that reducing tip speed and increasing no. of blades is the most effective way of reducing noise generation.

2. **Hydrodynamic** noise originates in the irregular flow around the hull as the platform passes through water. The fluctuating pressures associated with the turbulent boundary layer radiate noise directly into the water and excites vibrations in the hull structures which radiates as sound in water. In particular the flow noise, due to the turbulent flow over the body, can give an important listening degradation on its own platform sensors due to increase in self-noise the sensors. The decay of the hydrodynamic noise is very quick when the distance from the body increases and hence the disturbance in far field is reduced.
3. Ships will have numerous **machinery** such as diesel engines, generators, propulsion motors, compressors, pumps etc. both rotating and reciprocating systems for their propulsion and habitability. Most of the noise generated by machinery is concentrated at precise operating frequencies or their harmonics; thus called tonal noise or narrow band noise, and appear as narrow peaks in the spectra of platform's noise signature. As the operating conditions of the ship changes, different machinery will have different behaviors. The machinery associated with the main propulsion will generate noise with a higher frequency as the ship's speed increases, while many auxiliary machines, such as generators or pumps will not change their acoustical signatures. Due to the noise and vibration levels of the machines on board, besides a control on the noise source level, it is possible by proper study and application of acoustic treatments on all the internal propagation paths in order to reach the objective noise levels. Figure 1 is a schematic representation of the machinery radiated noise for a typical on board installation. The machinery vibration represent generally the most important contribution to the underwater noise, and requires a great attention in all the aspects that can affect the underwater noise.

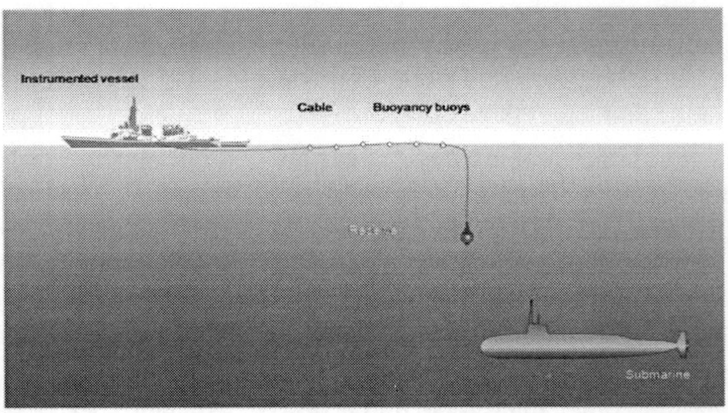

Fig. 1: Portable Ranges Set-up

Generally at low speeds the signature of the ship is dominated almost by tones from the Diesel Generator, propulsion machinery and other auxiliaries. At medium speeds ship's signature is dominated by main diesel firing rate harmonics, blade harmonics. At higher speeds it is dominated by wide band cavitation noise.

Propeller shaft and blade rate components lie in the frequency range 1–20 Hz and extend up to 100 Hz with multiple harmonics, while the ship machinery like generator, compressor etc extend up to 2 KHz. Propeller cavitation, turbulent flow and bubble generated sounds range from 100 Hz upwards with spectral slope inversely proportional to frequency. It has been reported in the literature that directivity showed reduced noise levels at bow and stern.

The radiated noise from War Ships and merchant vessels range between 100 and 200 dB ref 1micro pa at 1m depending on their speed and tonnage.

NOISE MEASUREMENTS AND DATA COLLECTION

Radiated noise signature measurements can be broadly grouped into two categories i.e Static and Dynamic. In Static ranges the vessel suspended from set buoys at specified location and depth. The measuring hydrophones are placed sufficiently far from the hull at different locations. Individual or set of machinery can run while recording the noise signature. Static ranges are highly useful in identifying the contribution of each machinery to the underwater noise and thus control measures can be implemented on erotic machinery. Ranges that are used to measure over all signature of the platform in dynamic condition are called Dynamic Ranges. In order to understand the acoustic characteristics of an individual ship, measurements should be made using a recognized technique, with the ship in a known configuration and at a known distance, so that representative source levels can be back calculated to a one metre reference level.

In order to estimate Sonar detection risk, the measurement points must be as close as possible to horizontal plane of the platform movement since the most of the energy detected by passive Sonar is radiated in this direction. To estimate mine detection risk one must measure underwater noise levels under the platform in additional to above.

Underwater radiated noise signature data of vessels in dynamic condition is generally collected by using Portable (Figure 1)/Mobile (Figure 2) ranges which are deployed in the specific identified area from special supporting vessel or fixed (stationary) permanent (Figure 3) ranges.

No doubt the fixed ranges are extremely expensive with complicated engineering structures .The various advantages and disadvantages of each range listed below (Table 1).

The main requirement for hydro acoustic test range is that at the point located at the required distance R from the platform the valid signal level in the frequency band of interest should be significantly higher (at least 6 dB) than the interference signal. When the signal/noise ratio more than 10 dB, the effect of interferences on the measurement results is negligible. It may be pertinent to mention that all the Navies use much of the same algorithm and procedure for underwater noise level measurements of any platform.

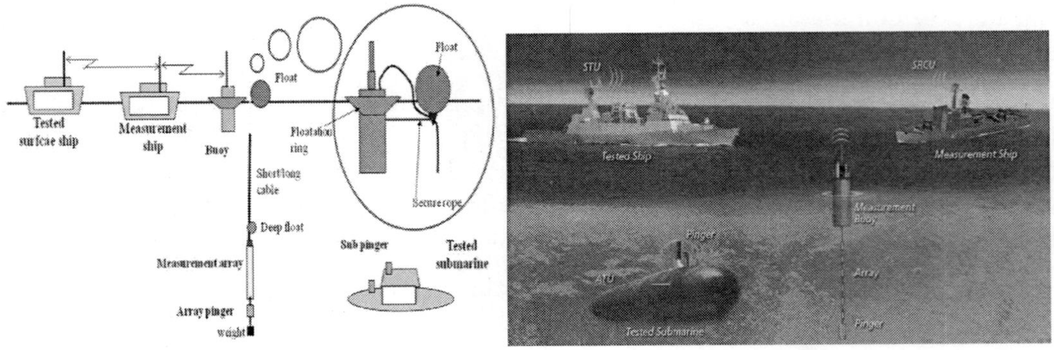

Fig. 2: Mobile Acoustic Range System

Table 1: Comparison of Portable and Fixed Ranges

Parameter	Portable range	Fixed Range
Accuracy in measurement	Fair due to splashing wave from the platform ship	Excellent
Repeatability	Poor due to inaccurate tracking and change in range location	Very good
Availability of ranges	Moderate	Very high
Reliability	Fair as damages may occur during laying and recovery	Excellent
Deployment	Anywhere	Only at specified site
Operating cost	High as logistic support is required	Moderate
Maintenance	Very easy	Easy

During ranging the platform moves evenly at a particular speed with specified necessary auxiliary machinery in a straight forward course relative to hydrophones. The hydrophones generates equivalent electrical signal on receiving the radiated sound of the platform. They also register background interference signal together with valid signal of the platform. The signals are analyzed either using linear averaging or exponential averaging techniques. Generally it is done for 8 sec average.

Most of the data was collected using portable ranges consisting of highly sensitive wideband hydrophone whose self noise is less than that of equivalent sea state zero, a low noise preamplifier and digital data recorder. Typical range diagram is in Figure 1 Some part of the data has been collected by conducting trials on permanent ranges shown in Figure 3 where the array of sensors are laid on concrete structures 2–3 m above sea bed and 50m away from each other. The system consists of junction box with associated electronics, data transfer cable, data recording and storage units.

Fig. 3: Permanent Ranges Setup

The data was analyzed using standard signal analyzer/software over the frequency range of interest with in the limitations of recording instrumentation.

EMPIRICAL MODEL FOR PREDICTION

It has become a common practice to formulate an empirical model for noise levels estimations by performing statistical analysis of measurements of large no. of ships when studying radiated noise signatures of either commercial or naval ships. RJ Urick and D Ross developed no, of empirical models for over all spectrum levels as a function of speed, displacement tonnage, number of blades and propeller tip speed. Whales and Heitmeyer more recently investigated a number of models based on measurement of 54 ships. Based on measurements on large no of ships, data collected through portable ranges and permanent ranges an empirical model has been developed and the predicted values are in close agreement with measured values.

ANALYSIS RESULTS AND OBSERVATIONS

From the analysis of radiated noise signatures of the ships , some of the observations are listed below.

1. Underwater radiated noise of ship increases with speed at particular rate around 1 dB/ knot, but beyond a particular speed (RPM) of the propeller, cavitation inception takes place and the noise increases abnormally by about 12–15 dB with little change in speed over a wide frequency range as shown in Figure 4

2. At low speeds the noise signature is dominated by tonal components generated by various rotating and reciprocating machinery and these tonals are dominated by broad band noise as the speed increases as shown in Figure 5
3. High pitched sound as shown in Figure 6 is generated when the vortex shedding frequency coincides with blade natural frequency; of course this occurs only at a narrow speed range of the vessel.
4. Noise levels are high at bearings 135 and 225 degrees of the vessel on estimations of directivity pattern as shown in Figure 7
5. Proper maintenance of vessel can result a reduction of about 5–6 dB over wide frequency range as shown in Figure 8
6. Noise signatures of different class of ships produce different noise spectra as shown in Figure 9 These signatures can be classified and identified based on various factors such as number of tonal components and their magnitude, spacing between the lines, slope of the spectrum etc.

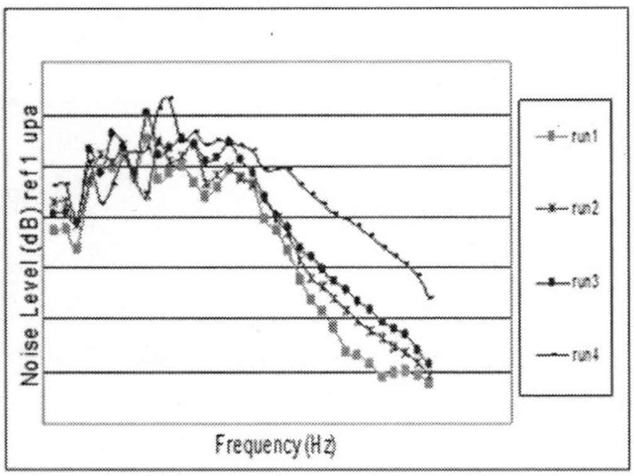

Fig. 4: Cavitation

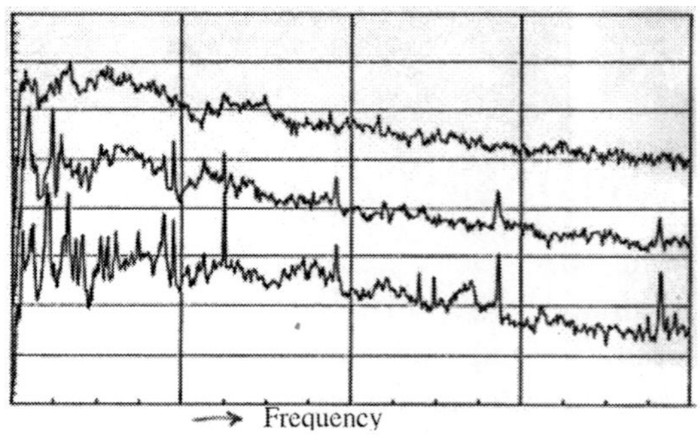

Fig. 5: Noise Level at Different Sped

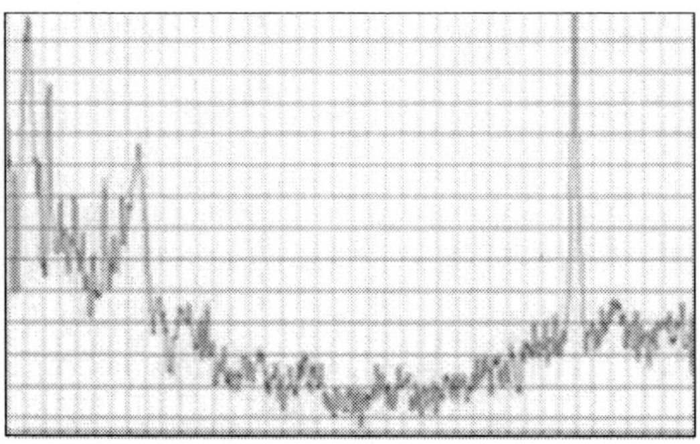

Frequency

Fig. 6: Propeller Singing

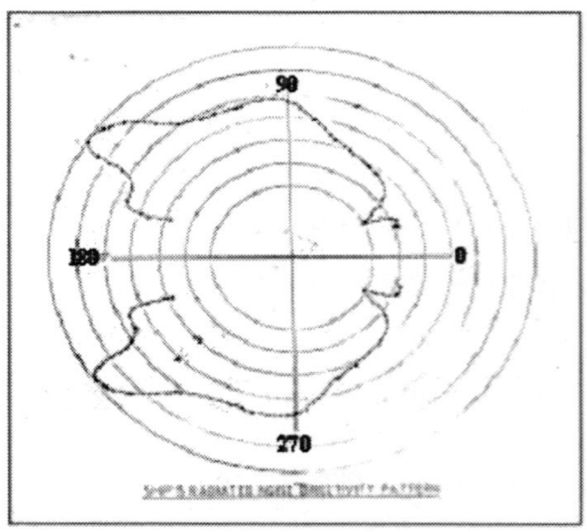

Fig. 7: Directivity Pattern

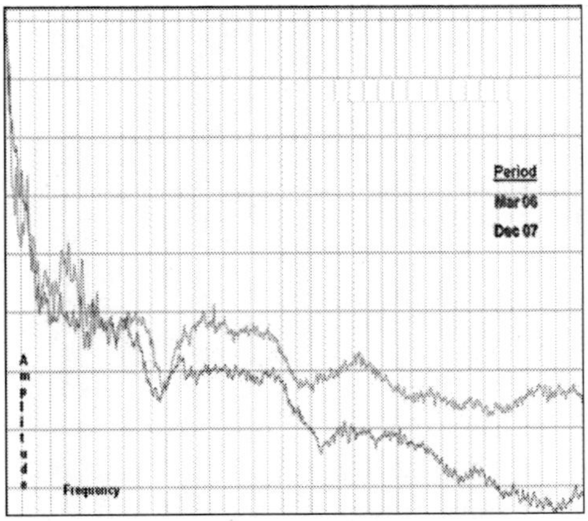

Fig. 8: Effect of Maintenance

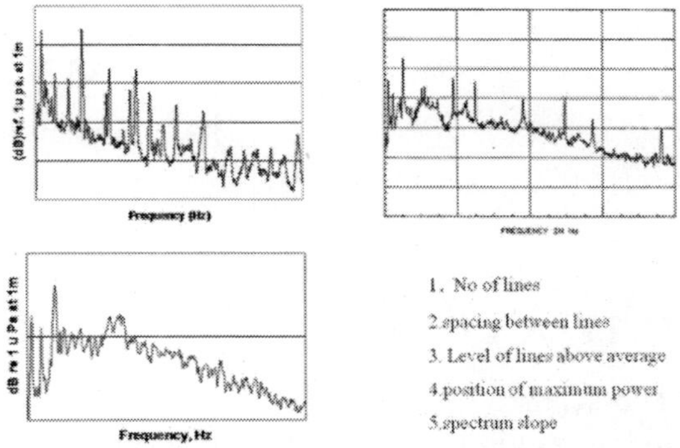

Fig. 9: Classification and Identification

CONCLUSIONS

The task of measurements of radiated noise of underwater vehicles look simple, but not so, because of hostile environment and uncontrolled environmental factors. Obtaining accurate and reliable data, of course is a challenging task.

ACKNOWLEDGEMENTS

The author expresses his deep gratitude to Shri. S.V. Rangarajan, Director, NSTL, Visakhapatnam for permitting to publish this paper. He also wishes to thank Shri. GV Krishna Kumar, Sc-'E', Shri AVS Chari, Sc-'G' & Shri KVVSS Murty, Sc-'G' of NSTL, Visakhapatnam for their support and encouragement during preparation of this paper.

REFERENCES

[1] Ross D., *Mechanics of Underwater Noise*, Pergamon, New York, 1976.

[2] Urick R.J., Principles of Underwater Sound, 3rd ed., McGraw-Hill, New York, 1983.

[3] Scrimger P., Heitmeyer R.M., "Acoustic source-level measurements for a variety of merchant ships", J. Acoust. Soc. Am., 89, 691–699, 1991.

[4] Wales S.C. and Heitmeyer R.M., "An ensemble source spectra model for merchant ship-radiated noise", J. Acoust. Soc. Am., 111, 1211–1231, 2002.

Advancements in Acoustics—Indian Scenario

B.V.A. Rao

KL University, Vijayawada, A.P.

INTRODUCTION

The science of acoustics, one of the oldest branches of physics, embraces the design of indoor environment to achieve acceptable levels of communication, comfort and privacy for the occupants. Acoustical issues include room shaping, reverberation control, speech clarity and Noise control. Virtually every space demands acoustic attention in order to function for its specified purpose like homes, offices, classrooms, factories, auditoriums, theaters and conference rooms.

Acoustical designing, problem solving needs technical expertise and experience in this field. It also has to be carried out with close association with other designing agencies and giving required flexibility to aesthetics.

Acoustics is an increasingly active subject in India. The awareness for sound control is growing by the day. With the industry growing at a rapid pace combined with innovative architectural designs, challenges in the acoustical design are increasing. In India, acoustics is a serous business today with plenty of acoustical research happening industrial as well as in architectural acoustics. NVIT is the criteria for modern automobiles. In addition, quieter technologies is the buzz word for all our industrial machinery.

INDUSTRIAL ACOUSTICS

Industrial acoustics primarily deals with noise control. With the pollution control board of India issuing gazette notification of sound levels to be emitted from machines, the design for quieter technologies is the need of the hour. Every major machine manufacturer is working towards manufacturing quieter machines. In this process, there are a lot of acoustical test cells and anechoic chambers are being constructed. NVH unit has become a compulsory department in big industries.

Major automotive companies in India like Tata Motor Ltd. Mahindra and Mahindra, Hero Honda, TVS Motor, Toyota Kirloskar Motors, Bajaj Auto, all have their own NVH test centres. The NVH test centres include engine test cells, dynamometer test cells and an anechoic chamber for noise labeling. Every machinery or appliance that have to be exported must carry a label declaring the sound power levels emitted by the product.

Appliances majors like the BPL, Videocon, etc also claim their own acoustical testing centres. GE BEL, GE ITC, Tatas have all invested in test cells for acoustical testing.

Many public sector companies have been involved in NVH testing. Institutions like National Aerospace Laboratories, Gas Turbine Research Establishment, Fluid Control Research Institute, Automotive Research Authority of India, etc have invested in huge Test Cells and Anechoic Chambers. We have the National Physical Laboratory in Delhi that has an anechoic chamber used for testing and certifying acoustical materials. IISc, the premier research institute of India has its own acoustical department with valuable research happening in vibro-acoustics, mufflers, aero and ocean acoustics. IIT Madras contributed significantly to machine dynamic problems and noise studies.

ARCHITECTURAL ACOUSTICS

The need for good acoustic within buildings has become a criterion. More and more people rely on acoustical consultants for designing acoustically good spaces, be it in modern day office spaces, conference rooms, auditoriums or home theaters.

The modern day office spaces are being designed and constructed keeping acoustics as an important criterion. With people spending more and more time in the offices, it is important to provide them with better work places to improve efficiencies. There are also international guidelines for the decibel level, noise criterion within office spaces, which are followed by many of the MNC's setting up offices in India. Hence office acoustical design has become a must in India. Also, with respect to Call-Centres, the design of acoustical barriers between seats has become important.

Most of the offices have video conferencing facilities installed in their conference rooms and training centres. Good acoustics with least distortions are a must and should be in these kinds of facilities.

Coming to auditorium designs. Any new auditorium being constructed has an acoustical consultant in its design team. The acoustical consultant is being involved from the concept stage of the auditorium construction. Older are being completely refurbished to accommodate for good acoustics. All the universities are inclined to improve the acoustics of their auditoriums, which have exited for a while now. Though majority of the auditoriums being constructed are for multi-purpose usage, there are a few auditoriums coming in places like Bangalore and Mysore that are being specially constructed for either drama, concerts or for speech purposes. The syndicate of Mysore University have decided to refurbish their syndicate hall to comply with the modern day standards by acoustically isolating the hall as well as providing good acoustics within the hall including electro acoustical facilities.

There are number of studios that have come up in India. With ever increasing television channels and radio stations, studio designing has become very important. Unlike yesteryears, companies opening up studios are very particular about acoustical consultants being involved with their studio design. Satellite radio station World Space recently set up 6 RJ studios in Bangalore and one in Hyderabad. The A R Rehman studio in Chennai can be considered as one of the best sound recording studios in India.

With the number of multiplexes growing in numbers rapidly in India, single screen theatres could be history soon. With four to twelve auditoriums constructed adjacent to each other with common walls in between, the acoustical design becomes very critical and important. This apart, the multiplexes are generally situated inside shopping malls make noise control measures inside the auditoriums even more significant. The new companies issue these certificates, only when the theatres comply with the acoustical parameters issued by these companies.

Home Theatres have become a life style statement. More and more houses insist in having a room dedicated for music and movies. Acoustical deign for such spaces have become popular and competitive.

Medical institutions and hospitals too are seeking acoustical consultants expertise in construction of their Audiometric Rooms. All India Institute of Speech and Hearing in Mysore have half a dozen audiometric rooms constructed. There are independent hearing clinics with audiometric rooms that have come up in large cities.

Acoustical design is widely used in HVAC installations too. Be it for offices, theatres or auditoriums, the chiller machines noise should not be carried through the ducts. This apart, apartments and buildings situated close to noisy points like a busy road near an airport are seeking the services of acoustical experts for solution from noise.

ACOUSTICAL MATERIALS

One finds a wide range of companies that are manufacturing acoustical materials like tiles and barriers. There are many materials like gypsum-based products, fibreglass wool based, rockwool based, mineral wool based, wood wool based, foam based and latest, polyster fibre base products. Various companies manufacture these products in India who would also provide acoustical design and to name few:

- Noise and Heart Control System, Bangalore
- Acoustics India Limited, Tiruchirappalli, Tamil Nadu
- Sound Wizard, Pondicherry
- Anutone Acoustics, Bangalore
- Rockwell India Limited, Mumbai
- Jal Mistry Consultants, Mumbai
- Sanjay Suri Consultants, New Delhi

- Mikasha International, New Delhi
- Air Control, Chennai
- Parimal Mehtha Consultants, Mumbai.

Considerable researching is done in these fields at Building Research Institute (Roorkee), NPL (Delhi) and many other private industries like Beardsell, Anutone., etc.

Classical examples of acoustic materials can be seen in the following illustrations:

Acoustical Analysis Shows Temple Transforms Echoes into Sound of Nature

Mayan tombs have been analyzed and researchers have found that they were designed to produce incredibly weird sound effects. If you stand in front of the staircase of the E1 Castillo pyramid and clap your hands, you'll hear an echo that sounds like the chirp of a bird. If you walk up the stones steps, it produces a flurry of echoes that sounds like rain falling into a bucket. Declercq's team has shown that the height and spacing of the pyramid's steps creates like an acoustic filter that emphasizes some sound frequencies, while suppressing others. But more detailed calculations of the acoustics shows that the echo is also influenced by other, and more complex factors, such as the mix of frequencies of the sound source.

Rani Gumpha is the current local name given to the largest structure in a group of excavated caves near Khandagiri, very close to Orissa's state capital, Bhubaneswar. The structure dates to about 2 to 3 century BC, the period just after Ashoka.

In Indian archaeological terms, it is a very early site, and represents an important transitional period in the country's history.Could this have been a theatre? Or at least, some kind of a performance space? The suggestion has been put forward before, notably by the art historian Percy Brown and Dhiren Dash, a theatre activist from the region. But the general explanation (offered by the archaeological Survey India and others) has been these were just another set of caves set apart for meditation.

Physical, Underwater and Atmospheric Acoustics (includes Ultrasound and Sonars)

60 to 70% of the publications at NSA conferences belong to this category, essentially application of ultrasonic to Sensors, Sonars and Chemical activity. IIT, Delhi; NOI, Goa; NPOL, Cochin; NSTL, Vizag are some other leading institutions.

Medical, Psycho and Bio Acoustics

Considerable work is going on in the field of Audiology, Cochlear implants and similar studies involving human health and consequent physiological and psychological effects for both low frequency and high frequency sounds. Good work is done at MMC, Chennai, AIIMH, New Delhi etc., They have also contributed to the Mitigation of Noise

Pollution and awareness of biological effects. There is a large population, who cannot understand the implications or consequences of this. After all noise is an invisible pollution, and what the eyes cannot see, the mind does not register it. In most of our industries, one can witness workers operating without using ear muffs and getting exposed to more than 100 to 120 dBA. Most of our carpenters are easily prone to duffness, so also drill operators

Establishment of Bio-Acoustic Analysis Facility at Wildlife Institute of India: Acoustical studies on *Pycnonotus* Bubuls

Bioacoustics is the study of animal acoustic signals, their production and perception mechanisms. It attempts to understand relationships between the structure of sounds and the nature of environment in which they are used. Proper documentation and recording of acoustics signals provide important base-line date for research in the field of systematic, behaviour, and play-back tools for survey and census. Establishment of bio-acoustic laboratories and animal sound archives are the fundamental needs to initiate such studies. Acoustic signals of Red-vented bulbul and Himalayan bulbul were recorded in their natural habitats during March 2002 to June 2003. Both species used their own vocal repertories.

Acoustic Technology by Japanese Team to Study Dolphins in Chilka

A seven-member Japanese team, led by Prof. Tamaki Ura University of Tokyo, has introduced an acoustic technology to observe the behavoiur and eco-system of rare endangered Irrawady Dolphins in Chilka Lake.

The team in association with the Chilka Development Authority, and experts from IIT, Delhi, and WWF India, conducted a pilot application of the study. Dr Ura mentioned that such acoustic study using the sound pulses of a specified signal has been carried out with the support of the Japanese Ministry of Education, Culture, Sports, Science and Technology. The Chilka system, he said, would be a major step for introducing and testing a day and night automated acoustic survey system for different species of endangered marine mammals. He hinted that a more ambitious survey of the Ganges River Dolphin would be soon carried out with the participating agencies using the new technology. Prof Ura said the pilot study carried out by the team revealed the presence of the Irrawady Dolphins at selected five pockets of the Chilka Lagoon. He said the team received many preliminary data on the concentration of the Dolphins at four points, which would be further analyzed at Tokyo to find out the underwater behaviour of the marine mammal.

The team claimed that the pilot study revealed that the Irrawady Dolphin moves at two meter per second.

Engineering and Machinery Acoustics (including transportation)

This is one of the largest contributors to our noise pollution in various forms including traffic covering both surface and air. India failed to meet global challenges in term of noise standards. Our machines produce more sound and vibration than their counterparts manufactures elsewhere. For the designers, this is the last priority. With more global inputs, things are getting better due to usage of suitable materials as well as manufacturing and assembling skills. Lot of attention is being paid to this aspect now to meet the challenges by going to quieter technologies. We have noise standards, but rarely these are strictly followed. Our industry appreciates both vibrations and noise diagnostics as a strategy for CBM, but very few organizations are serious about this implementation. For the past three decades considerable awareness has gone in this direction, yet we have to go a long way. IIT Madras, IIT Delhi, IIPE, many of our power plants and cement plants have contributed significantly to this novel technique. Academic Research involved considered such aspects as efficiency of sensors and instrumentation, inspection systems, and lastly economic benefits through cost model developments. This is one of the area in which I have personally taken interest in last three decades including introduction of courses like Diagnostic Maintenance in Condition Monitoring programmes. These were also highlighted in number of seminars and conferences. There are both passive and active techniques that are being researched in this quieter effort. IITs, IISc, Space and Defence Organizations have played a major role.

Music, Speech and Signal Processing (Electro Acoustics)

Next to Physical Acoustics, comes this field in our country with considerable contributions. This is essentially contributed by our electronic specialists from TIFR, Mumbai, IISc., Bangalore, CEERI (Pilani and Delhi), AIISH (Mysore) and many other centres throughout the country. There is a small group working on Musical Acoustics in Pune. The work involves speech analysis, speech processing and speech therapy. Most of our medical units dealing with ENT have good audiology sections and significant contributions.

IBM India Research Lab—Speech Technologies

The design and implementation of computer systems that can cover whole range of human-like interaction by using faces and voices is one of the challenging objectives of the Speech Group at the India Research Lab (IRL). The group aims to build state-of-the-art speech recognition/synthesis system for Indian languages and improve the existing techniques in general. IRL is currently pursuing activities in the areas of:

- Indian Language Desktop Speech Recognition
- Indian language Telephony Speech Recognition
- Indian-English Text-to-speech synthesis I.

ENVIRONMENTAL AND COMMUNITY NOISE (NOISE POLLUTION)

This is one of the major problems in the country due to lack of knowledge and education. This is a country that pays considerable weightage for religious, cultural and political ceremonies, where noise dominates. Noise legislations are there, framed by both CPCB and State PCB's. But hardly people pay attention. The industry, traffic and the community all contribute together to produce the Noise Pollution. I was one of the expert members to draft noise standards for the Ministry of Environment. It pains me to say the whole effort has gone waste, as no serious attention is paid at any level. Honking is anybody's business in this country, and when no honking is heard, the person thinks nothing is coming behind him or her. Movies have been made on this for screening in theatres, and many other steps were taken. During 1982–84, I served as Chairman for construction noise in Singapore introducing 5mn standard for Leq. There was an opposition, but it was Singapore! Discipline is lacking in our country.

RANDOM AND NON-LINEAR ACOUSTICS, JET NOISE, AERO ACOUSTICS AND CHAOS

These are highly specialized areas involving modern jet noises and their control (more due to aerodynamic effect). Some work was done at IIT Madras for producing 160–180 dB noise levels to impinge on rocket models to study their acoustic transmissibilities. These facilities have now been created at NAL Bangalore. Significant practical work was carried out both at IISc., Bangalore and NAL. Considerable work is being done on random noise, non linear acoustics and chaos theory relating to stabilities. I was involved in establishing NEF contours for the Meenambakkam airport in Chennai and extrapolate it to Pondicherry. This was done some 15 years ago and still Pondicherry airport has not yet been commissioned, though the airfield was put up at an enormous cost.

- Fluent announced coupling with LMS SYSNOISE for Acoustics Modeling on August 8, 2003. Lebanon, NH, USA and Leuven, Belgium Fluent Inc. and LMS International announced later the release of the world leading Vibro-Acoustics software LMS SYSNOISE Rev 5.6, which supports import of transient data from FLUENT 6.1, the leading computational fluid dynamic (CFD) software from Fluent. This allows results from FLUENT to be used for comprehensive computational aero-acoustics simulations within SYSNOISE Rev5.6, complementing FLUENT 6.1's native aero-acoustics modeling capabilities.
- Computational aero-acoustics is a critical technology for engineers in virtual prototyping environments, who want to predict and compare the aero-acoustics performance of proposed designs. By using computational tools to identify and rank sources of flow-induced noise, engineers are able to design for noise reduction early in their design cycle, enabling them to produce better products, while reducing the need for expensive physical testing, and shortening the overall time to market.

APPLICATION OF HOLOGRAPHY IN JET ACOUSTICS STUDIES

Source strength distribution on a jet boundary was obtained from measurements using the principle of acoustic holography. Measurements were conducted in an open field. Measurement of acoustics pressure on a cylindrical two dimensional contour located close to the vibrating jet boundary was used to obtain the acoustic source strength distribution of the sound field on to a cylindrical surface. A jet emanating from 5mm convergent nozzle was used for the holography experiments, assuming symmetry. Experimental results were compared with results obtained from holography.

LOW FREQUENCY SOUND AND PHYSIOLOGICAL EFFECTS

- In 1665, people in London heard the gun fire from English and Dutch men in the battle in North Sea.
- In 1883, volcanic eruption in Krakotoa from infrasonic waves traveled 5 times around the world.
- In 1933 Greenwood's famous novel 'Love on the Dole' mentions that 'the thump of the presses makes the workers feel sick'.
- In 1973 the 1st International conference on Informative was held in Paris.
- In 1980 the 1st International Conference on low frequency noise. Regrettably even there is widespread ignorance as to its quantitative and subjective effects, no convenient measurement unit to describe it and no definitive guidelines as to extent to which it should be controlled. Unfortunately people exhibit offering levels of individual sensitivity to low frequencies.

Tests conducted in Berlin on the effects of infrasound showed marked physiological disturbances such as elevated BP, ECG disorder and change in adrenaline. Females show earlier sensitivity than male.

Its effect on auditory systems such as NIHL and Tinnitus are well known. It has also a profound effect on non-auditory systems such as blood pressure, perspiration, heart rate, contraction of muscles, ischemic heart diseases, hypotension, hormonal secretions, physiological stresses, neurological disorders and epileptic seizures. Its effect on animals, sleep, foetus, visual acuity, loss of balance and whole body vibration are all well known.

Loudness sensation can be measured up to 1 Hz, and that produced by gasturbines in the range of 3.5 Hz has annoying characteristics. But usually the high frequency noises acting at the same time (like pop music) from a translator radius mask, the low frequency noises and sensitivity changes.

Examples of Low Frequency Noise:

- Silencing of a gas turbine test bed
- Telex rooms (around 17 Hz)

- Ventilating fan noise (20 Hz)
- Reciprocating type air compressors (20–31.5 Hz)
- Domestic boilers and industrial boiler plants (20 Hz)
- Diesel powered transport—with its low firing and of immense power is a source for low frequency noise. Ship noise spectrum contain both infrasonic and low frequency bands. Active control techniques help here.
- Low frequency noise and wind throb in cars (especially when windows are opened) act like a Helmholtz Resonator induces great SPL's.
- Buildings in the wind—telephone cables exposed to wind produces distinctive humming sound. The frequency calculated from strohaul's equation on windy days, it is difficult to manage children in schools. Some of the depression symptoms exhibited by dwellers in high buildings are noise induced.
- Distant guns—dull thud gun fire.
- Bristol hum—from gas turbines producing low frequency noise.

For Control of low frequency noise, 'A' network is not suitable as it is only in the audio range. They now recommend 'P' weighting. It falls at 12dB/octave. This has a disadvantage. Hence another weighted network 'N' is used, which falls at 6dB/octave. ISO is standardizing all these and coming out with reasonable standards.

Low frequency sound due to its low extenuation, its decay rate approximates to the simple inverse square law, as the long wavelengths involved causes the sound to spread in a spherical manner and covers a wide area.

CONCLUSION

Acoustics in India—yes, we create enough sound, as we like to be loudly heard. But we don't bother and we do not know even how to control it, when it is needed or parts damaged as in machinery. In Architecture, we overdesign than needed costing significantly. In countries like Australia (Queensland) and New Zealand, they have Ministries for Noise Abatement. In Germany, noise legislations are stringent. Noise maps are displayed in their town halls, curtailing the sound depending on the importance of those areas, and one cannot put up any noise unit that exceeds those values. Noise maps are prepared earlier for turn key projects, so that industrial noise affecting communities can be controlled.

In India, we have to go a long way to reach any of those norms or standards. Public awareness is still in its infancy here and the hazards involved limitless. We have each year NSA conferences, and how much of it, stresses on these aspects and the country gets thereby benefited has not been assessed. It is a cry in the wilderness. But we are sure one day our cries will be heard by future generations and a better sense dawns in the country.

ACKNOWLEDGEMENT

Many citations here have been drawn from sources earlier published in newspapers or magazines and other literature, which cannot be recollected. Special thanks and gratitude to all those authors. Considerable work also has been carried out by me and my colleagues, while I worked at IIT Madras. All of them are being acknowledged here. Appendix A gives some standards adopted. Appendix B gives the list of my involvement as consultancy projects, which will be briefly highlighted while presentation.

APPENDIX A

Acoustical Standards as Per IS

IS:2526–1963	Code of practice for acoustical design of auditoriums and conference halls.
1S:9736–1981	Glossary of terms applicable to acoustics in buildings.
IS:9000(PART 21)–1985	Basic environmental testing procedure for electronic and electrical items. Acoustic Noise Test.
IS:2264–1963	Preferred frequencies for acoustical measurements.
IS:9876–1981	Guide to the measurement of airborne acoustical noise and evaluation of its effects on man.
IS:4954–1968	Recommendations for noise abatement in town planning.
IS:7194–1973	Specification for assessment of noise-exposure during work for hearing conservation purposes.
IS:4242–1967	Method of measurement of acoustical noise emitted by ballasts for gaseous discharge lamps.
IS:6964–1973	Specification for octave, half-octave and third-octave band filters for analysis of sound and vibrations.
IS:10775–1984	Specification for body level hearing aids.
IS:1885(PART III/Sec 1)–1965	Electro technical vocabulary Acoustics
IS: 1885(PART III/Sec 2)–1966	Electro technical vocabulary Acoustical Electro-Acoustical System
IS: 1885(PART III/Sec 4)–1966	Electro technical vocabulary Sonics, Ultrasonic and Underwater acoustics
IS: 1885(PART III/Sec 5)–1966	Electro technical vocabulary Speech and Hearing
IS: 1885(PART III/Sec 6)–1967	Electro technical vocabulary Acoustical Instruments
IS: 1885(PART III/Sec 7)–1978	Electro technical vocabulary music
IS: 1885(PART III/Sec 8)–1974	Electro technical vocabulary Architectural Acoustics

APPENDIX B

Industrial Consultancies in Acoustics (by the author)

1. Impact noise studies through strain measurement on reactor domes ('76)
2. Dynamic characteristics of indigenous machine tools ('76)
3. Effect of machinery and noise vibration on neighbourhood for large Hotel complex ('76)
4. Acoustic tests on SLV 3 model—VSSC, ('77)
5. Dynamic and acoustic transmissibility test on rocket models, solar panels and antenna dishes ('77)
6. Acoustic tests on Solar panels and Antenna Dish for ASLV and SROSS—ISAC, Bangalore {'84}
7. Noise analysis and measurement of 1.3 MW induction motor—Jyoti Ltd., Baroda, 1985
8. Diesel Generator vibration and noise and foundation studies—India Cements, Sankar nagar and tube investments, Avadi (1986)
9. Airport noise survey—Ministry of Tourism, Govt. of Pondicherry
10. Development of NEF contours for the Madras Airport and its extrapolation to the proposed Pondicherry Airport for aircraft noise at critical locations
11. Acoustic design and treatment of community hall for CCI at Yerraguntla and Kallakurchi
12. Acoustical enclosure design for the gas turbines for BHEL, through SOL Engineers, Hyderabad and many other industries through Viprah Engineers, Coimbatore
13. Evaluation of the NEF contours for the proposed airport at Pondicherry, Ministry of Tourism (1987)
14. Acoustical treatment and reverberation of studies for a community hall, CCI, Yerranguntla (1987)
15. Noise measurement and analysis of a 1.3 MW Induction motor for Jyoti Ltd., Baroda (1987)
16. Vibration and Noise Studies on DG sets for India Cements, Sankarnagar and Tube Investments, Madras (1987)
17. Design of an acoustical enclosure for a pump-motor set for Beacon Weir, Madras (1987)
18. Environmental noise studies from exhaust fans for Kerala Laxmi Mills, Trichur (1988)
19. Refrigerator compressor noise studies for Godrej and Boyce, Bombay (1988)
20. Environmental noise studies from DG sets for Hydraulics India, Madras (1988)
21. Acoustic treatment for the Turf Club, Beard Sell ltd., Bangalore

22. Acoustic enclosure for gas turbine for BHEL, SOL Engineers, Hyderabad
23. Noise analysis of inlet silencer for coincidence effect, Anutone, Bangalore
24. Noise from A/C plant for Hotel Sangam, Tanjore
25. Noise from Diesel—Generator sets for number of industries and residential complexes and enclosure designs (Indo-American Hybrid, Ramco Systems, Raheja Complexes, etc.
26. Consultation on choice of pneumatic isolators for a anechoic tank for sonar studies – NPOL, Cochin
27. Study on a non-linear vibration/noise of an aircraft gas turbine—Kaveri Engine, GTRE, Bangalore
28. Design of a pneumatic isolated foundation bed for an accuracy of 0.03 μ–IBM, Bangalore.

Section–II

Engineering and Applied Acoustics

Optimal Design of an Acoustic Energy Harvester

T.S. Indraneel, Abhijit Sarkar and S. Narayanan

Department of Mechanical Engineering, Indian Institute of Technology Madras

ABSTRACT: *Acoustic energy harvester has been demonstrated in the literature to convert the acoustic energy to useful work. However, the output power and efficiency obtained in the experiments is reported to be very low. In this work, we study the energy harvesting system with an objective to improve its performance. Towards this end, the electroacoustic analogy is used to analyse the available design. A source ranking exercise is identified and principle component responsible for low power output of the harvester is identified. A numerical optimization technique is adopted to arrive at an optimal design of the energy harvester. The results based on classical optimization technique are validated against results obtained through genetic algorithm. The optimal design is observed to give a power output greater than the baseline design by 4 orders of magnitude.*

INTRODUCTION

With ever increasing demand of energy and growing scarcity of the conventional sources (viz. the fossil fuels), there is great demand to explore new sources of energy. The area of energy harvesting aims to tap the freely available ambient sources to produce usable energy. The objective is to develop an architecture of transducers which utilizes the available environmental conditions (such as vibration, ocean currents, sound, etc.) and generate usable power. At present, the technology has been demonstrated for low power applications for example in powering the sensors used in condition monitoring of remote installations such as off shore oil rigs. Powering these sensors becomes challenging since using dense wiring networks complicates the system. Moreover, using batteries for these devices is difficult since batteries have a limited lifetime and replacing them is often time consuming and expensive. Harvesting energy from the ambient environment is a viable solution in this case. These energy harvesting devices are helpful because they tap in energy from an unlimited source of energy and are very little on maintenance. Further, the ill- effects of pollution is absent.

BASELINE DESIGN

An acoustic energy harvester in the form of an eletromechanical energy harvester (EMHR) has been proposed in the work of Liu *et al.*[1, 2]. The schematic illustration of their set up is shown in Figure 1. It consists of a sound generator, a long tube, Helmholtz resonator which is attached with a compliant diaphragm, rectifying circuitry and the external load (the device which needs to be powered). A high amplitude tonal noise is produced by the sound generator. This sound is transmitted through a long tube. The sound propagation can be approximated to be in the form of one dimensional standing wave inside the long tube. The sound gets amplified due to the presence of the Helmholtz resonator. The amplified pressure fluctuations vibrate the composite diaphragm. The composite diaphragm is in the form of an annular piezo-electric material (PZT) attached to a silicon base. The piezoelectric material is attached to platinum electrodes and a layer of titanium oxide is used to bond the platinum with silicon. The deformation in the annular piezoelectric layer results in electric voltage across the layer. When the sound generator produces sound, the resultant output is that the piezoelectric material generates a voltage across it. The voltage thus produced has the same temporal form as the input sound. Appropriate electronic circuitry (in the form of a full wave rectifier and flyback converter) is employed to convert this alternating voltage to obtain a constant DC voltage and charge a battery source. The authors have demonstrated acoustic energy harvesting using sound source of 90–150 dB producing an output power of 10^{-12} to 10^{-6} Watts. The efficiency of conversion was also reported to be 10^{-2} at best.

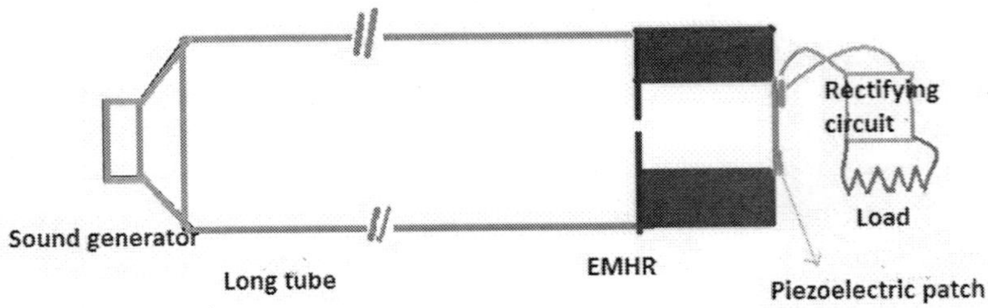

Fig. 1: Schematic Illustration of the Acoustic Energy Harvester

In the above work, the primary objective of the authors was to demonstrate the fabrication and working of an acoustic energy harvester. As reported by the authors, the performance of the device has scope for considerable improvement. In particular, the authors did not consider optimizing the design via impedance matching of the various analogous circuit elements. The present work aims at analysing this baseline design to study the performance characteristics namely output power. The objective is to understand the principal reasons for low values of output power. This in turn would facilitate improvements in the design.

For the analysis of the baseline design, we use the electroacoustic analogy as proposed in [1] is used. Using this analogy, the Helmholtz resonator is modelled as a series connection of a resistor (due to the viscous effect of air flow in the neck region) and an inductor (due to the inertia of the air mass executing oscillatory motion) connected in parallel with a capacitor (due to the acoustic stiffness of the cavity). Further, an equivalent circuit model of the piezoelectric patch backed by the compliant diaphragm (in the form of an annular ring) is used [2]. This is in the form of a series connected inductor, resistor, capacitor connection analogous to mechanical mass, damper and spring, respectively. The piezoelectric transduction is modeled as an electric transformer. The transduction coefficient is appropriately captured through the transformer ratio. This is further processed through a rectifier circuit and finally fed to external load which is in the form of the battery to be charged.

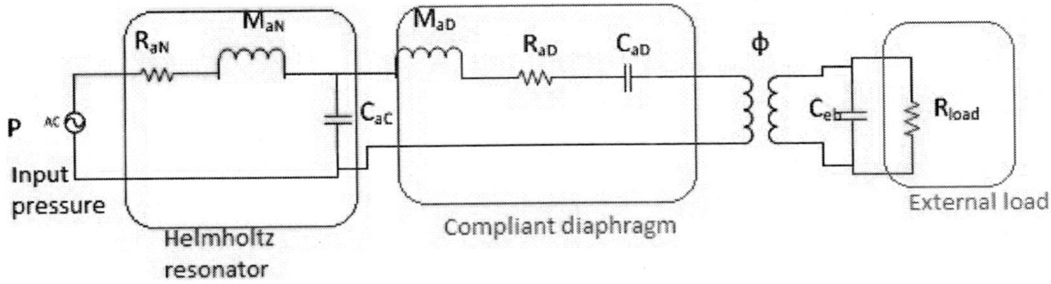

Fig. 2: Equivalent Circuit for the Acoustic Energy Harvester

The equivalent circuit for the design is shown in Figure 2. For a harmonic input acoustic pressure, the current and the voltage generated across each branch of the electrical circuit are calculated. Through these quantities, the power delivered at the load from the energy harvester can be computed. The electroacoustic lumped element analogy described above is based on the following assumptions:-

1. The acoustic wave propagation inside the long tube is in the form of plane waves. This limits the frequency of analysis to below the first cut-on frequency of the duct [4].
2. The compliant diaphragm undergoes a small deflection thus allowing for a linear model. Further, lumping this continuous vibrating system as a lumped element comprising of a fixed spring mass and damper is an approximation. In reality, the vibratory response of such systems will be a superposition of multiple modes in which case the above assumption will be erroneous.
3. The electric elements used also approximate the linear lumped element model. Further, it is assumed that the output load is of a purely resistive character.

SOURCE RANKING

Table 1 shows the dimensions of baseline design (as given in [1]). Using these parameters, the time-averaged power delivered across the load for a range of frequency is computed. The results of the analysis are shown in Figure 3. It was verified that the first peak is due to the resonance of the Helmholtz resonator, and the second peak is due to the resonance of the diaphragm. The peak power in this set up occurs at 1838 Hz corresponding to the first natural frequency and is of magnitude 3×10^{-11}W.

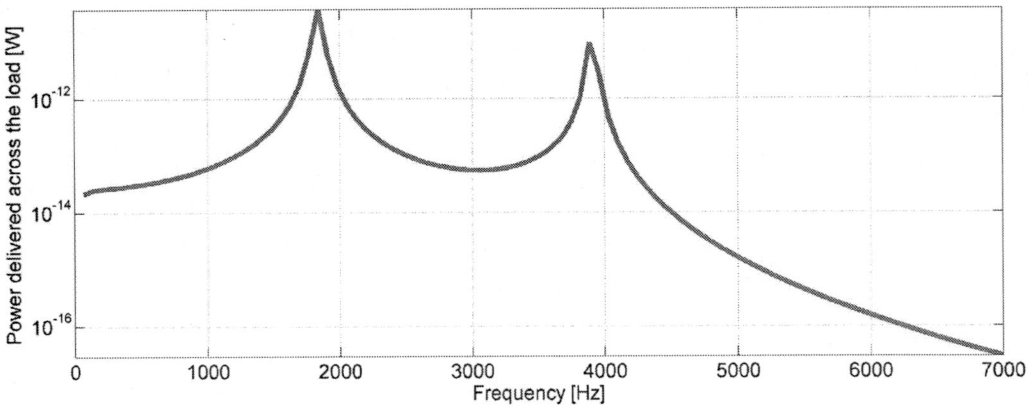

Fig. 3: Power Obtained Across the External Load Due to Acoustic Energy Harvesting as a Function of the Frequency of the Acoustic Input

Lower and upper bounds chosen for each design variable is also shown in the Table 1.

Table 1: Dimensions of the Baseline Design and the Optimal Design of the Acoustic Energy Harvester

Variable	Baseline design	Optimal design	Lower bound	Upper bound
Frequency	1838 Hz	1000 Hz	1000 Hz	4000 Hz
Length of neck	3.16 mm	1 mm	1mm	5 mm
Radius of neck	2.56 mm	5 mm	1mm	5mm
Volume of cavity	1950 mm^3	1950 mm^3	1000 mm^3	3000 mm^3
Outer radius of diaphragm	1.95 mm	3.02 mm	1 mm	5 mm
$\frac{R2}{R1}$	95	97.7	10	99
Thickness of Silicon	3μm	2.5μm	1μm	5μm
Thickness of PZT	6μm	9.99μm	1μm	10μm
Thickness of platinum	170 nm	10 nm	10 nm	500 nm
Thickness of $TiO2$	100 nm	10 nm	10 nm	300 nm
Power	3×10^{-11}W	0.3μW	-	-

To identify the elements that are major dissipators of energy, a source ranking exercise was carried out. In this exercise, the power dissipated by each of the elements was plotted against frequency. The power across any impedance can be calculated as the product of conjugate of the voltage across the impedance and the current that passes through it. The energy dissipated by each of the elements of the mechanism (viz. neck of EMHR, cavity, diaphragm and load) were plotted against frequency. This plot is shown in Figure 4. As seen in the graph, the major dissipaters of energy are the neck of the EMHR, and the compliant diaphragm. The cavity however, does not dissipate energy since it performs like a capacitor. The neck primarily dissipates power due to the viscous losses at the walls of the neck, and the diaphragm dissipates energy due to radiation. Thus, to improve the overall efficiency of the system, we have to primarily focus on redesigning the neck and the diaphragm.

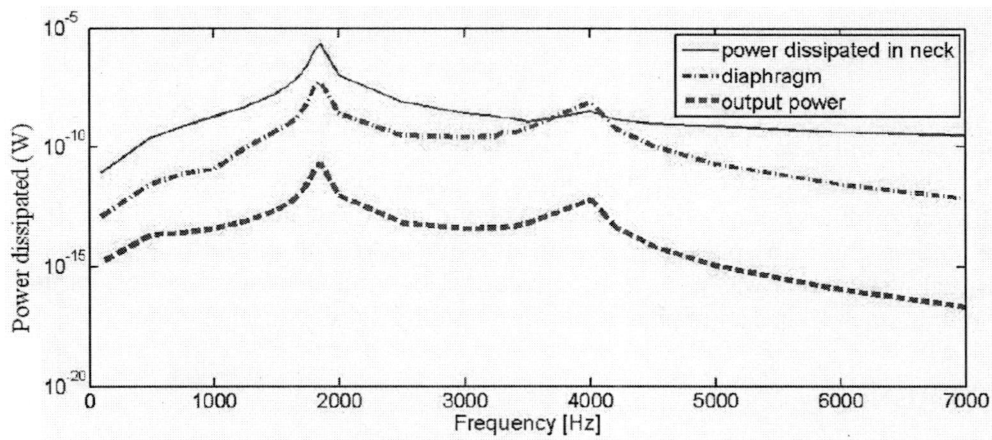

Fig. 4: Power Dissipated by Individual Elements of the Acoustic Energy Harvester

OPTIMIZATION

The availability of analytical expressions for each component of the equivalent circuit in terms of the geometric and material properties of the system enabled the usage of the classical optimization technique in order to arrive at a better design of the acoustic energy harvester. The objective of the optimization exercise was to get the maximum power output of the energy harvesting device. The design variables used for the optimization exercise are: input frequency, radius and neck of the Helmholtz resonator, volume of cavity, radius of the diaphragm, thickness of the piezoelectric layer, the platinum electrode, and the titanium oxide layers. The bounds on each of these variables were chosen such that the dimensions of the optimal design are comparable to the baseline design. These bounds are shown in Table 1. The *fmin* algorithm in MATLAB is used for this optimization exercise. The algorithm outputs a local optimum. The optimal values of the design variables are also shown in Table 1. The variable tolerance was chosen to be 10^{-10}, and the function tolerance was chosen to be 10^{-6}. As observed from Table 1, many of the design variables representing the dimensions of the neck and the thicknesses of the different layers namely $L_{neck}, r_{neck}, t_p, t_{pl}, t_{O2}$ at the optimal design assume either the upper or lower bound values prescribed. The results show that the power output obtained from the optimal design is 4 orders of magnitude greater than the baseline design. The former attains the better performance by superior impedance matching characteristics.

To verify if the optimum solution obtained is a global optimum, the optimization was effected using a genetic algorithm with suitable tolerances on the variables, objective function and the iteration limit. The genetic algorithm also gave a similar power output with the same geometry of the device. Therefore, the optimum reported is the global optimum.

CONCLUSIONS

In this work, we have arrived at an optimal design of the acoustic energy harvester as proposed by Liu *et al.* [1, 2]. Electro-acoustic analogy is used for the analysis of the system. Using the equivalent circuit model power dissipations occurring over each element of the energy harvester is calculated. It is concluded that the neck of the Helmholtz resonator is the principal dissipater of the power. An optimization exercise based on classical algorithims was used to minimize the dissipative losses. The results indicate the optimal design has an improved power output by 4 orders of magnitude.

REFERENCES

[1] F. Liu, S. Horowitz, T. Nishida, L. Cattafesta and M. Sheplak. A multi degree of freedom electromechanical helmholtz resonator. Journal of the Acoustical Society of America, 122(1), pp. 291–301, 2007.

[2] F. Liu, A. Phipps, S. Horowitz , K. Ngo, L. Cattafesta, T. Nishida and M. Sheplak. Acoustic energy harvesting using an electromechanical Helmholtz resonator. Journal of the Acoustical Society of America, 123(4), pp. 1983–1990, 2008.

[3] A. Suryanaryanan, N. Prasad, S. Horowitz, B. Homoeijer, B.V. Shanker, L.N. Cattafesta and M. Sheplak. Analytical electroacoustic model of a piezoelectric composite circular plate. AIAA Journal, pp. 2311–2317, 2006.

[4] L.E. Kinsler, A.R. Frey, A.B. Coppens and J.V. Sanders. Fundamentals of Acoustics. John Wiley and Sons, 2005.

Preliminary Study on Automotive Horn Sound Quality

B. Venkatesham, N. Pradeep[1], Goutham[1] and Jongdae Kim[1]

Department of Mechanical Engineering, Indian Institute of Technology Hyderabad, A.P.

[1]Electronics Engineering Department, Hyundai Motor India Engineering Pvt. Ltd., Hyderabad

E-mail: venkatesham@iith.ac.in

ABSTRACT: *Sound Quality (SQ) is a parameter that provides product differentiator in a competitive market. Automotive horn sound design is a compromise between the need to generate warning signals and also customized sounds for different manufacturers. Customized sound for a product will provide uniqueness to the manufacturer brand image. The sound quality study of automotive horn is a multi-disciplinary problem. The knowledge of electro-acoustic, musical acoustics and statistics will help the designers for providing optimized sound quality products.*

The aim of the current investigation is to develop a methodology for studying automotive horn sound quality. It has been discussed the methodology of recording horn sounds for Jury listening test. It has been prepared Jury listening Questionnaire. The recorded sounds are analyzed to calculate different objective parameters like loudness, sharpness, roughness, fluctuation of strength and tonality. Relative sensory pleasantness index considers as an aurally adequate psycho-acoustic index to explain pleasantness as the global behavior of horn sound. This value has been calculated from objective parameters.

INTRODUCTION

Human brain interprets the source of cause and provides instruction to react. It acts like a processor. This study will be a part of Psycho-acoustic study. Psycho-acoustic study is a science to read out human perception and acoustics. Every working product can create a specific sound. This sound may be pleasant or unpleasant based on human perception. Some sounds are generated for warning purposes. The purposes of these sounds are alert people and grab attention of people. These sounds convey some information. For example, alarm sound, ambulance siren sound, Factory siren sound, Police vehicle siren sound, and vehicle horn sound etc. The focus of the current study is vehicle horn sound. Vehicle horn sound is one of the major noise sources in urban noise pollution. People use horn very frequently countries like India. The major reasons for honking might be

• Traffic indiscipline (i.e., not following lane system) • Cross-Junctions • Hurriedness in people • Habitual	• Low confidence in driving • People walking on the road • Animals crossing road	• Vehicle Overtaking • Blind turns and Hair-pin turns • Genuine Warning signal

The automotive horn sound is more impulsive in nature and creates more annoyance to listeners. Sound Quality of Horn is necessary for reducing traffic noise annoyance.

Most of the automotive manufacturers are interested to create uniqueness to their Vehicle horn sound and should be more pleasant to listeners with proper warning. This motivates for a current study of Automotive Horn Sound Quality.

Sound Quality (SQ) study includes subjective test or Jury test and objective test. The purpose of subjective test is to collect human perception of sound. Objective test data can be used to calculate Sound Quality (SQ) metrics like loudness, sharpness, fluctuation of strength, Roughness and tonality etc., based on empirical equations. These two tests have independent results. The relation between these results can be established with a proper analysis. Existing literature related to automotive horn sound is in the references [1–4].

Guillaume Lemaitre [1–3] etc., from France studied extensively on the sound quality of car horns. They have discussed about the perceived quality of car horn sounds within a psycho-acoustical frame work. They studied timbre of car horns similar to the timbre of musical sounds. Their second paper provides insights into the design of new sounds suitable for European environment. Hai Ann ThiPhan [4] etc., From Japan conducted cross-cultural experiments between Vietnamese and Japanese to understand the annoyance from road traffic noise with horn sounds.

METHODOLOGY

The purpose of subjective test is to capture human perception about different sounds. This test can be conducted with actual product sounds in the field. It may not be possible to conduct test all times due to limitation of resources and complex interaction with other sounds. An alternative approach is to develop a procedure in the laboratory.

The main focus of the current study is as follows:

1. Psycho-acoustic analysis of horn sound data to estimate objective parameters,
2. A subjective test to identify subjective parameters, and
3. Data analysis to correlate subjective parameters with objective parameters.

It has been developed test setup for recording of horn sound. The recorded sound samples have been used in subjective test and objective test. Instrumentation, Mounting frame fabrication and test recording procedure details has been considered in test setup preparation. The subjective test has been conducted with recorded sound samples.

Objective data calculate SQ metrics and psycho-acoustic indices. Psycho–acoustic indices provide a global description of sound in terms SQ Quantities. The relations used in calculation of objective parameters are empirical in nature. Relative sensory pleasantness index considered as an aurally adequate indices for automotive horn sound.

Experimental Test Setup

The laboratory test setup development for recording horn sounds has been discussed in this paper. Binaural Head Set (BHS) instrument used to record horn sounds.

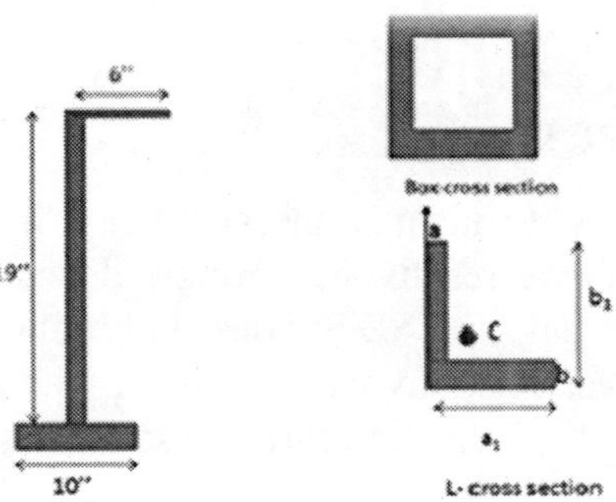

(a) (b)

Fig. 1: (a) Dimensions of Mounting (b) Cross-Sectional Details

Figure 1 shows the mounting frame dimensions with Cross-sections. These dimensions are estimated based on Horn sound standard and also the actual mounting position of Horn in a car. Different holes are provided on the horizontal arm of mounting frame to mount the hornin different locations. The mounting location is not changed in the current study.Table 1 shows the summary of horn sound sample recording test conditions.

Table 1: Recording Test Setup Details

Environment	*Open Ground (neglected wind noise effect)*
Recording duration	15 Sec with continuous Honking
Data recording details	24 bits, 44100 Hz sampling rate
Distance	2m from Horn in Front direction
Number of Horns	5 Manufactures, 10 Disc type, 4 Shell type total: 14
Tested Horns	14
Type of Horns	Disc and Shell type
Number of Trails	2

S.No.	1	2	3	4	5	6	7	8	9	10	11	12	13	14
Notation	M1DH1	M1DH2	M1SH1	M1SH2	M2DH1	M2DH2	M2DH3	M2DH4	M2SH1	M2SH2	M3DH1	M3DH1	M4DH1	M5DH1
Horn Type	Disc	Disc	Shell	Shell	Disc	Disc	Disc	Disc	Shell	Shell	Disc	Disc	Disc	Disc

Subjective Test

The subjective test consists of three major tasks. First one is a preparation of sound samples for listening with proper instrumentation; the second one is jury evaluation methods and the last one is jury evaluation questionnaire.

There are different types of Jury evaluation tests are available in the literature. Detailed explanations of Jury evaluation methods for automotive applications have been provided in reference [5]. Based on the problem objective, three types of tests are selected as part of jury evaluation methods. These are

1. Relevant Test (RT)
2. Semantic Differential Test (SDT)
3. Pair Comparison Test (PCT)

The purpose of relevant test is to identify the significant horn model, which people can recognize it as a car horn. These results are strongly depends on subject knowledge and experience. Semantic differential test (SDT) consists bipolar variables with seven verbal Interval scale. Each bi-polar variable may have a linking with another variable. The purpose of Pair Comparison Test is to check the subject response consistency and preferred choice.

Objective Data Analysis

Major psycho-acoustic metrics are Loudness, specific loudness, roughness, sharpness, and fluctuations of strength defined or calculated based on application. Each metric represents one type of sound behavior but global representation is possible with psycho-acoustic indices. For example, relative sensory pleasantness is a function of loudness, roughness, sharpness and tonality. It is used to describe global behavior of most of the sound quality applications. Even Horn Sound Quality also described with relative sensory pleasantness index. These indices are monaural (i.e., Calculations are conducted for each ear signal without combining signals). It generates two separate values for left and right channels.

Fastl [6] provides the detailed information about psycho-acoustic metrics and psycho-acoustic indices with formulas etc., Developed MATLAB [7] code based on these formulas to calculate objective data parameters for all recorded sound samples.

RESULTS AND DISCUSSION

Objective parameter metrics in relative percentage are shown in Box plot format in Figure 2. The Box plot data shows 25% to 75% quartile in a box format. The 1.5 time quartile range is shown as whisker length. Beyond whisker length represents outliers. Red mark shows the outlier points. Mean line shown with objective metrics.

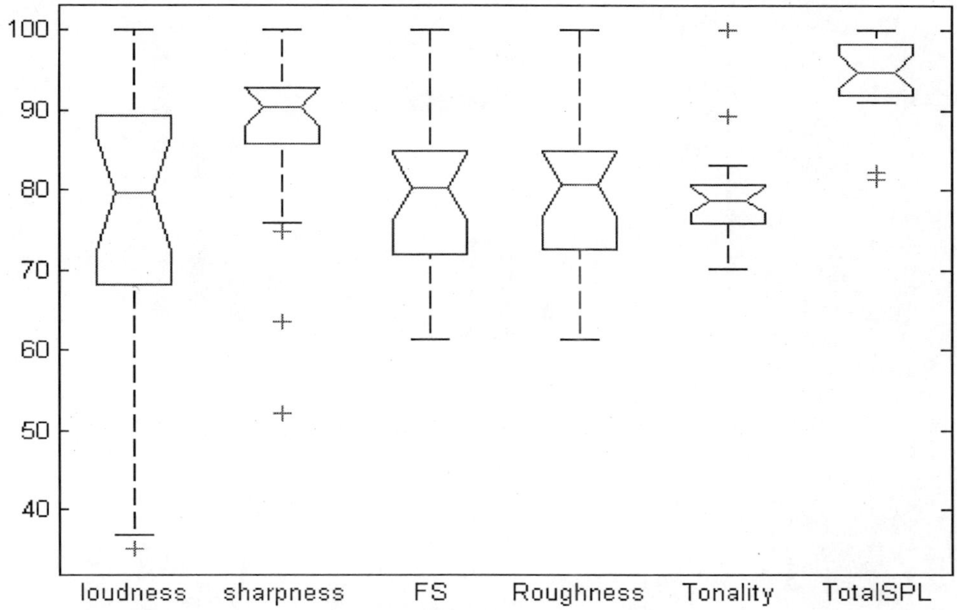

Fig. 2: Objective Data Analysis Results in a Box-Plot Representation

Total sound pressure level data have a smaller range variation and highest mean relative percentage. Fluctuation strength and Roughness values are proportional. Data with higher sharpness mean value and lower Roughness mean value can be interpreted as the dominance of higher order harmonics. Two horns tonality is shown as outliers in the Box-plot.

Relative sensory pleasantness index calculated for all horns considering M2DH3 horn as a reference sound sample for calculating pleasantness index. If the pleasantness index value is higher than reference horn sound pleasantness index value, those sound samples are pleasant. Pleasantness index values less than reference sound, then those horn sound samples are unpleasant.

Figure 3 shows the pleasantness index for objective data generated using BHS unit. It can be observed that shell horns, (M1SH1, M2SH1, M1SH2, M2SH2) are more pleasant compared to disc type horns. These results are useful to identify extreme pleasant or unpleasant horn sound samples with respect to reference sound.

The purpose of relevant test data is to identify car horn sound. Number of responses recognized as a car horn sound normalized with the number of responses for each horn and multiplied by 100 to represent as a percentage.Most of the subjects have knowledge associated with the disc horn as a car horn. A cognitive effect plays an important role in relevant test.

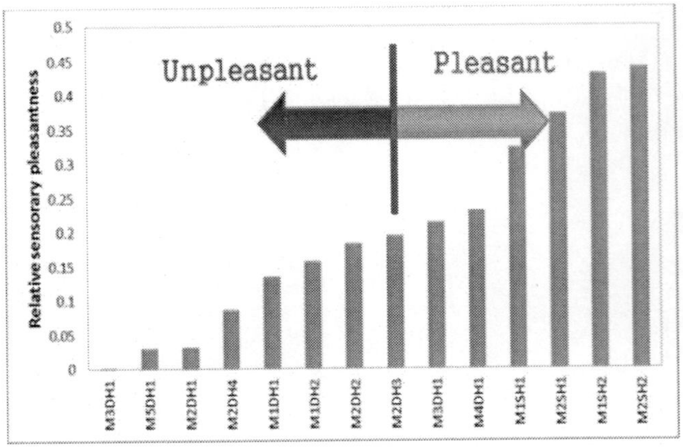

Fig. 3: Comparison of Different Horn Pleasantness Index

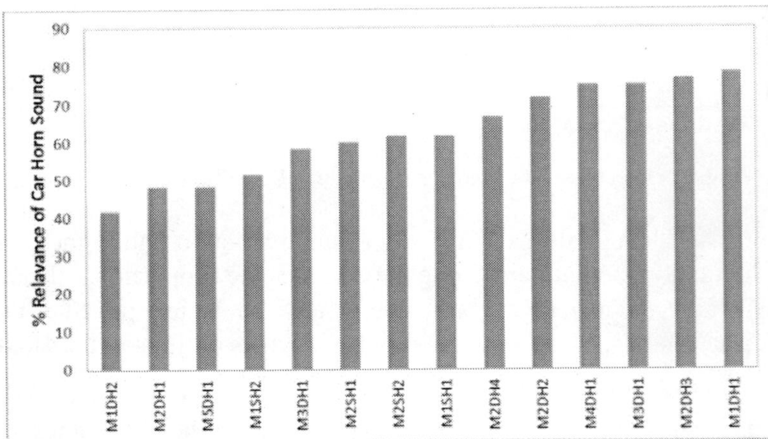

Fig. 4: Relevant Percentage of Different Car Horn Sounds

Table 2 shows a summary of pleasantness data results of objective and subjective data analysis. It should be observed that identification of pleasantness or unpleasantness for extreme psycho-acoustic parameter metrics is simple but it is difficult for other horn sound samples. The results depend on the instrumentation, cognitive effects, subjective knowledge and experience. There is a lot of research scope to understand all these parameters.Subjective test data of Relevant test and SD test could be used to develop a design chart to relate relevant percentage of car horn recognition to pleasantness.

Table 2: Summary of Pleasantness Data Results in Subjective and Objective Data Analysis

	M2DH1	M2DH2	M2DH3	M2DH4	M2SH1	M2SH2	M4DH1	M1DH1	M1DH2	M1SH1	M1SH2	M3DH1	M3DH2	M5DH1
Subjective data	Pleasant	Pleasant	Pleasant	Pleasant	Unplesant	Pleasant	Pleasant	Pleasant	Pleasant	Pleasant	Pleasant	Unplesant	Unplesant	Unplesant
Objective data	Unplesant	Unplesant	Unplesant	Unplesant	Pleasant	Pleasant	Pleasant	Unplesant	Unplesant	Pleasant	Pleasant	Pleasant	Unplesant	Unplesant

Figure 5 shows results from RT test and SD. A few comments based on subjective test results are:

- The different horn model has different values of percentage recognition and percentage of pleasantness.
- Shell horns have a higher pleasantness index and also some models have a good percentage of recognition as a car horn sound.
- These graphs might be used as a design optimization chart.

The pair comparison test has questions related to forced choice and pair comparison similarity. The forced choice response is in the form of binary input and pair comparison similarity response is in the form of interval scale. In a PC test, a forced choice method used to select a horn sound sample in a given pair. Based on this response, relations are established for different horn sounds. Associative laws are used to develop a relation at different stages. It has been observed that shell horn sound samples are more preferred choice than disc samples.

Based on subjective data analysis, the following observations are made

- Analysis of subjective data shows that disc type horns have recognized as car horn sound. This result might depend on subject knowledge and experience.
- Shell horns are more pleasant and preferred choice based on SD test and PC test results.

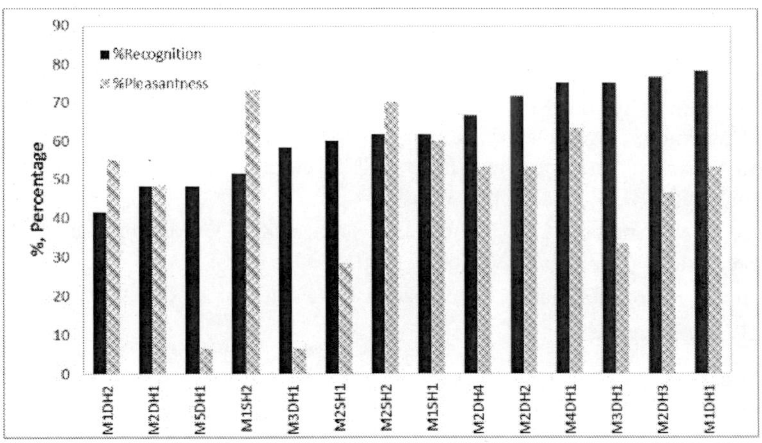

Fig. 5: Design Chart based on RT and SDT Test

SUMMARY

Preliminary study of sound quality analysis for automotive horn has been discussed in this paper.The experimental setup for sound recording has been discussed.

Some of the significant points of have been discussed in the current article are:

1. A comprehensive subjective test evaluation method has been developed to get the subject response related to automotive Horn. The three models of the Jury evaluation questionnaire provided basic information of subject's perception about horn sound.
2. It has been observed from subjective test data that soft, calm, relax and slow sounds are more pleasant.
3. It has been observed from the current investigation that shell horns are more pleasant.

ACKNOWLEDGEMENT

We would like to thank Hyundai Motors Engineering India Pvt. Ltd (HMIE) for financial support to conduct current investigation and Indian Institute of Technology Hyderabad for providing the required resources.

REFERENCES

[1] Guillaume Lemaitre, Boris Letinturier and Bruno Gaengel, "Model and estimation method for predicting the sound radiated by a horn loudspeaker-With application", Applied Acoustics 69 (2008) 47–59.

[2] G. Lemaitre, P. Susini, S. Winsberg, B. Letinturier and S. McAdams, "The sound quality of car horns: a psychoacoustical study of timbre", ActaAcustica united with Acustica 93 (2007) 457–468.

[3] G. Lemaitre, P. Susini, S. Winsberg, B. Letinturier and S. McAdams, "The Sound Quality of Car Horns: Designing New Representative Sounds", ActaAcustica united with Acustica 95 (2009) 356–372.

[4] HaiAnhThiPhan, Tsuyoshi Nishimura, Hai Yen ThiPhan, Takashi Yano, Tetsumi Sato, and Yoritaka Hashimoto, "Annoyance from road traffic noise with horn sounds: A cross cultural experiment between Vietnamese and Japanese", 9th International Congress on Noise as a Public Health Problem (ICBEN) (2008), Foxwoods, CT.

[5] Norn Otto, Scott Amman, Chris Eaton and Scott Lake, "Guidelines for Jury Evaluations of Automotive Sounds", Sound and Vibration (2001) 1–13.

[6] Hugo Fastl and EberhardZwicker, Psycho-acoustics Facts and Models, Springer Publications.

[7] MATLAB User Manual.

Acoustical Absorption Studies in Aqueous Lithium Salts

S. Vinothini, C.S. Prabhakaran, V. Aroulmoji[1] and G. Rajarajan[1]

Department of Physics, Mahendra Arts and Science College, Kalippatti, Namakkal, Tamil Nadu

[1]Centre for Crystal Growth and Nanotechnology, Department of Physics, Mahendra Engineering College, Mallsamudram, Namakkal District, Tamil Nadu

E-mail: grajarajan@gmail.com

ABSTRACT: *Ultrasonic absorption studies are carried out in aqueous lithium acetate and lithium chloride in the concentration range 0.1– 0.5 molar and in the frequency range 3–12 MHz. The ultrasonic absorption is measured using a pulsed power oscillator at 303 K. The measured and computed parameters viz., density, velocity, relaxation frequency, relaxation amplitudes A and B are tabulated. The variation in ultrasonic absorption is discussed on the basis of Hall's two-state model for liquid water.*

Keywords: Absorption, Velocity, Lithium Acetate, Lithium Chloride, Relaxation Amplitude.

INTRODUCTION

The earlier ultrasonic velocity and absorption studies [1–3] in aqueous electrolyte solutions of ammonium chloride, ammonium dihydrogen phosphate and ammonium oxalate at temperature 303 K and in di-hydrogen phosphates of sodium and potassium at temperatures 303 K, 313 K and 323 K, showed non-linear increase in the velocity with concentration and it is attributed to the increased cohesion among the water molecules due to the addition of electrolytes. Lower charge density anions (of relatively high compressibility) contribute more to the increasing rates of conversion between the structural forms of water.

When a solute either electrolyte, or non-electrolyte, or both, is added to water, it affects the structural equilibrium existing between hydrogen bonded clusters and monomers. Electrolytes dissolving in water have been classified as structure makers or breakers, depending on whether the existing structural equilibrium of water shifts to the hydrogen-bonded clusters or to the denser monomers. It has been reported that the ions with a low charge density termed as net structure breakers and the ions with a higher charge density termed as net structure makers [4, 5]. In continuation of our earlier work [6] on ultrasonic velocity studies in aqueous lithium acetate and lithium chloride, present ultrasonic absorption study in the frequency range 3–12 MHz is undertaken in order to understand these aspects more vividly and elucidate the relaxation parameters.

EXPERIMENTAL

All the chemicals used in the present study are of AR/BDH quality and are used as such without further purification. Solutions of aqueous lithium acetate and lithium chloride in the concentration range 0.1–0.5 molar are prepared by dissolving known amounts of the salt in double distilled water. The ultrasonic velocity is measured using an ultrasonic time intervalometer. The ultrasonic absorption is measured using pulsed power oscillator in the frequency range 3–12 MHz. The accuracy in the measurement of velocity and absorption is 0.001 ms^{-1} and ±1% respectively. The density and viscosity of the solutions are measured using *Specific Gravity Bottle* and *Ostwald's Viscometer*. The accuracy in the measurement of density and viscosity is 0.001 gm and ±0.01 sec. The temperature of the solutions is maintained at 303 K by circulating water from a thermostatically controlled water bath with an accuracy of ±0.1°C. The temperatures of the solution and the circulated water are noted by using dual terminal digital thermometer sensors by inserting them in the provided holes in the ultrasonic liquid cell for this purpose.

RESULTS AND DISCUSSION

The measured and computed parameters viz., density, ultrasonic velocity, relaxation frequency, relaxation amplitudes A and B for aqueous lithium acetate and lithium chloride are given in Tables 1 and 2 respectively. The variation of observed absorption with frequency is graphically represented in Figures 1 and 2 for aqueous lithium acetate and lithium chloride. The variation of absorption per wavelength with frequency is given in Figures 3 and 4. The solid lines in Figures 3 and 4 represent the relaxation frequency.

Table 1: Ultrasonic Relaxation and Related Parameters for Aqueous Lithium Acetate

c *molar*	ρ Kg/m^3	*u* *m/s*	f_r *MHz*	*A* $\times 10^{-15}$ $Npm^{-1}s^2$	B $\times 10^{-15}$ $Npm^{-1}s^2$
0.1	1019.69	1513.14	5.92	47.12	12.90
0.2	1020.00	1534.00	6.25	39.36	11.92
0.3	1021.60	1537.56	7.15	37.28	11.88
0.4	1022.89	1542.00	8.18	40.69	12.65
0.5	1028.79	1543.92	8.68	41.37	12.46

From the measured absorption coefficient (α), the observed absorption $(\alpha/f^2)_{ob}$ is calculated and it is fitted to conventional debye type single relaxation equation,

$$\left(\frac{\alpha}{f^2}\right) = \frac{A}{1+\left(\frac{f}{f_r}\right)^2} + B$$

where f, f_r, A and B represents frequency, relaxation frequency, amplitude of the relaxation and the contribution form frictional processes, e.g., shear viscosity, and also can include $(\alpha/f^2)_{ob}$ arising from other processes having relaxation frequencies at much higher frequencies than f_r respectively. The variation of observed absorption with frequency for fitted curve is shown in Figure 1A and 2A for aqueous lithium acetate and lithium chloride. The above parameters are computed using non-linear method proposed by D. W. Marquardt [7] and are given in Table 2.

Table 2: Ultrasonic Relaxation and Related Parameters for Aqueous Lithium Chloride

c *molar*	*ρ* *Kg/m³*	*u* *m/s*	f_r *MHz*	*A* $\times 10^{-15}$ $Npm^{-1}s^2$	*B* $\times 10^{-15}$ $Npm^{-1}s^2$
0.1	1014.50	1521.14	5.32	31.27	10.13
0.2	1016.59	1530.40	5.91	33.86	10.80
0.3	1017.39	1531.54	6.89	40.84	11.27
0.4	1019.50	1532.74	7.63	42.74	12.64
0.5	1020.04	1535.38	8.08	37.88	11.31

c = concentration; ρ = density; u = velocity; f_r = relaxation frequency; A & B = relaxation amplitudes

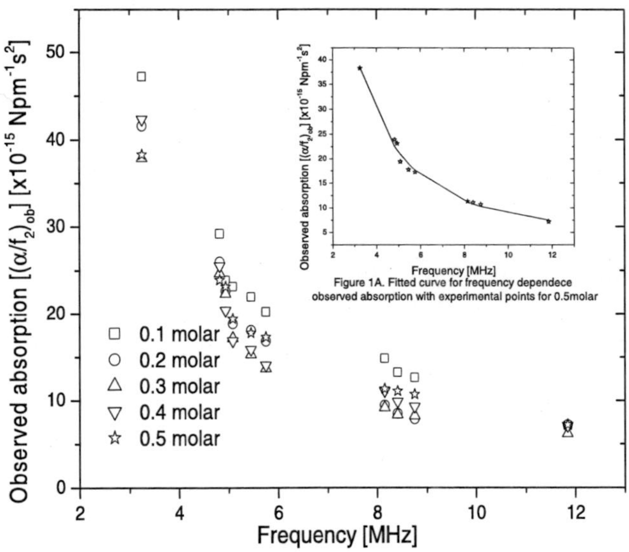

Fig. 1: Variation of Observed Absorption with Frequency for Aqueous Lithium Acetate

The ultrasonic absorption obtained in the frequency range 3–12 MHz is showing higher values at the lower frequencies and decreases as the frequency increases and thereby showing that the ultrasonic absorption is frequency dependent. The observed absorption is varies non-linearly as the concentration of the lithium acetate and lithium chloride is increased and is higher in lithium acetate. The computed parameters viz., relaxation frequency is increases as the concentration increases and is higher in lithium acetate than lithium chloride. The relaxation amplitudes A and B varies non-linearly as the concentration increases and is higher in lithium acetate than lithium chloride. The cation Li^+ being the same in these electrolytes, the difference in the variation of observed absorption may be mainly due to the anions (namely acetate and chloride ions) provided the solvent-solvent interactions are similar. Hence, this difference in the observed absorption may be attributed to the difference in charge densities of the anions [8].

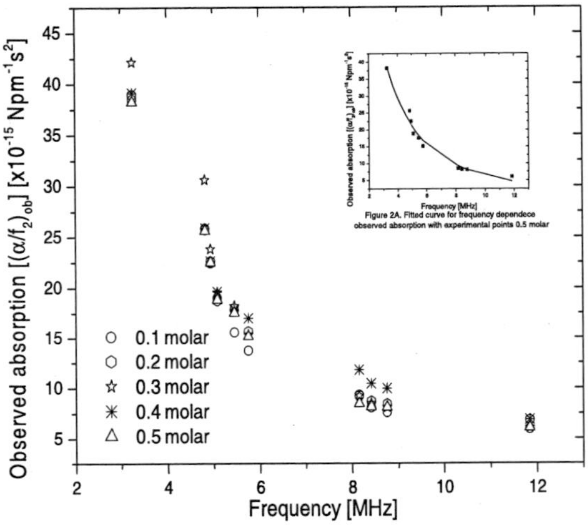

Fig. 2: Variation of Observed Absorption with Frequency for Aqueous Lithium Chloride

This variation of excess absorption as a function of concentration of electrolytes may be explained as follows. The addition of an electrolyte to a polar solvent can be having two effects [9]. The ions act as acceptors and they can compete with the protons for the lone pair of electrons on the oxygen. This leads to the formulation of salvation sheath around the ion. The ion may possess a large volume and hence a small effective charge density. The equilibrium between the two structural forms is disturbed. Both these effects contribute leading to a change in the acoustic absorption.

The variation in the excess absorption can also be explained by Hall's two state model of water [10]. According to this model, water is considered to be mixture of two molecular species. One of them has open ice like structure while the other has more densely packed structure. The addition of electrolytes to water may shift the equilibrium between the two configurations, resulting in reduction of excess absorption of the solution. This may result in a reduction of potential barrier between the two structural forms of water molecules particularly in the vicinity of the ions. This would reduce the relaxation time and may lead to the decrease of excess absorption at frequencies well below the relaxation frequency [11].

From the Figures 3–4, it is seen that the absorption per wavelength for the aqueous lithium acetate and lithium chloride increases with increasing frequency and reaches a maximum value at a particular frequency called as 'relaxation frequency (f_r)' and then decreases with further increase in frequency. This gives the information that all the systems studied show a single relaxation behavior in the frequency range 3–12MHz. The variation of absorption per wavelength with frequency follows a similar trend with increasing concentration of electrolytes.

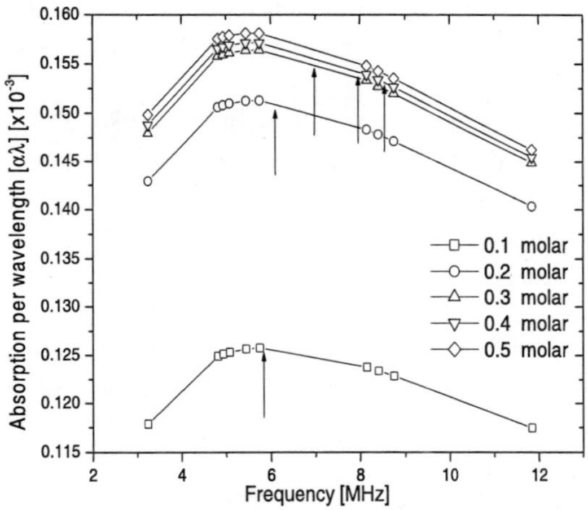

Fig. 3: Variation of Absorption Per Wavelength with Frequency for Aqueous Lithium Acetate

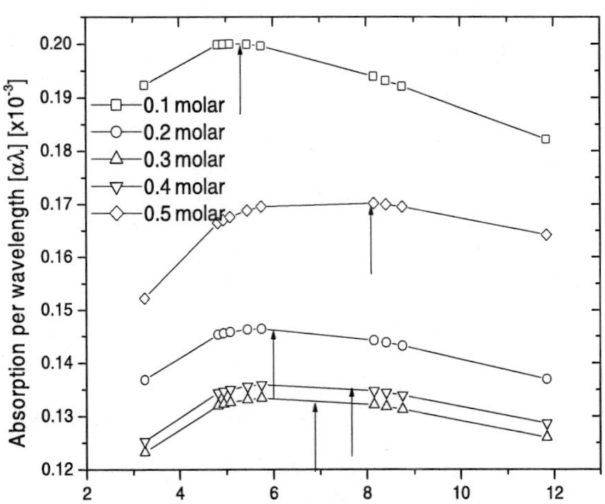

Fig. 4: Variation of Absorption Per Wavelength with Frequency for Aqueous Lithium Chloride

In conclusion, it can be seen that the change in the variation of observed absorption value with increasing concentration of electrolytes may be due to the difference in their charge densities, which contribute more to the increasing rates of conversion between the two forms of water. From this ultrasonic absorption study it is inferred that acetate ions have a greater structure breaking effect than chloride ions.

REFERENCES

[1] G. Ravichandran, A. Srinivasa Rao and T.K. Nambinarayanan, *Ind. J. Pure and Appl. Phy.*, 29 (1991), 792.

[2] G. Ravichandran and T.K. Nambinarayanan, Acoustic Letters, 19 (1996), 245.

[3] S. Kalaiselvi, M. Phil., thesis submitted to Pondicherry University, (2003).

[4] D.V. Jahagirdar, B.R. Arbad, Smt. C.S. Patil and A.G. Shankarwar, *Ind. J. Pure and Appl. Phy.*, 38 (2000) 645.

[5] T.K. Nambinarayanan and A. Srinivasa Rao, Acustica 53 (1983) 264.

[6] G. Rajarajan, P. Revathy and T.K. Nambinarayanan, *Indian J. Physics* 77B (2003), 677.

[7] D.W. Marquardt, *J. Soc. Ind. and Appl. Math.*, 11 (1963) 431.

[8] R. Barret, R. Beyer and F. McMamara, *J. Chem. Phys.* 14 (1946), 608.

[9] H. Ratajczak and W. J. Orville Thomas, *Molecular Interactions,* Vol. 3 (John-Wiley and Sons: New York) 1982.

[10] L. Hall, *Phys. Rev.,* 73 (1948) 775.

[11] W.P. Mason, *Physical Acoustics,* Vol. II Part A (Academic Press: London) 1965.

Non-Destructive Evaluation of SS304L Welded Material

S.V. Ranganayakulu, M. Premkumar and R. Gowtham

Centre for Non Destructive Evaluation, Guru Nanak Institutions, Technical Campus Ibraimpatnam, Hyderabad, Andhra Pradesh

E-mail: nayakulu@rediffmail.com

ABSTRACT: *The field of Nondestructive Testing (NDT) is a very broad, interdisciplinary field, that plays a critical role in assuring the structural components and systems, perform their function in a reliable and cost effective fashion. Welded sample of SS304L is tested by various NDT techniques such as Magnetic Particle Inspection, Liquid Penetrate test and Ultrasonic Flaw Detection in the present investigation studies. Testing procedure and methodology for detection of flaw is clearly discussed in the present studies. In Ultrasonic testing we detected two discontinuities (SD = 25.34 BP = 32.38 D = 1.69 and SD = 34.05 BP = 43.40 D = 8.94). In Liquid Penetrant test, no surface defects were found. In Magnetic Particle Inspection, no surface and sub-surface defects were found.*

Keywords: *Non-destructive Testing, Ultrasonic Testing, Liquid Penetrant Testing, Magnetic Particle Inspection.*

INTRODUCTION

Basic Principles of Non-Destructive Testing (NDT)

Non-destructive testing is the use of physical methods which will test materials, components and assemblies for flaws in their structure without damaging their future usefulness.NDT is concerned with revealing flaws in the structure of a product. It, however, cannot predict where flaws will develop due to the design itself.

Types of NDT Methods Used

Dye penetrant testing, Magnetic particle testing, Ultrasonic testing.

SAMPLES INTRODUCTION

- SS304L (Figure 1) is an austenitic Chromium-Nickel stainless steel offering the optimum combination of corrosion resistance, strength, and ductility. These attributes make it a favorite for many mechanical switch components. The low carbon content reduces susceptibility to carbide precipitation during welding.

Nominal Composition

- Chromium 18.2% Silicon .5%
- Nickel 8.5% Carbon .015%
- Manganese 1.6% Iron Balance

EXPERIMENTAL METHODS

- Ultrasonic testing
- Magnetic particle inspection
- Liquid penetrant testing

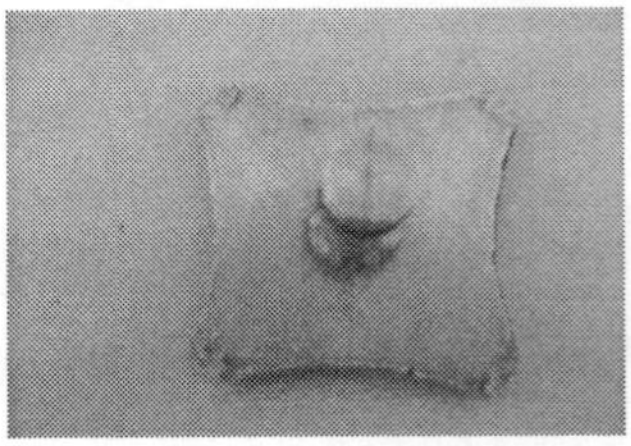

Fig. 1: SS304L Component Used for Testing

Ultrasonic Testing

Apparatus used: ultrasonic flaw detector, probes, 2T oil, V1 & V2 blocks.

Machine calibration procedure on V1 block with Normal probe and V2 block with Angle Probe:

- Machine is calibrated using four methods to understand the working condition
  - (a) Horizontal linearity
  - (b) Vertical linearity
  - (c) Sensitivity
  - (d) Resolution

Process of Testing

- After calibration of equipment and probes, we have to clean the surface of the job on which we are performing test.
- Here our testing piece is SS304L welded, so by using angle probe we can get better results.
- Apply the couplant on surface of the job to fill the gap between probe and surface.
- Now start the movement of the probe in which direction the whole welded region get detect.
- With 45degrees probe of frequency 2MHz, defects are detected at 1/2V, 1V, 11/2V sound path according to machine displayed values.
- By that we can detect any type of discontinuities which are in different orientation.

Drawing of DAC

- Draw DAC curve (Figure 2) from reference block in three paths, 1/2V, 1V, 11/2V paths. We can get automatically 50%20% curve.

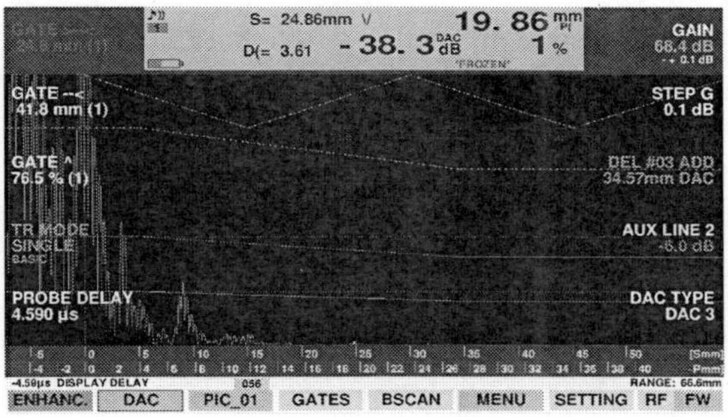

Fig. 2: DAC Curve

After testing the component we got two discontinuities (Figures 3 & 4) which are figured below:

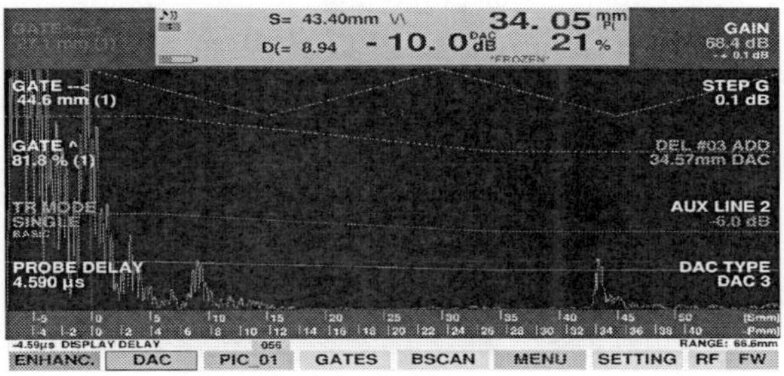

Fig. 3: Discontinuity-1
SD = 25.34 BP = 32.38 D = 1.62

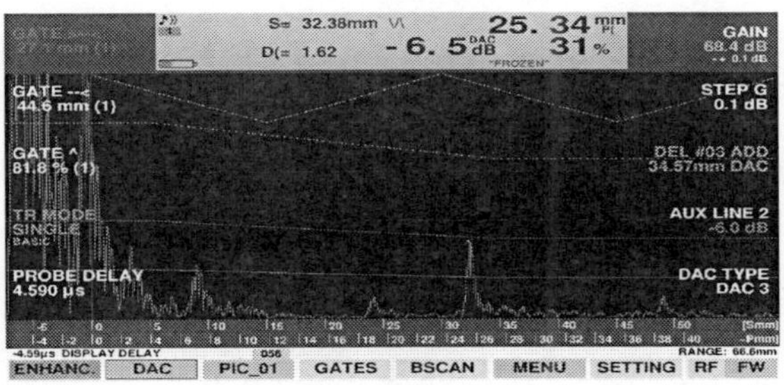

Fig. 4: Discontinuity 2
SD = 34.05 BP = 43.40 D = 8.94

Discontinuities which are get detected at 11/2V sound path there defect size is less than the reference reflector (Where SD-Surface Distance, BP-Beam Path, D-Distance)

Liquid Penetrate Testing

- Liquid penetrate examination is a method for the detection of discontinuities which are open to the surface of metal and materials.

- When the surface irregularities might mask the indications of unacceptable discontinuities, the surface shall be prepared by grinding or machining or any other suitable method.

Chemicals for Examination

- The cleaner used for the surface cleaning.
- Penetrant used shall have a contrast color which can be seen.
- Developer, when it dries form a white coating.

Procedure

Initially clean the surface with PMC Cleaner, because it shall be free of any dirt, grease, lint, rust, paint, oil or any other matter that could interfere the examination by preventing the penetration.

- The PMC Dye penetrant is applied by spraying after the surface is dried. For spraying aerosol cans is used.
- The penetrant is allowed to retain wet on the part for minimum 10 minutes (Dwell time).
- After the required dwell time, the excess penetrant remaining on the surface is removed by wiping, with a clean cloth wetted lightly with the cleaner.
- The developer is applied over the surface in such a manner to assure complete coverage of the part of with a thin layer of developer.
- Discontinuities at the surface will be indicated by bleed out of the penetrant, i.e. reddish indications on white background.

Magnetic Particle Inspection

- Magnetic particle examination is used for detecting the presence of discontinuities at surface or near the surface (sub-surface) of ferromagnetic materials such as cracks and other discontinuities like lack of fusion, incomplete penetration etc.
- Its sensitivity is greatest for surface and sub-surface (up to 6 mm from top) discontinuities and diminishes rapidly with increasing depth of discontinuities below the surface.

Equipment

- Here we are following two types of methods:
- Yoke and Prod technique.

Procedure

- For both techniques we are using electricity to magnetize the material, for that AC type of current is used for surface and DC type of current is used for subsurface.

- Generally, Yoke method is used for small length jobs, Prod method is used for industry purpose; by this we can detect large area.
- Here for prod method same the position of the prods, but depends on prod distance section thickness current generation varies.
- Both methods should repeat 2–3 times by changing the position of equipment for coverage total area.
- Dry magnetic powder/ wet water mixed magnetic particles used as inspection medium.
- Then the particles will attract on the surface of piece as discontinuity orientation.

RESULTS AND DISCUSSION

- Two internal discontinuities are detected through Ultrasonic Testing, discontinuities are showed in Figures 3 and 4.
- No surface and sub-surface defects are get detected through Magnetic particle testing.
- No surfaces discontinuities are get detected through Liquid Penetrant testing.

ACKNOWLEDGEMENTS

Authors are grateful to AICTE for carryout research under Research Promotion Scheme **(Project No: 8023/RID/RPS-58/ (POLICY-III) (pvt)/2011-2012).** Gratefully acknowledgements due to **Sardar G.S. Kohli**, Vice Chairman, and "**Dr H.S. Saini**, Managing Director, Guru Nanak Institutions, Ibrahimpatnam, for cooperation and encouragement.

REFERENCES

[1] G.A. Georgeou, "Non-Destructive Testing and Evaluation of Metals", *Material Science and Engineering*, Vol. III.

[2] P.P. Nanekar and B.K. Shah, "Characterization of Material Properties by Ultrasonics" *BARC News letter*, Issue No. 249.

[3] D.S. Forsyth, R.D. Leblanc, A. Fair, A. Maslouhi and A. Moreau, "Practical Ultrasonic Techniques for assessment of Heat-Damage in 7050 Aluminium alloy" NDT.net, November 1997, Vol. 2.

[4] Anish Kumar, B.K. Choudary, K. Laha, T. Jayakumar, K. Bhanu Sankara Rao and Baldev Raj, "Characteristics of Microstructure in 9% Chromium Ferritic Steels using Ultrasonic measurements", *Trans. Indian Inst. Met.* Vol. 56, No. 5, October 2003.

[5] "ASTM E-709-01 Standard Guide for Magnetic Particle Examination and ASTM E-1444-01 Standard Practice for Magnetic Particle Examination, ASTM, 100 ", Barr Harbor Drive, West Conshohocken, PA, 19428, USA.

[6] Chedister, William C., "Evaluation of Magnetic Gradients for Magnetic Particle Testing", *Materials Evaluation*, Vol. 60, No. 9, September, 2002, pp. 1083–9.

[7] Nikolai P. Migoun and Nikolai V. Delenkovsky, "The Ways of Penetrant Testing Applicability for Rough Surfaces" *17th World Conference on Nondestructive Testing*, 25–28 Oct. 2008, Shanghai, China.

[8] Nielson, D.C. and Thompson, J.G.H., "Evaluation of Liquid Penetrant Systems, Materials Evaluation", Vol. 33, No. 12, December 1975, pp. 284–292.

Non-Destructive Evaluation of Mild Steel Material

S.V. Ranganayakulu, R. Gowtham and M. Premkumar
Centre for Non Destructive Evaluation, Guru Nanak Institutions, Technical Campus, Hyderabad, Andhra Pradesh
E-mail: nayakulu@rediffmail.com

ABSTRACT: *This paper gives about orientation of defect present in mild steel material (welds).The defect is analyzed by non destructive methods namely ultrasonic method, magnetic particle and liquid penetrant method. First method trace out internal defect. Last two methods implies surface and sub- surface defect in the mild steel material. Dimensional measurement of welds 300 mm × 300 mm×10 mm (Length × Breadth × Height). Mild steel material consist of coarse grain structure and submerged arc welding. The pulse echo of ultrasonic method has been implemented in weld to trace out of defect along with DAC (Distance Amplitude Correction). Dry penetrant method is give out surface defect and Magnetic particle method through electromagnetic yoke is used along with Ferromagnetic particle give accurate surface and sub-surface defect in welds .Results has been achieved through Ultrasonic method that two defect are traced out at SD = 25.34 BP = 32.38 D = 1.69 and SD = 34.05 BP = 43.40 D = 8.94 respectively. Such that defect are within limits of DAC curve .similarly liquid penetrant and magnetic particle gives no defect such that mild steel material is accepted according to acceptance standards.*

Keywords: Defect, Welds, Penetrant, DAC, Pulse Echo, Electromagnetic Yoke.

INTRODUCTION

Today's mild steel is heavily dependent on the means of fast non destructive quality inspection of the welds joints parts together. This paper will present you three non-destructive methods suitable for welds. The aim is to raise discussion on the applicability of NDT methods namely.

Ultrasonic Method

Ultrasonic method is used as NDT-Method to evaluate the integrity of welds. The principle is to employ high frequency acoustic wave to probe the inspected sample. As acoustic wave penetrates sample, the wave is attenuated or reflected by any change in the density of the material. By observing the returned signal many characteristics can be determined.

Dye Penetrant

This is suitable for detection of cracking and porosity in welded joints and surface of the welds. The principle is that the surface of the sample is coated with penetrant in which a fluorescent dye is dissolved. The penetrant is pulled into surface defects by capillary action. After a waiting period to insure that the dye has penetrated into the cracks, the excess penetrant is cleaned from the surface of the inspected part. A developed, a white powder, is sprayed over the part. This lifts the penetrant out of the defect and the dye indication can be identified the defect. Less than one hour is usually required as an inspection time. The method is a lot cheaper compared to radiography or ultrasonic, but can only defect external defects.

Magnetic Particle

Magnetic particle method can be used for identification of surface or sub-surface. The principle is that the sample is magnetized by dusting magnetic particle over it. A surface defect will form a magnetic anomaly, attracting and holding magnetic particle and thus giving a visual indication of the defect. The evaluation time is typically few minutes. The sample must be ferromagnetic and therefore this technique cannot be used on most stainless steels. This method also is a lot cheaper compared to radiography or ultrasonic, but like dye penetrant, it only can defect external defects.

SAMPLE PREPARATION

Mild steel material (Welds) is a submerged arc welding type which consists of coarse grain structure due to annealing heat treatment is implemented. Reflector of 2 dia hole is made to the adjacent to the welds. The tolerance for hole diameter drilled is 4 mm in deep.

Table 1: Chemical Composition of the Studied Steel (weight %)

Chemical Composition (%)									
C	*Mn*	*P*	*S*	*Si*	*Cu*	*Al*	*N*	*Ti*	*Sn*
0.04	0.193	0.017	0.0241	0.20	0.002	0.02	0.0044	0.002	0.008

Table 2: Mechanical Properties of the Studied Steel

Hardness	*Cupping*	*Yield Strength*	*Tensile Strength*	*Ys/Ts*	*Elongation*
43 HRB	11.3	218MPa	325 MPa	0.67	45%

EXPERIMENTAL STUDY

Ultrasonic Method

Ultrasonic weld inspections are typically performed using a straight beam transducer in conjunction with an angle beam transducer. A straight beam transducer, producing a longitudinal wave at normal incidence into test piece, is first used to locate any defects at HAZ area. Scanning with straight probe no defect is located.

The second step in the inspection involves using an angle beam transducer to inspect the actual weld. Angle beam transducer to inspect the actual weld. Angle beam transducer uses the principle of refraction and mode conversion to produce refracted shear or longitudinal waves in the test material.

During the process of scanning the surface of the material around weldment with transducer. Refracted sound wave will bounce off a reflector in the path of the sound beam. With proper angle beam techniques, echoes returned from the weld zone and weld surface may allow determining the location and type of discontinuity. With 45 degree probe of frequency 2 MHz defect are detected at 1/2V, 1V and 1⅓V sound path according to machined displayed values.

Similarly DAC curve is constructed from peak amplitude responses from reflectors of equal area at different distance in the same material. A-scan echoes are displayed at peak amplitude of each signal is marked on the flaw detector screen. Therefore, two defects are located as follows:

Defect-1 There is a defect which is getting detected at 1V sound path there defect size is less than reference reflector.

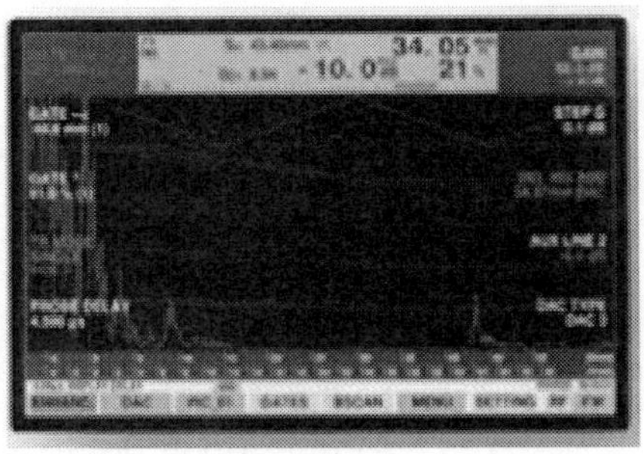

Defect 1

SD = 25.34 mm, BP = 32.38, D = 1.27 mm
SD-surface distance, BP-beam path & D-depth

Defect-2 There is a defect which is getting detected at 1½ V sound path there defect size is less than reference reflector.

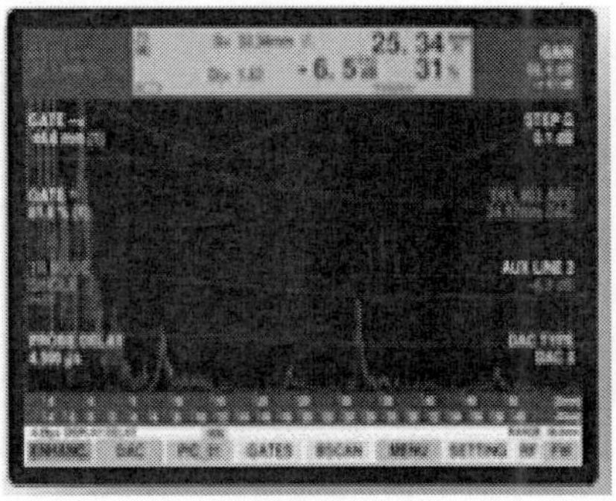

Defect 2

SD = 34.05 mm, BP = 43.40 mm, D = 4.94 mm

Liquid Penetrant Method

In order to perform liquid penetrant method on weld plate. The weld plate is cleaned with PMC cleaner sprayed to remove the dust and dirt. Wiped with a lint free cloth and PMC dye penetrant is sprayed near the weld area and 20 minutes dwell time is given.

It is wiped off and then PMC developer is sprayed near weld area and 10 minutes dwell time is given.

Finally no defect was observed on the weld plate.

Interpretation on HAZ (Heat Effected Zone) and Whole Surface Area

No defect is traced out on weld surface by liquid penetrant test.

Magnetic Particle Method

With this method the ferromagnetic powder is usually applied from shaker, bulb or blower and follows these steps. Apply magnetized current yoke placed on the weld plate.

The powder should float to the magnetized area so that particle will be attracted to any flux leakage.

Inspection on HAZ Area and Whole Weld Surface Area

No defect has been detected.

RESULTS AND DISCUSSION

According to ultrasonic testing method mild steel material consists of 3 defects were detected. These defects are crossed 50% DAC. So, the weld piece is rejected. No significant cracks, wormhole or other surface deformities were observed in the mild steel material produced by magnetic particle and liquid penetrant method.

CONCLUSIONS

Non-Destructive testing techniques via: ultrasonic testing, magnetic particle testing and liquid penetrant method were successfully conducted on welds produced at various parameter combinations. It can be concluded that the magnetic particle testing of the welds is not the best technique as it was found that although all the welds passed this test and appeared as defect-free welds, the ultrasonic testing technique successfully detected the defects present in the welds and can be said to be appropriate in this regard.

ACKNOWLEDGMENTS

Authors are grateful to AICTE for carryout research under Research Promotion Scheme *(Project no: 8023/RID/RPS-58/(POLICY-III) (pvt)/2011-2012).* Gratefully acknowledgements due to “**Dr H.S. Saini**, Managing Director and **Sardar G.S. Kohli**, Vice Chairman, Guru Nanak Institutions, Ibrahimpatnam for cooperation and encouragement.

REFERENCES

[1] Atlas of Some Steel Casting Flaws as shown by Non Destructive Testing, Steel Castings Research and Trade Association, Sheffield, England, 1968.

[2] Recommended Procedure for the Ultrasonic Examination of Steel Castings, Steel Castings Research and Trade Association, Sheffield, England, 1971.

[3] ASTM A609, “Standard Specification for Longitudinal Beam Ultrasonic Inspection of Carbon & Low Alloy Steel Castings”.

[4] ASME Boiler & Pressure Vessel Code, Section V, T524.2, “Angle Beam Examination of Steel Castings.

[5] R/D Tech Operating Manual, TC-5700, October, 1998.

[6] .ASNT Handbook, 2nd edition, Vol.4, ET, R. McMaster, P. McIntire, M. Mester, 1986.

[7] . ASNT Handbook, 2ndedition, TU, Vol.7, A. Bikes, R. Green, P. McIntire, 1991.

Section–III

Speech, Hearing and Music

Categorical Perception of Voice Onset Time (VOT) and F2 Continuum by Kannada and Hindi Speaking Individuals

T. Jayakumar, Vijay Kumar Narne[1] and M.V. Pragati Rao

Department of Speech-Language Sciences, All India Institute of Speech and Hearing, Mysore

[1]Department of Audiology, All India Institute of Speech and Hearing, Mysore

E-mail: jayakumar82@gmail.com

ABSTRACT: *The present study aimed to check the perception of synthetic speech continuums in native Hindi and Kannada speakers. Two synthetic speech continuums varying in place of articulation (onset of F2) and voicing (Voice Onset Time) were presented to the speakers of the two language groups. Both the language groups perceived the continuums varying in F2 similarly. But the Hindi speakers required a larger VOT to categorize a sound as being voiced. The implications of these findings are discussed.*

Keywords: Synthetic Speech, Speech Perception, Logistic Regression, Cross Language, Categorical Boundary.

INTRODUCTION

Categorical perception is a well researched phenomenon in the field of speech perception. VOT (Voice Onset Time) has been found to be distinctive in many languages like English (Lisker and Abramson, 1972), Hindi, Punjabi and Bengali (Lahiri, 1980). Several Indian studies (Usha Rani and Savithri 1989, Vinay Rakesh and Savithri 1989) have focused on assessing VOT as a major cue for voicing in Indian language. There is a dearth of literature in assessing onset of F2 as a cue for place of articulation in Indian languages. In India, most of the research in this area has focused on perception of synthetic speech continuum created by using analysis by synthesis method. There has been no published literature where completely synthetic speech was used to check perception in different populations. The present study focused on creating completely synthetic speech continuums for F2 and VOT and checked the perception of these continuums in native Hindi and Kannada speakers.

METHOD

Stimulus and Participants

Synthetic speech continuums were created using Klatt synthesizer in MATLAB (McLennan, 2000). The synthesizer consists of a Simulink model which incorporates the Klatt synthesizer (Klatt, 1980). The input parameters to the Simulink model were varied using MATLAB and the synthetic speech continuums were generated. The synthetic continuums varied in onset of F2 (second formant) and VOT (Voice Onset Time. In the F2 continuum, onset of F2 was varied from 800 Hz to 1900 Hz in steps of 100 Hz to create 12 stimuli. In Indian languages, there is a lead VOT (prevoicing component) in the voiced syllables such as /da/ or /ba/. This was generated synthetically using MATLAB and was added to the base syllable. The VOT continuum consisted of 12 stimuli where VOT varied from –50 msec to +5 msec in 5 msec steps. The lag VOT stimulus of +5 msec was generated by delaying the aspiration noise which excites the first formant filter (also known as F1 cutback). The generated continuums were then perceptually validated by performing the goodness of stimuli test on 10 individuals with normal hearing. A pilot experiment was also conducted where identification scores were obtained for the end points of the continuums. E.g. for the VOT continuum stimulus 2 (0 msec VOT) and stimulus 12 (–50 msec VOT) were consistently identified by all the participants of the pilot experiment as /ta/ and /da/ respectively. Similarly, in the F2 continuum stimulus 1 (onset of F2 at 800 Hz) and stimulus 12 (onset of F2 at 1900 Hz) were identified as /pa/ and /ta/ respectively.

Participants consisted of 30 individuals who were divided into two groups of 15 native speakers of Hindi and Kannada respectively. All the participants were in the age range of 18–30 years. None of the participants had any history of speech, language and hearing disorders.

Stimulus Presentation

Alvin software (Hillenbrand and Gayvert, 2005) was used to present the stimulus to the participants. Binary forced choice identification was performed. The participant's task was to choose whether the stimulus sounded like one of the two options on the screen. In the identification task, 10 iterations of the 12 stimuli in each continuum were randomly presented to the listeners. The participants were asked to click on the option on the screen as fast as possible without sacrificing accuracy. The response of the participant for both voice and place categories was calculated using the following formula: Percentage identification = (Obtained response for the token/total number of iterations *100). This percentage of identification for each stimulus in each of the participant was noted down.

RESULTS AND DISCUSSION

From the identification experiment, for each of the stimuli scores were obtained in the form percentage of identification. The results for F2 continuum and VOT continuum will be discussed separately.

F2 Continuum

The results were obtained as percentage of times identified as /pa/. The following Figure 1a shows the average of the identification scores for the Hindi and Kannada group. Maximum likelihood procedure was used to fit logistic regression curve to this average data. This was implemented using MATLAB. From the identification curve, fifty percent crossover, lower limit of phoneme boundary (LLPB), upper limit of phoneme boundary (ULPB) and phoneme boundary width (PBW) was obtained (Doughty, 1949) Fifty percent crossover is the point at which 50% of the subject's response corresponds to the one category. Lower limit of phoneme boundary width is the point along the acoustic cue continuum where an individual identified voiced (voiceless) or bilabial (place) stop 25% of the time and upper limit of phoneme boundary width defined as the corresponding point of the identification of the unvoiced (voiced) or palatal (place) cognate 75% of the time. Phoneme boundary width was determined by subtracting the lower limit from upper limit of boundary width.

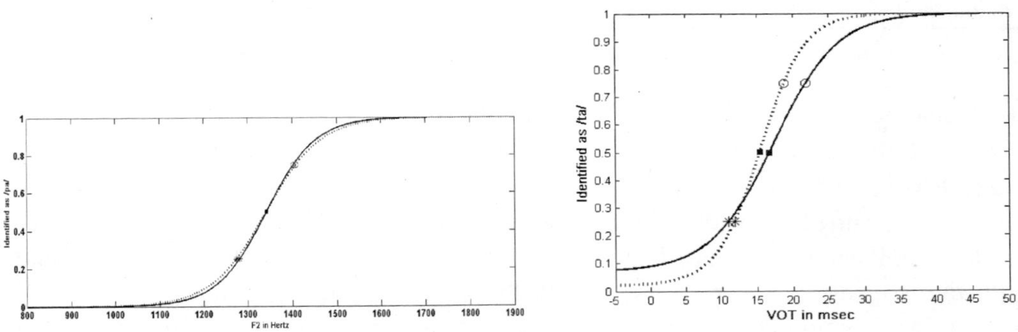

(a) (b)

Fig. 1(a) & (b): Logistic regression curves fitted for the average identification functions of Hindi participants (solid line) and Kannada participants (dotted line) for the F2 continuum and VOT continuum. Black filled squares show the fifty percent crossover points while the unfilled circle and the asterisk show ULPB and LLPB respectively.

From the Figure 1a it can be seen that differences do exist in the fifty percent crossover points, ULPB for the two language groups. Then, the maximum likelihood procedure was used to fit logistic regression curve for the identification data of each of the participant. To determine goodness of fit for the curves generated Nagelkerke estimated R^2 values were calculated for each participant as recommended by McMurray and Spivey (1999). Multivariate analysis of variance (MANOVA) was used to compare LLPB, 50% crossover point, ULPB and PBW across the two language groups. No significant difference was noted between the two groups on any of the variables. Mean, standard deviation in parenthesis, F value and level of significance are shown in the Table 1.

Table 1: Mean, SD, F& P-values in Kannada and Hindi Speakers for Parameters of Identification Curve and Nagelkerke Estimated R^2 Values for VOT and F2 Continuum

Parameters	*F2 Continuum*				*VOT Continuum*			
	Hindi Mean (SD)	*Kannada Mean (SD)*	*F (2,28)*	*P value*	*Hindi Mean (SD)*	*Kannada Mean (SD)*	*F (2,28)*	*P value*
LLPB in Hz	1287.76 (104.4)	1299.38 (85.79)	0.11	0.74	–12.25 (5.92)	–13.85 (4.15)	0.74	0.40
50% point in Hz	1332.06 (88.12)	1345.18 (83.22)	0.18	0.68	–16.42 (4.68)	–15.82 (3.98)	0.14	0.71
ULPB in Hz	1377.65 (84.73)	1391.80 (91.25)	0.19	0.66	–20.45 (4.86)	–17.74 (4.25)	2.63	0.12
PBW in Hz	89.89 (74.28)	92.42 (59.75)	0.01	0.92	8.20 (5.50)	3.88 (2.64)	7.50	0.01*
Nagelkerke estimated R^2	0.94	0.95	–	–	0.87	0.93	–	–

Note: *significance at 0.05 level.

Hindi and Kannada speakers identified all the stimuli of the F2 continuum similarly. According to Ohala (1983), the various places of articulation for plosives in Hindi are bilabials (p, b), velar (k, g), dentals (t̪, d̪) and retroflex (ʈ, ɖ). Similar classification is also present in Kannada (Upadhyaya, 1972). Since is a dearth of published F2 studies in production, it can be postulated that since the place classification of plosives is similar for both the languages, the speakers of both the languages perceive stops varying along a place of articulation continuum similarly. From the above data it is also clear that both Hindi and Kannada speaker perceived stimulus 1 and stimulus 12 as /pa/ and /ta/ with an accuracy of greater than 90%. The crossover boundary or 50% crossover point occurred between stimulus 6 and 7 (closer to stimulus 6) in both the language groups. A relatively narrow phonemic boundary width in both the language groups suggests that the listeners did not have much confusion while perceiving the stimuli.

VOT Continuum

Similar to the F2 continuum, the percentage identification scores were obtained for each stimulus in the VOT continuum. Average percentage identification was calculated and logistic regression curves were fit to see the average trend of identification for both the language group. Figure 1b shows the average of the identification scores for the Hindi and Kannada group for the VOT data. The Figure 1b also shows the average fifty percent crossover points, LLPB and ULPB values. Similar to the data analysis in the F2 continuum, maximum likelihood procedure was used to fit the identification data for each of the participants. From the curves generated, fifty percent crossover points, LLPB, ULPB and PBW was determined. The parameters were defined in the same manner as described in the previous section

From the Figure 1b it can be seen that there are differences in the fifty percent crossover points, ULPB between the two language groups. To determine goodness of fit for the curves generated Nagelkerke estimated R^2 values were calculated for each participant as recommended by McMurray and Spivey (1999). The range of the Nagelkerke estimated R^2 values were from 0.46–0.97 for the Hindi group and 0.81–0.97 for the Kannada group. MANOVA was used to compare LLPB, 50% crossover point, ULPB and PBW across the two language groups. The table shows the descriptive statistics and the F values from MANOVA for each of the variables.

It can be observed from the table that only the phonemic boundary width differs significantly across the two groups. The phonemic boundary width is significantly larger for the Hindi group than the Kannada group. This implies that Hindi speakers require a larger VOT (of up to 8.2 ms) to categorize a sound as being voiced. Very few published production studies exist to throw more light on the mechanisms of perception. Savithri and Santosh (2003) compared voicing parameters of plosives in 15 Indian languages. On closely observing the mean production data for VOT, it was found that Dravidian Languages (Kannada, Kodava, Telugu, Tulu) had smaller VOT values than the Indo-European languages (Assamese, Bengali, Gujarati, Hindi, Marathi, Oriya, Rajasthani) in the word initial position. From the production data, it might be extrapolated that since Hindi speakers had larger VOT in the voiced syllables in the word initial position, they also required a larger VOT to perceive a sound as being voiced. It can also be postulated that since Hindi speakers had a significantly larger phonemic boundary width and shallower phoneme identification curves they had greater confusions in perceiving voicing. This also implies that they had a weaker categorical distinction between /ta/ and /da/ (Manis *et al.*, 1997).

ACKNOWLEDGEMENT

Author would like to thank the Director, All Indian Institute of Speech and Hearing (AIISH) and the participants of the study. It is part of ARF project.

REFERENCES

[1] Doughty, J.M. (1949). The effect of psychophysical method and context on pitch and loudness functions. *Journal of Experimental Psychology, 39*(5), 729–745.

[2] Hillenbrand, J.M. and Gayvert, R.T. (2005). Open source software for experiment design and control.. *Journal of Speech, Language, and Hearing Research, 48*(1), 45–60.

[3] Klatt, D. (1980). Software for a cascade/parallel formant synthesizer. *Journal of the Acoustical Society of America,* 67(3), 971–995.

[4] Lahiri, A. (1980). Voicing contrast in four-category stop languages. *Journal of the Acoustical Society of America, 67*(S1), S51–S51.

[5] Liberman, A.M., Harris, K.S., Hoffman, H.S. and Griffith, B.C. (1957). The discrimination of speech sounds within and across phoneme boundaries. *Journal of Experimental Psychology,* 54(5), 358–368.

[6] Lisker, L. and Abramson, A.S. (1972). The voicing dimension: Some experiments in comparative phonetics. In proceedings of the Sixth International Congress on phonetic sciences. Prague. 1967 (pp. 563–567) Prague: Academia.

[7] Manis, F.R., Mcbride-Chang, C., Seidenberg, M.S., Keating, P., Doi, L.M., Munson, B. and Petersen, A. (1997). Are speech perception deficits associated with developmental dyslexia? *Journal of Experimental Child Psychology,* 66(2), 211–235.

[8] McLennan, S. (2000). Klatt Synthesizer in Simulink. Retrieved June 29, 2012, from http://www.shaav.com/professional/linguistics/klatt.pdf

[9] McMurray, B. and Spivey, M.J. (1999). The Categorical Perception of Consonants: The Interaction of Learning and Processing. *Proceedings of the Chicago Linguistics Society*. pp. 205–219.

[10] Ohala, M. (1983). Segments and Features. In Jain, D. (Ed.), Aspects of Hindi phonology (pp. 1–2). Patna. Motilal Banarasidass.

[11] Savithri, S. and Santosh, M. (2003). Voicing contrast in Indian languages- Acoustic measurements. *Journal of Acoustic society of India,* 31, 122–127.

[12] Upadhyaya, U. (1972). Description of Kannada speech sounds. In Biligiri, H. (Ed.), Kannada phonetic reader, (pp. 15–20). Mysore. Central Institute of Indian Languages.

[13] Usha Rani and Savithri S.R. (1989). Temporal Perceptual Cues of Velar and Bilabial stop consonants. Unpublished master's dissertation, University of Mysore, Mysore, India.

[14] Vinay, R. Savithri S.R. (1990). Cross-language study of velar and bilabial stop perception. Unpublished master's dissertation, University of Mysore, Mysore, India.

Search for Spectral Cues of Emotion in Hindustani Music

Anirban Patranabis, Kaushik Banerjee, Ranjan Sengupta, Nityananda Dey[1] and Dipak Ghosh

Sir CV Raman Centre for Physics and Music, Jadavpur University, Kolkata

[1]ITC Sangeet Research Academy, Tollygunge, Kolkata

ABSTRACT: *The foremost element of Hindustani Music (HM) is Rasa—the finest emotional behavior, outcomes from a perfect rendition of any raga that touches the hearts of world music lovers. In this work we attempt to find out the spectral cues of emotion in six ragas. As per literature and research findings, out of these six ragas three are of positive emotion (arousal) and rest three are of negative emotion (valence). Keeping this as standard, our objective is to find out the spectral cues of emotion from the sung signals in these six ragas. Features that communicate emotion are extracted from the music as a function of time. Our study agrees with the fact that emotion is a time varying quantity.*

INTRODUCTION

Most of us feel that music is closely related to emotion or that music expresses emotions and/or can elicit emotional responses in the listeners. Every known human society is characterized by making music from their very beginnings. One fascinating possibility is that, although different cultures may differ in how they make music and may have developed different instruments and vocal techniques, we may all perceive music in a very similar way. That is, there may be considerable agreement in what emotion we associate with a specific musical performance [1].

In HM, the artist aims at expressing the *rasa* (emotional flavour) associated with the ragas. There is a well established system of ragas and *talas* (rhythms) that form its edifice. Ragas can be described a melodic framework with specific tones with a well defined ascending (*aroha*) and descending (*avroha*) structure as well as unique phrases (*pakkad*). Each raga in the HM is conventionally assigned to a corresponding *rasa* (emotion) and is also known to consistently evoke more than one emotion. As per treatise, nature of raga depends on *Vadi-Samvadi* and combination of notes (*chalan)* and that could create more than one Rasa from a single raga. Earlier study indicates that the following parameters are related to the emotional expression of music: pitch or F_0 (i.e., the strongest cycle component of a waveform), intensity or loudness, attack or the rapidity of tone onset, tempo or the velocity of music, articulation or the proportion of sound-to-silence, timing or rhythm variation, timbre or high-frequency energy of an instrument/singers formant [1].

In this work we have taken the aalap part of six ragas (eliciting contrasting emotions- joy and sorrow), each of 1min duration, played by an eminent sitar player for analysis. These ragas chosen for analysis are Darbari Kannada, Adana and Mian-ki Malhar and Chayanat, Jayjayanti and Basant. The first three are under the category of pathos or *Karuna* and the last three expresses the mode of joy/romanticism. It is expected that the associated emotion with these ragas should be within the sound clips and therefore analysis of the sound spectra of each clip might reveal the spectral cues of the associated emotions. Here we did a study to find the spectral cues of emotion from the analysis of the raga clips.

EXPERIMENTAL PROCEDURE

For our analysis only the *aalap* parts were selected from each raga. The digitization of the signal was done at the rate of 44100 samples/sec (16 bits/sample) in mono channel. Signal files, stored as 'wav files', are thus selected for analysis, which constitute the database. Each signal taken is of one minute duration. Pitch was extracted at 10 milliseconds interval using open source software Wavesurfer of KTH, Stockholm. Cool edit software of Syntrellium corporation was used for further signal processing.

METHODOLOGY

Since all the six signals are from a single sitar with single artist and so timbre variations are fixed. All the six signals are normalized to 0db and hence intensity or loudness and attack cue are not being considered. Extraction of notes from the pitch profile needs the knowledge of the tonic used by the player. A skilled musician was requested to listen to the signal files one after another to detect the position of tonic Sa in the signal file. The finding of the ratio-intervals is done by first dividing the smoothed pitch values for each song by the pitch value of the 'Sa', tonic of that raga rendering. This gives the frequency ratios for each pitch data. From the ratio data, steady state sequences are created with all consecutive pitch in a sequence, which is terminated when $| x_i + 1 - M | > M/30$ where $M = (1/i) \Sigma x_i$. If the duration of any sequence were less than that of a certain minimum value (60 milliseconds for this experiment) then the sequence is rejected. Elements of these sequences are considered as suitable candidate data for this analysis [2]. Whenever the ratio is less than 1 it is multiplied by a factor of 2 and when it is greater than 2 it is divided by 2. This effectively folds all pitch data into the middle octave. These are now distributed in 1200 bins of one-cent width each. The peaks of these distributions for each song are purported to be indicative of the sruti positions for that raga rendering (shown in Figure 1). After getting pitch value of Sa we have calculated the pitch value of other notes from standard sruti model, thus each note is being found out from the along with their duration. Time gap between two successive notes is termed here as silence. Silence to note duration is an important feature for emotion determinant (shown in Table 1).

Musical features can be considered either global or local. Global features are measured over an entire selection of music and are not time-varying. Local features are measured over small sections of time [1]. It is established that the elicited emotion from different segments from the same raga has some specificity, i.e. it might be that different segments from the same raga show different emotions. Also the emotional response from the segments of a raga does not generally correspond to those prescribed in Indian treaties [3]. We took the help of some experienced musicians for identifying the emotional phrases in the music signal along with their time intervals, based on their feelings. These are shown in Table 1. We attempted to match the results of our analysis with this subjective information.

Table 1: Time Intervals of Emotional Phrases in Different Ragas

Raga	*Duration (secs)*
Adana	0–28.379
Darbari	04.234–09.928 and 24.174–53.963
Mian-ki-Malhar	04.795–48.845
Jayjayanti	07.352–12.401, 19.323–27.4 and 48–end
Basant	09.861–end.
Chayanat	03.737–12.537 and 27.805–59.349.

Since emotion is a time-varying quantity, in this paper, we tried to analyse the whole signal as well as the small segments in the each signals as marked by the musicians as emotion locations.

RESULTS AND DISCUSSION

After the extraction of pitch data from each signal clips, it is smoothed and the steady state was detected. Figure 1 shows the frequency distribution of the duration of steady states of raga Adana and raga Basant (1min clip each) spread over 1200 bins of 1-cent interval. The peak shows the ratios, which correspond to different sruti positions utilized in the respective raga. Each peak corresponds to a note utilized in the raga and the height of the peak denotes the number of times the note was hit in the clip. The relationship between the height and the distance between the peaks may give some indication about the nature of the raga and the emotion associated therein. Similar distributions were done from the signals of Adana, Darbari Kannada, Mian-Ki-Malhar, Jay-Jayanti, Basant and Chayanat and the rendered ratios were found out. These distributions showed the preferred notes in the raga used by the artist.

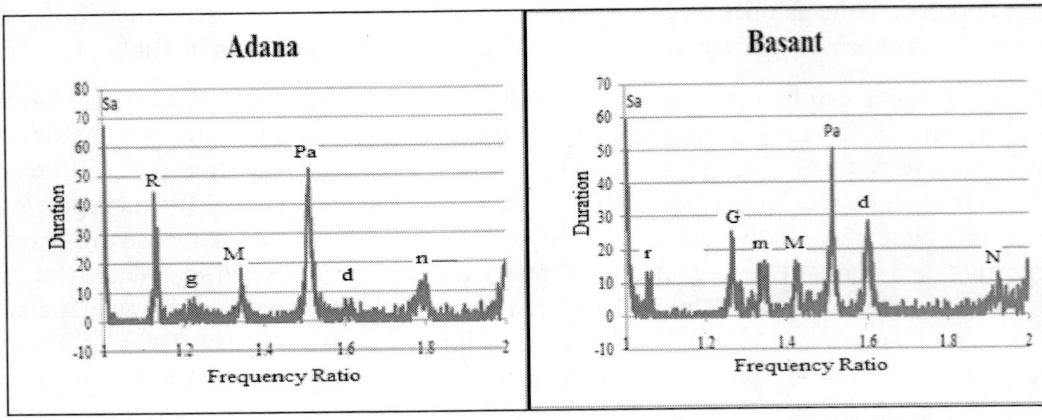

Fig. 1: Two Out of Six Distributions are shown over 1200 Bins of 1 Cent Width

Analysis was also done with pitch pattern behavior/harmonic behavior of each signal, in which we have made a distribution of pitch/harmonic shift [1] of each peaks of LTAS from the spectral centroid. In all six distributions, the best fit polynomial curve is of order six and the R^2 values were noted.

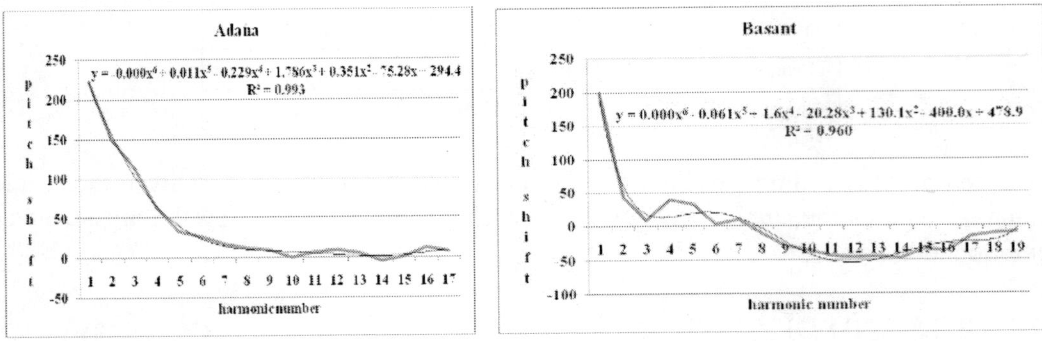

Fig. 2: Two Out of Six Distributions are shown with Best Fit Trend Line and R^2 Value

The distribution reveals the fact that the partials are more harmonically distributed around the spectral centroid. The figure also shows that pitch shift decreases linearly with the harmonic number for the ragas Adana, Mian ki Malhar and Darbari Kannada which may be an spectral attribute of negative emotion (valence). The ragas Basant, Chayanat and Jayjayanti showed a complex nature of the variation of pitch shift with harmonic number showing an indication of positive (arousal) emotion in the signals. It is also observed that the R^2 value is higher for signals of positive emotions and lower for signals of negetive emotions.

The note sequences in each signal were found out and also their durations were measured [2]. The silence zones in the signals were detected and duration of silence was measured.

Silence to note ratio was measured. This is done globally for the entire signal and also over small sections of time interval 10 seconds each. It is understood that this ratio when closer to one reveals a very slow tempo and hence the emotion may be sad/devotion while this ratio is very less than one reveals a fast tempo and hence the emotion may be joy/happiness [1].

Table 2: Ratio of Silence to Note Durations Local and Global

	Local						*Global*
Time → in sec,	*0–10*	*10–20*	*20–30*	*30–40*	*40–50*	*50–60*	*0–60*
Adana	0.5337	0.5801	0.5105	0.6497	0.8353	0.8817	0.8817
Darbari	0.4864	0.8106	0.7296	0.6890	0.8512	0.9322	0.9322
Mian-Ki Malhar	0.5493	0.6016	0.6801	0.5232	0.6801	0.6801	0.6801
Jayjayanti	0.7884	0.5709	0.8972	0.6525	0.5709	0.7069	0.8972
Basant	0.8898	1.0593	0.8051	0.9322	0.7627	1.0169	1.0593
Chayanat	0.6288	0.4353	0.4111	0.5079	0.7014	0.5321	0.7014

Now comparing Tables 1 and 2 we find the emotion cues and are marked in italics and underlined in Table 2. This is an objective evaluation with the help of subjective information. Comparing the above two tables it may be concluded that the silence to note ratio is higher for signals of positive emotions while the ratio is lower for signals of negative emotions. Another interesting conclusion is that a piece of music may be undeniably emotionally powerful, and at the same time be experienced in very different ways by another person who hears it. Different part of a signal reveals different emotions.

Tempo is the speed or pace of a musical piece. It may be measured as the numbers of notes per unit time [4]. Studies indicate an association between fast tempo and happiness or excitement. Slow tempo may be associated with sadness or serenity [5]. Measurement is done for the entire signal (globally) and then over the small sections of time interval 10 seconds each and is shown in Table 3. Comparing this with the emotion revealing phrases in the signal, as labeled by a musician (Table 1), it may be inferred that the emotion revealing part of Jayjayanti, Basant and Chayanat have higher tempo than the remaining part while Adana, Darbari Kannada and Mian ki malhar have lower tempo than the remaining part.

Comparing with Table 3, data matched with the musician's opinion are marked in italics and underlined in Table 4. Comparing these two tables it may be concluded that the tempo is higher for signals of positive emotions while tempo is lower for signals of negative emotions. But considering the whole signal at a time, we find very little difference.

Table 3: Tempo—Local and Global

	Local						*Global*
Time → in sec,	*0–10*	*10–20*	*20–30*	*30–40*	*40–50*	*50–60*	*0–60*
Adana	23	25	22	28	36	38	172
Darbari	12	20	18	17	21	23	111
Mian-Ki-Malhar	21	23	26	20	26	26	142
Jayjayanti	29	21	33	24	21	26	154
Basant	21	25	19	22	18	24	129
Chayanat	26	18	17	21	29	22	133

CONCLUSION

1. Emotion in musical signal is time varying and so in order to find the emotional cue in the signal spectra we should concentrate on the smaller segments of signal.
2. Keeping timbre and loudness constant, spectral cues of emotion might lie in the silence to note ratio and other features like pitch/harmonic shift and tempo.
3. Musical signal with positive and negative emotions will have higher and lower tempo respectively.
4. Silence to note ratio is higher for signals of positive emotions while this ratio is lower for signals of negative emotions.
5. Pitch/harmonic shift decreases linearly with the harmonic number for ragas of negative emotion (valence) while a complex nature of the variation of pitch shift with harmonic number indicates positive (arousal) emotion.

REFERENCES

[1] Mark David Korhonen, "Modeling Continuous Emotional Appraisals of Music Using System Identification", A thesis presented to the University of Waterloo, for the degree of Master of Applied Science in Systems Design Engineering, Waterloo, Ontario, Canada, 2004.

[2] A.K. Datta, R. Sengupta *et al.*, "A methodology of note extraction from the song signals", Proc. Frontiers of Research on Speech and Music (FRSM-2006), 9–10 January, 2006.

[3] Alicja Wieczorkowska, Ashoke Kumar Datta, Ranjan Sengupta,Nityananda Dey and Bhaswati Mukherjee, "On Search for Emotion in Hindusthani Vocal music", Adv. in Music Inform. Retrieval, SCI 274, pp. 285–304, Springer-Verlag Berlin Heidelberg 2010.

[4] P.G. Hunter, E.G. Schellenburg and U. Schimmack, "Mixed affective responses to music with conflicting cues". Cognition and Emotion 22: 327–35, 2008.

[5] S.O. Ali and Z.F. Peynircioglu, "Intensity of emotions conveyed and elicited by familiar and unfamiliar music" Music Perception: An Interdisciplinary Journal 27: 177–182, 2010.

Effect of Harmonium Usage in Hindustani Music Performances: A Signal Processing Approach

Kaushik Banerjee, Anirban Patranabis, Ranjan Sengupta and Dipak Ghosh
Sir CV Raman Centre for Physics and Music, Jadavpur University, Kolkata

ABSTRACT: *In this paper we attempted to find out the influence of harmonium usage (ET scale) on the raga renderings in Hindustani music. The pitch period was extracted from 17 renderings of 12 eminent vocalists of Hindustani music, for two different ragas. Only the steady pitch periods were used for analysis. The approach was to study the duration distribution of steady pitch states against the ratio of each note in an octave. Prominent peaks indicate preferred ratios. Attempt was made to see whether these ratios obey ET scale or Indian 22 sruti intervals. The result shows that most of the vocalists follow a mixture of both systems with a preference of any one depending on the raga.*

INTRODUCTION

Hindustani music (HM) has had some uniqueness and so it has been highly esteemed by the rest of the world. In HM, swara (note) is conceived not merely as a sound of fixed pitch position, but as the entire tonal range between itself and its previous swara. Though this interval can be theoretically divided into infinitesimal parts, the ancient musicologists believed that only a limited number, not exceeding four, sounds could be distinctly cognized by the ear in a swara-interval. These cognizable sounds are known as Srutis and the interval, which separated one swara from the next, was measured in terms of Srutis. Sruti was thought of both as the least audible interval between two sounds, as well as the sounds themselves, which were separated, by such an interval. The total number of Srutis was fixed unambiguously at 22 in ancient, musical treatises [1]. The tradition of teaching HM is an oral one. It is taught directly by the guru to the disciple, without the use of textual material. The basics starting from voice culture to the understanding of notes and Srutis are taught through direct oral communication accompanying with only tanpura. It was showed that the use of *Srutis* varies with *gharana*s as well as with the *raga*s and has also a student- teacher relationship [2]. It is likely that each *gharana* may have its own sense of aesthetic satisfaction of a *raga* and the use of *Sruti* may reflect these. It is also known that musicians feel that a particular note sounds better when placed relatively low for a particular *raga* while it may have to be raised much higher for some other *raga*. For example some say that *komal* re needs to be placed high in Todi but low in Bhairav. In the later part of the British rule in India, western musical instrument Harmonium (using ET scale) was introduced in HM in about middle of 19th century. The instrument was much nourished and developed in the eastern part of Indian, especially in Kolkata [3].

This paper presents the *Sruti* position as revealed from the performances of contemporary and older generation singers. Regarding the use of Srutis by the singers, substantive and distinctive preferences to different Srutis in two different ragas (Bhairav and Todi) are studied particularly in the context of the usage of Harmonium (ET scale).

EXPERIMENTAL DETAILS

17 renderings on two *raga*s namely Bhairav and Todi, sung by 12 singers from different *gharana*s of Hindustani music, collected from different archives, were taken for the study. For our analysis only slow tempo parts (mostly *Vilambit Vistar*) of each sample was selected from each *raga*. The digitization of the signal is done at the rate of 44100 samples/sec (16 bits/sample) in mono channel. Signal files, stored as 'wav files', are thus selected for analysis, which constitute the database. Pitch was extracted at 10 milliseconds interval using open source software Wavesurfer of KTH, Stockholm. Extraction of notes from the pitch profile needs the knowledge of the tonic used by the player. A skilled musician was requested to listen to the signal files one after another to detect the position of tonic Sa in the signal file. Cool edit software of Syntrellium corporation was used for further signal processing. The finding of the ratio-intervals is done by first dividing the smoothed pitch values for each song by the pitch value of the 'Sa', tonic of that raga rendering. This gives the frequency ratios for each pitch data. From the ratio data, steady state sequences are created with all consecutive pitch in a sequence, which is terminated when $| xi + 1 - M | > M/30$ where $M = (1/i) \Sigma xi$. If the duration of any sequence were less than that of a certain minimum value (60 milliseconds for this experiment) then the sequence is rejected. Elements of these sequences are considered as suitable candidate data for this analysis [4]. Whenever the ratio is less than 1 it is multiplied by a factor of 2 and when it is greater than 2 it is divided by 2. This effectively folds all pitch data into the middle octave. These are now distributed in 1200 bins of one-cent width each. The peaks of these distributions for each song are purported to be indicative of the Sruti positions for that raga rendering.

RESULT AND DISCUSSION

Aim of this study is to examine closely the calculated Sruti intervals [5] for individual performers, which reveals discreet well-defined finer structures of their Sruti position. It also reveals that the sounds having two different pitches in the same *Sruti*-interval have the same note value and are identical in this perceptual sense. However in terms of consonance with another sound, these two may behave quite differently. One may be in consonance while the other may not. In this sense they are perceptually differentiable. This differentiability demands cognizance at least in modal music [6].

A Euclidean distance measurement technique (eqn. 1) is used to find the nearest neighbour of 22 Sruti intervals from that of ET scale.

$$d(p,q) = \sqrt{(p_1 - q_1)^2 + (p_2 - q_2)^2 + ... + (p_i - q_i)^2 + ... + (p_n - q_n)^2} \quad ... (1)$$

As per the calculated minimum distance i.e. the nearest neighbour, values closer to 22 Srutis are marked in italics in Table 1 and 2 while values closer to ET scale are in normal font.

Table 1: *Sruti* Positions Obtained from Raga Todi for Nine Artists

Ratio (22)	*Artist a*	*Artist c*	*Artist d*	*Artist f*	*Artist g*	*Artist w*	*Artist x*	*Artist y*	*Artist z*
sa	1.000	1.000	1.000	1.000	1.000	1.000	1.000	1.000	1.000
r1			*1.051*					*1.050*	
r2	1.059	1.061		1.050	1.059	1.060	1.061		1.061
r3									
r4			*1.158*			1.123		*1.136*	
g1	1.186			1.185	1.189		1.185		*1.179*
g2		*1.200*							
g3									
g4						*1.262*		*1.256*	
m1									
m2								1.333	1.338
m3		*1.413*	1.417			*1.412*	1.420	1.416	*1.401*
m4	1.428			1.427	1.431				
pa	1.497	1.498	1.501	*1.508*	*1.514*	*1.509*	1.488	*1.506*	*1.507*
d1					1.575				
d2	1.582	*1.597*	1.591	1.583			1.586		*1.597*
d3									
d4						1.681			
n1									
n2									
n3								*1.819*	
n4	1.910	1.903	1.893	1.903	*1.926*	*1.920*	1.890	1.888	*1.920*

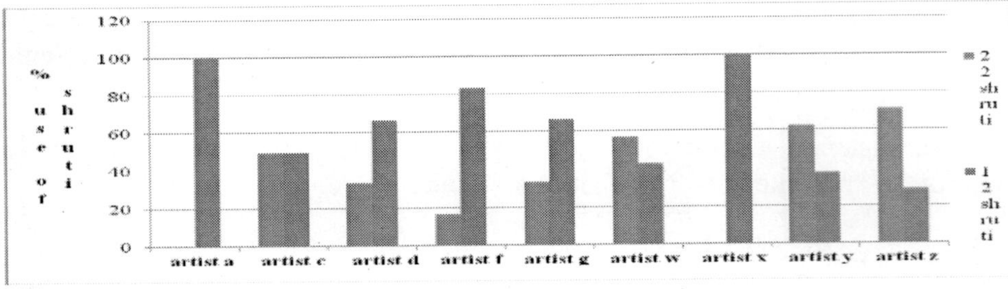

Fig. 1: Usage of Sruti by Nine Artists in Raga Todi

Usage of 22 Srutis (microtonal intervals) are perceptually distinguishable than that use of ET scale. And according to some people 22 Srutis are more pleasant to listen. Possibly the frequent use of harmonium as an accompaniment for some artists make their ratios mostly matched with the ET scale.

Table 2: *Sruti* Positions Obtained from Raga Bhairav for Eight Artists

Ratio (22)	*Artist A*	*Artist C*	*Artist D*	*Artist F*	*Artist G*	*Artist H*	*Artist I*	*Artist J*
sa	1.000	1.000	1.000	1.000	1.000	1.000	1.000	1.000
r1		*1.043*			*1.047*	*1.053*	*1.043*	*1.049*
r2	1.061		1.058	1.059			*1.078*	
r3								
r4								
g1								
g2		*1.209*						
g3								
g4	*1.265*		*1.263*	*1.263*	*1.266*	*1.258*	*1.263*	*1.261*
m1		*1.313*					*1.327*	
m2	1.334		1.336	1.335	1.333	1.330		1.332
m3						*1.411*		
m4								
pa	*1.506*	1.498	*1.512*	*1.517*	*1.510*		1.494	1.497
d1		*1.562*				1.583		1.571
d2	1.593		1.596	*1.601*	1.586		1.588	
d3								
d4								
n1								*1.709*
n2					1.764			
n3			*1.795*					
n4	1.906	1.869	1.906	*1.914*	1.875	1.896	1.896	1.886

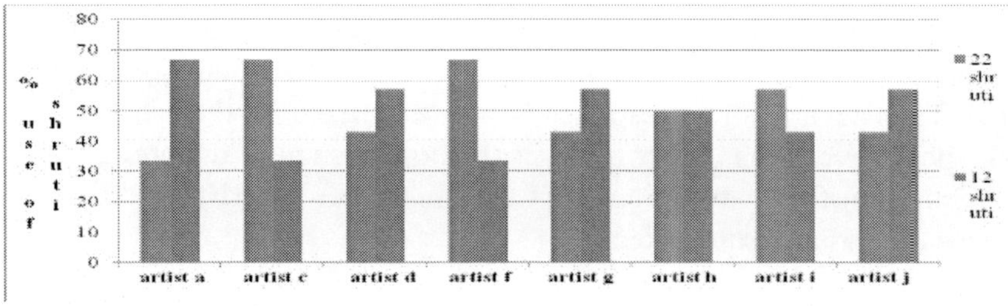

Fig. 2: Usage of Sruti by Eight Artists in Raga Bhairav

It is being observed from the tables and Figures 1 and 2 that some musicians are using ET scale ratios and some musicians are using mixed ratios both 22 srutis and ET scale.

Presently this trend is increasing in HM and a recent survey on the assessment of tuning in HM shows the predominance of the use of ET scale [7]. Such inclination of HM towards equal-tempered scales will gradually loose the significance of *shruti*s due to the influence of foreign musical elements like harmonium. This might lose the cognitive influence of HM regarding many musical qualities such as mood, consonance or timbre particularly the 'feelings', which is the heart of HM.

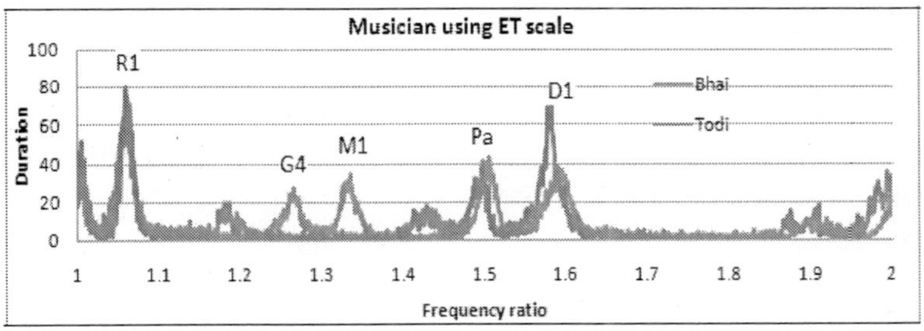

(a)

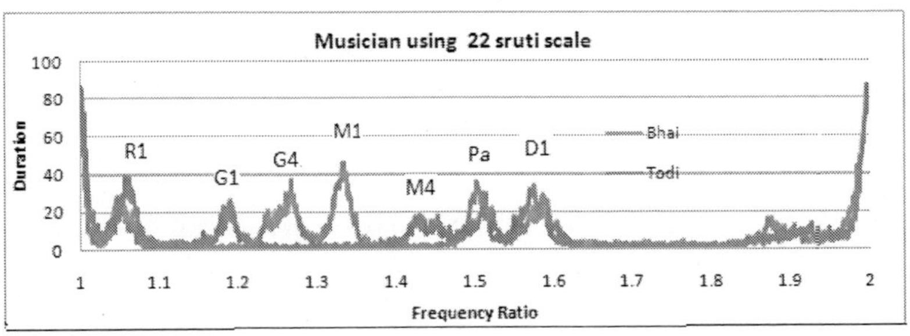

(b)

Fig. 3(a) & (b): Frequency Distribution of Two Renderings of *raga* Bhairav and Todi Spread over 1200 Bins of 1-*cent* Interval. The Peaks can be Easily Identified with the Ratios, Which Correspond to Different *Sruti* Positions Utilized by the Artist in the Rendering

Figure 3 (a&b) shows the frequency of occurrence of the duration of ratio of the steady pitch periods and Sa (tonic) folded to middle octave and distributed over 100bins of one cent interval. Thus each peak corresponds to a note utilized by the artist in the raga. The height of the peak gives the number of times the note is sung in the processed signal. It is being observed from the distributions that the singer (a) uses ET scale mostly and the singer (b) uses the Indian Sruti system.

CONCLUSION

Influence of harmonium have enabled the vocalists of HM to deviate from the older 22 Sruti interval system and at present most of the musicians either follow a mixed system of ET and 22 Sruti intervals or purely ET scale. There is also a preference of the system on the raga.

REFERENCES

[1] Sangitaratnăkara of Săraingadeva (Vol. II): Text and English translation by R.K. Shringly and Prem Lata Sharma, published by Munshiram Manoharlal Publishers Pvt. Ltd. 1st published on 1989, Delhi.

[2] Asoke Kumar Datta, Ranjan Sengupta,Nityananda Dey and Dipali Nag,"Experimental Analysis of Srutis from performances in Hindustani music", Monograph published by Scientific Research Department, ITC Sangeet Research Academy, 2006, ISBN 81-903818-0-6

[3] http://en.wikipedia.org/wiki/Harmonium#India

[4] A.K. Datta, "A methodology of note extraction from the song signals", Proc. Frontiers of Research on Speech and Music (FRSM-2006), 9–10 January, 2006.

[5] A.K. Datta, R. Sengupta *et al.*, "Sruti usage by old and contemporary singers in Khyal: An objective approach", *Proc. Frontiers of Research on Speech and Music (FRSM-2004),* 8–9 January, 2004, Annamalai University, Tamil Nadu, India.

[6] A.K. Datta, R. Sengupta *et al.*, "Objective analysis of srutis from the vocal performances of Hindustani Music using Clustering Algorithm",Published in EUNOMIOS—An open online journal for Theory, Analysis and Semiotics of Music, http://www.eunomios.org/, dt. 25.6.2011.

[7] Joan Serr`a, Gopala K. Koduri, Marius Miron and Xavier Serra, "Assessing the tuning of sung Indian classical music", Int. Soc. for Music Information Retrieval Conf. (ISMIR), Miami, USA, 2011.

Processing for Emotions Induced by Hindustani Instrumental Music

Ranjan Sengupta, Anirban patranabis, Kaushik Banerjee, Tarit Guhathakurta, Samabrata Sarkar, Gautam Ganguli[1], Atanu Biswas[1], P.K. Saha[1] and Dipak Ghosh

Sir C V Raman Centre for Physics and Music, Jadavpur University, Kolkata
[1]Bangur Institute of Neuroscience, Kolkata

ABSTRACT: *Music has an intriguing ability to evoke powerful emotions without having such an intrinsic link. However, establishing exactly how music communicates emotion is a topic of much debate. The present study concentrates on the signal analysis of two ragas (Chayanat and Darbari Kannada) played by an eminent sitar player as well as performing listening tests and EEG analysis on naïve listeners. An attempt was made to correlate the changes of EEG signal characteristics due to emotion with the note sequence of the ragas.*

INTRODUCTION

Western classical music which is based on harmonic relation between notes versus the melodic mode (raga) structures in the Hindustani music system (HM) within the rhythmic cycle music may demand qualitatively different cognitive engagement. Each raga in the HM is conventionally assigned to a corresponding rasa/emotion is known to consistently evoke certain emotions [1]. The artist exploits ones creativity and elaborates the raga well within the melodic framework to bring out the rasa or the emotion. The human brain, which is one of the most complex organic systems, involves billions of interacting physiological and chemical processes that give rise to experimentally observed neuro-electrical activity, which is called an electroencephalogram (EEG). The use of EEG signals as a vector of communication between men and machines represents one of the current challenges in signal theory research. The present study concentrates on the signal analysis of the aalap part of two ragas (eliciting contrast emotions, i.e., joy/romantic and sorrow/pathos) played by an eminent sitar player and subsequently conducting listening tests with the raga clips along with the EEG analysis to study the emotion as processed in the brain.

EXPERIMENTAL DETAILS

Two ragas chosen are for our analysis are Chayanat (romantic/joy) and Darbari Kannada (pathos/sorrow). Each of these sound signals was digitized with sample rate of 44.1 kHz,

16 bit resolution and in a mono channel. From the complete playing of the ragas, segments of about 3mins and 40 seconds were taken out for analysis for each raga. Detection of pitch was done by using the open source software package WAVESURFER of KTH, Stockholm. The tonic (Sa) was extracted from the signal by the help of a musician and then the note sequence was found out. Signals processing was performed to find out the tempo, pitch sift and note distribution of the emotion revealing part. From the sequences the phrases which contain the emotion as identified by a musician was played and subjected to listening tests by 30 listeners. The listeners' opinions were collected in the score sheet as in [1]. The listeners were asked to assign each raga clip to only one emotional category, if possible. Statistical elaboration of results in work was performed taking into account the choices of each listener. The data from the responses found in the score sheet was accumulated in a matrix where columns represented the number of the sound clips and the rows represented the listeners. Each value in the matrix gives the number indicating the emotion using the following code values:

1 anger, 2 joy, 3 sorrow, 4 heroic, 5 romantic, 6 serenity, 7 devotion, 8 anxiety, 9 nil emotion

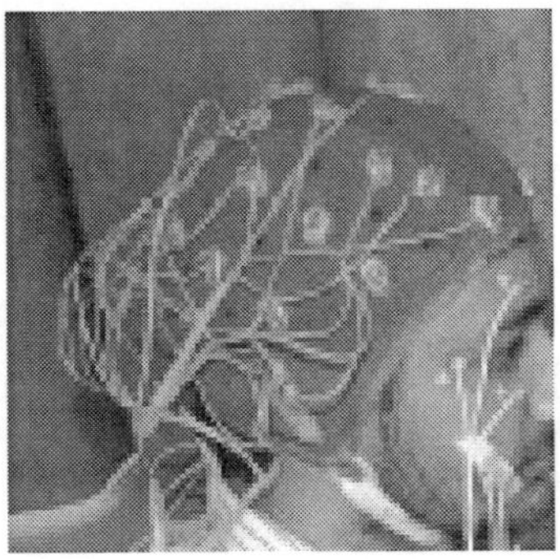

EEG was done subsequently to record the brain-electrical response of two such listeners. Each listener was prepared with an EEG recording cap with 19 electrodes (Ag/AgCl sintered ring electrodes) placed in the international 10/20 system. Impedances were checked below 5 kOhms. The EEG recording system (Recoders and Medicare Systems) was operated at 256 samples/s recording on customized software of RMS. The data was bandpass-filtered between 0.5 and 35Hz to remove DC drifts and suppress the 50 Hz power line interference. Each listener was asked to be seated comfortably on a chair in a shielded measurement cabin. He was also asked to close his eyes. A sound system (Logitech R _ Z–4 speakers) with very low S/N ratio was set up in the measurement room that received input from outside the cabin. After initialization, a 16mins 20secs recording period was started, devoting 3minutes each at the start, 3minutes at the end and 3minutes in between the signals because of the relaxation from the preceding signal period. Each signal length was 3mins. 40 secs. Markers were set at start, signal onset/offset, and at the end of the recording.

RESULT AND DISCUSSION

After extracting note duration and transition time for the whole signal, it was observed a negligible difference between the two signals. Then with the help of expert musicians we have sorted out the portions of each raga signal where the emotion of that signal reveals. The notation of that part of the signals for each raga is shown below:

SS RR GG M DPMP M, GMRGMP, RPR, GMNDP, P, PNṠṘṠṠ, ṠnD DnPP, RGMNDMP, R—GMPMR, snD, nSRS.

The phrase containg emotion as identified from the raga Chayanat sound clip by a musician. The time span of this phrase is 2.56.646–03.23.229 secs in the complete signal.

C—SRG -, C—MPD -, C—NNṠ -, C—Ṡ-, D-, NṠṘ-Ṙ, C—ṠṘĠ-Ġ ṀṘ - , C—DNṘṠ C—D - D - N - P,

The phrase containg emotion as identified from the raga Darbari Kannada sound clip by a musician. The time span of this phrase is 03:17.496–03:39.550 secs in the complete signal.

Extraction of the note duration of the above portions the distribution is shown in Figure 1. In Darbari most dominating note is Re while in Chayanat it is MA and Pa.

Tempo is the speed or pace of a musical piece. Studies indicate an association between fast tempo and happiness or excitement [2]. Slow tempo may be associated with sadness or serenity Measurement is done for the entire signal and then over the small sections of time interval 30 seconds each and is shown in Figure 2. Comparing Figure 2 with the emotion revealing phrases in the signal, as labeled by a musician, it may be inferred that the emotion revealing part of Chayanat have higher tempo than Darbari. But considering the whole signal at a time, we find very little difference. Analysis was also done with pitch pattern behavior/harmonic behavior of each signal, in which we have made a distribution of pitch/harmonic shift of each peaks of LTAS from the spectral centroid.

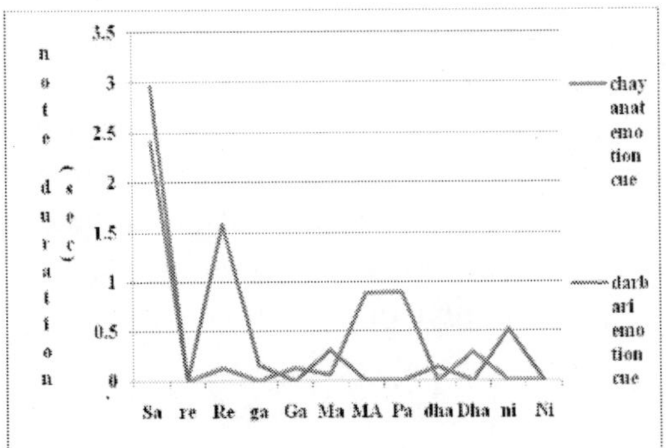

Fig. 1: Note Distribution of Emotion

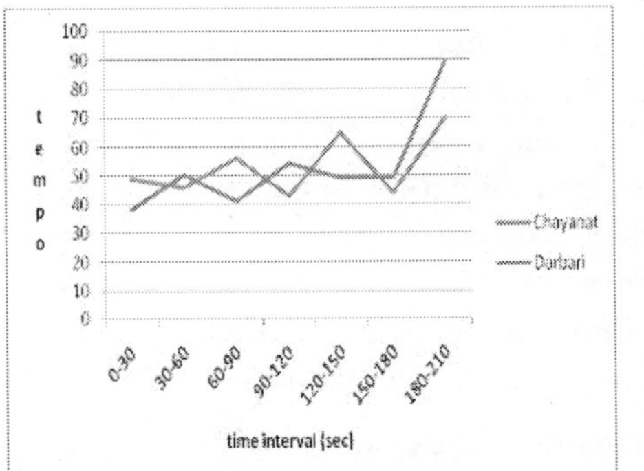

Fig. 2: Tempo at 30 Sec Interval Revealing Part

In both the distributions, the best fit polynomial curve is of order six and the R^2 value is shown in the Figure 3. The figure reveals the fact that the partials are more harmonically distributed around the spectral centroid. The figure also shows that pitch shift decreases linearly with the harmonic number for the ragas Darbari Kannada which may be an spectral attribute of negative valence emotion. The raga Chayanat showed a complex nature of the variation of pitch shift with harmonic number showing an indication of positive valence emotion in the signal.

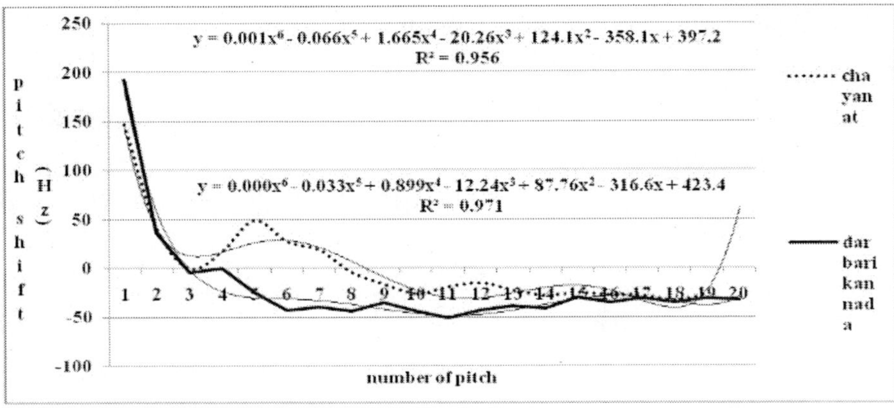

Fig. 3: Distributions of Pitch Shift are Shown with Best Fit Trend Line and R^2 Value

Table 1: Listener's Emotional Response in Percentage for the Ragas Chayanat and Darbari Kannada.

Raga	Anger	Heroic	Joy	romantic	serenity	devotion	Sorrow	Anxiety
Chayanat	3.571	3.571	**32.14**	**32.143**	7.1429	3.57143	10.71	7.1429
Darbari	3.448		6.897		20.69	13.7931	**44.83**	10.345

It is observed from Table 1 that 45% of the listeners' pointed out sorrow for raga Darbari and 32% each pointed joy or romantic for raga Chayanat, which is quite interesting. 71.4% listeners indicated positive valence emotion for raga Chayanat and the rest indicated negative valence emotion. It is said in the music treatises that there is an element of pathos in the romantism of raga Chayanat. 28.6% of negative valence emotion might reflect this. In raga Darbari 90% listeners correctly indicated negative valence emotion. Thus it can be said that timbre effects very little in the recognition of emotion in Hindustani music. Figure 4 below presents the distribution of emotional responses in circular graphical form.

Two listeners were subjected to EEG while listening to music. They were asked to sit in a relaxed manner in a shielded measurement cabin and the electrode cap was properly fitted on their head. The ground electrode was they properly fitted and adjusted. After settings were adjusted then recording of EEG was started without music. After 3 minutes

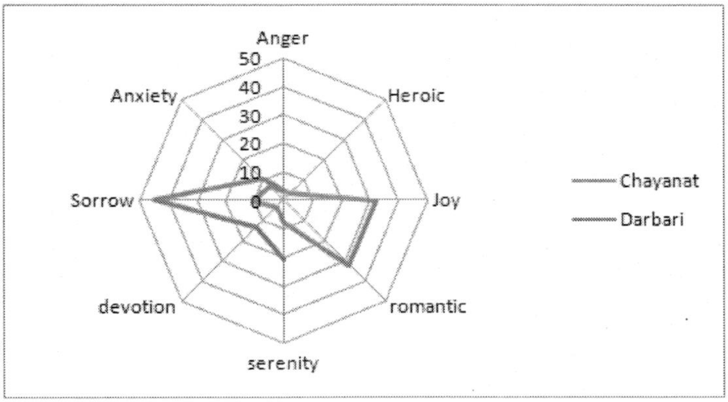

Fig. 4: Emotional Responses of the Two Raga Sound Clips

the raga Chayanat was played in the speaker. Then raga Darbari Kannada was played and EEG was continued. After that EEG was continued for 3 mins. The output from each electrode was observed and it was seen that there is a marked variation of electric potential with the change of the ragas. The frequency content of the EEG signal is a fundamental thing for appraising the EEG. Most of the valuable information is hidden within the frequency of 10 Hz. In general, EEG rhythms are classified into four basic types: (i) delta (1/2–4 cycles per second), (ii) theta (4–7 cycles per second), (iii) alpha (8–13 cycles per second) and (iv) beta (13– 40 cycles per second). Variation in frequency bands clearly shows the changes of functional and emotional states of listener. This type of arousal effect was also observed in the tanpura sound induced EEG [3].

Figure 5 shows the frequency analysis for 2 seconds interval FP2–F4 electrodes. The distribution shows an increase in alpha activity while the music listening to music. It is observed from the figure that in no music condition the theta activity increases. After removal of the first music clip (raga Chayanat) we see that the theta activity is on the rise. When the second music clip (raga Darbari Kannada) was removed then again theta predominates alpha. It is said that alpha pattern appears when in wakefulness where there is a relaxed and effortless alertness.

It was interesting to observe that the alpha activity was very high during 2.6–2.8 secs for the music clip (raga Chayanat) and nearly maximum during 3.2 secs- 3.4 secs for the clip (raga Darbari Kannada). This is well within the note sequences as containing emotion of that raga mentioned above. It is being observed that almost no discrimination regarding the contrasting emotions is being observed from the frequency analysis of the EEG potential structure. Both the signals produce arousal in the brain mapping of both the listeners though the variations are different between the listeners. Figure 6 shows the frequency mapping of 2 secs interval for a particular listener when he was listening to raga Chayanat. The alpha predominance is very well observed.

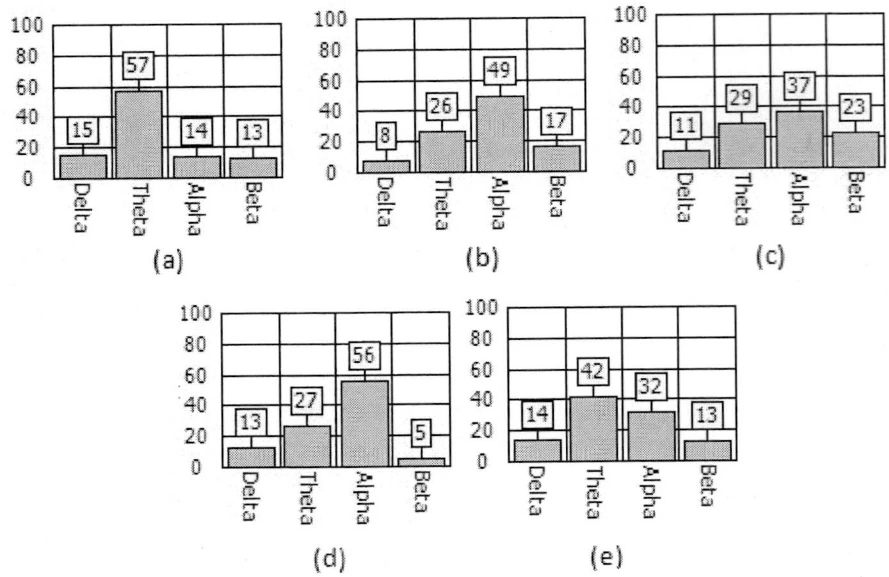

Fig. 5: (a) Relaxed Mode—Before Music Started (b) Raga Chayanat was Established (c) After Raga Chayanat was Stopped (d) Raga Darbari Kannada was Established (e) After Raga Darbari Kannada was Stopped

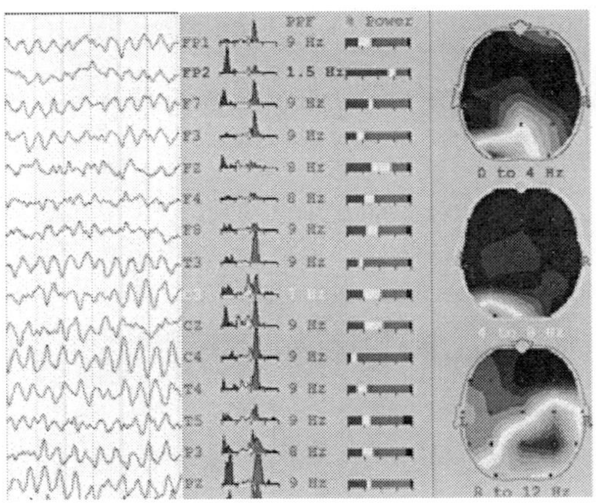

Fig. 6: Frequency Map of 2 secs. While the Second Listener was Enjoying Raga Chayanat

CONCLUSION

Emotion in music clip (raga rendering) is embedded in some specific note sequences of the clip. Arousal occurs to a listener while listening to the music clip particularly when that sequence in being played. Our present EEG analysis is unable to distinguish between contrasting emotions. As evidenced by the listening test, emotion in the Hindustani raga music is timbre independent and mainly concentrated on the particular note sequences, note to note transitions (meends) and other ornaments. Listening test result shows that naïve listeners can distinguish between contrast emotions. More rigorous analysis of EEG data with different types of music needs to be done to frame a robust algorithm to have a thorough knowledge of music processing in the brain. This is our future work.

REFERENCES

[1] Alicja A. Wieczorkowska, Ashoke Kumar Datta, Ranjan Sengupta, Nityananda Dey and Bhaswati Mukherjee, "On Search for Emotion in Hindusthani Vocal Music", Adv. in Music Inform. Retrieval, SCI 274, pp. 285–304, Springer-Verlag Berlin Heidelberg 2010.

[2] S.O. Ali and Z.F. Peynircioglu, "Intensity of emotions conveyed and elicited by familiar and unfamiliar music" Music Perception: An Interdisciplinary Journal 27: 177–182, 2010.

[3] Matthias Braeunig, Ranjan Sengupta and Anirban Patranabis, "On Tanpura Drone and Brain Electrical Correlates", LNCS 7172, pp. 53–65, Springer-Verlag Berlin Heidelberg 2012.

Vowel Duration in the Geminate Contexts: Observations from Oriya Speaking Children and Adults

Venkat Raman Prusty and Lakshmi Venkatesh[1]
MediCiti Institute of Medical Sciences, Ghanpur, Andhra Pradesh
[1]Dr. S.R. Chandrasekhar, Institute of Speech and Hearing, Bangalore

ABSTRACT: *Temporal characteristics of vowels in geminate contexts (phonological long vowels) were observed through acoustic analyses of vowels occurring in word pairs of Oriya language. Productions of Oriya speaking typical developing children in the age range of 3–14 years and young adults (25–35 years) in response to recorded model were subjected to acoustic analyses to measure the duration of vowels. Results revealed a developmental trend such that the duration of all geminate vowels differed across the different age groups with no difference was observed between male and female participants in each age group. Findings are discussed in terms of the age at which children's productions reached adult like durations and comparison with duration of singleton production of vowels reported in an earlier study.*

Keywords: Development, Oriya, Acoustic Analysis.

INTRODUCTION

Phonemes are converted into phonetic units during the process of speech production through the realization of different rules. Durational properties observed at the level of phonetic output are the result of both segmental and suprasegmental factors. Variations due to intrinsic (segment specific) and extrinsic (contextual) specifications may arise from maturation, predisposition and cognitive learning. An analysis of developmental changes helps understanding aspects of normal maturation. Quantifying effects of language dependent factors are helpful in identification of abnormal forms of productions in development.

Vowel length is contrastive in many languages with varied language-specific realizations of the length contrast. Several languages (e.g., Sanskrit, Kannada, Japanese, Hungarian etc) from around the world make a phonemic distinction between long and short vowels. Long vowels are distinct phonemes from short vowels leading to doubling the number of vowel phonemes in a particular language. Long vowels in Japanese for example are analyzed as two same vowels or a vowel + the pseudo-phoneme /h/ (e.g., short /i/ and long /ii/ or /ih/). Languages such as Japanese or Finish are referred to as "Geminate vowel" languages which allow a long vowel (a phoneme) to be analyzed as sequences of short vowels [1]. In Finnish, both consonant length and vowel length contrasts are phonemic and are independent of each other. For example, words such as *taka* /taka/ "back", *takka* /tak ː a/ "fireplace", *taakka* /ta ː k ː a/ "burden", and so forth are different, unrelated words. Ladefoged defines geminates as "long consonants (or vowels) that can be analyzed as double consonants (or vowels)" [2]. In several languages, gemination is not productive in the creation of phonemes and only occurs only in allophonic variation.

Gemination is contrastive in Japanese or Finnish as seen from examples above, whereas it is not contrastive at the phonemic level and only occurs at lexical boundaries in English. The "Geminate Vowel" languages are in contrast to several "Tense-Lax" languages such as English, in which the "length" contrast is realized by quality as well as duration [3].

Oriya is an Indo-Aryan language spoken by about 31 million people in state of Orissa and some areas of neighboring states. The features of this language which makes it marked include the vowel [ɔ] without a front vowel counterpart and the vowel /a/ considered as secondary cardinal vowel. Table 1 shows the classification of vowel sounds in Oriya in terms of tongue height, tongue root advancements and mouth opening [4].

Table 1: Classification of Vowel Sounds in Oriya

Feature	*Front*	*Central*	*Back*
Close (High)	i		u
Half Close (High Mid)	e		o
Half Open (Low Mid)			ɔ
Open (Low)		a	

Investigations of acoustic characteristics of vowels in geminate contexts have been limited to languages having geminate vowels (arising out of suffixation or compounding) or a long vowel. Oriya is a language which has neither genuine geminate vowels nor long vowels. Several languages such as Oriya which have large number of words ending in a vowel have phonological long vowels or vowels which occur in geminate context in running speech. Investigations of the developmental changes in production of such vowel contexts may help in understanding the complex process of phonemic- phonetic-perceptual- cognitive interaction during production.

An earlier study [5] reported preliminary data on the temporal characteristics of six vowels (/ɔ/, /a/, /i/, /u/, /e/ and /o/) in Oriya. Developmental changes occurred in duration of vowels; duration of vowels decreased with increase in age for all vowels in Oriya in the context of voiced and voiceless post-vocalic stop consonant for both male and female speakers. Duration of vowels in the post-vocalic voiceless stop context was lower than in the voiced stop consonant context. Such context based differences were present from the age of three years and developed with age. The current study was aimed at investigating developmental changes in duration of Oriya vowels in geminate contexts (phonologically long vowels). Effects of age, gender and vowel type on vowel duration were also targeted for analysis.

METHOD

Native Oriya speaking children (40 males and 40 females) of 3 years to 14 years and young adults (10 males and 10 females) were included based on following criteria as reported by parents in case of children and self in case of the adult group: (1) Oriya as their primary language, (2) typical development of speech and language, (3) no prior enrollment in speech or language intervention (4) normal hearing status, and (5) absence of a history of neurological and/or psychological disorder. Exposure to English and Hindi was limited. The structure and function of oral articulators were assessed to be within normal limits. The first five age groups of children were separated by an interval of one year and next three groups by an interval of two years for precise analysis of developmental changes in the very young children.

Frequently used bi-syllabic words were paired to form utterances of the form- $CVCV_1$ – V_1CV, where V1 was the target vowel to be analyzed resulting in the production of the target vowel in a geminate form or phonologically long vowel $CVCV_{long}CV$. Stimuli comprised of six word pairs judged to be familiar or most familiar by a group of 10 Oriya speakers. Word pairs judged as unfamiliar by any of the listeners were not included. The six word pairs included /kad̪ɔ-ɔd̪a/, /iʈa-aʈ̪ɔ/, /pɔʈi-iʈa/, /babu-upi/, /goʈe-eka/ and /ɔʈo-od̪a/ representing the six vowels /ɔ:/, /a:/, /i:/, /u:/, /e:/ and /o:/ respectively.

Productions of these words pairs by a native female speaker of Oriya were recorded and presented in random order as model for imitation by the participants. Both children and adult participants produced three attempts of each word pair in response to the recorded model presented via a computer. Feedback regarding accuracy of production was not provided. Production of the word pairs of all participants was recorded and analyzed with PRAAT (version 5.0.33). Vowels were identified based upon the regularity of the vertical striations in the time domain waveform and simultaneous display of formants on the spectrograph. The onset of vowel was marked as the cycle after the burst portion of the consonant and offset was marked as the last cycle of periodicity on the acoustic waveform. The duration of the vowel segment as defined by the onset and offset of the vowel was measured for each production after confirmation by highlighting and playback by the experimenter.

RESULTS AND DISCUSSION

Two-way repeated measures ANOVA was performed on the duration of geminate vowels with age groups and gender as between subject factors and vowels (six vowels) as within subject factors. There was a main effect for age groups ($F(8, 69) = 73.582$, $p<0.01$) indicating that the duration of vowels produced by children differed among the different age groups. The main effect for gender was not significant. A main effect was also found for geminate vowel type ($F(1, 69) = 95.049$, $p<0.01$) indicating that there was a statistically significant difference in the duration of the six geminate vowels. No interaction effect existed for geminate vowel types across groups, indicating that the differences observed across vowel types were similar in all age groups. As evidenced from Figure 1, the duration of geminate vowels were longest in age group I (3–3; 11) and shortest in the adult group.

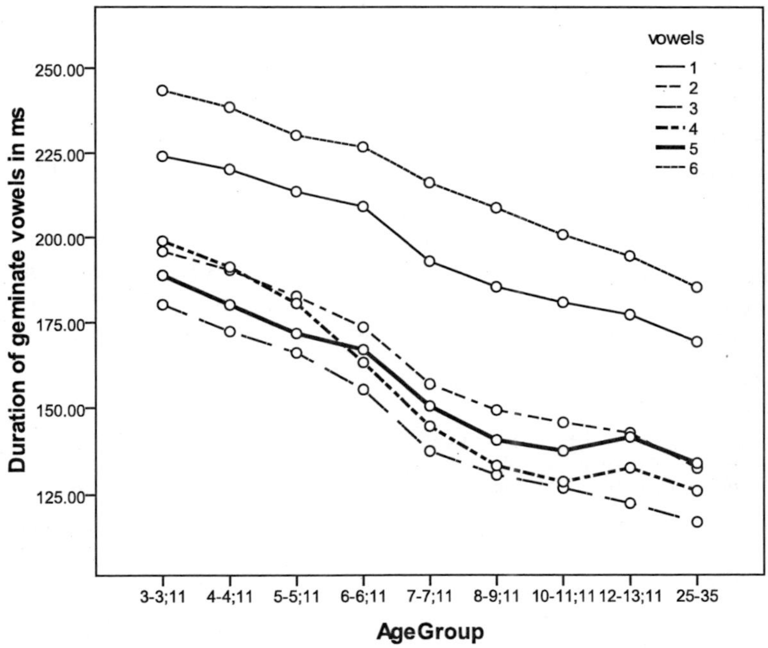

Fig. 1: Mean Duration of All Six Geminate Vowels Produced by Participants in the Nine Age Groups

Developmental Effects in Germination

Post hoc analyses between durations produced by participants in the different age groups showed that there was no difference among children for the production of geminate vowel /a:/, /i:/ and /o:/ in the age groups of 3-to-4, 4-to-5 and 5-to-6 years. Duration of vowels /u:/ and /e:/ showed no difference in the 3-to-4 and 4-to-5 years of age. Duration of vowel /ɔ:/, was not different in children in the age range of 3-to-4, 4-to-5, 5-to-6, and also 6-to-7 years of age. The duration produced by children in the age group of 7-to-8 years was significantly different from those of the children of younger age groups, but did not show any difference from children of 8-to-10 years and 10-to-12 years. For production of vowels /i:/, /u:/ and /e:/, there was no difference in the durations of vowels produced by children in the age groups of 8-to-10, 10-to-12, 12-to-14 years and those produced by the adult group indicating children reached adult like productions by 8-to-10 years of age. For vowels /ɔ:/ and /a:/, the adult group did not vary significantly from children of 10-to-12 years; for vowel /o:/, the adult group did not vary significantly from children in the age groups of 12-to-14 years. The age at which children's productions reached adult like durations differed by vowel type. It could be inferred that the early acquisition of specific characteristics of vowels involves an interaction between factors specific to languages, and related to articulatory mechanism.

Geminate Contexts Versus. Singleton Context

Mean duration of each of the six vowels in geminate vowel as produced by speakers of the adult group are shown in Table 2 in comparison to the mean vowel duration in singleton post-vocalic voiced consonant context (SVD) and post-vocalic voiceless consonant context (SVL) as reported in an earlier study [5]. Vowels in singleton production and geminate context were not compared statistically as the structure of words taken differed in both syllable size and syllable structure and this might affect the durations of vowels [6, 7]. Although statistical analyses were not performed, it can be seen that the duration of geminate vowel was longer than singleton vowels.

Table 2: Vowel Duration in Geminate Context and Singleton Contexts

Vowel	*Singleton Context: Post Vocalic Voiced Consonant*	*Singleton Context: Post Vocalic Unvoiced Consonant*	*Geminate*
/T/	110.7	77.3	169.3
/a/	122.2	81.2	132.4
/i/	93.3	62.1	117
/u/	98.1	67.2	125.9
/e/	99.1	70.5	133.8
/o/	102.3	73.4	185.2

Similar findings have been reported in study on gemmination in Lebanese Arabic [8], where geminate vowels were longer than singleton vowels and vowels were longest when they were phonologically long vowels. A study of durational characteristics of Hindi vowels and consonants in continuous speech found that the duration of geminate vowel /e/ was 87% longer than the singleton vowel production [9].

The current study provided preliminary data on the temporal characteristics of vowels in Oriya. The data added to the much needed corpus of acoustic characterization of vowels and consonants in Indian languages. Such data may serve useful in analysis of speech of children with speech sound disorders as well as adults with neurological disorders. The values of vowel duration may inform models for speech recognition and synthesis for Oriya language. It was evident that the early acquisition of vowel specifications involves an interaction between language specific features and articulatory predispositions associated with phonetic context.

REFERENCES

[1] Shibatani, M. (1990). The Languages of Japan. Cambridge: Cambridge University Press.

[2] Ladefoged, P. 2006. A course in Phonetics. Boston: Thomson Wadsworth.

[3] Durand, J. (2005). Tense/lax, the vowel system of English and phonological theory. In Headhood, Elements, Specification and Contrastivity, 77–97.

[4] Ramaswami, N. (1994). Common linguistic features in Indian languages: Phonetics. Central Institute of Indian Languages, Mysore.

[5] Prusty, V. R. and Venkatesh, L. (in submission). Duration of vowels in Oriya: A developmental perspective.

[6] Nagamma R.K. (1988). The influence of plosive consonant voicing and aspiration on duration of vowels in Telugu and Hindi. *Journal of Institution of Electronics and Telecommunication Engineers, Special Issue of JIETE on Speech Processing, New Delhi, 34, No. 1.*

[7] Nagamma R.K. (1985). Phonetic conditioning of Duration of vowels and consonants in Telugu. Osmania Papers in Linguistics, 11, 54–83.

[8] Khattab, G. (2007). A phonetic study of gemination in Lebanese Arabic. *Proc. 16th ICPhS*, Saarbrücken, 153–158.

[9] Samudravijaya, K. (2003). Durational characteristics of Hindi phonemes in continuous speech. Technical Report, TIFR (April 2003).

Comparing Acoustic Parameters of Conversational and Clear Speech during Reading in Kannada Speaking Individuals with Parkinson's Disease

Lakshmi Venkatesh and Ritu Mary Mathew
Dr. S.R. Chandrasekhar, Institute of Speech and Hearing, Bangalore

ABSTRACT: *The current study compared acoustic parameters of reading samples produced in clear speech and conversational speech conditions by individuals with Parkinson's disease (PD), grouped into three groups of PD (early, middle and late PD) depending on disease progression. A listener judgement task validated that instructions given to persons with PD to speak clearly indeed resulted in differences in speech production with a large proportion of listener's judgments indicating a preference for clear speech condition in comparison to conversational speech condition. Acoustic analysis of reading passages demonstrated significantly increased standard deviation of Fo, standard deviation of intensity as well as greater duration (reduced articulatory rate) in the clear speech condition in comparison to conversational reading. Further differences among the three groups of participants with PD in acoustic parameters reveal trade-off's in the acoustic measures possibly reflecting differences in physiological capacities with the progression of the disease.*

INTRODUCTION

Hypokinetic dysarthria associated with Parkinson's disease (PD) is considered to be a perceptually distinguishable motor speech disorder associated with basal ganglia control circuit pathology, characterized by a progressive loss of dopamine in the brain. Clinical characterization of hypokinetic dysarthria emphasizes in rank order; monopitch, reduced stress, monoloudness, imprecise consonants, inappropriate silences, short rushes, harsh voice, continuous breathiness, pitch level disturbances, and variable rate [1]. Few studies of speakers with dysarthria have shown that clear speech can have an important effect on intelligibility [2]. Clear speech includes speaking in a deliberate manner, characterized by exaggerated articulation, reduced rate, and increased loudness relative to habitual speech. A recent study examining the acoustic characteristics of clear speech by individuals who have hypokinetic dysarthria due to PD suggested that such an intervention may have the potential to improve intelligibility for some speakers with dysarthria [3]. Acoustic analysis revealed that group of individuals with PD did use some of the clear speech strategies, as indicated by significantly decreased articulation rate, increased mean fundamental frequency (MeanFo) and standard deviation of speaking fundamental frequency (Fo-SD) in clear speech condition in comparison to conversational speech condition during reading and monologue [3]. Studies are necessary to describe production features of clear speech in individuals with dysarthria, so that those production changes that has greatest impact on intelligibility can be emphasized in intervention.

There is increasing evidence that acoustic/perceptual features that characterize the conversational-to-clear speech transformation may vary across languages [4]. There is some research emerging on acoustic and perceptual characterization of properties of clear speech in Indian languages. For example, clear speech in Kannada was characterized by longer vowel durations and decreased speaking rate, higher mean values of pitch, increased amplitude in spectra of consonants and fricative [5]. Research on clear speech and conversational speech in individuals with dysarthria speaking different languages would determine language specific acoustic/perceptual features as well as provide additional support for benefit if any of clear speech over conversational speech. The purpose of this study was to evaluate the differences in clear speech and conversational speech conditions in a reading task among individuals with dysarthria due to PD using acoustic analysis and perceptual judgements.

METHOD

A total of 13 speakers with hypokinetic dysarthria due to PD (10 Male and 3 Female; Mean age 61.23 years; mean disease duration; 7.38 years) participated in the study. Participants were grouped into three groups of PD depending on the Hoehn-Yahr Scale of disease progression such that there were four participants each in the early and middle stages and five participants in the late stages of the development of PD. The severity of dysarthria increased with stage. Participants were native Kannada speakers who were able to read and write Kannada. Participants with evident linguistic disturbances were excluded from the study.

Stimuli comprised of a standard reading passage in Kannada. Participants were asked to read in two speaking styles; conversational speech condition and clear speech condition. For the conversational style, the talkers were instructed to read the passage as they would normally speak to someone familiar. For the clear speaking style, the talkers were instructed to read the passage as clearly and precisely as possible as if the experimenter was having trouble hearing them, or in understanding the language or as if there was noise in the background. Both speaking conditions were elicited through the use of instructions, and the participants' speech was audio-recorded as they read the passage.

Acoustic analysis of reading passage was performed by means of a commercially available software (Motor Speech Profile/Computerized Speech Lab CSL 4300; Kay Elemetrics Corp., USA) which allowed auditory playback of marked segments. The initiation and termination of the reading passage was marked by means of cursors, thereby eliminating the background noise from the sample. Voiced period marks were placed throughout the sample. Acoustic measurements included mean fundamental frequency (meanFo), standard deviation of fundamental frequency (SD-Fo), mean intensity (mean-Int) and standard deviation of intensity (SD-Int), measured across the entire length of the passage. In addition, the total time taken to read the entire passage was also noted. Speech samples comprising of reading passage in both speaking styles produced by participants were given to listeners for comparison to indicate their preference for either of the styles or none in terms of ease of understanding and clarity of speech.

RESULTS AND DISCUSSION

Perceptual Judgments

As seen from Table 1, listeners indicated that clear speech was better in terms of ease of understanding and clarity of speech in comparison to the conversational speech for the task of reading passage. Out of the total 130 responses, 92 listener judgments (70.77%) showed a preference for clear speech. A small number of listener judgments 13 (10%) indicated preference for conversational speech condition over clear speech. 24 (18.46%) judgments reported no preference or suggested that both conditions were perceived similar in terms of ease of understanding and clarity. 67.5% (27) and 62.5% (25) of the listener judgements were for clear speech condition in the early and middle PD groups respectively.

Table 1: Percentage and Number of Listener Preferences as Clear/Conversation or Same

Reading Passage		*Listener Preference*		
		Clear	*Conversation*	*None (same)*
Early	% (Total–40)	67.5 (27)	10 (4)	22.5 (9)
Middle	% (Total–40)	62.5 (25)	10 (4)	25 (10)
Late	% (Total–50)	80 (40)	10 (5)	10 (5)
Overall	% (Total–40)	70.77 (92)	10 (13)	18.46 (24)

An increased number of preferences for clear speech (80%) observed in the speech of participants with severe dysarthria associated with the late stage of PD in comparison to early and middle stages of PD suggested that clear speech instructions resulted in changes in speech production among these participants which were detected by listeners to contribute to ease of understanding and clarity in light of increased severity of speech deficits.

Acoustic Parameters

Wilcoxon signed rank test used to compare the acoustic measures on clear and conversation condition across the 13 participants revealed a significant difference in total duration of reading passage between clear (Mdn = 30.62 sec) and conversation speech condition (Mdn = 25.06 sec; z = 3.181, p = 0.001). A significant difference was seen for SD-Fo of the reading passage between clear (Mdn = 26.59) and conversation speech condition (Mdn = 22.17; z = 3.181, p = 0.001), as well as for SD-Intensity of reading passage between clear (Mdn = 7.89) and conversation speech condition (Mdn = 7.27; z = 2.901, p = 0.004). The measures of meanFo and mean-intensity did not differ between reading of passage in the clear condition and conversation condition.

When prompted to use clear speech, participants in the current study consistently took longer duration to read the passage in comparison to conversational speech condition. Interestingly, this was observed across all participants in the study. These results are similar to those reported earlier in terms of a decrease in the articulation rate [3]. The increase in duration for the clear samples may possibly be due to slower speaking rate. Previous research has found that non-impaired speakers produce slower articulation rates when asked to speak clearly [6, 7]. A slower rate of speech has been reported in clear speech due to the wide range of acoustic/articulatory adjustments, inclusive of increased stress and rhythm, hyper articulation of phonemic segments, vowel lengthening, and insertion of long duration and more number of pauses to increase the intelligibility [7, 8].

No significant differences were observed in mean Fo and mean-intensity values between the clear and conversation conditions across speakers. The mean Fo data in the current study are an average of male and female participants' Fo data. Interestingly, the findings of meanFo in conversational reading (159 Hz) are consistent with values reported in an earlier study [4] for conversational reading among participants with PD (157Hz). However, participants in the current study did not increase their mean Fo in the clear condition as seen in the study by Goberman and Elmer [4]. Earlier research on unimpaired speakers has also found higher speaking Fo in clear speech condition compared to conversational speech associated with increased laryngeal tension in clear speech production [7]. Participants in the current study ranged widely in the degree of progression of PD and resulting speech impairment. The lack of differences may have resulted from the variations in the participants. The results of the current study are consistent with several previous findings that speakers with PD already demonstrate a higher speaking Fo that is significantly higher than that of controls [9] and not all participants may be able to raise their further raise their Fo to produce clear speech.

Although there were no difference in the mean values of Fo and intensity, the standard deviation of Fo and standard deviation of intensity were significantly higher in the clear condition in comparison to the conversation condition. Increased SD-Fo in clear speech is consistent with earlier studies of comparison of clear and conversational speech among speakers with PD [4] as well as those in non-impaired speakers [7]. Compared to controls, PD speakers have been shown to have decreased speaking variability in Fo as well intensity [9] which has been attributed to loss of prosody resulting in perceptions of monopitch and monotonous speech. The results of the current study suggest that all speakers with PD were possibly attempting to increase the range of Fo and intensity across the entire reading passage in order to compensate for the loss of prosody in clear speech condition.

The ratio of clear to conversational speech samples, for all acoustic measures were taken to examine the extent of difference in the acoustic measures in both conditions across the participants (Table 2). All groups demonstrated a ratio higher than one for the duration measure reflecting the increased time or slower articulation time for clear speech in comparison to conversation speech. However, in comparison to middle and late groups of PD, the early PD group showed a greater ratio (Mdn = 1.44) probably indicating that they relied on the strategy of slowing down speech rate when directed to speak clearly.

Table 2: Ratio of Acoustic Measures of Clear Versus Conversational Reading

Groups		*Total Duration Ratio*	*Mean F0 Ratio*	*F0- SD Ratio*	*Mean Intensity Ratio*	*Intensity- SD Ratio*
Early PD	Median	**1.444**	0.980	1.073	1.032	1.072
	Mean (SD)	1.441 (0.01)	0.973 (0.01)	1.077 (0.01)	1.036 (0.03)	1.092 (0.082)
Middle PD	Median	1.082	**1.159**	**1.260**	1.003	1.122
	Mean(SD)	1.079 (0.03)	1.149 (0.08)	1.353 (0.255)	1.004 (0.022)	1.118 (0.103)
Late PD	Median	**1.151**	0.961	**1.157**	**0.987**	1.090
	Mean (SD)	1.171 (0.04)	0.972 (0.04)	1.188 (0.104)	0.945 (0.09)	1.114 (0.158)

Note: Measure differentiating each group with other groups is marked in bold.

Examination of the meanFo ratio values by group shows that all participants in the early group showed reduced Fo in the clear condition (meanFo ratio less than 1). Participants in the late group showed inconsistent patterns. However, all participants in the middle stage showed increased meanFo in the clear condition in comparison to conversational reading (meanFo ration greater than 1). No such unique differences among groups were seen for the measures of Fo-SD and intensity-SD. All participants demonstrated an increase in variability in Fo and intensity in the clear condition. This change occurred to the same extent in participants of the three groups. Examination of mean intensity ratios revealed a distinct pattern in the productions by the later group of speakers with PD such that the intensity reduced in the clear speech condition in comparison to the conversation speech condition (mean intensity ratio less than 1). The participants in the late group appears to have achieved the difference between the two conditions with reductions in rate of speech (although not to the extent done by the early group) and increased Fo-SD which was accompanied with decrease in mean intensity. Such differences in the acoustic measures across the three groups of speakers with PD point to possible differences in strategies utilized by the speakers to achieve the perceived changes in the clear speech condition. These changes may well be defined by the physiological capacities and possible trade-off's among the speech systems with the progression of the disease. Detailed analysis of unique contribution of acoustic measures to perceived changes is required to further analyze the differences among the groups. Findings of the current study both perceptual and acoustic measures suggest that clear speech production may hold promise as a simple approach to improving speech production in individuals with PD.

REFERENCES

[1] J.R. Duffy (2005). Motor speech disorders. Substrates, Differential diagnosis and Management. osby.

[2] D.R. Beukelman, S. Fager, C. Ullman, E. Hanson and J.A. Logemann (2002). Impact of speech supplementation and clear speech and the intelligibility and speaking rate of speakers with traumatic brain injury. Journal of Medical Speech Language Pathology, 10 (4), 237–242.

[3] A.M. Goberman and L.W. Elmer (2004). Acoustic analysis of clear versus conversational speech in individuals with Parkinson disease. Journal of Communication Disorders, 38, 215–230

[4] R. Smiljanic, and A.R. Bradlow (2005). Production and perception of clear speech in Croatian and English. Journal of Acoustical Society of America, 118, 1677–1688.

[5] P. Kumar (2008). Comparing Acoustic Parameters of Conversational and Clear Speech in Kannada. Unpublished Master Dissertation, Mangalore University, Mangalore, India.

[6] A.R. Bradlow, N. Kraus and E. Hayes (2003). Speaking clearly for children with learning disabilities: Sentence perception in noise. Journal of Speech Language, and Hearing Research, 46, 80–97.

[7] M.A. Picheny, N.I. Durlach and L.D. Braida (1986). Speaking clearly for the hard of hearing II: Acoustics of clear and conversational speech. Journal of Speech and Hearing Research, 29, 434–446.

[8] R.M. Uchanski, S.S. Choi, L.D. Braida, C. Reed and N.I. Durlach (1996). Speaking clearly for the hard of hearing. IV: Further studies of the role of speaking rate. Journal of Speech and Hearing Research, 39, 494–509.

[9] E.J. Metter and W.R. Hanson (1986). Clinical and acoustical variability in hypokinetic dysarthria. Journal of Communication Disorder, 19, 347–366.

Hearing Profile of Indian Light Music Orchestra Musicians

Aswin Kumar K.V., Aparna Ravichandran[1], Venkat Raman Prusty[2] and Lakshmi Venkatesh[3]

National Hearing Care, Brisbane, Queensland, Australia

[1]AYJNIHH, SRC, Secunderabad, Andhra Pradesh

[2]MediCiti Institute of Medical Sciences, Ghanpur, Andhra Pradesh

[3]Dr. S.R. Chandrasekhar Institute of Speech and Hearing, Bangalore

ABSTRACT: *The present study aimed at assessing effect of exposure to Indian light music orchestra music on the hearing status of musicians. Measures included measurement of sound levels, assessing hearing status by using pure tone audiometry, loudness discomfort level and oto-acoustic emissions and self-reported questionnaire related to hearing sensitivity, localization, subjective judgments, and quality of hearing, tinnitus and hyperacusis. Data from 40 light music orchestra musicians of 20–35 years of age in four groups: high strings, low strings, percussion and other) were compared with those from 40 non-musicians of same age with no exposure to noise. Results were analyzed and discussed based on instrument type, frequency and ear (left or right). Possible effects of exposure to noise among musicians in orchestra are reported and preventive measures to protect hearing are discussed.*

Keywords: Light Music Orchestra Musician, Hearing Status.

INTRODUCTION

Noise is the new age or the latest generation hazard, which has become the inevitable part of the civilized world brought by industrial evolution. An adverse effect of noise is defined as a change in the morphology and physiology of an organism that results in impairment of functional capacity, or an impairment of capacity to compensate for additional stress, or increases the susceptibility of an organism to the harmful effects of other environmental influences (International Program on Chemical Safety, WHO, 1994). Among reported effects of noise like annoyance, depression, anxiety, distraction, exhaustion, one of the effects is also on hearing mechanism, generally described in terms of Temporary threshold shift (TTS), Permanent threshold shift (PTS) and Acoustic Trauma.

Sources of noise may be broadly classified into industrial and non-industrial [1]. Noise is possibly generated interestingly form recreational purposes. Music, though composed and performed for purposes of aesthetic pleasure, religious or ceremonial or entertainment, have been recognized as potential causes of noise induced hearing loss. The sound levels required to make the audience hear music from a distance are such that even with moderately long rehearsals and concert durations, the defined risk limits are exceeded with the Leq of about 94 dB (A) [2]. The average sound levels in the Indian orchestras were found to be 93.1 dB (A) and 102.45 dB (C) [3].

The underlying pathology occurs with earliest changes in the outer hair calls with loss of stereocilia followed by deformation, swelling and disintegration of the cell body, deterioration of the supporting cells such as pillar cells, Deiter's cells and Hensen's cells with progressive severity of injury, and at later stages the atrophying inner hair cells and the cochlear neurons [4].

Hearing requirements for professional musicians are high, requiring accurate pitch, loudness, and timbre perception. Diagnosis of NIHL has often been limited to the measurement of hearing thresholds; while musicians along with complaint of reduced hearing sensitivity, report other problems like tinnitus, hyperacusis, distortion and binaural diplacusis [5].

This study aims at investigating whether professional musicians of Indian light music orchestras should be treated as a special group with regard to hearing, noise, and noise related hearing problems and related expressions like hyperacusis and tinnitus. It also aimed at determining the sound levels in the Indian light music orchestra.

METHOD

The study was carried out in two phases such as phase-I- Measurement of sound levels of music during orchestral performance and phase II- Audiological evaluation of orchestral performers.

Phase I included measurements of the sound level during a concert on an open stage, with public address (PA) system positioned facing towards the audience. The musicians had feedback speakers placed beside them. A calibrated integrating Sound Level Meter (SLM; Quest technologies 1900) with half inch microphone on a tripod stand at a distance of 3 feet from the musical instruments. A-weighting network was selected in order to suppress the responses of the high and low frequencies and the response time was set to "Slow" to reduce the likelihood of an overestimation of sound levels and to smooth out the noise fluctuations.

Phase II included audiological evaluation. (40 light music orchestra musicians in age range of 20–35 years, divided into four groups: high strings (HS), low strings (LS), percussion (PC) and other (OT); compared with 40 non musician adults (80 ears) of same age with no noise exposure. Subjects in both the groups underwent various test procedures like otoscope evaluation, Pure tone audiometry (Grason Stadler-61 clinical diagnostic audiometer) to estimate the hearing sensitivity, discomfort/loudness perception levels, matching of tinnitus, Immittance audiometry (Madson Maico immittance), Otoacoustic Emissions (Otoread screener- DPOAEs) and a self-report questionnaire (related to hearing sensitivity, localization, subjective judgments of own hearing capacity, questions about rating quality of hearing, tinnitus and hyperacusis.)

RESULTS AND DISCUSSION

Phase I

Results of phase I indicated the L_{Peak} was 124.67 dB(A) which is higher than the level (95 to 112 dB(A)) that was reported in earlier studies **[6]**. The present findings of L_{eq} exceed the 'maximum permissible limits of noise to the worker' prescribed by Factories Act (1997), which recommends a maximum of 91 dB(A) exposure for a 2-hr time-weighted average.

Table 1: Average Values of L_{eq}, L_{Max}, L_{Min} and L_{Peak} Across 1/3rd Octave Band Frequencies

Frequency (Hz)	*L_{eq} (dB(A) SPL)*	*L_{Max} (dB(A) SPL)*	*L_{Min} (dB(A) SPL)*	*L_{Peak} (dB(A) SPL)*	*Frequency (Hz)*	*L_{eq} (dB(A) SPL)*	*L_{Max} (dB(A) SPL)*	*L_{Min} (dB(A) SPL)*	*L_{Peak} (dB(A) SPL)*
12.5	49.05	50.95	48.3	79.325	630	84.85	85.85	82.7	122.475
16	50.625	51.575	48.7	81.975	800	86.625	88.05	84.375	123.45
20	55.975	56.625	54.425	96.025	1 k	90.225	91.525	87.625	124.2
25	56.625	57.325	55	96.45	1.25 k	91.05	92.4	87.82	124.67
31.5	61.6	61.925	59.5	101.425	1.6 k	89.85	91.15	87.575	124.55
40	58.275	58.625	56.8	98.4	2 k	82.25	83.625	79.55	119.825
50	62.3	63.2	60.95	102.2	2.5 k	80.825	82.4	78.5	118.775
63	61.525	62.875	59.475	100.1	3.15 k	79.225	80.975	75.55	118.325
80	62.625	63.525	60.5	102.65	4 k	78.625	79.575	75.575	116.5
100	68.125	69.5	66.925	105.375	5 k	74.85	76.225	71.75	110.35
125	72.3	72.7	70.65	111.175	6.3 k	72.4	74	70.275	104.175
160	73.425	74.45	71.95	112.125	8 k	70.925	72.625	68.375	108.75
200	74.4	75.025	72.725	113.175	10 k	69.825	71.25	67.375	106.15
250	76.575	77.125	74.75	115.275	12.5 k	68.95	70.675	66.325	105.275
315	80.45	81.2	77.275	119.3	16 k	65.4	66.9	63.425	102.7
400	82.525	83.775	80.275	121.275	20 k	62.875	64.35	60.825	100.65
500	84.575	85.375	81.775	122.425					

Phase II

Group mean, standard deviation, minimum and maximum values of hearing thresholds for test frequencies 0.25 to 8 kHz for both the ears of group I were compared with that of group II. It was observed that the mean hearing thresholds of individuals in group I were poorer (i.e. >20 dB HL) at 4, 6 and 8 kHz when compared to other frequencies. Maximum values of hearing thresholds were poorer at 2, 3, 4, 6 and 8 kHz for both the ears. The mean hearing thresholds of group II were less than 15 dB HL.

A GLM repeated measures analysis over the hearing thresholds per ear at all frequencies showed that there is a significant difference between group I and group II ($F (1, 78) = 214.885, p < 0.001, \eta^2 = 0.73$). There was no significant difference between the right and left ears ($F (1, 78) = 3.600, p > 0.05$). The instrument played by the musicians affected the distribution of hearing thresholds ($F (1, 36) = 7.043, p = 0.001, \eta^2 = 0.57$). Post-hoc Tukey's HSD test showed that the hearing threshold of LS was significantly better than the hearing threshold of HS, OT and PC groups.

Effects in Loudness Perception: The Loudness Discomfort levels (LDL) of group I ranged from 86 to 110 dB HL across the test frequencies in both the ears. The mean LDLs ranged from 95.9 dB HL at 6 kHz to 102.3 dB HL at 0.5 kHz in right ear and from 96.05 dB HL at 6 kHz to 102.05 dB HL at 0.5 kHz in left ear. The LDLs of group II ranged from 94 to 110 dB HL across the test frequencies in both the ears. The mean LDLs ranged between 103.4 dB HL at 8 kHz and 107.4 dB HL at 0.5 kHz in right ear and between 102.7 dB HL at 8 kHz and 107.4 dB HL at 0.5 kHz in left ear. A GLM repeated measures analysis over the LDLs per ear at all frequencies showed that there is a significant difference between group I and group II ($F (1, 78) = 51.05, p < 0.001, \eta^2 = 0.69$) and no significant difference between the ears. There was no significant difference between the instrument groups ($F (3, 36) = 0.143, p > 0.05$) and no significant difference between the right and left ears across the groups. Thus for the further analysis, all the four groups were considered as a single group. There was also no significant difference found for the right and the left ear ($F (1, 273) = 2.71, p > 0.05$). A main effect was found for the frequencies ($F (6, 273) = 7.905, p < 0.001, \eta^2 = 0.54$) suggesting that the LDLs obtained across the frequencies differed significantly.

Tinnitus: As tinnitus matching can only be performed in case the tinnitus was present at the time the testing, 14 (35%) musicians participated in this test. Tinnitus was more often localized in midline (6, 42.9%) when compared to right (4, 28.6%) and left (4, 28.6%). On an average the sensation level of the tinnitus was 7 dB SL, but it ranged from 2 to 12 dB SL. The pitch of tinnitus was matched with pure tones of frequencies between 0.25 and 8 kHz. Seven (50%) participants indicated the pitch of their tinnitus to be 8 kHz. Three (21.4%) participants indicated a pitch of 6 kHz. Two (14.2%) participants indicated a pitch of 4 kHz. Among the remaining 2 participants, 1 (7.2%) indicated the pitch to be 3 kHz and the other 1 (7.2%) indicated a pitch of 2 kHz.

Effect in OAE Responses: The mean DP amplitudes in group I ranged from 15 dB SPL at 2 kHz to 4.2 dB SPL at 5 kHz in right ear and from 14.5 dB SPL at 2 kHz to 4 dB SPL at 5 kHz. The mean DP amplitudes ranged between 23.7 dB SPL at 2 kHz and 14.3 dB SPL at 5 kHz in right ear and between 23.1 dB SPL at 2 kHz and 14.3 dB SPL at 5 kHz in left ear. There is a significant difference between group I and group II ($F(1, 78) = 113.9$, $p < 0.001$). No significant effect for ears ($F(1, 78) = 0.112$, $p > 0.05$) and no significant interaction effect ($F(1, 78) = 0.510$, $p > 0.05$) was observed which shows. There was no significant difference between the instrument groups ($F(3, 36) = 0.809$, $p > 0.05$) and the ears across all the four groups. When checked for frequencies as between subject factor, there was no significant difference found for the right and the left ear ($F(1, 156) = 0.496$, $p > 0.05$). A main effect was found for the frequencies ($F(3, 156) = 59.768$, $p < 0.001$) suggesting that the DP amplitudes obtained across the frequencies differed significantly.

Self-Report Questionnaire: Majority of musicians (29, 72.5%) reported healthy ears without any reduced hearing sensitivity. Nine (22.5%) musicians reported a reduced hearing sensitivity in both the ears, while 2 (5%) of the musicians reported reduced hearing sensitivity in the right ear. No participant reported a hearing loss in the left ear. When the musicians were compared across the four groups in reporting the reduced hearing sensitivity, they differed significantly ($\lambda^2(6) = 13.70$, $p = 0.03$).

CONCLUSION

The sound levels in the Indian light music orchestras are greater than the permissible noise levels prescribed by Factories Act (1997). Professional musicians of the Indian light music orchestras should be treated as a special group with regard to hearing, noise, and noise related hearing problems. The instrument type played by the musicians is responsible for different patterns of hearing damage. Music predominantly affects the hearing of musicians at 3, 4, 6 and 8 kHz with a pronounced notch at 6 kHz. Tinnitus and hyperacusis seem to have much impact on the professional practice of the musicians, also their prevalence are suggestive for the involvement of NIHL. As the musicians are exposed to high sound levels at their work place and as they are at high risk for developing NIHL certain measures could be undertaken to protect their hearing. Continuing education about the risks of intensive sound exposure, with the emphasis on the possible development of tinnitus and hyperacusis, is warranted. Musicians should receive a hearing test regularly. Also, they could be considered for hearing conservation program.

It was concluded that though music is wanted sound listened with pleasure and cannot be compared to industrial noise, since music can also be characterized by its intensity, duration and spectral content, it may induce damage to the inner ear resulting in hearing impairment to different extent depending on the individual susceptibility to noise.

Thus this study gave a foundation and essential at measuring the sound levels present in the Indian light music orchestra and indicated that professional light music orchestra musicians should be treated as a special group with regard to hearing, noise, and noise related hearing problems.

REFERENCES

[1] Trivedi, P.R. and Raj, G. (1997). Noise pollution. New Delhi: Akashdeep Publishing House.

[2] Jansson, E.V. and Karlsson, K. (1982). Sound levels and long-time spectra recorded within the symphony orchestra and risk criteria for hearing loss. STL—Quarterly Progress and Status Report, 23(1), 31–48.

[3] Satish (2002). Hearing profile in orchestral performers. Masters dissertation. Unpublished data. University of Mysore.

[4] Siebenmann, F. and Yoshii, U. (1908). Demonstration Von experimentellen okustischen Schadigungen des Gehororganes. Verh Dtsch Otol Ges, 17, 114–118.

[5] Jansen, E.J.M., Helleman, H.W., Dreschler, W.A. and de Laat, J.A.P.M. (2008). Noise induced hearing loss and other complaints among musicians of symphony orchestras. International Achieve of occupational Environmental Health.

[6] McBride, D., Gill, F., Proops, D., Harrington, M., Gardiner, K. and Attwell, C. (1992). Noise and the classical musician. British Medical Journal, 305, 1561–1563.

Performance Analysis of DCT and DFT Based Speech Coding

Shijo M. Joseph and P. Babu Anto
School of Information Science and Technology, Kannur University, Kerala
E-mail: Shijomjose71@gmail.com

ABSTRACT: *Speech compression system focuses on reducing the amount of redundant data while preserving the integrity of the signals. Speech compression algorithms are generally classified in to three categories. They are parametric coders, hybrid coders and transform's coders. This paper analyzes the effectiveness of different transform technique on Indian regional language. Speech data base used for this experiment is created in Malayalam, one of the South Indian languages. To remove insignificant coefficient from the feature vector we apply hard thersholding technique on Discrete Cosine transformed coefficient. Lossless run length encoding algorithms is used to encode the remaining coefficients. Performance of this experiment is evaluated in terms of NRMSE, SNR, PSNR, MOS and percentage of compression occurred. It has been found that Discrete Cosine Transforms (DCT) is superior to Discrete Fourier Transforms (DFT).*

Keywords: Discrete Cosine Transforms (DCT), Discrete Fourier Transforms (DFT), Run Length Encoding (RLE), Mean Opinion Score (MOS), Compression percentage (CP).

INTRODUCTION

In information theory, data compression or source coding is the process of encoding information using fever bits than the original one. Data compression is useful because it helps to reduce consumption of expensive resources such as memory, HDD and transmission bandwidth, etc. [1]. On the down side of encoding and decoding process demands some amount of computation. The design of data compression schemes therefore involves trade off among various factors including the degree of computation the amount of distortion and computational resource required to compress and uncompress the data [2].

Compression algorithms are classified in to lossless compression and lossy compression. Lossless algorithms are usually exploiting statistical redundancy in the input data. Perfect reconstruction is possible in lossless compression without any loss of fidelity. But in the case of lossy compression there are many trade off where by some degradation in the quality of data will occur as part of compression [3]. There is mechanism to measure the efficiency of compression without which an effectives rating of compression techniques may not be possible. There are many parameter used to evaluate the effectiveness of data compression [4]. The main area of data compression is speech, image and multimedia processing.

Speech is a very unique signal by the nature due to many reasons. The most prominent factor is that the speech is a non stationary signal. This makes the speech signal hard to analyze and model. Second factor is that speech signal contains not only the information part but also many information like emotion, health condition, age and gender of speaker etc [5]. This make analyses and synthesis of speech signal more complex. There are many type of algorithms are available to process speech signal. Each one has its own merit and demerits. So far no algorithms have 100% perfection in analyses and synthesis of speech signal. Therefore demand for perfect algorithm is still exist in the area of compression.

Speech coding or speech compression is a process of obtaining a compact representation for the speech signals, for the purpose of efficient transmission over band limited wired or wireless channels and also for efficient storage. In the modern digital world speech coding is an essential part of telecommunication and multimedia application. Speech coding algorithms enable to carry out more calls per channel which leads to reduce the cost of communication [6]. The main areas of speech coding are telecommunication, satellite communication, video conferring, and storage media.

This paper deals with a well known DFT and DCT transforms (lossy) along with run length encoding (lossless) and evaluate its effectiveness in continuous Malayalam speech coding. Different quantitative (SNR, PSNR, NRMSE) and qualitative parameters are used to evaluate the performance of the system and compared with others. The system developed proved to be very efficient in terms of compressing speech files. This paper is organized as follows: A brief introduction of speech compression is presented in section 1. Section 2 gives overview about Discrete Fourier Transforms (DFT) and Discrete Cosine Transforms (DCT). Section 3 furnishes the details of data base created for this experiment. Section 4 discusses the methodology used in this experiment. Finally Section 5 presents the simulation and result and concludes this paper with remarks in section 6.

DIFFERENT TRANSFORM TECHNIQUES

Discrete Fourier Transforms (DFT)

The fast Fourier transform (FFT) is a discrete Fourier transform (DFT) algorithm which reduces the number of computations needed for N points from $2N^2$ to 2N logN, where log is the base 2 algorithm. A DFT is used in Fourier analysis of signal. It transforms the signal in to frequency domain. DFT decompose a set of values in to components of different frequency values [7]. The way it does this is defined by the following equation [1]

$$X(f) = \int_{-\infty}^{\infty} x(t) \bullet e^{-2j\pi ft} dt \qquad \dots (1)$$

and inverse FFT can be describe in equation 2.

$$x(t) = \int_{-\infty}^{\infty} X(f) \bullet e^{2j\pi ft} df \qquad \dots (2)$$

In this equation 't' stands for time 'f' for frequency and 'x' denotes the signal at hand 'X' denotes the signal in the frequency domain.

Discrete Cosine Transforms (DCT)

The Discrete is in a class of mathematical operations that includes the well known fast Fourier transforms (FFT) and many others. The basic operation performed by these transforms is to take a signal and transform it from one type of representation to another. DCT transformation identifies pieces of information in the signal that can be effectively "thrown away" without seriously compromising the quality of speech signal or image. The formula for one-dimensional DCT is shown in equation 1 with its partner the IDCT shown immediately below equation 4.

Discrete Cosine transform (DCT) of 1-D sequence b[n] of length N is

$$\mathrm{B[m]} = \left(\frac{2}{\mathrm{N}}\right)^{\frac{1}{2}} c_m \sum_{m=0}^{N-1} b[n] \cos\left[\frac{(2n+1)m\pi}{2N}\right] \qquad \dots (3)$$

where m = 0, 1, … N–1

Similarly the inverse transformation IDCT is defined as,

$$\mathrm{b[n]} = \left(\frac{2}{\mathrm{N}}\right)^{\frac{1}{2}} c_m \sum_{m=0}^{N-1} B[m] \cos\left[\frac{(2n+1)m\pi}{2n}\right] \qquad \dots (4)$$

In equations (3) and (4), c_m is defined as

$$\mathrm{C_m} = \left(\frac{1}{2}\right)^{\frac{1}{2}} \text{ for m = 0 or } \mathrm{C_m} = 1 \text{ for m} \neq 0$$

In DCT the coefficient B[0] which is directly related to the average value of the time-domain block, is termed as the DC coefficient, and the remaining coefficients are called AC coefficients [8].

SPEECH DATA BASE CREATION

Aim of this section is to give details of the speech data base created for this experiment. For this part of the experiment continuous speech samples were collected from six hundred individuals of different age and gender. First and second sentence of the Indian National pledge is selected for the creation of database. This sentence is uttered in Malayalam (one of the regional languages). The first sentence consists of three words and has average duration of 1.5 seconds and second utterance contains five words and has average duration of 2.4 seconds. Each word contains vowels, consonants, fricatives and diphthongs. The Malayalam sentences and its corresponding International Phonetic Alphabet (IPA) are given in Table 1. Speech samples were recorded using unidirectional USB microphone with following specifications: frequency response 100–10,000 Hz and sensitivity: 62 dBV/µBar –42 dBV/Pa +/-d 3dB. 'Goldwave' the software at a sampling rate of 8 KHz (4 KHz) band limited was used for capturing and processing the speech signal. The speakers were native speakers of Malayalam and free from speech disabilities. The recorded speech samples are trimmed by removing prefix and suffix silence from the utterances and it is labelled and stored in the data base for further processing.

Table 1: Speech Sentence and its IPA Format

Malayalam	1. ഇന്ത്യ/എന്റെ/രാജ്യമാണ്. 2. എല്ലാ/ഇന്ത്യക്കാരും/എന്റെ/സഹോദരി/സഹോദരമാരാണ്.
English	1. India/ente/rajamanu. 2. Ella/indiyakarum/enete/sahothari/sahotharamranu.

SPEECH COMPRESSION

The experimental module consists of three phases they are, transforming the signal, thersholding the coefficients and encoding the remaining coefficients. We utilize Discrete Cosine transforms and Discrete Fourier transform to transform signal in to its corresponding domain. The transformed signal reveals relevant information. This process extracts information from the signal that is the essential for signal compression. In this experiment DCT and DFT decompose the signal only. These coefficients are used to create feature vector.

Feature vector is undergoing a dimensionality reduction process known as thersholding. Thersholding is a procedure which takes place after decomposing the signal. This algorithm is a lossy algorithm since the signal cannot be reconstructed exactly [9]. Coefficients below this threshold level are zeroed. There are two types of thersholding techniques are used, soft thresholding and hard thresholding. Mathematically these two thersholdings are represented by the following equations. Equation 6 represent the hard thersholding and equation 7 represent the soft thresholding. In this experiment we use hard thersholding.

$$D_H^{\lambda}(w) = \begin{cases} \text{W} & \text{for all } |\text{w}| > \lambda \\ 0 & \text{otherwise} \end{cases} \quad \dots (6)$$

$$D_s^{\lambda}(w) = \text{sgn}(w)\max(0; |w| - \lambda \quad \dots (7)$$

The third phase of this experiment is encoding threshold signal by lossless Run length Encoding (RLE) algorithm. In this implementation consecutive zero valued coefficient are encoded with two bytes. One byte is used to specify a staring string of zero and the second bytes keeps track of the number of successive zeros. Figure 1 illustrating the compression process involved in compression of speech signal. Inverse process is applied to reconstruct the speech signal. Performance of the encoding and decoding process are evaluated in a number of quantitative and qualitative analyses.

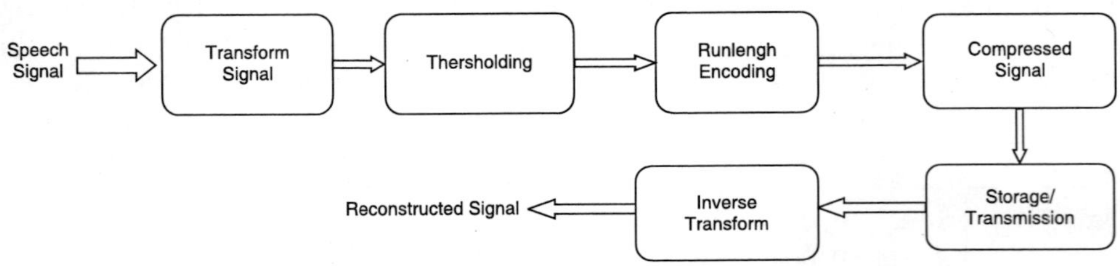

Fig. 1: Speech Coding Using Transforms Technique

Performance Evaluation

In this section we talk about the performance of the synthesized signal in terms of five parameters [9]. The following are the parameters we compared in addition to mean opinion score (MOS).

- Compression Percentage (CR)
- Signal to Noise Ratio (SNR)
- Peak Signal to Noise Ratio (PSNR)
- Normalized Root Mean Square Error (NRMSE)

The results obtained for the above quantities are calculated using the following formulas:

Compression—Percentage

This is the simplest way to measure the performance of compression algorithm. Compression–percentage can be calculated by the following formula:

$$\text{CR} = (1\text{-}(\text{Compressed-size/raw-size})) \times 100 \quad \dots (8)$$

Signal to Noise Ratio (SNR)

This value gives the quality of reconstructed signal, higher the value the better and it is calculated by,

$$\text{SNR} = 10\log_{10}\left(\frac{\sigma_x^2}{\sigma_e^2}\right) \qquad \ldots (9)$$

σ_x^2 is the mean square of the speech signal and σ_e^2 is the mean square difference between the original and reconstructed signals.

Peak Signal to Noise Ratio (PSNR)

This value indicates the ratio between the peak signal and noise within the reconstructed signals. Higher value is better. PSNR is calculated by the following formula,

$$\text{PSNR} = 10\log_{10}\frac{NX^2}{\|x-r\|^2} \qquad \ldots (10)$$

In this equation N is the length of the reconstructed signal, X is the maximum absolute square value of the signal x and $\|x\text{-}r\|^2$is the energy of the difference between the original and reconstructed signals.

Normalized Root Mean Square Error (NRMSE),

$$\text{NRMSE} = \sqrt{\frac{x\left(x(n)-r(n)\right)^2}{\left(x(n)-\mu_X(n)\right)^2}} \qquad \ldots (11)$$

x(n) is the speech signal, r(n) is the reconstructed signal, and $\mu_X(n)$ is the mean of the speech signal.

Mean Opinion Score (MOS)

Subjective evaluation by listeners is still a method commonly used in measuring quality of reconstructed speech. MOS provides a numerical indication of the perceived quality of received media after compression and/or transmission. The MOS is expressed as a single number in the range 1 to 5. Where 1 is the highest perceived quality and 5 is the lowest perceived quality. When taking subjective test, listeners focus on the difference between the original and reconstructed signal and rate it. The MOS is generalized by averaging the result of a set of standards, subjective tests, where a number of listeners rate reconstructed speech signals. In our experiment the MOS rate was excellent in DCT experiment and it was good in DFT based experiment.

RESULTS

Continuous Malayalam speech compression is implemented by different transform techniques along with thersholding and run length encoding. In these experiments the qualitative and qualitative performances are measured in terms of NRMSE, SNR, PSNR, Compression Percentage and MOS.

The results obtained in these experiments are tabulated in Table 2 and corresponding measures are graphically illustrated in Figure 2. Average measures of DFT experiments

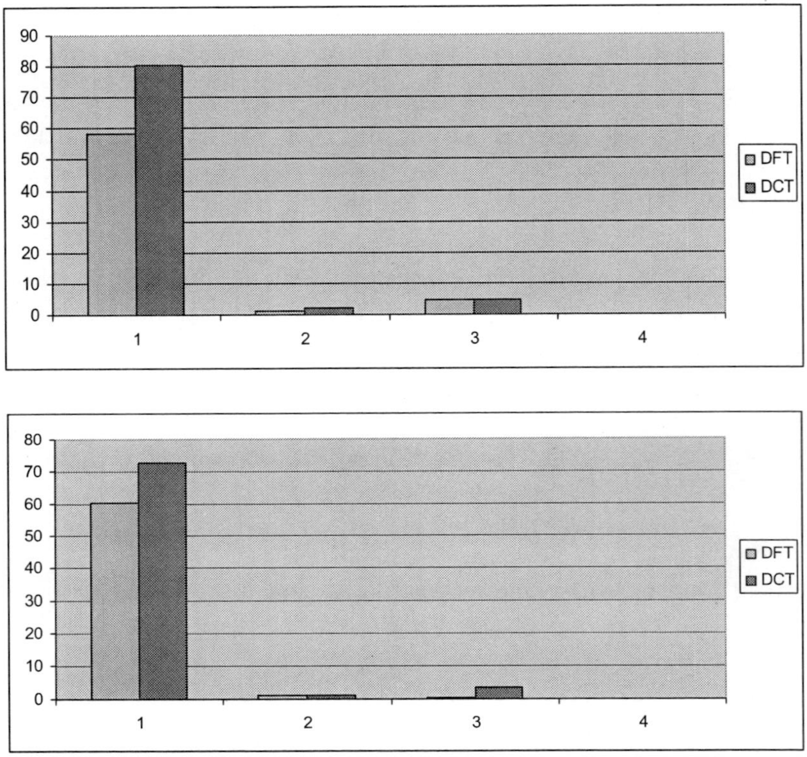

Fig. 2: DFT vs. DCT Performance Comparison Based on First and Second Sentence

Table 2

Sentence 1	*DFT*	*DCT*	*Sentence 2*	*DFT*	*DCT*
CP%	58.4662	80.4025	CP%	60.312	72.87
SNR	1.1148	2.17112	SNR	1.00657	1.01972
PSNR	4.6556	4.7269	PSNR	2.5001	3.5161
NRMSE	0.16503	0.16058	NRMSE	0.0187	0.0178

of first and second sentence are SNR is 1.0606, PSNR is 3.57788, NRMSE is 0.0918 and compression percentage is 59.3891. Similarly in DCT experiments the obtained average values are SNR is 1.59542, PSNR is 4.1215, NRMSE is 0.08918 and compression percentage is 76.6366.

CONCLUSION

The main aim of speech coding is to represent speech signal with minimum number of bits for storage or transmission. This paper discusses the efficiency of well known transform based speech coding on Indian regional language. From the stated results it is proved that, Discrete Cosine Transform (DCT) perform better than the DFT in Indian regional languages.

REFERENCES

[1] Mark Nelson and Jean Loup Gailly, "The Data Compression Book", 2nd Edition, BPB publications 2003.

[2] Khalid Sayood, "Introduction to Data Compression", Morgan Kaufmann Publishers 2nd edition 2005.

[3] David Salomon "Data Compression, the Complete Reference', Springer International Edition, 4th Edition, 2011.

[4] Lawrence Rabiner, Biing Hwang Juang, "Fundamentals of Speech Recognition", Pearson Education, 2009.

[5] Thomas F. Quatieri, "Discrete-Time Speech Signal Processing Principle and Practice', Pearson Publication 8th edition 2012.

[6] K.P. Soman, K.I Ramachandran, "Insight in to wavelets from theory to practice", Prentice-Hall of India Private limited, 2nd Edition. 2008.

[7] G. Rajesh, A. Kumar and K. Ranjeet, "Speech compression using different transform techniques", Second IEEE International Conf. on Computer, and Communication Technology, pp. 146–151, 15–17 Sept. 2011.

[8] M. Satya Sai Ram, P. Siddaith and Madhavi Latha, "Speech Coding and Recognition", world academy of Science and Technology 2009.

[9] Shijo M. Joseph, Babu Anto P. "Orthogonal and Biorthogonal wavelet based speech coding" International Conference on Communication Computing and Security (ICCCS 2012) ELSEVER Proceedings, ISBN 978-1-4503-0464–1, 06th Oct 2012.

Speech Rhythm in Individuals with Stuttering

M. Santosh, Priyanka Parakh and V. Sahana
All India Institute of Speech and Hearing, Manasagangothri, Mysore

ABSTRACT: *The objective of this study was to investigate speech rhythm in persons who stutter. Ten adults who stutter and ten age and gender matched controls participated in the study. Participants in both the groups were native speakers of Kannada. Five sentences were constructed in Kannada and the participants were asked to read out these sentences. Using Praat software, two kinds of rhythm metrics were calculated–pairwise variability indices and interval measures. Pairwise variability indices include normalized Pairwise Variability Index for vocalic intervals (nPVI–V) and raw Pairwise Variability Index for consonantal intervals (rPVI–C). Interval measures include percentage of vocalic intervals, standard deviation of the vocalic and consonantal intervals, and coefficient of variables for vocalic and consonantal intervals. Significant difference ($p < 0.05$) between the two groups was present for all the interval measures except coefficient of variables for vocalic intervals. However, there was no significant difference ($p > 0.05$) between the groups for Pairwise variability measures (nPVI–V and rPVI–C). The durational differences between the successive stressed and unstressed segments are taken into consideration in pairwise variability indices. The findings of the present study indicate that few measures of speech rhythm are deviant in persons who stutter while the others indicate that they follow the same rhythm pattern.*

INTRODUCTION

Stuttering is a developmental disorder of speech generally characterized by repetitions, prolongations and silent blocks. It is estimated that 1% of the world's population stutters. Stuttering usually starts between 2 and 4 years of age and in 50% of the children, condition persists. Although stuttering is one of the extensively researched topics in the field of speech pathology, till today, its cause remains unknown. Previously, several attempts have been made to understand the underlying mechanisms behind stuttering. Numerous studies have shown timing irregularities in stutterers' fluent speech and report that stutterers have longer voice onset times (VOT), longer voice initiation times (VIT), longer voice termination times (VTT), and longer vowel durations (Agnello, 1975; Agnello, Wingate, and Wendell, 1974 among others) than nonstutterers. However, few other studies (Cullinan and Springer, 1980; McFarlane and Shipley, 1981; Venkatagiri, 1982) did not find any timing abnormalities when compared to nonstutterers. One reason for the differences in the results may be different response modes employed (phonatory gestures, phonation of speech-like sounds, phonation of words, etc.) and differences in the severity of stuttering.

Many other studies have investigated the regularities in the timing of speech and non-speech events. Some studies have found stutterers to be more variable than fluent peers (Boutsen, Brutten, and Watts, 2000), but others have found no differences between groups (Max and Yudman, 2003; Zelaznik, Smith, and Franz, 1994). Most of these studies have investigated the temporal patterning or rhythm at the segmental level. Although stuttering is disorder in the rhythm of speech, to date, surprisingly no study has been attempted quantifying speech rhythm in sentence level in stuttering individuals. One reason for this could be difficulty with the quantification of speech rhythm.

Over the years, studies in the speech rhythm have mainly focused on classifying languages based on the isochrony principle. Isochrony principle states that languages can be either stress-timed or syllable timed. A third type of rhythm, mora timed was also proposed by Ladefoged (1975). Traditionally languages were classified into different rhythmic classes based on perception. However, with the development of quantification measures like pairwise variability index (PVI), more work is done at the production level. Pairwise variability index (PVI) calculates the pattering of successive vocalic and intervocalic (or consonantal) intervals showing how one linguistic unit differs from its neighbor. Two measures are derived using PVI; normalized pairwise variability index (nPVI) and raw pairwise variability index (rPVI). Both these measures help in computing the variability of successive measurements. Both these measures have been successfully used to classify different languages as stressed-timed, syllable timed or mora-timed. As these durational measures can be used to quantify variability in successive acoustic phonetic intervals, it will be interesting to see if stutterers differ from their fluent peers on these different measures of durational variability. Such evidence will fill remaining gaps in the field's knowledge and help us understanding the mechanisms underlying stuttering. Hence, the aim of this study was to investigate (1) whether stutterers differ from non-stutterers on the measures of speech rhythm.

METHOD

Subjects: Ten male adults with stuttering (AWS) in the age range of 17–30 (mean age–21.27 years) years and ten age and gender matched controls (AWNS) (mean age–20.9 years) participated in the study. Participants in both the groups were native speakers of Kannada. Adults who stuttered reported to have developmental stuttering and no history of other neurological, psychological and communication problems.

Procedure: Three tasks were recorded from each participant. First was reading standardized 300- word reading passage in Kannada. Second task was conversation between the second author and client. Third task was reading five sentences. First two tasks were used to calculate severity of stuttering and third task for measurement of rhythm. Participants were given set of sentences, and were instructed to silently read and rehearse. Once they were well versed with the material, recording of these sentences was done. They were instructed to read the sentences in their normal loudness and comfortable rate. They were asked to read the same sentences if they made any mistake while reading the sentences. The control participants also read the same set of sentences in the manner described above.

Analysis: First, the recorded sentences were edited by removing the silences between the words (Grabe and Low, 2002) and also omitting the disfluencies in AWS. Using wideband spectrogram, vocalic and consonantal intervals were measured for all edited sentences. The duration between the onset and the offset of the vowel or cluster of vowels was considered as the vocalic duration. Similarly, the duration between the onset and the offset of the consonant or a cluster of consonants was taken as the consonantal duration. The rules that governed the segmentation of the sounds were based on former studies on rhythm analysis and acoustic analysis (e.g., Peterson and Lehiste, 1967). The boundary between the vocalic and the consonantal interval was indicted by a significant drop in the amplitude and a break in the second formant structure. The identification of consonant boundaries was made easy by the cues obtained from the spectrogram which was associated to the manner of production of that sound. Apart from this standard criterion, a few other factors were considered to obtain consistency in the measurements. As it is difficult to determine the boundaries of approximants reliably, they were taken as a part of vocalic interval (White and Mattys, 2007). Pre-vocalic glides were included in the consonantal duration and the post vocalic glides were included in the vocalic duration (Ramus *et al.*, 1999). Aspiration following the stop release was taken in the consonant duration (White and Mattys, 2007). Even though there was significant lengthening noticed in the phrase final position, it was not excluded from the analysis (White and Mattys, 2007). Further, it was difficult to mark the glottal sound /h/, and hence it was also taken as a part of vocalic interval for uniform analysis.

Subsequent to measuring the vocalic and the consonantal durations, the various measures under the two kinds of rhythm metrics, that is, the interval measures and pairwise variability indices were derived. For calculations of these metrics, the first pair of vocalic and consonantal interval was eliminated to control variations in the speech characteristics related to the abrupt or sudden start of the utterances. The variables calculated under interval measures were percentage of vocalic intervals (Perc V), standard deviation of the vocalic intervals (Delta V), standard deviation of the consonantal intervals (Delta C), variable coefficients of vocalic intervals (VarcoV) and variable coefficients of the consonantal intervals (Varco C). In pairwise variability indices, raw pairwise variability for the consonantal intervals (rPVI-C) and normalized pairwise variability index for the vocalic intervals (nPVI-V) were calculated. The average of each of the rhythm metrics for each participant was calculated and these values were subjected to further statistical analysis. The formulas used to compute different rhythm metrics are provided in appendix.

RESULTS

As the goal of this experiment was to obtain the differences in rhythm metrics between persons who stutter and fluent speaking individuals, independent sampled t-test was done to compare the two groups. The mean values of different rhythm metrics were higher for AWS than control speakers except for nPVI. For nPVI, control speakers had slightly higher mean values compared to AWS. Results of independent 't' test showed statistically significant difference ($p < 0.05$) between the groups for all the interval measures except variable coefficient of consonantal intervals (percentage of vocalic intervals, standard deviation of the vocalic and consonantal intervals and variable coefficient of vocalic intervals). However there was no significant difference ($p > 0.05$) between the two groups for the pairwise variability index measures (nPVI–V and rPVI–C).

Table 1: Mean, S.D, 't' Values, df Values and 'p' Values for Different Rhythm Metrics in AWS and Control Groups

Parameters	*Groups*	*Mean*	*S.D*	*'t' Values*	*df*	*p*
nPVI	AWS	84.7973	4.41	0.258	18	0.799
	Control	85.9120	12.90			
rPVI	AWS	0.0645	0.083	−1.162	18	0.260
	Control	0.0339	0.003			
PercV	AWS	72.6371	2.56	−2.748	18	0.013*
	Control	69.9350	1.768			
DeltaC	AWS	0.0386	0.086	−2.199	18	0.041*
	Control	0.0321	0.0034			
DeltaV	AWS	0.1667	0.0255	−2.116	18	0.049*
	Control	0.1432	0.024			
Varco V	AWS	49.7421	4.95492	−3.494	18	0.003*
	Control	43.5944	2.532			
Varco C	AWS	87.3082	9.125	0.900	18	0.380
	Control	84.3082	3.146			

AWS-adults who stutter, nPVI-normalized pairwise variability index, rPVI-raw pariwise variability index, PercV—percentage of vocalic intervals, DeltaC—standard deviation of consonant intervals, DeltaV—standard deviation of vocalic intervals, Varco V—variable coefficients of vocalic intervals, Varco C—variable coefficients of consonant intervals. * indicates significance at 0.05 level

DISCUSSION

Here in our study, we investigated the sensitivity of pairwise variability indices and few other additional interval measures in differentiating speech rhythm of AWS and control speakers. Results showed significant difference between two groups only for selected interval measures (PercV, Delta V, Delta C, Varco V). Results highlight that selected rhythm metrics can be objectively used to document the rhythm abnormalities in people who stutter. However, there was no significant difference between the two groups for PVI measures. The durational differences between the successive vocalic and intervocalic segments are considered in pairwise variability index. Pairwise variability index (PVI) calculates the pattering of successive vocalic and intervocalic (or consonantal) intervals showing how one linguistic unit differs from its neighbor (Low, 1998). The values of PVI indices of AWS were similar to control participants. This suggests AWS are indistinguishable from control participants if we measure their rhythm using PVI. The results support similar observations for non-speech sounds (Max and Yudman, 2003) and suggest that individuals who stutter are not deficient with their isochronous rhythm timing if we document using PVI measures.

CONCLUSION

This study documents the rhythm abnormalities in persons with stuttering. As there is a mixed literature regarding the notion of stuttering as a disorder of speech rhythm and timing abnormalities, the findings of the present study also is similar, i.e., a few measures indicating that the speech rhythm characteristics are deviant in persons with stuttering while the others indicating that they follow the same rhythm pattern. The results have positive implications for speech language pathologists. Selected activities can be taken up to teach rhythm to patients with aprosodia considering the fact that one knows the type of rhythm in that language.

REFERENCES

[1] Agnello, J. (1975). Voice onset and voice termination features of stutterers. In L. M. Webster and L. C. Furst (Eds.), *Vocal tract dynamics and dysfluency*. New York: Speech and Hearing institute.

[2] Agnello, J., Wingate, M.E. and Wendell, M. (1974). Voice onset and termination times of children and adult stutterers. *Journal of the Acoustical Society of America, 56,* S62.

[3] Boutsen, F.R., Brutten, G.J. and Watts, C.R. (2000). Timing and intensity variability in the metronomic speech of stuttering and nonstuttering speakers. *Journal of Speech, Language and Hearing Research, 43,* 513–520.

[4] Cullinan, W.L. and Springer, M.T. (1980). Voice initiation and termination times in stuttering and non stuttering children. *Journal of Speech and Hearing Research, 23*, 344–360.

[5] Grabe, E. and Low, E.L. (2002). Durational variability in speech and the rhythm class hypothesis. In N. Warner and C. Gussenhoven (Eds.), *Papers in laboratory phonology 7* (pp. 515–546). Berlin: Mouton de Gruyter.

[6] Ladefoged, P. (1975). *A course in Phonetics.* New York: Harcourt Brace Jovanovich.

[7] Max, L. and Yudman, E.M. (2003). Accuracy and variability of isochronous rhythm timing across motor systems in stuttering versus non-stuttering individuals. *Journal of Speech, Language, and Hearing Research, 46,* 146–163.

[8] McFarlane, S.C. and Shipley, K.G. (1981). Latency of vocalization onset for stutterers and non-stutterers under conditions of auditory and visual cueing. *Journal of Speech and Hearing Disorders, 46,* 307–312.

[9] Peterson, G.E. and Lehiste, I. (1960). Duration of syllable nuclei in English. *Journal of the Acoustical Society of America, 32,* 693–703.

[10] Ramus, F., Nespor, M. and Mehler, J. (1999). Correlates of linguistic rhythm in speech signal. Cognition, 73, 265–292.

[11] Venkatagiri, H.S. (1982). Reaction time for "s" and "z" in stutterers and non-stutterers: A test of discoordination hypothesis. *Journal of Communication Disorders, 15,* 55–62.

[12] White, L. and Mattys, S.L. (2007). Calibrating rhythm: First language and second language studies. *Journal of Phonetics, 35,* 501–525.

[13] Zelaznik, H. N., Smith, A., Franz, E.A. (1997). Differences in bimanual coordination associated with stuttering. *Acta Psychologica, 96,* 229–243.

APPENDIX

Formulaes Used

1. Normalized pair-wise variability index of Vocalic interval (nPVI-V):
   nPVI-V = 100 × (k = 1∑m-1 | (dk – dk+1)/(dk + dk+1)/2|)/(m-1)
2. Raw pair-wise variability index of the Consonantal interval (rPVI-C):
   rPVI-C = (k=1∑m-1 |dk – dk + 1|)/(m–1)
3. Percentage of Vocalic interval: %V = ∑V/∑ (V + C)
4. Delta C: ▲C = Standard deviation of C
5. Delta V: ▲V = Standard deviation of V
6. Variable coefficient of consonantal Interval (VarcoC): VarcoC = ▲C/Average C
7. Variable coefficient of Vocalic Interval (VarcoV): VarcoV = ▲V/Average V

Abbreviations Index

m = total number of intervals
d = duration of "kth" interval
V = duration of vocalic interval
C = duration of consonant interval

Consonant Classification Using Time/Frequency Domain Features and Decision Directed Acyclic Graph Support Vector Machine

T.M. Thasleema and N.K. Narayanan

Department of Information Technology, Kannur University, Kannur, Kerala

E-mail: thasnitm1@hotmail.com; nknarayanan@gmail.com

ABSTRACT: *This paper discusses two novel feature extraction approaches to model Malayalam Consonant-Vowel CV) speech unit waveform. To extract the time domain feature coefficients Reconstructed State Space (RSS) based State Space Point Distribution (SSPD) Parameters are used. To develop the feature vectors a scatter plot namely State Space Map (SSM) is generated from RSS plot. SSPD parameters are generated from the SSM plot. To extract the frequency domain feature parameter, Multi Resolution Analysis (MRA) approach using Wavelet Transform based Normalized Wavelet Hybrid Feature (NWHF) parameter is used. In order to identify the credibility of these two parameters, statistical classification approach namely Decision Directed Acyclic Graph Support Vector Machine (DDAGSVM) is used. From the experimental study it is reported that RSS models are able to differentiate each consonant speech unit compared with MRA approach in speaker independent environments.*

INTRODUCTION

In the past few decades, significant progress has been made by researchers in the area of speech and language processing. In the present work we are discussing two different feature extraction techniques namely State Space Point Distribution (SSPD) parameter and Normalized Wavelet Hybrid Feature (NWHF) for the recognition of Malayalam Consonant-Vowel (CV) speech units based on the Decision Directed Acyclic Graph Support Vector Machine (DDAGSVM) statistical pattern classifier. Simulation experiments are carried out by dividing the CV speech database into five different phonetic classes namely unaspirated, aspirated, nasals, approximants and fricatives.

SSPD PARAMETER EXTRACTION USING RECONSTRUCTED STATE SPACE APPROACH

In dynamical system approach, by embedding a signal into adequately high dimensional space, a topologically equivalent to the original state space structure of the system generating the signal is formed [1][2]. This embedding is known as Reconstructed State Space (RSS). The time evolution within the RSS traces out a trajectory pattern referred to as its attractor which is a representation of the dynamics of the underlying system [3]. Since the attractor of an RSS captures all the relevant information about the underlying system, it is an efficient choice for signal analysis, processing and classifications. The RSS approach proposed here has the advantage of extracting both linear and non-linear aspects of the entire system.

According to Takens embedding theorem, a RSS for a dynamical system can be produced for a measured state variable S_n, n = 1, 2, 3,..., N via method of delays by creating vectors given by,

$$\mathbf{s}_n = [s_n\ s_{n+\tau}\ s_{n+2\tau}, \ldots, s_{n+(d-1)\tau}] \qquad \ldots (1)$$

where d is the embedding dimension and τ is the time delay value. The row vector **sn** defines the position of a single point in the RSS. To completely define the dynamics of the system and to create a d dimensional RSS, corresponding trajectory matrix is given as,

$$S_d = \begin{bmatrix} s_1 & s_{1+\tau} & \cdots & s_{1+(d-1)\tau} \\ s_2 & s_{2+\tau} & \cdots & s_{2+(d-1)\tau} \\ \cdots & \cdots & \cdots & \cdots \\ s_N & s_{N+\tau} & \cdots & s_{N+(d-1)\tau} \end{bmatrix} \qquad \ldots (2)$$

A speech signal with amplitude values can be treated as a dynamical system with one dimensional time series data. Based on the above theory, this study investigates a method to model a RSS for Malayalam consonants through the use of time delayed versions of original scalar measurements. Thus a trajectory matrix $\mathbf{S}_1$ with embedding dimension d = 2 and τ = 1 can be constructed by considering the speech amplitude values $\mathbf{s}_n$ as one dimensional time series data. Thus $\mathbf{S}_1$ is given as,

$$S_1 = \begin{bmatrix} s_1 & s_2 \\ s_2 & s_3 \\ \cdots & \cdots \\ s_{N-1} & s_N \end{bmatrix} \qquad \ldots (3)$$

For every consonant speech signal a trajectory matrix is formed with embedding dimension d = 2 and time delay τ = 1.

The State Space Map (SSM) for the Malayalam consonant CV unit is constructed as follows. The normalized N samples values for each CV unit is the scalar time series s_n where n = 1, 2, 3,…, N. For every consonant speech signal a trajectory matrix is formed with embedding dimension d = 2 and time delay $\tau = 1$. Now the scatter plot SSM is generated by plotting the row values of the above constructed trajectory matrix by plotting s_n versus s_{n+1}. Figure 1 shows the SSM for the first consonant sound /ka/.

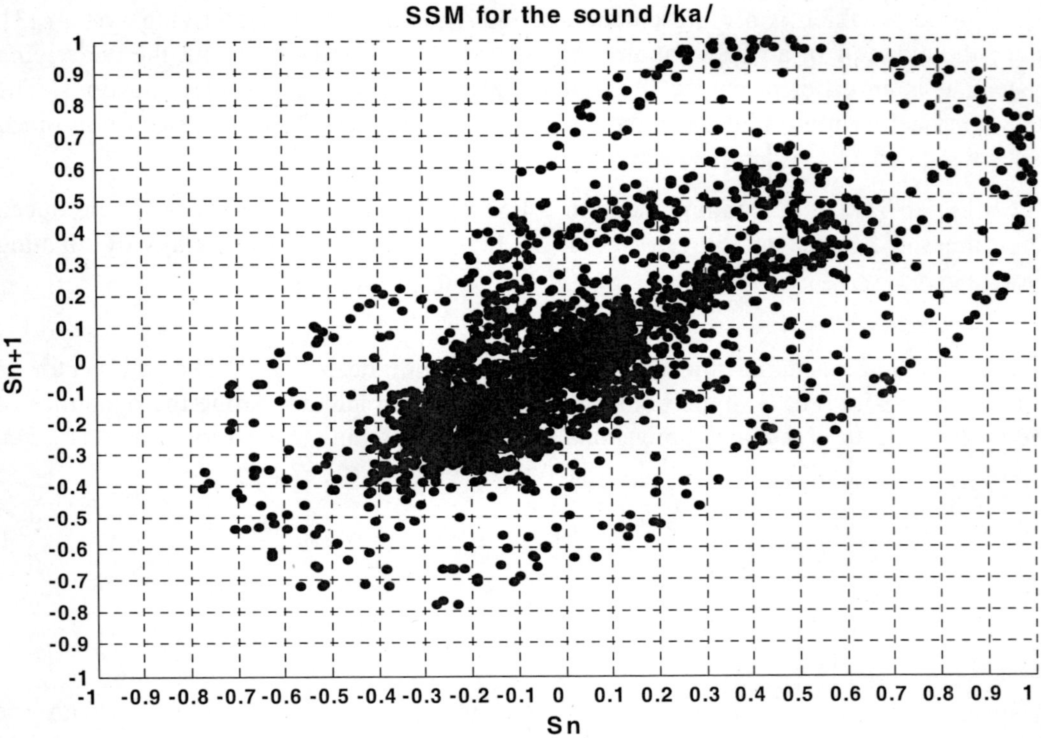

Fig. 1: Scatter Plot (SSM) for the Sound /ka/ with d = 2

In Automatic Speech Recognition (ASR), selection of distinctive features is certainly the most important factor for the high recognition performance. Present study uses non linear feature extraction technique called State Space Point Distribution (SSPD) from their SSM. For this purpose the SSM of the speech unit is divided into grids with 20 × 20 boxes. The box defined by co-ordinates (–1, 0.9), (–0.9, 1) is taken as box 1 and box just right side to it as taken as box 2 and so on in the x-direction with the last box being (0.9, 0.9), (1,1) is taken as box 20. The process is repeated for all the rows and boxes are numbered consecutively for the 400 boxes. The SSPD for each pattern is calculated by estimating the number of points distributed in each of these 400 boxes. This can be mathematically represented as follows.

The reconstructed SSPD parameter for location 'i' in two dimensions can be defined as,

$$(SSPD)_i = \sum_{n=1}^{N} f([s_n, s_{n+1}], i)$$

where $f([s_n, s_{n+1}]), i) = 1$, if SS point defined by the row vector $[s_n, s_{n+1}]$ is in the location 'i',

0, otherwise

Using this information the SSPD plot is plotted by taking the box number along x-axis and the number of points in each box along y-axis. The SSPD plot for the first Malayalam CV sound /ka/ is given in Figure 2.

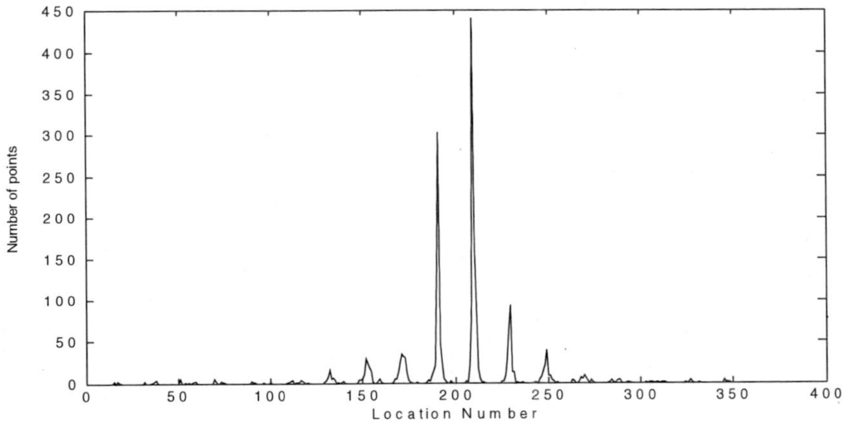

Fig. 2: SSPD Plot for the Sound /ka/

The SSM and the corresponding SSPD plot obtained for different speaker shows the identity of the sound so that an efficient feature vector can be formed using SSPD. The feature vector of size 20 is estimated by taking the average distribution of each row in the SSPD graph.

NWHF PARAMETER EXTRACTION USING WAVELET TRANSFORM

Certain ideas of wavelet theory appeared quite a long time ago [4]. Over the last decades wavelet analysis has turned to be a standard technique in the areas of geophysics, meteorology, audio signal processing and image compression [5][6][7]. Wavelet transform can be defined as the transformation of the signal under analysis into another representation which presents the signal in a more useful form [8]. The Discrete Wavelet Transform (DWT) which is based on subband coding is a fast computation of wavelet transform. The main advantage of DWT over Fourier Transform (FT) is the multi resolution analysis of signals with localization on both time and frequency. In the present work we utilize these characteristics of wavelet transform using two major wavelet decomposition techniques namely Classical Wavelet Decomposition (CWD) and Wavelet Packet Decomposition (WPD) for CV speech unit recognition.

The process for extracting NWHF feature vector is described below.

In the first step, the sound signal is made to undergo recursively to decompose into k^{th} level of resolutions; therefore the approximation coefficient matrix at this level is a sufficiently small representative of the original sound signal and carries enough information content to describe sounds characteristics coarsely.

Let A_k represents this approximation matrix at decomposition level k, which can be written as,

$$A_k = \begin{bmatrix} A_{1,1} & A_{1,2} & \cdots & A_{1,n} \\ A_{2,1} & A_{2,2} & \cdots & A_{2,n} \\ \cdots & \cdots & \cdots & \cdots \\ A_{k,1} & A_{k,2} & \cdots & A_{k,n} \end{bmatrix} \qquad \ldots (4)$$

Then, the first component of NWHF feature vector v_1 is,

$$v_1 = \bigcup_{i=1}^{k} \bigcup_{j=1}^{n} \{A_{i,j}\} \qquad \ldots (5)$$

In the second step, for Wavelet Packet Decomposition, decompose each sound using k^{th} level of resolutions for the best level of wavelet packet decomposition tree. The first coefficient matrix at the best level tree contains enough information to represent the given input consonant CV speech unit without loss of much speech features. Let m represent mean of one row vector in the coefficient matrix then the WPD feature vector v_2 is given by

$$v_2 = m_i, \ \mathrm{i} = 1, 2, \ldots, \mathrm{m} \qquad \ldots (6)$$

(number of rows in the best level coefficient matrix)

In the third step we combined v_1 and v_2 to fusion CWD and WPD coefficients,

$$V = \bigcup_{i=1}^{2} \{v_i\} \qquad \ldots (7)$$

Then the final feature vector F after z-score normalization is given as,

$$F = \frac{V - \mu(V)}{\sigma(V)} \qquad \ldots (8)$$

Thus an efficient feature vector is extracted using the proposed NWHF vectors.

The feature vector of size 20 is estimated by concatenating CWD and WPD features for hybrid feature vectors The NWHF vector for different speaker shows the identity of the same sound so that an efficient feature vector can be formed using NWHF vectors.

CLASSIFICATION

Pattern recognition can be defined as a field concerned with machine recognition of meaningful regularities in noisy or complex environments [9]. Nowadays pattern recognition is an integral part of most intelligent systems built for decision making. In the present study reports the performance of widely used approach for pattern recognition problems namely Support Vector Machines (SVMs).

SVMs are mainly used for binary classifications. For combining the binary classification into multiclass classification a relatively new learning architecture namely Decision Directed Acyclic Graph (DDAG) is used. For N class problem, the DDAG contains, one for each pair of classes. DDAGSVM works in a kernel induced feature space and uses two class maximal margin hyperplane at each decision node of the DDAG. The DDAGSVM is considerably faster to train and evaluate comparable to other algorithms. The present study proposes an SVM based recognition system for Malayalam CV speech unit recognition. The support vectors consist of small subset of training data extracted by the DDAGSVM algorithm.

All the simulation experiments are carried out using Malayalam CV speech unit database, uttered by 96 different speakers. We used 8 kHz sampled speech signal which is low pass filtered to band limit to 4 kHz. We divide the dataset into training and test set which contains first 48 samples for training and next 48 for testing. Thus training and test set contains total of 1728 samples each. The recognition accuracies obtained for Malayalam CV speech database for the five different phonetic classes using SSPD and NWHF parameters based on DDAGSVM algorithm are tabulated in Table 1.

Table 1: Experimental Results Using SSPD and NWHF based on DDAGSVM Classifier of 5 Classes

Class	*Recognition Accuracy*	
	NWHF	*SSPD*
Unaspirated	76.8	84.15
Aspirated	79.9	83.63
Nasals	83.6	96.23
Approximants	81.9	94.92
Fricatives	84.2	91.43
Average	**81.28**	**90.07**

CONCLUSION

This paper describes two feature extraction techniques viz., State Space Point Distribution (SSPD) and Normalized Wavelet Hybrid Feature (NWHF) for the classification of Malayalam CV speech unit database in speaker independent environments. Statistical pattern classification using Support Vector Machine based Decision Directed Acyclic Graph Support Vector Machine (DDAGSVM) algorithm has been presented for the classification study. The recognition accuracies are calculated and compared using SSPD and NWHF parameters and it is observed that the statistical learning approaches performs well using time domain based SSPD parameters. Classification of Malayalam consonants inherent with all the vowels are some of our future research direction.

REFERENCES

[1] E. Ott (1993). *Chaos in Dynamical Systems*, Cambridge University Press.

[2] G.L. Baker and J. Gollub (1996). *Chaotic Dynamics: An Introduction*, Cambridge University Press.

[3] Michael T. Jhonson, Rchard J Povinalli, Andrew C Lindgren, Jinjin Ye, Xiaolin Liu and Kevin Indrebo, (2005), "Time Domain Isolated Phoneme Classification using Reconstructed Phase Space", IEEE *Trans. on Speech and Audio Processing*, Vol.13, No. 4, pp. 458–466.

[4] A. Grossman, J Morlet and P Gaoupillaud, Cycle octave and related transforms in seismic signal Analysis, *Geoexploration,* 23(1), (1984) 85–102.

[5] C. torrence and G P Compo, "A practical guide to Wavelet Analysi"s, *Bull. Amer. Meteorological. Soc*, 79(1), (1998) 61–78.

[6] Hongyu Liao, Mrinal Kr and Bruce F Cockburn, "Efficient Architectures for 1-D and 2-D Lifting Based Wavelet Transform", *IEEE Trans. on Signal Processing*, 52(5), (2004) 1315–1326.

[7] S. Mallat, *A wavelet Tour of Signal Processing, The Sparse Way* (NewYork: Academic, 2009).

[8] K.P. Soman and K.I. Ramachandran, *Insight into Wavelets, from Theory to Practice* (Prentice Hall, India, 2005).

[9] Duda, R.O. and Hart, P.E., *Pattern Classification and Scene Analysis* (Wiley Inter Science, New York, 1973).

Voice Range Profile Characteristics in Mothers of Children with Hearing Impairment and Mothers of Children with Normal Hearing

D. Rekha, H.S. Shwetha and R. Rajasudhakar

All India Institute of Speech and Hearing, Manasagangothri, Mysore
E-mail: rekha.d22@gmail.com; shwethahs.srinivas@gmail.com

ABSTRACT: *The present study investigated the voice range profile (VRP) in mothers of children with hearing impairment. There have been many studies published by considering different types and levels of singers and different professional voice users using voice range profile (VRP). A very few studies were addressed the voice features of mothers' who have children with hearing impairment. Due to sensory loss, mothers' of children with hearing impairment need to voice out at louder level in day to day conversation and also train their children at higher loudness level. Two groups of participants participated in the study. Group I consisted of mothers of children with hearing impairment and Group II consisted of mothers of children with normal hearing. All the participants produced one complete voice range profile that is, by gliding /a/ vowel from the lowest to the highest possible pitch at first and secondly, from the lowest to the highest possible loudness/intensity. Results indicated no significant difference in the VRP curves between the two groups. The measured parameters such as minimum fundamental frequency (min F0), maximum fundamental frequency (max F0), range of fundamental frequency (range of F0), minimum intensity (min I0), maximum intensity (max I0) and range of intensity (range of I0) between the two groups were relatively similar. Though there was no statistical difference between the two groups, the range of fundamental frequency (range of F0) was lesser in group I compared to group II. The standard deviation of range F0 was higher in group I compared to group II. The reason for this can be attributed to duration of training given by mothers of children with hearing impairment. The mothers of children with hearing impairment have just started the training and hence the vocal usage is not significantly different from that of mothers of children with normal hearing.*

Keywords: Phonetogram, Vocal Range, Fundamental Frequency, Hearing Impairment.

INTRODUCTION

Voice is defined as the laryngeal modulation of the pulmonary air stream which is then modified by the configuration of the vocal tract (Brackett, 1971). It is an auditory perceptual term, which means the audible sound produced by the larynx, which embodies such parameters as pitch, loudness, quality, and variability. Aronson in 1980 explained about the meaning of voice disorders. He says, 'A voice disorder exists when quality, pitch, loudness or flexibility differs from the voice of others of similar age, sex and cultural group.' Voice of a person can be assessed objectively. One of the tools to quantify the vocal function is voice range profile (VRP).

Voice Range Profile (VRP) is a graphical representation that reflects a speaker's ability to produce maximum and minimum vocal intensities at fixed percentages of his maximum frequency range of phonation under controlled conditions of vowel production and mouth opening. Voice range profile or the Phonetogram plots the dynamic range (dB A) as a function of fundamental frequency (Hz), documenting the extreme possibilities of voice, which is relevant for voice classification. Voice range profile has its unique ability to detect subtle changes in vocal function (Benninger and Murray, 2006). It is a two-dimensional image or graphical representation of phonation/singing/shouting sample in the range of frequency in the X-axis and range of amplitude in the Y-axis. The resulting plot is an ellipse-shaped frequency-intensity profile, with dimensions expressed in semitones. It assesses both the fundamental frequency and intensity of an individual's absolute minimum and maximum capabilities.

Measurement derived from the voice range profile appear to be valuable for the assessment of normal and pathological voices since they reflect not just the patient's vocal frequency and intensity capabilities, but also interaction between these two variables. This information is used by the voice clinician for both diagnostic and therapeutic purposes. Phonetogram measure makes it easier to tract the areas of vocalization where there is a reduced control (break of register) or a physical problem such as vocal polyps (Heyley *et al.*, 1998). There are many studies which have been done to know the VRP in different types and levels of singers. Sulter, Schutte and Miller (1995) evaluated the gender and training differences in phonetogram in 224 subjects. The authors reported that the female singers produced louder phonations and had larger frequency range when compared to male singers. Also, trained singers had larger frequency range and larger area of phonetogram. Furthermore, Studies have been published in professional voice users using VRP. Le Borgne and Weinrich (2002) examined the phonetogram changes in trained singers over nine month period of vocal training. The authors considered 21 vocal music students and analyzed their vocal profile range. The authors reported a significant difference on frequency range and minimum vocal intensity across frequency levels and no significant difference for maximum vocal intensity across frequency levels after vocal training. Girish and Rajasudhakar (2012) studied voice range profile in different levels of Carnatic singers. They found a significant difference between junior singers and vidwath grade singers in terms of maximum F0 and range of F0. In addition, the VRP metrics/contours/areas depicted in the study hints as a continuum offering information about poor to excellent sides of continuum in terms of vocal range in singers. The authors concluded that it can be used as an objective tool to identify the singing talent and training progress.

Mothers of children with communication disorders have to train their children much more than the mothers of normal hearing children, especially in children with hearing impairment. In this process they use their vocal energy, most of the time, at louder level in a prolonged manner. There are no studies which has empirically documented the performance of mothers of children with hearing impairment on voice range profile. Also, it is interesting to study the VRP characteristics of mothers of children with hearing loss, as these mothers use their voice maximally in training their children compared to mothers of children with other communication disorders and normal hearing children.

AIM OF THE STUDY

The study aimed at investigating the voice range profile characteristics in mothers of children with hearing impairment. Hypothesis set forth in the present study is that there is no difference between the voice range profile parameters in mothers of children with hearing impairment and mothers of children with normal hearing.

METHOD

Participants: 21 participants were considered for the study. The participants were divided into two groups, namely group I and group II. Group I included 13 mothers of children with hearing impairment whose hearing loss varies from severe to profound degrees and Group II included 8 mothers of children whose hearing sensitivity within normal limits. The age range of participants in both the group was between 23 to 36 years. All the participants are non-working, non-singers and free from upper respiratory tract infection at the time of recording. Table 1 depicts the details of the participants.

Table 1: Information of Participants

S. No.	*Groups*	*Total*
1.	Mothers of children with Hearing Impairment (*Group I*)	13
2.	Mothers of children with Normal Hearing (*Group II*)	08
Total No. of Participants		**21**

Procedure: Voice Range Profile (VRP) of computerized speech lab (CSL, 4500 model) from Kay PENTAX (New Jersey, USA) was employed. The microphone (SHURE, SM-48) was used, where the distance between the microphone to the mouth was kept constant (10 cm) for all participants. The microphone was connected to an attenuator in order to cut down the hardware internal amplification at the time of recording. Participants were made to sit comfortably in a quiet room. Each subject was instructed to phonate /a/ vowel from the possible lowest loudness to the possible highest loudness and from the possible lowest pitch to the possible highest pitch that they can. They were asked to carry out this task freely with many trails.

Analysis: The following parameters were extracted from VRP statistics of VRP program. The parameters measured were, (a) minimum fundamental frequency (min F0), (b) maximum fundamental frequency (max F0), (c) range of fundamental frequency (range of F0), (d) minimum intensity (min I0), (e) maximum intensity (max I0), and (f) range of intensity (range of I0). The mean (M) and standard deviation (SD) of these parameters were calculated and compared between the two groups. MANOVA was administered to find the statistical significant difference between the groups using SPSS 16 software for the measured dependent variables.

RESULTS AND DISCUSSION

The results are discussed by comparing two groups of mothers. The mean and standard deviation of measured six parameters were calculated using descriptive statistics. Table 2 shows the mean and standard deviation of VRP profiles of mothers of children with hearing impairment and mothers of children with normal hearing sensitivity.

Table 2: Mean and standard deviation (SD) of VRP values across the two groups

S No.	*Parameters*	*Group I*		*Group II*	
		Mothers of Children with Hearing Impairment		*Mothers of Children with Normal Hearing Sensitivity*	
		Mean	*SD*	*Mean*	*SD*
1	Min F0	185.50 Hz	26.34 Hz	170.26 Hz	34.33 Hz
2	Max F0	440 Hz	92.61 Hz	505.87 Hz	75.31 Hz
3	Range of F0	254.42 Hz	99.85 Hz	336.10 Hz	82.05 Hz
4	Min I0	59.08 dB	4.11 dB	56.75 dB	1.66 dB
5	Max I0	87.23 dB	3.51 dB	88.63 dB	4.53 dB
6	Range of I0	28.15 dB	6.24 dB	31.88 dB	4.61 dB

It was found that VRP parameters such as Max F0, range of F0, Max I0, and range of I0 were reduced in group I (Mothers of children with HI) compared to group II (Mothers of children with normal hearing). Also, other parameters like Min F0 and Min I0 were higher in group I compared to group II. These differences in VRP values were subjected to statistical analysis for group differences. One way MANOVA results revealed that there was no statistical significant difference found between the two groups on the measured VRP parameters. Results of MANOVA are as shown in Table 3.

Table 3: Results of MANOVA Analysis

Parameters	*Min F0*	*Max F0*	*Range of F0*	*Min I0*	*Max I0*	*Range I0*
F	1.317	2.862	3.764	2.291	0.626	2.113
Sig	0.265	0.107	0.067	0.147	0.439	0.162

Mothers of cochlear implanters' reported to speak to their implanted child (mother-infant interaction) with higher F0, F0 range and F0 variability (Bergeson, Miller, and McCune, 2006). In another study by Lam and Kitamura (2010) found no difference in mean F0, F0 range and duration of vowels in mother-infant (who is a hearing impaired) interactions. The results of the present study may not directly relate with the above studies. But, the infant-directed speech by mothers had higher acoustic voice parameters. Contrary to this, the present study found reduced fundamental frequency and intensity parameters in VRP among mothers of children with HI. Figure 1 illustrates an example of VRP curves/contours of a mother from each group.

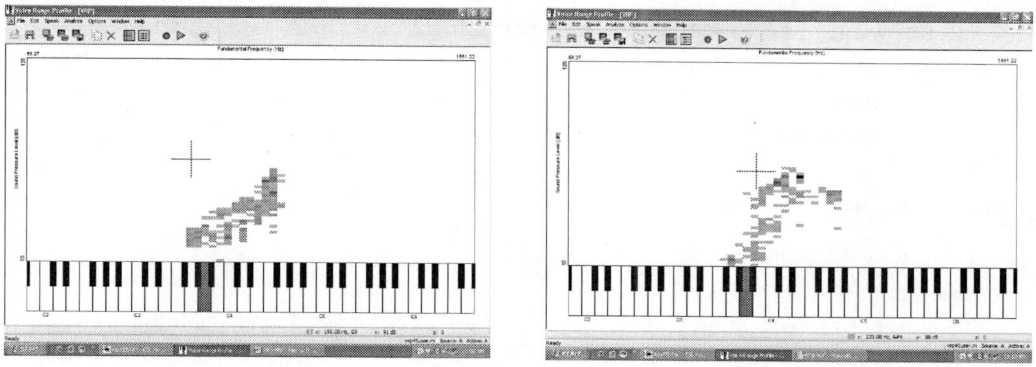

(a) (b)

Fig. 1: VRP Curves/Contours of a (a) Mother of Children with Hearing Impaired and (b) a Mother of Children with Normal Hearing

Though there was no statistical difference between the groups, the range of F0 was lesser in group I (254 Hz) compared to group II (336 Hz). The reduced range of F0 value in group I would be because of inflexibility of vocal folds to vibrate at two pitch levels, higher and lower. The standard deviation (SD) of range of F0 was higher in group I (100 Hz) compared to group II (82 Hz). Increased SD of range of F0 could be attributed to rough and unstable voice. Other possible reason for the increased SD of range of F0 could be impaired vocal folds movements, which is a symptom of fatigue (Hammarberg, 1986).

CONCLUSIONS

The present study investigated the voice range profile in mothers of children with hearing impairment and mothers of children with normal hearing. The present study did not find any significant difference on VRP between the two groups. Hence, the null hypothesis formed in the present study is accepted. The reasons for this can be attributed to duration of training given by mothers of children with hearing impairment. The mothers of children with hearing impairment teach approximately 3 to 4 hours per day. These parents enrolled in speech therapy just a month before the data collection (voice recording) of this study where the vocal usage would not significantly used much to bring about any alteration or different from that of mothers with normal hearing children. Furthermore, less number of participants participated in the present study could be the other possible reason for the obtained unexpected results. Future studies are warranted in the similar line by considering different duration of training (can be in months, years) and more number of subjects that would help in counseling the mothers of children with hearing impairment about voice care planning vocal hygiene measures.

ACKNOWLEDGEMENT

The authors would like to thank Dr. S.R. Savithri, Director, All India Institute of Speech and Hearing, Mysore for permitting us to carry out this research. The authors also thank HOD, Department of Speech-Language Sciences for permitting us to avail the lab facility and thanks are due to the subjects who participated in the study.

REFERENCES

[1] Aronson, A.E. (1980). Clinical Voice Disorders: An interdisciplinary Approach, Thieme-Stratton, New York.

[2] Benninger, M.S. and Murray, T. (2006). *The Performers' voice*.Plural Publishing, Oxfordshire OX14 1 EA, United Kingdom.

[3] Bergeson, T.R., Miller, R.J. and McCune, K. (2006). Mothers' speech to hearing-impaired infants and children with cochlear impants. *Infancy*, 10, 221–240.

[4] Brackett, I.P. (1971). *Parameters of voice quality*. In L.E. Travis (Ed). *Handbook of Speech Pathology and Audiology*. New Jersy: Prenice-Hall, Inc. 441–464.

[5] Girish, K.S. and Rajasudhakar, R. (2012). Voice Range Profile in different levels of Carnatic Singers. *Proceedings of the International Symposium on Frontiers of Research on Speech and Music*. 166–170.

[6] Hammarberg, B. (1986). Perceptual and acoustical analysis of dysphonia. [Doctoral dissertation]. Stockholm, Sweden: Huddinge University Hospital.

[7] Heylen, L.F.L., Mertens, F., Bodt, M.D., Pattyn, J., Croux, C. and Van de Heyning, P.H. (1998). Evaluation of the vocal performance of children using a voice range profile index. *Journal of Speech, Language and Hearing Research*, 41, 232–238.

[8] Lam. C. and Kitamura. C. (2010). Maternal Interactions with a Hearing and Hearing-Impaired Twin: Similarities and Differences in Speech Input, Interaction Quality and Word Production. *Journal of Speech, Language and Hearing Research, Vol.* 53, 543–555.

[9] Le Borgne, W.D. and Weinrich, B.D. (2002). Phonetogram changes for trained singers over a nine-month period of vocal training. *Journal of Voice,* 16(1), 37–43.

[10] Sulter, A.M., Schutte, H.K. and Miller, D.G. (1995). Difference in phonetogram features between male and female subjects with and without vocal training. *Journal of Voice*, 9(4), 363–367.

Speech Intelligiblity in School Class Rooms: A Case Study of Schools in an Indian Town

Ar.G. Muthu Shoba Mohan
Department of Architecture, Mohamed Sathak School of Architecture, Kilakarai

INTRODUCTION

The chief function of a school classroom is instruction, and instruction comes and likely will continue to come, by word-of-mouth and listening. The school classrooms are established to promote learning, which is by speaking and listening. In elementary school classrooms, most learning involves listening to the teacher and to each other. Therefore, acoustics is one of the most important physical properties that determine how well the school classroom building can serve its primary function. Thus the exclusion of noise and reverberation are indispensable in adapting classroom to the function of oral instruction in elementary school classrooms.

SCHOOLS IN INDIAN TOWN

Despite the existence of guidelines for school and classroom noise, and a body of research on the effects of noise on children and teachers in the classroom, there is relatively less information on noise levels outside the classrooms and its influence within the classrooms (Dockrell and Shield 2002). The acoustics of school classrooms in India are also influenced by climatic factors. In a warm-humid climate, where air-conditioning and tightly closed enclosures for the classrooms are not in practice the noise inside classrooms is highly influenced by the external noise as the windows and doors are kept open.

The Speech Intelligibility is influenced by the background-noise, reverberation time and the distance from the source to the listener. In the classrooms of the warm-humid climate regions the possibility of increased background noise in the classrooms due to the intrusion of noise through the open windows and doors along with the noise from fans remains higher than that in the enclosed classrooms. As the background noise is a fundamental parameter for determining the intelligibility in classrooms, the assessment of speech intelligibility in such classrooms becomes necessary. It is necessary to assess how far the typical Indian classrooms satisfy the requirement of better learning environment for school children.

SPEECH INTELLIGIBILITY IN CLASSROOMS

Speech Intelligibility (SI), which quantifies the quality of verbal communication, is a major concern regarding the acoustical characteristics of classrooms. SI is the percentage of speech material presented to an average listener which is correctly transmitted from a speaker to a listener. Non-optimal classroom design, acoustical conditions and SI can result in impaired verbal communication between a teacher and a student, with detrimental effects on learning. Both Bradley (2007) and Hodgson (2002) have carried out experimental and theoretical studies to investigate the relationship between background-noise, Reverberation Time and Speech Intelligibility in the classroom.

STUDY AREA

The schools evaluated are from an educationally advanced town Nagercoil located in the southern part of India, in the Kanyakumari District, which is one of the 32 districts in the state of Tamil Nadu, having 808 schools out of which 522 are primary and middle schools with about 190,000 students. The locations of the schools in relation to noise sources vary; some are adjacent to busy roads, others are set back from the road, separated from the roadside by playgrounds; many are located away from main roads, on side streets. It was therefore necessary to survey schools subject to a wide range of noise levels. Hence the study area was classified under three categories as: (i) Schools located near public roads (Noisy-sites); (ii) Schools located in Housing sites; and (iii) Schools located in Quiet sites.

METHODOLOGY

A total of 120 classrooms in 25 schools in the three sites were studied. The surveyed schools were attended by children from the 1st to the 8th grades (6–14 years old) of the fundamental educational system, which corresponds to primary and middle schools. Internal and external noise surveys were carried out around the schools to provide information on typical sound levels and sources to which children are exposed while at school. Many of the schools surveyed were exposed to noise from road traffic; the average external noise level was measured outside each school. Detailed internal noise surveys were carried out in the classrooms together with classroom observations to verify the acoustical conditions of the classrooms for the development of teaching learning process. External noise levels were compared with internal levels to determine the influence of external noise on the internal noise environment of the classrooms.

Reverberation time was measured in the classrooms when the classrooms were empty and when they were occupied by the students. The software ClassTalk was used to calculate also the RT in the classrooms. The speech intelligibility parameters like Signal to noise ratio, Speech Intelligibility and Speech Transmission Index (STI) at different locations in the classrooms were computed and discussed. The means to improve the speech intelligibility are also discussed and presented.

SOURCES OF NOISE IN THE CLASSROOM

The noise in a classroom is made up of external noise which is transmitted through the building envelope and open windows and doors, plus internally generated noise, as a result of which children in school may be exposed to noise from a wide variety of sources. The predominant external noise source, particularly in urban areas, is likely to be road traffic (Shield and Dockrell 2002) although aircraft noise may also affect many schools, with fewer schools exposed to railway noise.

In addition to external noise transmitted through the building facade, open windows and doors to a classroom, noise inside a classroom may include noise from teaching equipment (computers, projectors and so on), noise from building services (fan) in the classroom and noise transmitted through the walls, floor and ceiling from other parts of the school.

It was found that the dominant source of noise in a school classroom was the noise generated by the pupils themselves as they took part in a range of classroom activities. Class activity noise is the noise created by the occupants of the room whilst working. It is a background noise issue which has not been adequately addressed in classroom acoustics studies.

NATIONAL BUILDING CODE, INDIA (NBC, 2005)

NBC 2005 gives recommended values for background-noise levels and reverberation times in teaching spaces for schools. It recommends a value of 40 to 45 dB for background noise, and 35 to 45 dB for libraries and the proposed RT value of 0.75 s for occupied class and a higher value of 1.25 s for the empty classroom. These RT values correspond to an octave band mid frequency of 500 Hz. It is stated that it is preferable to have same limit for other frequencies.

MEASURED NOISE LEVELS

Background Noise Level

The background noise (BN) was measured in the three conditions in an occupied classroom in the three zones, namely Noisy-site, Housing and Quiet sites. The average sound levels in the occupied and unoccupied classes for the three zones are shown in Table 1.

Table 1: Average Sound Levels in Classrooms

Schools at	*Average Sound Levels in Unoccupied Classrooms dBA*	*Average Sound Levels in Occupied Classrooms dBA*
Quiet site	46.2	54.7
Housing site	49.7	58.0
Noisy site	52.8	61.2

External Noise Survey

Measurements made at the curb side of the road helped to characterize the study area as Noisy (located on public roads), partly noisy (located on Housing sites) and Quiet (located in remote areas). The influence of external noise on the existing school planning gives an insight into the acoustic effectiveness of the classrooms. There has not been any particular planning for the schools surveyed and it has been as per the availability of land and other factors that influenced the school planning.

MEASUREMENT OF REVERBERATION TIME

The traditional parameter which has been studied in evaluating the effect of room acoustics is Reverberation Time (RT). Reverberation Time is defined as the time (in seconds) it takes for the sound from a source to decrease in level by 60 dB (Sabine 1964) after the source has stopped. Reverberation Time was measured (Table 2) in furnished unoccupied classrooms which had a volume and a seat capacity for the three sites of Noisy, Housing and Quiet.

Table 2: The Average Values of RT in Occupied and Unoccupied Classrooms

Distribution of RT at 1 kHz in Occupied Classrooms		*Distribution of RT at 1 kHz in Unoccupied Classrooms*	
RT Values	*% of Classrooms*	*RT Values*	*% of Classrooms*
1.25 and above	62.5%	More than 1.0	5.0%
0.8 to 1.25	23.2%	0.76 to 1.00	21.7%
0.6 to 0.8	6.8%	0.61 to 0.75	16.7%
0.6 and below	7.5%	0.41 to 0.6	46.6%
		0.4 and below	10.0%

The stipulated value of RT according to NBC 2005 in unoccupied classrooms should be 1.25s or less at the frequency of 500 Hz and in the occupied classes the permitted value of RT is 0.75s. On comparison with the calculated values from Table, it is seen that only in 36% of classes, the RT is less than 1.25 s and 0.75 s.

SPEECH INTELLIGIBILITY

For good Speech Intelligibility (SI) the learning space should have low noise levels and minimum reverberation (Table 2). The Speech Intelligibility and the parameters influencing it are to be computed using the software ClassTalk (Murray Hodgsn 1994) and the results are discussed and presented. Speech Intelligibility in percentage at a listener position in a classroom is determined by two main room-acoustical factors: a)

Signal-to-Noise Level Difference (SNA), the difference in decibels between the total A-weighted Speech-signal Level (SLA) and the ambient-background-noise level (BGNA), both at the listener's ears (Hodgson 2002), b) Reverberation. The quqlitative ratings ofSpeech intelligibility is given asexcellent etc as stated below.If SI is more than 98% it israted as excellent; 96% to 98% is very good, 93 to 95% is good, 88% to 92% is fair 80 to 87% is poor and below 80% is poor.

COMPUTATION OF SI AT VARIOUS LOCATIONS IN CLASSROOM

In the classrooms under study, the BN in Noisy, Housing, Quiet zones of classrooms varied from 51.6 to 61.2 dB A, 48.8 to 57.5 dB A and 45.3 to 54.2 dB A respectively. The RT varied from 0.5 to 0.7 s. The background noise has great influence on Speech Intelligibility.

Table 3: SI Values in Schools of Three Sites

Quality of SI	*Range of SI*	*Quiet site %*	*Housing site %*	*Noisy site %*
Excellent	>98%	–	–	–
Very Good	96–98	–	–	–
Good	93–96	8	–	–
Fair	88–93	44	36	16
Poor	80–87	40	28	8
Bad (unacceptable)	<80	8	36	76

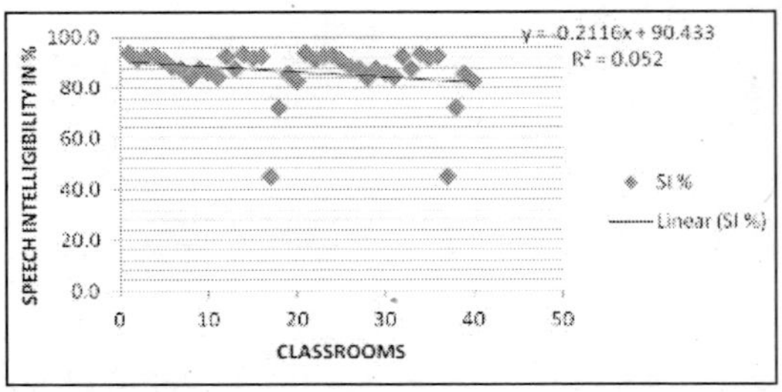

Fig. 1: Speech Intelligibility in Classrooms at Quiet Sites

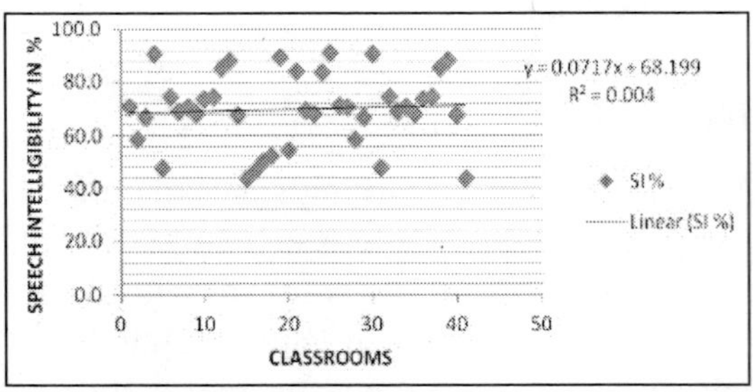

Fig. 2: Speech Intelligibility in Classrooms at Noisy-Sites Summary and Conclusions

The Speech Intelligibility in classrooms is the most important factor to attain efficient learning in classrooms, and to fulfil the function of the classroom In classrooms at Noisy-sites the Speech Intelligibility level was less than acceptable in many classrooms and in classrooms of Housing site it was better than Noisy-site but worse compared to Quiet site. The BN and RT stipulated by Indian code NBC 2005 have been found to be higher than the international stipulations. With the BN of occupied silent class of about 9 dB A more than the unoccupied BN as suggested by NBC 2005, the Speech Intelligibility in the rear rows was still less than acceptable. Hence the stipulation of the BN should be for occupied silent class a value as suggested by NBC 2005 45 dBA.

REFERENCES

[1] Bradley, J. 2007 Classroom Acoustics to support student learning: Encyclopedia of language and Literacy development London, pp. 1–7.

[2] Dockrell, J. and Shield, B., Children's and teacher's perceptions of environmental noise in classrooms.Proceedings. Institute of Acoustics, 2002.

[3] M. Hodgson, R. Rempel and S. Kennedy, "Measurement and prediction of typical speech and background noise levels in university classrooms during lectures, J. Acoust. Soc. Am. 105, 226–233, 1999.

[4] Hodgson, M. "UBC-Classroom acoustical survey," Can. Acoust. 22, 3–10, 1994.

Lexical Tone Patterns in Manipuri Language

Usha Shastri and U. Ajith Kumar
All India Institute of Speech & Hearing, Mysore
E-mail: shastri.usha@gmail.com

ABSTRACT: *There is a discrepancy in the established number of lexical tones in Manipuri, a tonal language, as well as the terminology used to describe them. Hence, the present study attempts to describe the characteristics of lexical tones in Manipuri using acoustical analysis and it perception by native speakers. Monosyllable minimal pair words contrasting in tone were recorded from three native speakers of Manipuri language. Pitch values extracted for each word were used to identify different tone contours based on the configuration of the pitch contour, initial pitch and final pitch. Results revealed four pitch contours namely, level, falling, rising and falling-rising. In addition, individual differences were noticed in the configuration of pitch contours obtained for the same word across speakers. However, for each speaker, each minimal word pair contrasting in tones were always distinguishable using pitch contour, initial pitch and final pitch of the pitch contour. The meaning of all the words was identified correctly above chance level by four native speakers. It is possible that normalization of the lexical tones produced by different speakers was done by the listeners.*

INTRODUCTION

Suprasegmental features of speech include variations in fundamental frequency, amplitude and duration. Certain languages of the world make use of variations in fundamental frequency or pitch to distinguish meaning of the words which have same segmental features. These languages which make use of phonologically contrastive variations in pitch at the word or syllable level are called tonal languages. Manipuri is one such tonal language belonging to Kuki-Chin group of the Tiebeto-Burman sub-family, spoken predominantly in Manipur, a North-eastern state of India (Chelliah, 1997; Singh, 2000; Singh, 1975).

Though it is established that Manipuri is a tonal language, the number of lexical tones are not well characterized in this language till date. There is inconsistency in the identified number of lexical tones in Manipuri, as well as the terminology used to describe them. Two to six tones have been identified in Manipuri language (Chelliah, 1992, 1997; Radhakrishnan and Savithri, 2005; Singh, 1975; Singh, 2000). Singh (1975) reported that Manipuri language shows a three way tonal contrast namely, falling, rising, and level. However, this identification of tones was not based on acoustical analysis. Chelliah

(1992, 1997) suggested that low and high are the two tones in Manipuri language. Chelliah analyzed minimal tone pairs produced by a native Manipuri adult male speaker. Based on the differences in the initial pitch of the word obtained by pitch tracking, existence of low and high tones was shown. Pitch contour was not given as much importance as that given to the initial pitch. Singh (2000) also reported of two tones in Manipuri, however, they were named, level and rising tones. Additional description of these tones is not available.

Radhakrishnan and Savithri (2005) recognized six tones in Manipuri and labeled them as rising, falling, level, rising-falling, falling-rising, and rising-falling-rising tones. Minimal tone pairs spoken by an adult female who was native speaker of Manipuri was used for the acoustical analysis. Highest Fo, lowest Fo, transition of tone, transition duration and speed of transition of each tone was measured. In addition, perceptual discrimination scores of Manipuri minimal tone contrast words by native speakers of Manipuri ranged from 87% to 100% with an average of 97.5%.

Investigators who performed acoustical analysis of Manipuri tones (Chelliah, 1992, 1997; Radhakrishnan and Savithri, 2005) have studied only single speaker data. Acoustic data from more than one native speaker of Manipuri might better characterize the tones in Manipuri (Radhakrishnan and Savithri, 2005). Also perceptual identification of Manipuri minimal tone contrast words by native speakers of Manipuri has not been studied. Investigating the identification of lexical tonal pairs is required to supplement the information obtained from acoustical analysis of lexical tones. Hence, the present study aimed to investigate the lexical tones in Manipuri language using acoustical analysis and its perception by native speakers of Manipuri.

METHOD

Participants and Material/Stimuli

Three native speakers of Manipuri language, two females and one male participated in the study. The participants ranged in age from 21 to 25 years. Four native speakers of Manipuri participated in the perception study. Seven minimal pairs of monosyllabic words contrasting in tone were selected from the previous study (Singh, 2000). These seven pairs forming 14 words were recorded from three native speakers of Manipuri language. They were required to say each word with particular meaning. All the words were spoken by each speaker in isolation. Each word was spoken thrice at different instances were recorded using a unidirectional microphone using Praat software (Boersma and Weenink, 2012) at a sampling frequency of 44,100 Hz, quantization level set at 16 bit and was stored to a personal computer. The recording was done in a sound treated room.

Procedure

Acoustical Analysis

Acoustical analysis was done using Praat software. Fundamental frequency values were extracted at 1 ms interval for each word using autocorrelation. In order to compare the pitch contours across the words, the duration of each word was normalized by transforming the unit of time from millisecond to percentage (Howie, 1976). Thus, fundamental frequency values were plotted in graph against time (in percentage) at 10% interval for each word and for each speaker separately to obtain pitch contours. Within a speaker, the pitch contours obtained for each word across three repetitions was consistently similar and hence, fundamental frequency values across three repetitions were averaged. Thus obtained pitch contours were used to characterize the lexical tones in Manipuri language.

Perceptual Analysis/Perception of Minimal Tonal Pairs

To investigate the perception of these minimal tonal pairs by native speakers of Manipuri, identification of the meaning of the recorded words were done. In this experiment, 42 words (14 words*3 repetitions) spoken from each speaker were presented in random order to a native speaker of Manipuri through headphones. The task was to write the meaning of the word heard. For one listener, same task was repeated by presenting all the 42 words spoken by same speaker thrice at different instances to ensure the reliability of the response. Percent correct identification scores were calculated for each listener identifying meaning of the words spoken by each speaker.

RESULTS

Acoustical Characteristics of Lexical Tones in Manipuri

Contours of lexical tones were identified for each speaker separately from the average pitch contour from three iterations of the single word. From the averaged pitch contours of 14 words spoken by Speaker 1 (S1) three contours were obtained namely level (6 words), falling (7 words) and rising (1 word). Visualization of pitch contours of words spoken by Speaker 2 (S2) revealed three contours namely level (6 words), rising (2 words), and falling-rising (6 words). Pitch contours of Speaker 3 (S3) showed four contours which could be labeled as level (3 words), falling (3 words), rising (2 words), and falling-rising (6 words). S1 and S3 shared same pitch contour for five words. However all the speakers produced each pair of words in contrasting pitch. The pitch contour obtained for each of these 14 words, for each speaker is given in Figure 1.The words, their meaning, and the average pitch contours obtained for each of the speakers is given in Table 1.

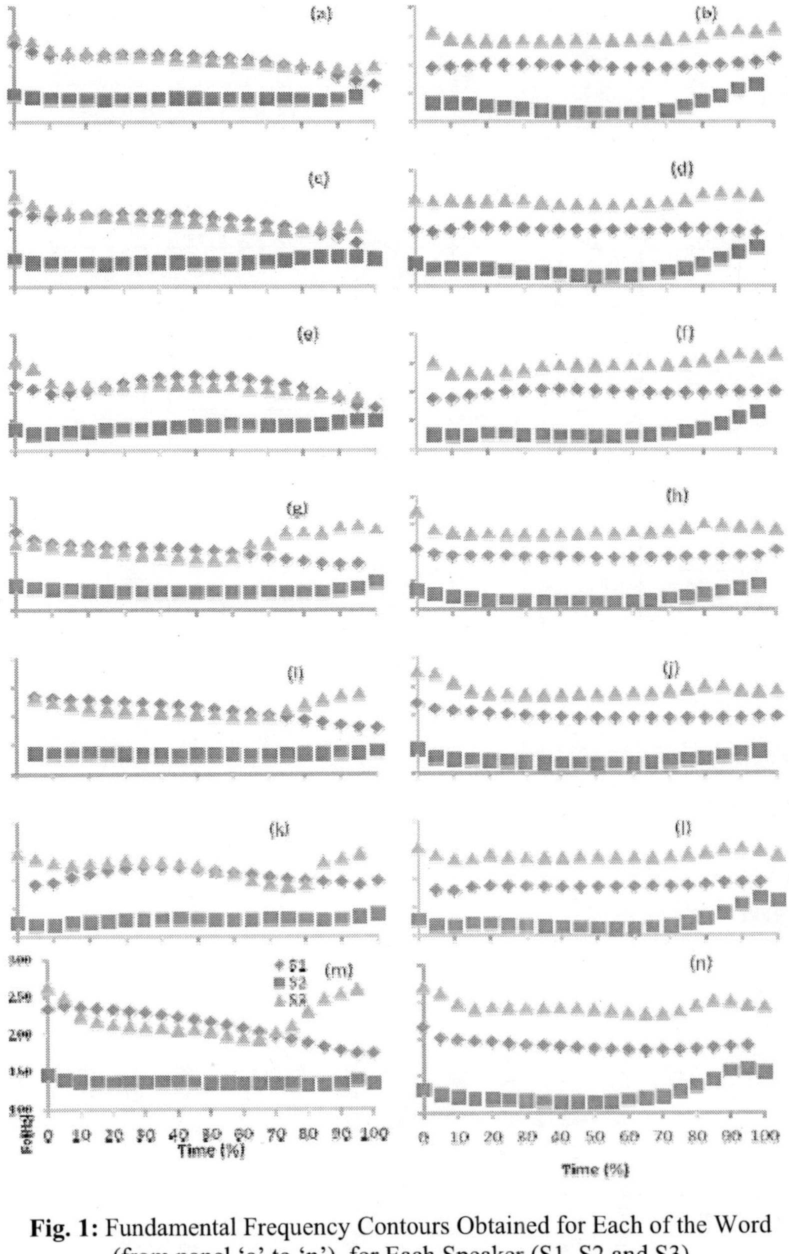

Fig. 1: Fundamental Frequency Contours Obtained for Each of the Word (from panel 'a' to 'n'), for Each Speaker (S1, S2 and S3)

From the Figure 1 and Table 1 it can be seen that basic three pitch contours level, falling and rising was identified in all the three speakers. An additional contour of falling-raising was present only in speaker 2.

Table 1: Different pitch contours obtained for different Manipuri words spoken by three native speakers of Manipuri. (S1-Speaker1, S2-Speaker 2, S3-Speaker 3, L-level, F-falling, R- rising, FR- falling-rising)

Sl. No.	*Manipuri Word*	*Meaning in English*	*S1*	*S2*	*S3*
1.	I	Blood	F	L	FR
2.	I	Thatch	R	FR	F
3.	Un	Skin	F	L	F
4.	Un	Ice	L	FR	FR
5.	Mi	People	F	R	F
6.	Mi	Spider	L	FR	R
7.	ka	Mosquito	F	L	FR
8.	ka	A game	L	FR	R
9.	kho	Leg	F	L	FR
10.	kho	Stream	L	FR	L
11.	lo	Spear	F	L	FR
12.	lo	Fishing net	L	FR	L
13.	tho	Door	F	L	FR
14.	tho	Bridge	L	R	L

Perceptual Analysis

Words spoken by S1 were identified correctly in meaning 90% times. Correct meaning identification scores for words spoken by S2 and S3 were 66.67% and 71.43% respectively. Overall percent correct identification of each of the word averaged across speakers is given in Figure 2.

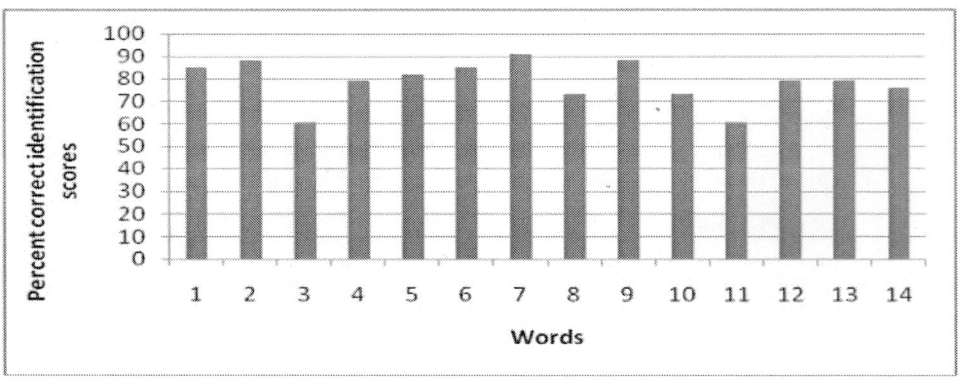

Fig. 2: Percent Correct Identification of 14 Words Averaged Across Three Speakers. The Words Corresponding to Each Number can be Seen from Table 1

From Figure 2 it can be observed that none of words were identified correctly 100% of the times. The percent correct identification scores ranged from 60.60% to 90.90% for each word. Thus, none of the words were identified below chance level of 50%.

DISCUSSION

From the results of the present study, four pitch contours were identified in Manipuri language namely, level, falling, rising and falling-rising. It is interesting to find that though the pitch contours were different for most of the words spoken by three speakers, the meaning could be identified well above chance level for all the words. Also, it is being noticed that in each minimal pairs, all the three speakers produced the words with distinguishing pitch contour. This contrasting pitch could have helped the listeners to identify the meaning of the word. In addition, the reason for not obtaining 100% correct identification scores could be because of absence of contextual cues as the words were presented in isolation. It is possible that normalization of the lexical tones produced by different speakers was done by the listeners.

CONCLUSION

Present study attempted to characterize the lexical tones in Manipuri language. It was found that four pitch contours were reliably identified. Three to four pitch contours were identified in each speaker. Though pitch contours produced by each speaker was different, listeners could normalize the lexical tones and could correctly identify the meaning of the word well above chance level.

REFERENCES

[1] Boersma, P. and Weenink, D. (2012). Praat: Doing phonetics by computer (Version 5.3.32) [Computer program]. Retrieved from http://www.fon.hum.uva.nl/praat

[2] Chelliah, S.L. (1992). Tone in Manipuri. In M. Ratliff, and E. Schiller (Eds.).*Papers from the First Annual Meeting of the Southeast Asian Linguistics Society* (pp. 65–85).

[3] Chelliah, S.L. (1997). *A grammar of Meithei.* Berlin: Mounton de Gruyter.

[4] Howie, J.M. (1976). *Acoustical studies of Mandarin vowels and tones.* Cambridge: Cambridge University Press.

[5] Radhakrishnan, S. and Savithri, S.R. (2005). Perception of Manipuri tones by native and non-native speaker. Students research at AIISH Mysore (Articles based on dissertations at AIISH), 4, 95–111.

[6] Singh, C.Y. (2000). *Manipuri grammar.* New Delhi: Rajesh Publications.

[7] Singh, I. (1975). *Manipuri phonetic reader.* Mysore: Central Institute of Indian Languages.

Frequency of Occurrence of Phonemes in Kannada: A Comparative Study between Two Dialects

N. Sreedevi, M.D. Vikas and Smitha K. Nair

Department of Speech Language Sciences, All India Institute of Speech and Hearing, Mysore

E-mail: vikasmdslp@gmail.com

ABSTRACT: *Kannada has a very complex range of regional, social and stylistic variations: the Mysore dialect, the coastal dialect (Mangalore), the northern dialect (Dharwar) and Kalaburgi dialect (Upadhyaya, 1976). The number of phonemes and their frequency of occurrence vary across languages. This information is essential for several research and clinical purposes. Several studies in different languages like English, Hindi, Oriya, Malayalam and Kannada calculated the frequency of phonemes using written source of materials. The purpose of the present study was to obtain the frequency of occurrence of various phonemes in two dialects of Kannada using conversation samples. The conversation samples were transcribed using IPA transcription. Further the transcribed data was analyzed using the SALT software for obtaining frequency of phonemes. The results show that /a/ was the most frequently occurring phoneme among vowels followed by /I/, /u/, /e/, /a:/ whereas among consonants /n/ was most frequently occurring followed by /r/, /d/, /l/, /t/, /g/, /k/ and /v/ in both the dialects of Kannada. The results obtained will aid in preparing speech corpora for several research purposes and also help audiologists and speech language pathologists in developing and updating the existing test material for evaluating various communication disorders and also for selection of treatment targets in such population.*

INTRODUCTION

The frequency of occurrence of phonemes is in research since 1930s. The studies of phonemes were mainly concentrating on the written context whereas Mines, Hanson and Shoup (1978) used conversation in an interview and the frequency of occurrences of phonemes were listed in descending order/**a, n, t, i, s, r, l, d, ɛ /.**

On similar lines, initial studies on Indian languages available were by Bhagwat (1961) in Marathi and Ghatage (1964) in Hindi and Malayalam. The sources were mainly concentrated on written context in all these studies. Ranganatha (1982) considered different written sources to calculate the frequency of occurrence of phonemes in Kannada. The results revealed that vowel /**a**/ was most frequently occurring followed by **/I/, /u/, /n/, /d/, /a:/.** The least frequently occurring phonemes were /**p^h**/, /**g**/, /**ɖ^h**/, /**k^h**/, /**ʈ^h**/. Considering the written sources, Jayaram (1985) studied the frequency of occurrence of phonemes in Kannada and listed out a series of vowels and consonants by rank order of their frequency of occurrence and the order was **/a, I, n, r, u, d, ɔ, e, t, l/**.

There is dearth of studies about the frequency of phonemes in Indian languages. These studies would provide a database for developing speech materials for assessment and selecting treatment targets for various communication disorders. The previous studies in Indian languages have considered only written source of materials including the earlier studies in Kannada (Ramakrishna, 1967; Ranganatha (1982); Jayaram, 1985) but there exist differences in phoneme frequency among spoken and written source of data (Sandoval *et al.*, 2008). Hence the present study was planned to obtain the frequency of occurrence of phonemes in conversational speech samples in Kannada.

METHOD

Participants

Participants included were 21 native speakers of Mysore dialect divided into 5 groups, each group consisting of 3 to 4 participants and 15 native speakers of Mangalore dialect of Kannada divided into another 5 groups consisting of 3 participants in each group. Participants in the age range of 20 to 50 years who did not have any clinical history of speech, language, hearing or any neurological problems and with a minimum of 10 to 12 years of education in Kannada medium were selected for the study.

Instrumentation

A digital recorder (Olympus WS 100) was used for recording the conversation samples.

Procedure

The data was collected through conversation in controlled natural environments which lasted for about 25 to 30 minutes of duration. Participants were not given any particular topic about the conversation. Any current topic of interest was initiated by the participants themselves. They were instructed to avoid using words from other languages but they were not restricted from using commonly used borrowed English words (E.g.: Bus, ticket, phone, car etc). A total of 10 sessions (5 of Mysore dialect, 5 of Mangalore dialect) of general conversation were recorded for the study.

Data Analysis

The conversation samples were transcribed using IPA transcription. The raw data obtained was analyzed using the software SALT RV version 9 for the frequency count. SALT (Systematic Analysis of Language Transcripts), is a computer based program designed to help in analyzing and interpreting language samples from one or more speakers during a communicative interaction. A database of Kannada phonemes was prepared and saved in the editable standard wordlists of SALT software. The Salt software compares the database and the loaded phoneme file and provides the phoneme count. Sample of each recording was subjected to inter judge and intra judge reliability measures. A reliability index of 87% and 94% were obtained for inter judge and intra judge reliability respectively.

RESULTS

The present study aimed to obtain the frequency of occurrence of phonemes in conversational speech in Mysore and Mangalore dialect of Kannada and to compare findings across the two dialects. The results of phoneme count obtained using SALT software were tabulated.

Figure 1 shows the phoneme counts obtained from the ten recording sessions. A total of 69,624 phoneme counts accounted for the Mysore dialect data and 75,803 for the Mangalore dialect from conversation samples of five groups of participants in both the dialects. Using descriptive statistics, the mean and standard deviations of the frequency of occurrence of vowels and consonants were calculated.

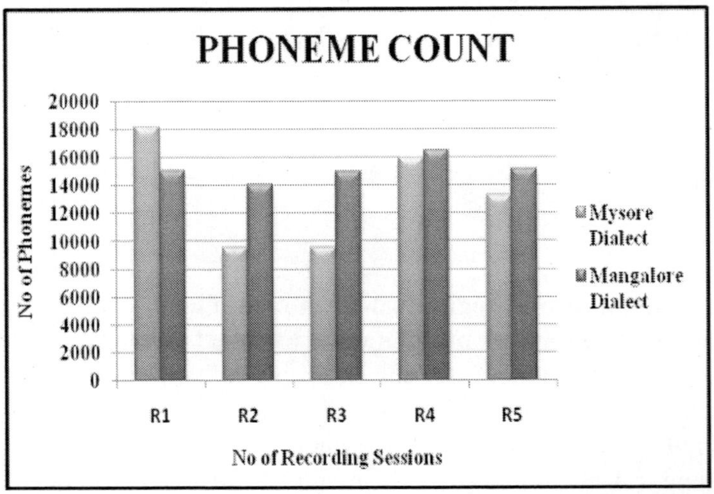

Fig. 1: Phoneme Count of Recordings from Mysore and Mangalore Dialects

Figure 2 represents the mean frequency of vowels and consonants calculated from all the 10 samples. The vowel percentage in Mysore and Mangalore dialects were 44.3% and

43.8% respectively and consonant percentage was 55.7% in Mysore dialect and 56.1% in Mangalore dialect of Kannada.

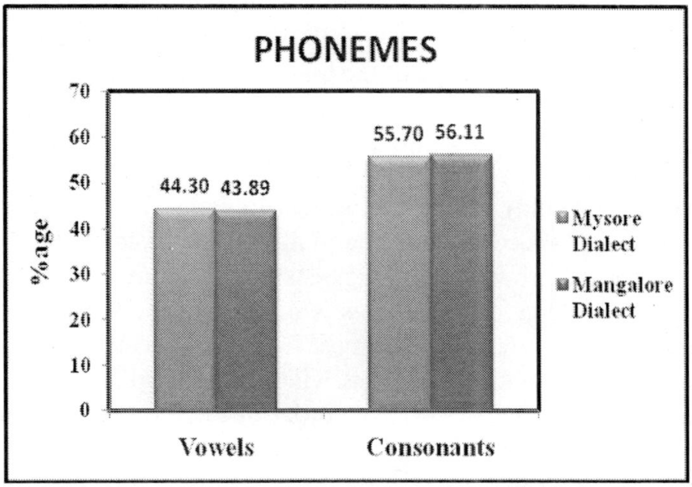

Fig. 2: Mean Frequency of Occurrence of Vowels and Consonants in Percentage

Figure 3 represents the mean percentage of the most frequently occurring phonemes. Vowel /a/ occurred 13% of the total data in Mangalore dialect and 14.1% of total data in Mysore dialect. The ten most frequently occurring phonemes in discending order are **/a/, /n/, /i/, /e/, /a:/, /r/, /d/, /l/, /t/, /u/** in Mysore dialect and **/a/, /n/, /i/, /u/, /e/, /a:/, /r/, /v/, /d/, /l/** in Mangalore dialect of Kannada.

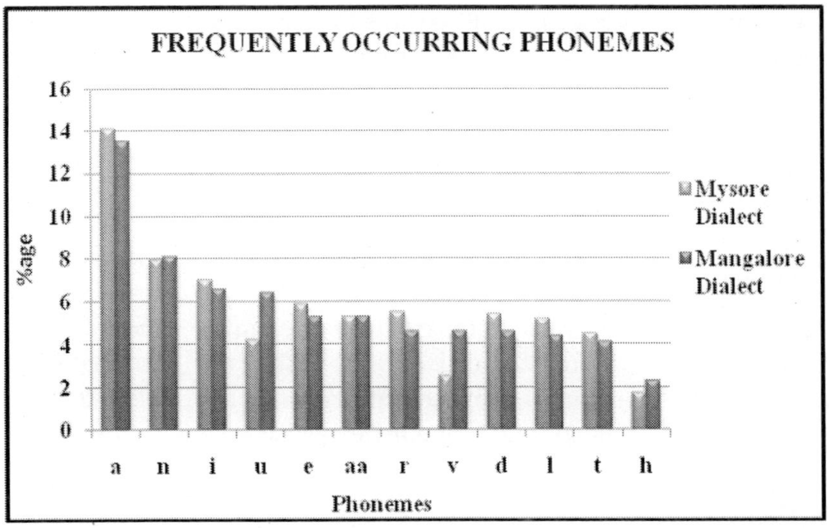

Fig. 3: Mean Percentage of Frequency of Occurrence of Frequently Occurring Phonemes

In both the dialects, vowel **/o/** occurred for 2% whereas phonemes /s/, /j/ occurred for less than 2% and **/p/, /tʃ/, /ʃ/, /i:/, /o:/** occurred less than 1% each of the total data. Consonants **/ŋ/, /pʰ/, /ɖʰ/** did not occur in the ten recorded conversational samples though they are present in the Kannada phoneme system. The aspirated phonemes were rarely seen. Diphthongs **/ai/ and /au/** occurred for less than 1% of the total data.

DISCUSSION

The results of the present study had several salient findings. The results indicated that the Kannada conversational speech (both the dialects) consists of around 55–56% of consonants and 44–45% of vowels. This suggests that the consonants are used more than vowels in our day to day conversation. This is in consonance with the earlier reports of several languages. In the present study, the most frequently occurring phonemes constituted 70–72% of the total data in both the dialects where as in English it was 47% of the total data (Mines, Hanson and Shoup, 1978). The order of the frequently occurring phonemes in the present study is in contrary to the reports of Ramakrishna (1967) in written Kannada. Though he found that vowel /**a**/ occurred most frequently, the predominantly occurring consonants were /**r**/, /**d**/ and /**t**/.

In the present study, /**n**/ was found to be the most occurring consonant in both the dialects of Kannada. Followed by /n/, consonants **/r/, /d/, /l/, /t/** occurred most frequetly in Mysore dialect whereas **/r/, /v/, /d/, /l/** in Mangalore dialect. The consonants /h/ and /v/ had a higher frequency of occurrence in Mangalore dialect. The differences in the order and phoneme frequency were minimal as both the dialects selected belonged to southern part of Karnataka. The findings of Jayaram (1985) in written Kannada also showed /**n**/ as the most frequently occuring consonant but was followed by **/r/, /d/, /t/** and **/l/** indicating a difference in the sequence of the frequently occurring phonemes across conversation and written samples.

CONCLUSIONS

The frequency of occurrence of different phonemes in Kannada for two dialects (Mysore and Mangalore) was determined using conversational samples and the most frequently occurring phonemes are **/a/, /n/, /I/, /l/, /e/, /d/, /a:/, /u/, /v/, /r/, /t/** which constituted 70–72% of the total data. There were no major differences in the order of the frequency of occurrence of phonemes in both the dialects. The information obtained in this study is applicable not only in the diagnostic and therapeutic aspects of communication disorders but also in the area of linguistics and automatic speech recognition which are a few to mention.

ACKNOWLEDGEMENTS

We extend our sincere gratitude to Prof. S.R. Savithri, Director, and Prof. Y.V. Geetha, Head, Department of Speech Language Sciences, All India Institute of Speech and Hearing for permitting us to carry out this study. We also thank all the participants of our study for their kind cooperation. This study is a part of the ARF project titled "Frequency of occurrence of phonemes in Kannada".

REFERENCES

[1] Jayaram, M. (1985). Sound and Syllable distribution in written Kannada and their application to Speech and Hearing. *Journal of All India Institute of Speech and Hearing,* 16, 19–30.

[2] Mines, M.A., Hanson, B.F. and Shoup, J.E. (1978). Frequency of occurrence of phonemes in conversational English. *Language and Speech,* 21, 221–41.

[3] Ramakrishna, B.S. (1962). *Some aspects of the relative efficiencies of Indian languages: a study from information theory point of view.* Bangalore: Indian Institute of Science.

[4] Ranganatha, M.R. (1982). *Morphophonemic analysis of the Kannada language.* Mysore: CIIL.

SECTION–IV

Underwater Acoustics

A Study on the Effect of Sea Surface on Different Underwater Signals

R.P. Raju and P. Balasubramanian
Naval Physical and Oceanographic Laboratory, Kochi

ABSTRACT: *Active SONARs employ different signals like Continuous wave (CW) and Linear Frequency Modulated wave [LFM] for detection, range-bearing estimation and classification of targets.*

During propagation, these signals interact with sea-surface and sea-bottom. In an earlier paper[1] [Acoustic waves (Proceedings of NSA-2011), Chapter 41, pp. 392–40], the effect of the sea bottom on the shape and spectrum of the signals was brought out. In this paper, the effect of sea-surface on the shape and spectrum of the propagating signals is analyzed. A Pierson-Moskovitz sea-surface spectrum is employed to generate different surface realizations corresponding to different wind speeds. A Parabolic Equation Model based on Split-step Pade approximation is used to simulate the channel impulse response and received time series for different signals. This study will be highly useful for SONAR engineers in designing the SONARs.

INTRODUCTION

Active SONARs use different kinds of acoustic waveforms to detect the targets. The propagation of these signals is modeled to predict the range of SONAR corresponding to the prevailing ocean conditions. The modeling of acoustic wave propagation in ocean involves modeling of interaction of the acoustic wave with the medium and the ocean boundaries. Sea surface and bottom constitute the boundaries of ocean. In most of the cases the root mean square wave height of the sea surface is considered to calculate the reflection coefficient of the sea surface using empirical formulae. In this paper the undulations of the sea surface corresponding to different wind speeds are being generated using Pierson–Moskovitz (P-M) spectrum. A study on the effect of undulated sea surface on acoustic wave propagation is done using a parabolic equation (PE) model based on split-step Pade approximation.

METHODOLOGY

The PE model is a wave-theory based approach to formulate and solve problems of acoustic propagation in a medium. PE approach involves reducing the elliptic Helmholtz wave equation to parabolic differential equation by paraxial approximation[2]. The parabolic equation can be solved by various methods.

A broadband PE model based on split-step Pade approximation[2] which can include coordinates of the undulated sea surface as an input has been implemented. The sea surface scattering is included in this PE model by using conformal mapping technique[2]. Also, the sea surface has been generated from P-M wave number spectrum by using methodology outlined by Thorsos, I E, *et al.*[3] (1988). This PE model is used to generate channel impulse response and received time series.

The effect of sea surface on different acoustic waveforms like Linear Frequency Modulated (LFM) pulse and Continuous wave (CW) pulse is evaluated by computing the replica correlator[4] outputs.

CASE STUDY AND RESULTS

The effect of sea surface on acoustic signals like LFM and CW pulse in two different ocean environments is studied. The summer and winter ocean environment parameters (Table 1) are fixed and sea surface is varied for three wind speeds. The received time series is generated for a range of 5000 m. The received time series is subjected to replica correlation. Replica correlation is a good tool to evaluate the change in shape and spectrum of a transmitted signal after propagation. In this case study the effect of sea surface will be evaluated using replica correlation, as only sea surface is varied for two different ocean environments. The received time series and replica correlator outputs for LFM signal are presented in Table 2. The received time series and replica correlator output for CW pulse is presented in Table 3.

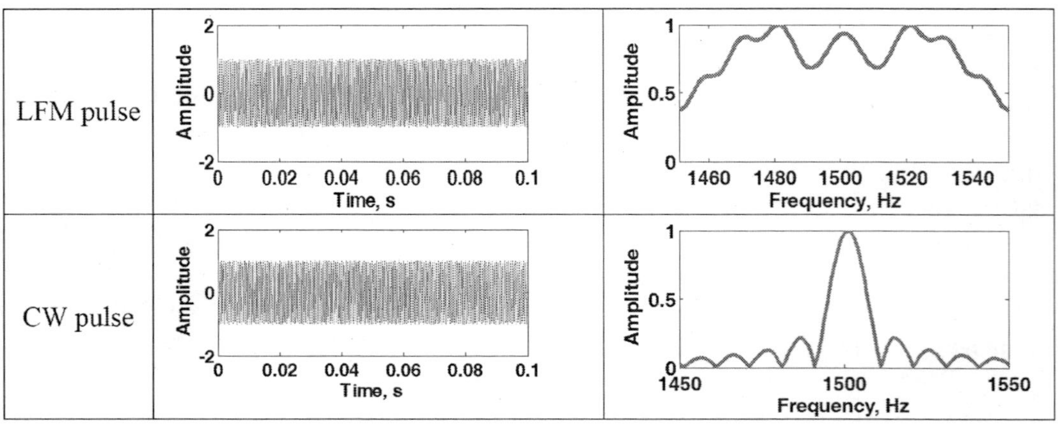

Fig. 1: Transmitted Signal (Time series and frequency spectrum)

Table 1: Ocean Environment

Environment	*Depth of Water Column, m*	*Depth, m*	*Sound Speed in Water, ms^{-1}*	*Sound Speed in Sediment, ms^{-1}*	*Density of Sediment, $Kg\ m^{-3}$*	*Attenuation in Sediment, $dB\ \lambda^{-1}$*
Summer[4]	100	0	1494.0			
		20	1494.3			
		50	1490.6	1560	1800	1.00
		100	1491.3			
Winter[4]	100	0	1490.0	1560	1800	1.00
		100	1492.0			

Propagation Model used : Broadband PE Model (Split-step Pade approximation)
Sea surface realization : P-M spectrum for wind speeds of 0, 10, 20 ms^{-1}
Source depth : 50 m
Receiver Depth, Rd : 10 m and 75 m, Range: 5000 m
Waveform : 0.1 s LFM signal with a bandwidth of 100 Hz centered at 1500 Hz. (Figure 1)
0.1 s CW pulse of frequency 1500 Hz. (Figure 1).

The plots in Table 2 and 3 clearly indicate the distortion of the transmitted signal. The LFM and CW signals are stretched to approximately 0.2 s. The replica correlator outputs in Table 2 indicate that the received signals at a depth of 75 m are more correlated to the transmitted LFM signal than the ones received at 10 m. Also, in winter environment at 10 m depth the maximum correlation value increases as the wind speed increases from 0 to 20 ms^{-1}. The maximum correlation at 10 m receiver is very low and the pulse may not be detected in a noise-limited environment after replica correlation. The rough sea surface leads to higher propagation losses[4] and lower pulse distortion in the given environments. In case of flat sea surface, the trade-off is between low propagation losses and higher correlation in signals at 75 m depth. And in case of Rd at 10 m the trade-off is between high propagation losses and lower correlation in received signals.

In case of CW pulses it is observed from Table 3 that there is not much change in the maximum correlation value (0.5–0.6) between received and transmitted signal due to change in the sea surface. But, the shape of the signal changes after propagation.

Table 2: Received Time Series and Replica Correlator Outputs for LFM Signal

<table>
<tr><th rowspan="2">Environment</th><th colspan="3">Wind speed, ms⁻¹</th></tr>
<tr><th>0</th><th>10</th><th>20</th></tr>
<tr><td rowspan="2">Summer; Rd = 10 m</td>
<td>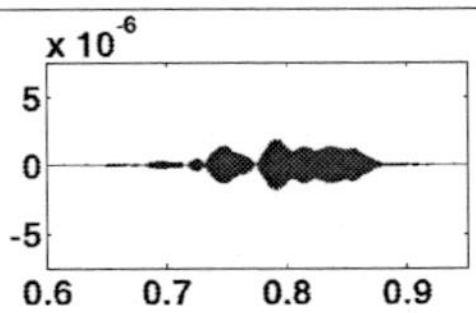</td>
<td>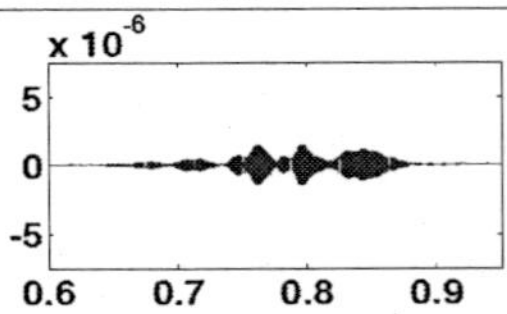</td>
<td>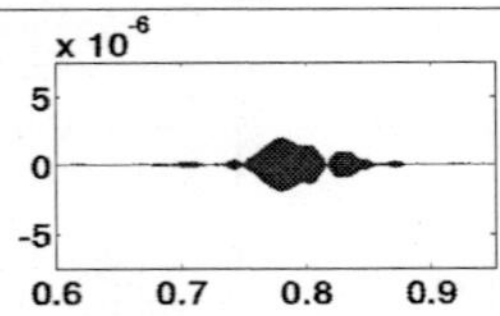</td></tr>
<tr><td>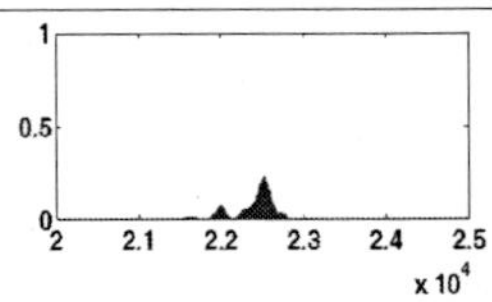</td>
<td>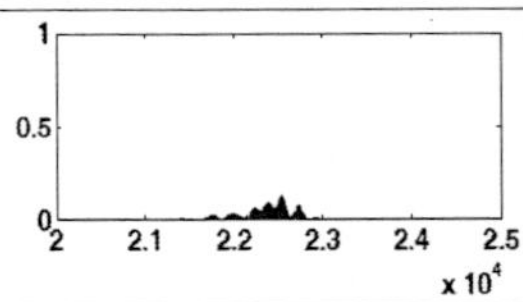</td>
<td>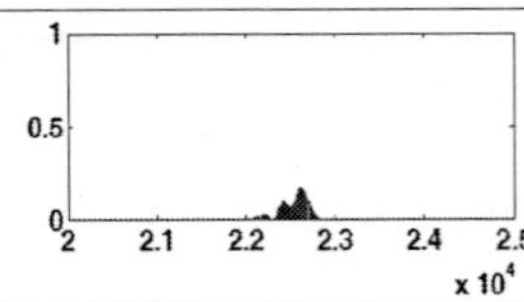</td></tr>
<tr><td rowspan="2">Summer; Rd = 75 m</td>
<td>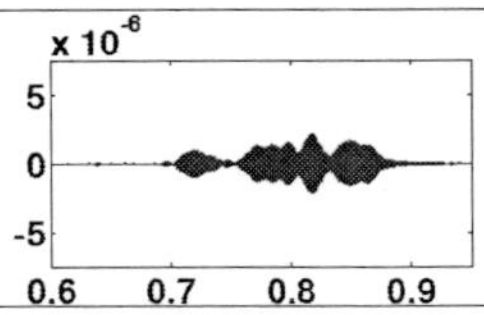</td>
<td>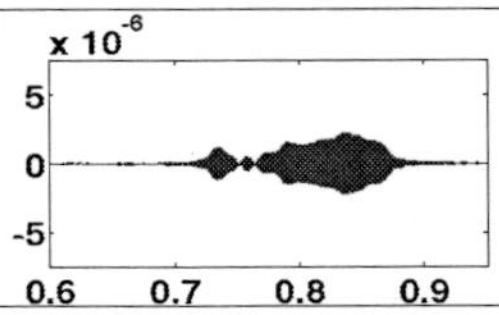</td>
<td>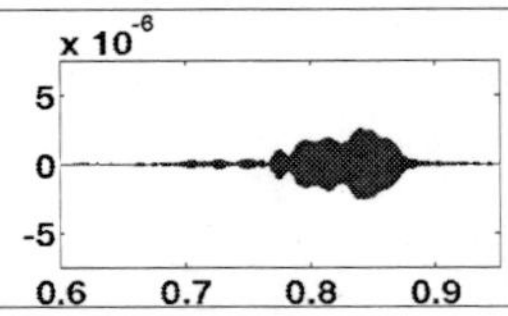</td></tr>
<tr><td>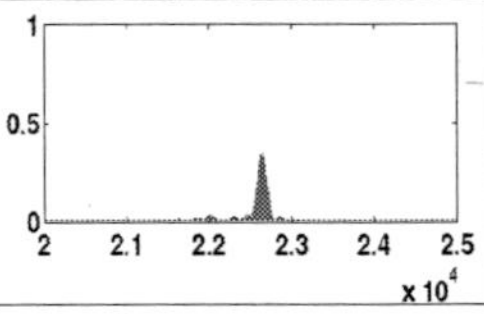</td>
<td>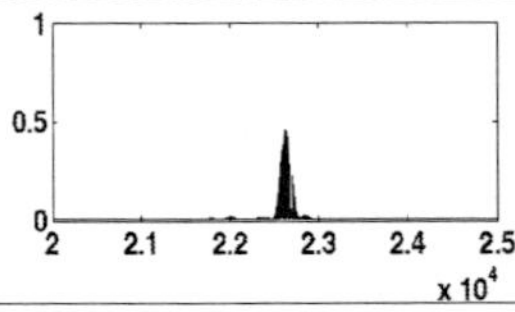</td>
<td>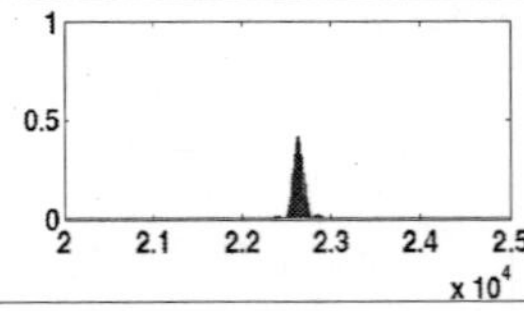</td></tr>
<tr><td rowspan="2">Winter; Rd = 10 m</td>
<td>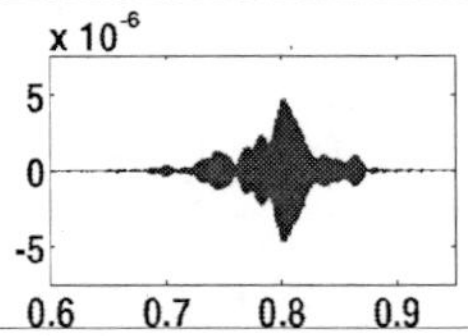</td>
<td>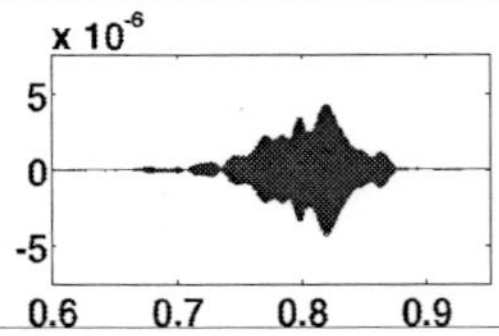</td>
<td>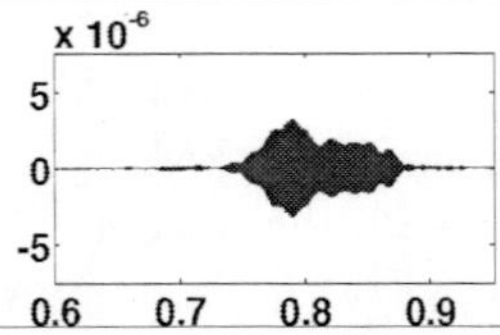</td></tr>
<tr><td>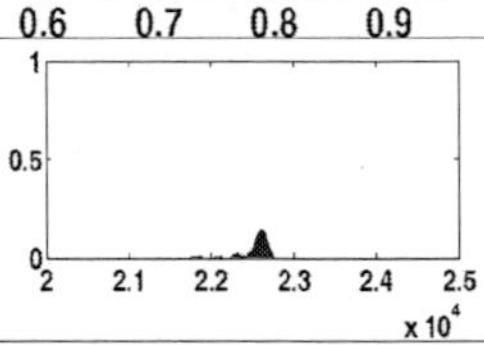</td>
<td>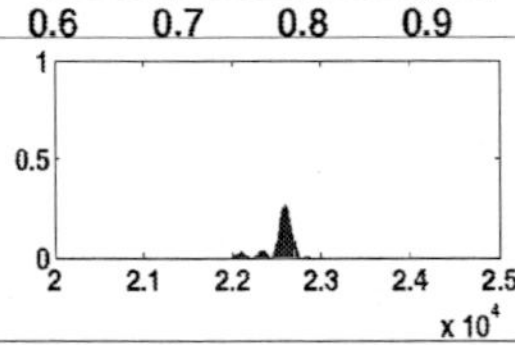</td>
<td>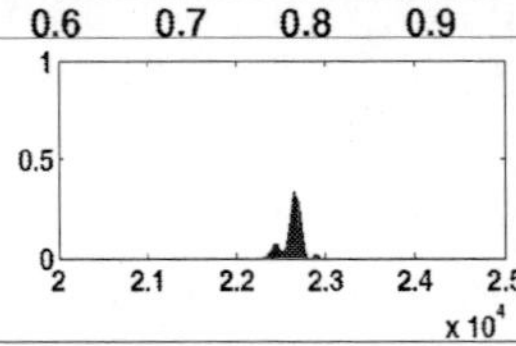</td></tr>
<tr><td rowspan="2">Winter; Rd = 75 m</td>
<td>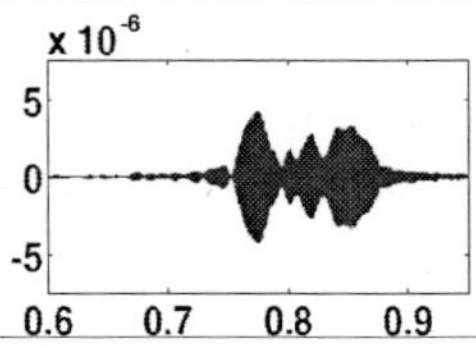</td>
<td>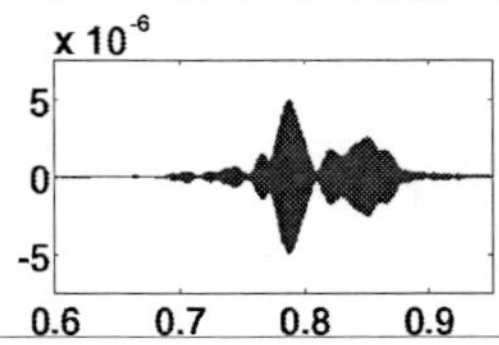</td>
<td>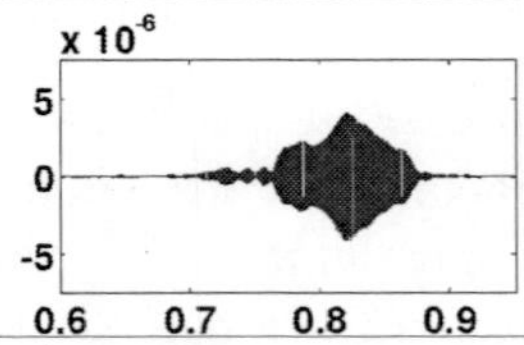</td></tr>
<tr><td>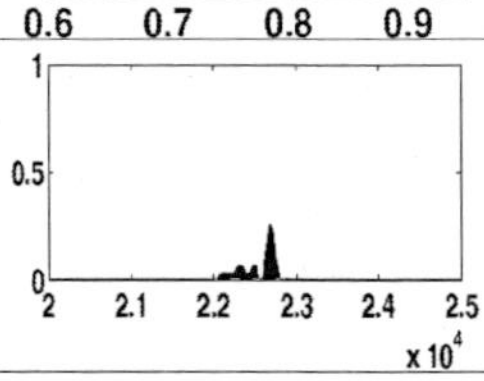</td>
<td>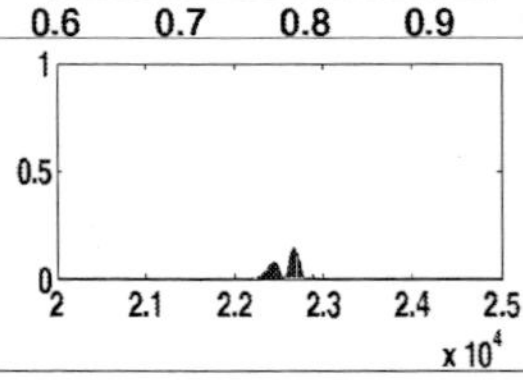</td>
<td>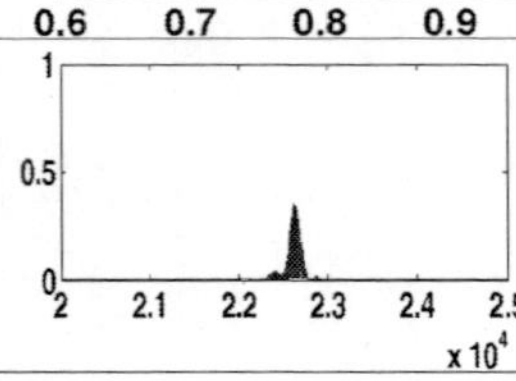</td></tr>
</table>

Table 3: Received Time Series and Replica Correlator Outputs for CW Pulse

Environment	*Wind speed, ms^{-1}*		
	0	*10*	*20*
Summer; Rd = 10 m	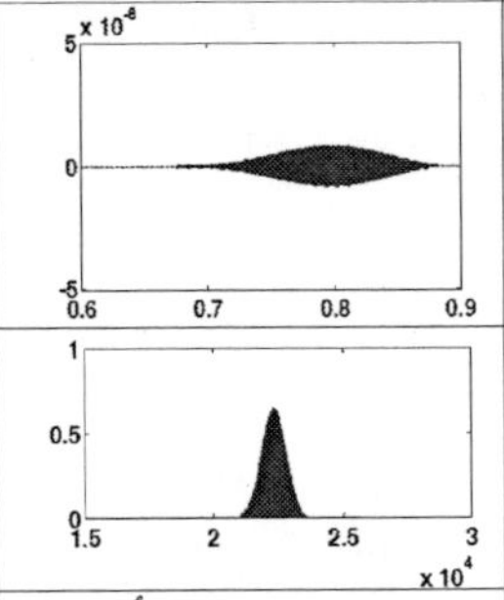	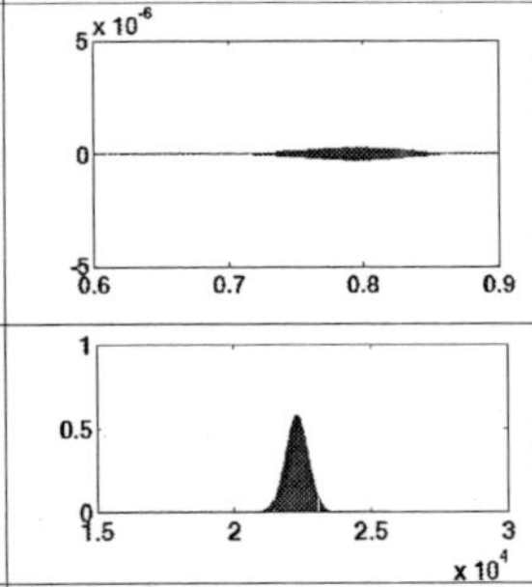	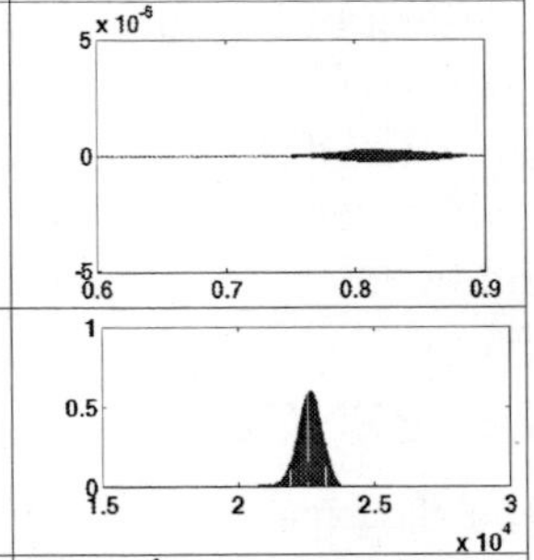
Summer; Rd = 75 m	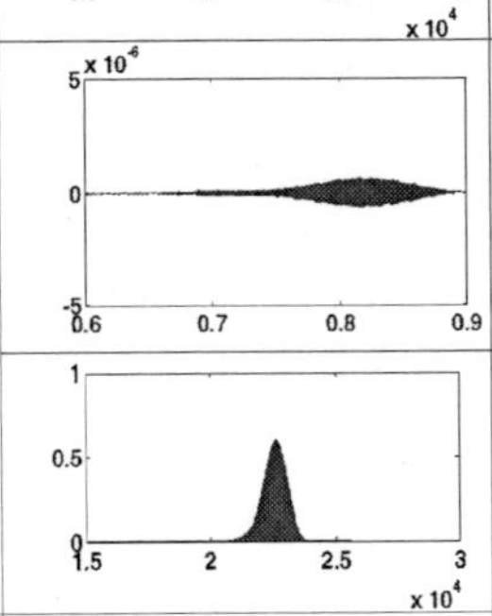	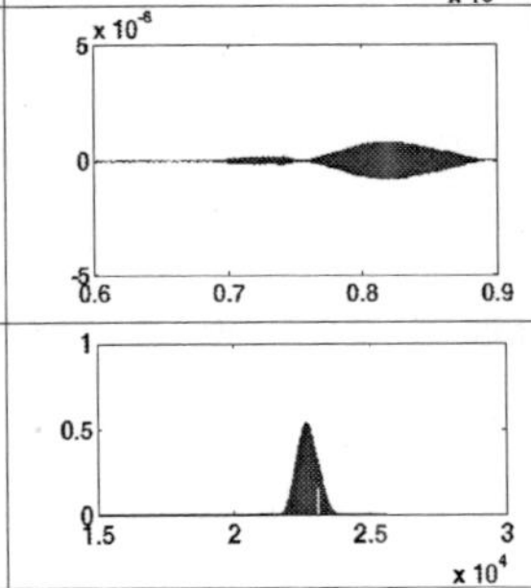	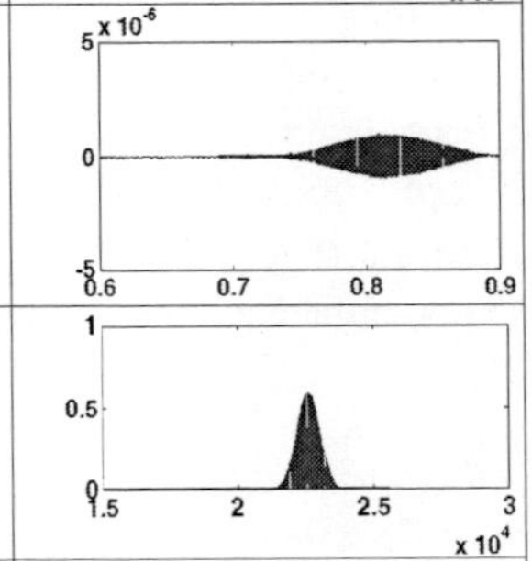
Winter; Rd = 10 m	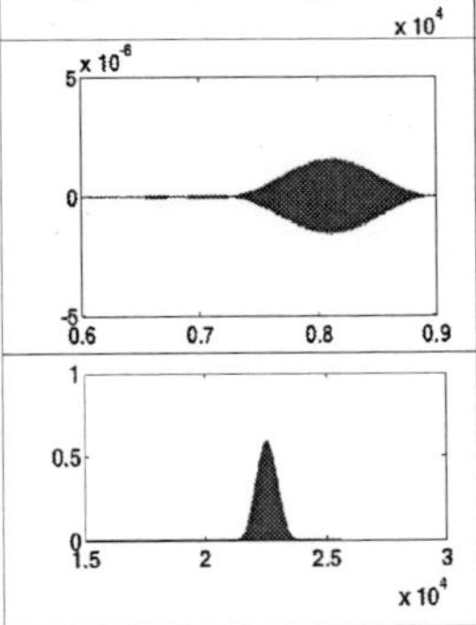	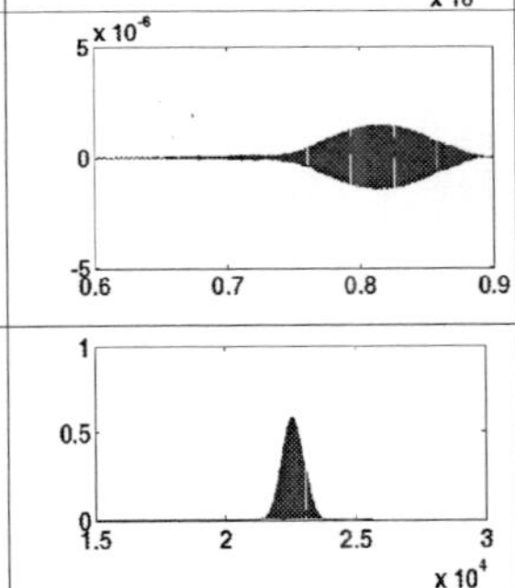	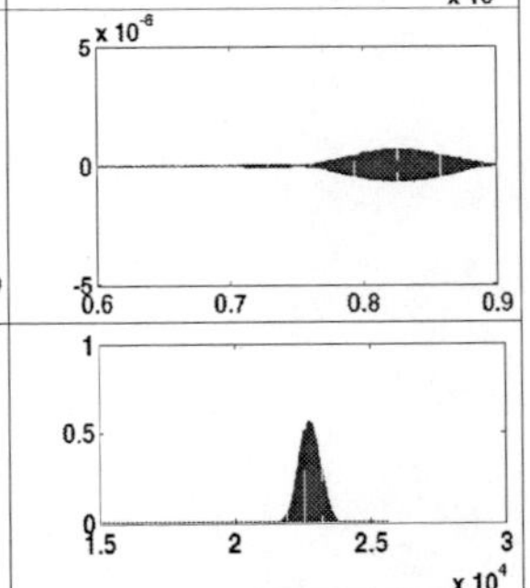
Winter; Rd = 75 m	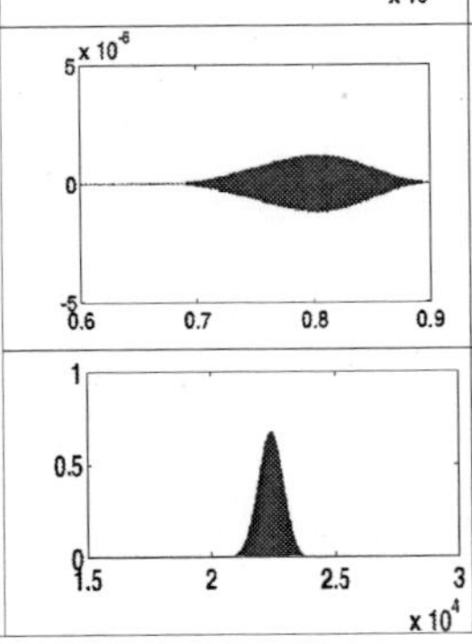	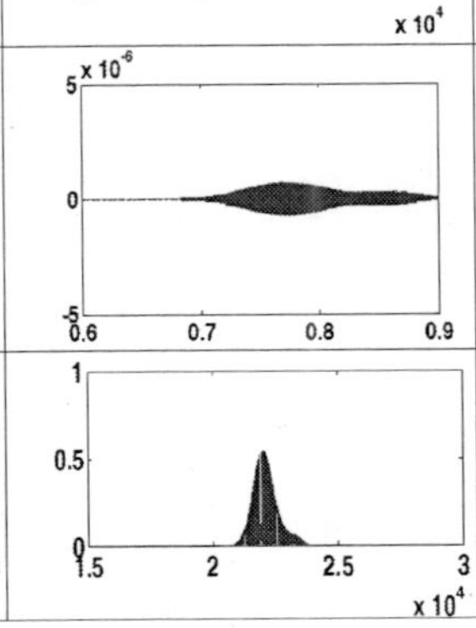	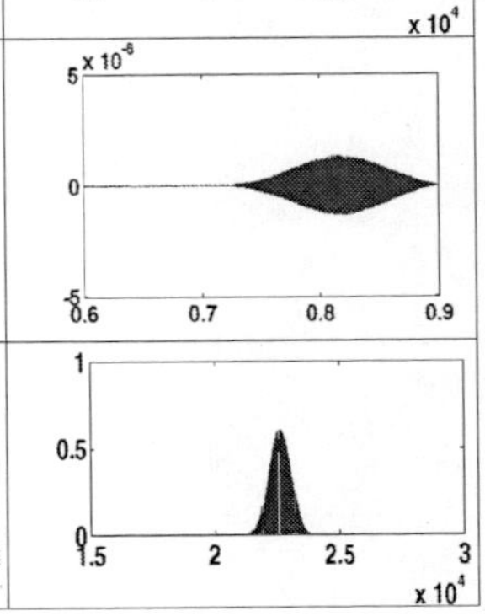

CONCLUSION

A broadband PE model based on split-step Pade approximation has been implemented. The effect of sea surface on acoustic pulse is clearly shown for the cases considered. The distortion in the pulse due to one-way propagation is characterized by low correlation values. The distortion in transmitted signal has to be considered during the implementation of detection techniques like Replica correlation, as it fails to be effective in realistic ocean conditions.

REFERENCES

[1] Balasubramanian, .P., *et al.*, pp. 392–401, Chapter 41, Acoustic waves (Proceedings of NSA-2011).

[2] Collins, M.D., "A split-step Pade solution for the Parabolic Equation Method", JASA, 93, 1736–1742 (1993).

[3] Thorsos, E.I., "Acoustic scattering from a "Pierson-Moskovitz" sea surface", JASA, 88(1), 335–349 (1990).

[4] Miles, D.A., *et al.*, "Forward scattering of pulses from a rough sea surface by Fourier synthesis of parabolic equation solutions", JASA, 114(3), 1266–1280 (2003).

Application of Estimation Techniques for Spatial Arrival Times of Simulated Multipath LFM Signals in a Typical Shallow Water Noisy Environment

R.V. Subba Rao and A. Unnikrishnan

Naval Physical and Oceanographic laboratory, DRDO, Thrikkakara, Kochi, Kerala

ABSTRACT: *In ocean, a transmitted acoustic signal can arrive at a receiver after travelling through multiple paths. The extraction of arrival times and amplitudes of these multipath components in the received composite signal is essential in many applications like ocean characterization through environmental inversion, underwater communication systems, and data telemetry. This work is about the comparison of two established methods for estimation of arrival times in a simulated experiment for a typical noisy environment in Arabian Sea shallow waters. The first method is the simple matched filter (MF) which gives optimum estimates of arrival time and amplitude if there is a single path but which introduces significant errors for a signal with overlapped components due to multipath. The second procedure is the maximum likelihood estimation (MLE) method that requires formulation of a 2N parameter estimation problem for N multipath. In this paper, for low Signal-to-Noise Ratio (SNR) the claimed advantage of ML estimation over MF method is investigated.*

Simulated realizations of received LFM signals were generated with multipath and noise characteristics for two typical sound speed profiles (one upward refracting and the other downward refracting from World Ocean Atlas 2009) in a shallow water site of east Arabian Sea. The ray model BELLHOP and the time series synthesis algorithm VIRTEX were used for this purpose. These models are well suited for simulating the signals in the range of frequencies 6 kHz to 10 kHz. The results were studied for a source placed 5 meters below the surface and receiver at 40 meters depth, in a water column of 70 meters. The white Gaussian noise was added to the received signal and the estimation algorithms were evaluated through Monte Carlo simulations. The range of SNR over which the processing has to be shifted to computationally intensive MLE technique from simple MF is indicated.

INTRODUCTION

Underwater acoustic systems operate in shallow water environments where the sound signals arrive at the receiver after traveling through many distinct paths from the same transmitter. Extraction of the closely spaced but different arrival times of these multipath signal components is helpful in applications like source localization, environmental inversion, and communication systems. The deterministic signal processing techniques of matched filter (MF) and the inverse filter (IF) are used for this purpose. However, their performance deteriorates with reduction in the signal-to-noise ratio (SNR) and with increase in the overlap between the multipath components. The maximum likelihood estimation (MLE) technique was applied to this problem in late 70's to improve the accuracy of arrival time estimates [1]. Since then many variants of MLE and other state space models have been introduced [2].

In this work simulation of pulse propagation in shallow east Arabian Sea waters has been carried out using the ray model BELLHOP and VIRTEX pulse synthesis tool box [3]. Subsequently noise was added to the signals using pseudo random generators. The MF and MLE techniques have been applied on such received signals in noise to estimate the arrival times of multipath components. As the range between the transmitter and the receiver increases, the SNR decreases and at some point the performance of simple MF deteriorates to unacceptable levels, and the more computationally intensive MLE has to be used. The region of SNR for this 'shift over' to be effected in the given environmental conditions has been discussed in the section on results.

THEORY AND BACKGROUND

When a sound signal of the form *s(t)* is transmitted into the ocean channel, the signal received by a distant receiver can be approximated by a linear superposition of multiple replicas that have travelled through different paths. If the transmitter and receiver were stationary, and the water column temporally static, then the received multipath signals would differ only in their arrival times and amplitudes. For sufficiently large source receiver separations, the multipath signals are discrete and finite as given by the set of eigenrays in a ray model,

$$S(t) = \sum_{i=1}^{N} A_i \, s(t - \tau_i) \qquad \dots (1)$$

Where A_i and τ_i are the amplitude and arrival time for the ith multipath respectively and N is the total number arrivals through multipath. The transmitted signal *s(t)* is normalized i.e. $\int_0^T s^2(t)\,dt = 1$, and T is the maximum possible duration of the elongated received signal.

The signals are received against the background of ambient noise in ocean. The ambient noise has log frequency spectrum that varies inversely with frequency. In simulations the noise *n(t)* is assumed to be white Gaussian with spectral density $N_0/2$ dB/Hz, which is reasonable over the limited signal band width. The received signal *r(t)* is written as

$$r(t) = \sum_{i=1}^{N} A_i\, s(t-\tau_i) + n(t) \qquad \dots (2)$$

In MLE technique the estimation of N sets of amplitudes A_i and arrival times τ_i are obtained by formulating a *2N* parameter estimation problem. The maximum likelihood estimate of the parameter vector $[A_i, \tau_i]$ is obtained by maximizing the joint probability density function of a set of observed random variables with respect to $[A_i, \tau_i]$. The likelihood ratio for *r(t)* is formulated as,[1]

$$\Lambda[r(t), \{A_i, \tau_i\}] = \frac{e^{\left\{\frac{2}{N_0}\int_0^T r(t) \sum_{i=1}^{N} A_i\, s(t-\tau_i)\, dt\right\}}}{e^{\left\{\frac{1}{N_0}\int_0^T \left[\sum_{i=1}^{N} A_i\, s(t-\tau_i)\right]^2 dt\right\}}} \qquad \dots (3)$$

This equation is maximized in a two step process. Initially the following equation (4), is obtained through manipulation of eq. (3). In the first step, the vector $[\tau_i]$ is numerically searched over a range of values $[\tau_i]_{min}$ to $[\tau_i]_{max}$, that maximizes eq. (4),

$$\underset{\{\tau_i\}}{\mathrm{Max}}\left[\mathbf{\Phi}^{T}\Lambda^{-1}\mathbf{\Phi}\right] \qquad \dots (4)$$

where $\mathbf{\Phi}^T = \left(\int_0^T r(t)\, s(t-\tau_i)\, dt, \dots, \int_0^T r(t)\, s(t-\tau_N)\, dt\right)$ and Λ is an NxN matrix given by $\lambda_{ij} = \int_0^T s(t-\tau_i)\, s(t-\tau_j)$

In the second step, the values of arrival times vector $[\tau_i]$ so obtained are substituted in the following equation (5) (also obtained through manipulation of eq (3)), to get values of $[A_i]$

$$\mathbf{A} = \Lambda^{-1}\mathbf{\Phi} \qquad \dots (5)$$

The other deterministic method of MF is a straightforward implementation of cross correlation of *r(t)* with *s(t)* in time domain.

SIGNAL AND ENVIRONMENT

The linear frequency modulated (LFM) pulse is an appropriate signal for the investigation of the multipath separation algorithms. The propagation characteristics of LFM pulses are immune to minor perturbations in source receiver positions. So using the data obtained with these pulses, generalized conclusions can be drawn about the influence of environment vis a vis the source-receiver configuration, e.g. a source placed at mid thermocline-depth. The simulations in this work were carried out using a single type of LFM pulse, with the frequency swept from 6 kHz to 10 kHz in 256 milliseconds. The width of the pulse is chosen as a compromise between the short pulse requirement for accurate arrival-time determination and the large duration required for long distance propagation. Many of the underwater communication systems operate around this band of frequency. In Figure 1(a), sample transmitted pulse is shown and its frequency spectrum in Figure 1(b). In Figure 1(c) the received signal through multipath channel as simulated by VIRTEX is given and in Figure 1(d) the same signal with noise added for 0dB SNR is shown.

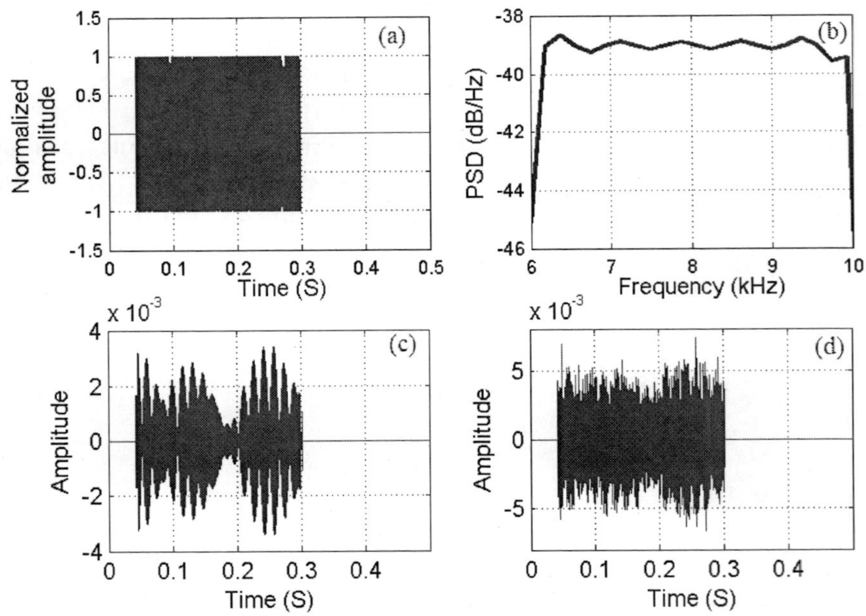

Fig. 1: (a) Transmitted LFM Pulse (b) Power Spectral Density of the Transmitted Pulse (c) Synthesized Received Pulse at 0.9 km and (d) Received Pulse with 0 dB SNR

A location in east Arabian Sea with a water column depth of 70 m and with a relatively flat bottom within a radius of 5 km has been selected from public domain ETOPO [4] data base and the monthly averaged sound speed profiles for January and May have been taken from the World Ocean Atlas 2009,[5] shown in Figure 2. The sound source is positioned at a depth of 5m below the sea surface and the receiver is located at a depth of 40 m at a range of 900 meters. The source has a vertical beam width of $\pm 10^0$ and the receivers are omnidirectional.

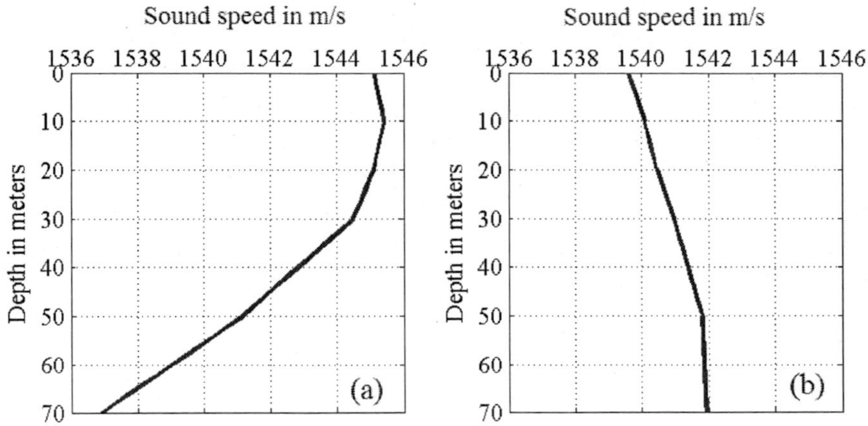

Fig. 2: Sound Speed Profiles from WOA2009 for the Months of (a) May (summer season), (b) January (winter season)

RESULTS AND CONCLUSIONS

Initially the received signal without noise has been processed using MF and MLE and the results are shown for comparison in Table 1. There are four multipath components in the signal for both May and January profiles. The first two arrivals are closely spaced in May (~0.26 ms) than those in January (~0.36 ms).

Table 1: The Arrival Times are Measured with Respect to First Arrival. MF and MLE are Applied on Signal Data Without Noise

Sound Speed Profile	*Method*	*1st Arrival (ms)*	*2nd Arrival (ms)*	*3rd Arrival (ms)*	*4th Arrival (ms)*
May	Ray arrival	0	0.2595	3.8759	4.4784
	MF	0	0.3481	3.9348	4.5309
	MLE	0	0.3125	3.9583	4.3750
January	Ray arrival	0	0.3629	2.5327	3.3067
	MF	0	0.3644	2.5518	3.3316
	MLE	0	0.4167	2.5000	3.3333

The performance of MF depends not only on the signal characteristics but also on the processing steps like envelop extraction and peak finding. Envelop extraction of correlated output was done using Hilbert transform and the peak finding was performed through detection of zero crossings of the derivative of envelop and amplitude thresholding of the envelop.

The performance of MF and MLE (Figure 3), have been generated with five Monte Carlo realizations. The mean square error is calculated as the mean of squared differences between each of the four arrivals at a particular SNR and the four arrivals calculated without noise.

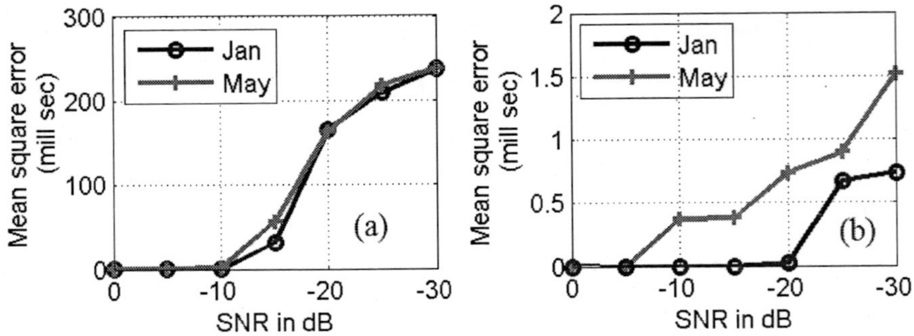

Fig. 3: Performance of (a) MF Processing (b) MLE Technique for Various SNR. (Note that Y axis scalings in (a) and (b) are different)

In Figure 3(a), for SNR values below -10dB there is a total deterioration of performance using MF. From Figure 3(b), which gives the performance of MLE, it can be concluded that useful results can be obtained even to a low SNR value of -30 dB, where the mean square error is around 1ms only. The sound speed profile of January has resulted in larger separation between the arrivals for the particular source receiver configuration and so both the MF and MLE performed better for January signal data compared to that of the May at same SNR.

ACKNOWLEDGEMENT

The authors thank the Director, NPOL for the computational facilities provided to carry out this work and granting the permission and encouragement to publish the results.

REFERENCES

[1] Ehrenberg *et al.*, "Signal processing techniques for resolving individual pulses in a multipath signal", J. Acoust. Soc. Amer, vol. 63, no. 5, pp. 1861–1865, June 1978.

[2] Zoi-Heleni Michalopoulou *et al.*, "Gibbs sampling for time-delay-and amplitude estimation in underwater acoustics", J. Acoust. Soc. Am. 117 (2), pp. 799–808, February 2005.

[3] http:// www.HLSresearch.com/oalib - as on 25 Oct. 2012.

[4] ETOPO1 can be found at the following address as on 25 Oct 2012 http://www.ngdc.noaa.gov/mgg/global/global.

[5] World Ocean Atlas 2009 in MATLAB format at the following address as on 25 Oct. 2012 http://909ers.apl.washington.edu/~dushaw/WOA09.

Measured and Modeled Acoustic Transmission Loss at a Deep Water Site off the West Coast of India

M. Padmanabham and P.V. Nair

Naval Physical and Oceanographic Laboratory, Kochi

E-mail: tsonpol@vsnl.com

ABSTRACT: *Acoustic data was collected at a deepwater site off west coast of India to measure transmission loss (TL). The water depth was around 2000 m. The field measured TLs were compared with TL obtained using geometric spreading law with attenuation and also compared with ray theory (BELLHOP) based propagation model predictions. Details of the BELLHOP model were also presented. Relevant environmental data were also collected during the experiment. Sound speed profiles of the water column indicate the presence of shallow sonic layer (~25 m).*

*High TL was noticed for receiver depths 45 and 75 m, since these were placed below the sonic layer whereas the source was within the sonic layer. However minimum TL was noticed at two ranges (15 and 28 km) due to low grazing bottom bounces. In addition two low TL zones were also noticed: one near to source region and other one at 23 km range. All these phenomena are well reflected in BELLHOP model predictions. The measured TL matched with the geometric spreading (17*log (range)) law with attenuation for the receiver depth 15 m only which is due to the receiver in the sonic layer. Whereas for other receiver depths 45 and 75m, measured values were much higher than the geometric spreading law.*

Keywords: Transmission Loss, Deep Water, Sound Speed Profile, Bottom Bounce.

INTRODUCTION

In sonar equations, transmission loss (*TL)* is an important parameter in sonar design and estimation of sonar ranging distance. Therefore, the performance of a sonar system then depends on the TL. An acoustic signal traveling through the ocean becomes distorted due to multipath effects and weakened due to various loss mechanisms. The TL in an underwater acoustic channel depends on the ocean environment and its boundaries. As the ocean environment is temporal/spatially variable, the best way to understand the characteristics of transmission loss is to measure it in sea trials. It is possible to make a theoretical estimation of TL, but clearly this cannot replace the at-sea experiment. A database of measurement results always forms the basis for sonar range prediction.

Therefore, we have conducted an experiment to collect acoustic data in deep waters of southwest coast of India to measure TL at different ranges and three frequencies. The field measured TLs were compared with geometric spreading law with attenuation and ray theory (BELLHOP) based propagation model predictions.

THE EXPERIMENT AND DATA ANALYSIS

Two ships were engaged in the experimental setup for TL measurements: One to transmit the signals from a fixed location and one to receive the signals at different ranges. The detailed schematic experimental setup is shown in Figure 1.

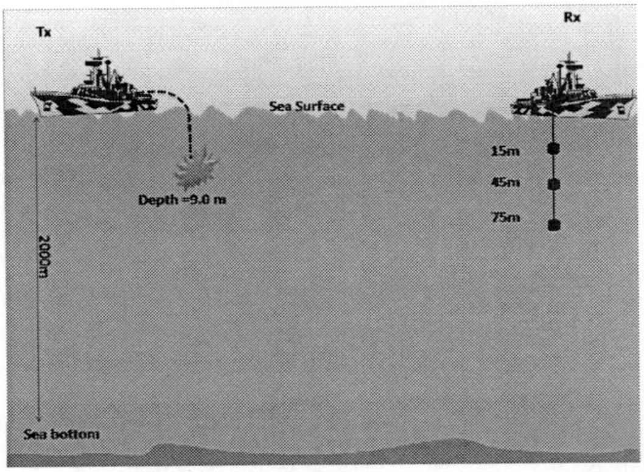

Fig. 1: Schematic Experimental Setup for TL Measurements at Sea

This experiment was conducted at a location, where the water depth was around 2000 m for a maximum range of 30 km. At the transmitting platform CW pulses of frequencies 1000 Hz and 1600 Hz were transmitted from a depth of 10 m. In the receiver platform, the data recorder and hydrophones were tested prior to the experiment. A three element hydrophone array was prepared such that the hydrophones were at depths of 15, 45, and 75 m. These hydrophones were lowered from the quarter deck of the ship. The data was recorded using digital recorder with a sampling frequency of 24 kHz. The pulses (measured in volts) recorded by the hydrophone elements, were separated at each range. Received signals were modified to incorporate hydrophone sensitivity and amplifier gain corrections and to convert to pressure unit of micro Pascal. Later, they were passed through a band-pass filter of width 50 Hz to improve the signal-to-noise ratio to estimate received level. The TL values are calculated by subtracting source level from the received level.

Sound Speed Profile (CTD): Along with acoustic measurements, CTD (Conductivity, Temperature, Depth) profile was also collected to determine sound speed profile at the location. The calculated sound speed profiles using CTD data is plotted in Figure 2. (*Left*

panel full profile—Right panel zoomed profile). The figure clearly indicates the presence of shallow sonic layer (~25 m).

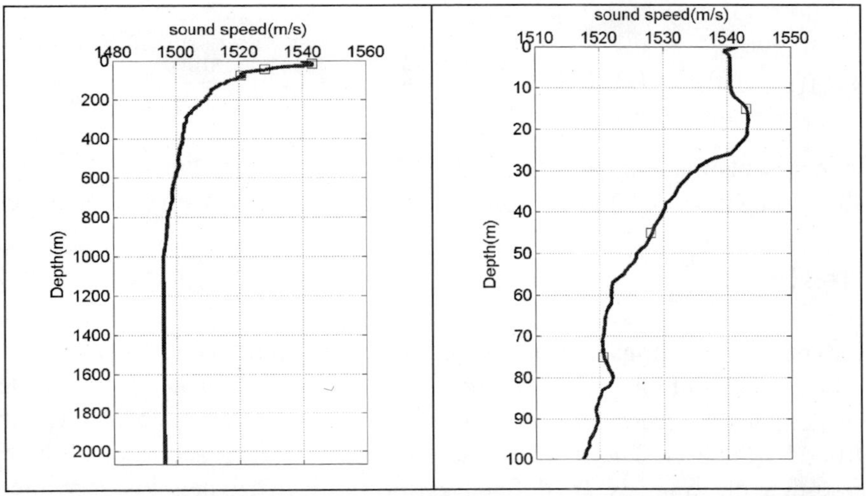

Fig. 2: Sound Speed Profile at the Experimental Site. Left Panel Full Profile and Right Panel Zoomed Profile. The Square Boxes Indicate Position of Hydrophones in the Water Column

BELLHOP ACOUSTIC PROPAGATION MODEL

Ray theory methods are still used extensively in the operational environment where computational speed is a critical factor and environmental uncertainty poses much more severe constraints on the attainable accuracy. Furthermore, much of the insight derived from studying ray theory is important in interpreting the results of other models. Therefore an Ocean acoustic propagation model based on ray theory, BELLHOP (http://oalib.hlsresearch.com/modes/acoustictoolbox/at.zip) was utilized to compute the TL. BELLHOP is designed in order to perform two-dimensional acoustic ray tracing for a given sound speed profile c (z) in ocean waveguides with attenuation or variable absorbing boundaries. The calculation of acoustic pressure is based on the theory of Gaussian beams. Ray tracing requires the solution of the ray equations to determine the ray coordinates. Amplitude and acoustic pressure requires the solution of the dynamic ray equations, which are described in [3]. For a system with cylindrical symmetry, the ray equations can be written as [5],

$$\frac{dr}{ds} = c\xi(s), \quad \frac{d\xi}{ds} = \frac{1}{c^2}\frac{dr}{ds}$$

$$\frac{dz}{ds} = c\zeta(s), \quad \frac{d\zeta}{ds} = \frac{1}{c^2}\frac{dc}{dz}$$

Here, r(s), z(s) represent the ray coordinates in cylindrical coordinates and s is arc length of the ray, c is sound speed, and the pair [$\xi(s)$, $\zeta(s)$] represents tangent along the ray. Initial conditions for, r(s), z(s), $\xi(s)$, $\zeta(s)$ are,

$$r(0) = r(s),\ z(0), = z(s),\ \xi(0) = \frac{\cos(\theta\ s)}{c(s)}\ ,\ \text{and}\ \zeta(0) = \frac{\sin(\theta\ s)}{c(s)}$$

where θs represents the launching angle, (r(s), z(s)) source position and c(s) is the sound speed at the source position. The coordinates are sufficient to obtain the ray travel time:

$$\tau(s) = \tau(0) + \int_0^s \frac{ds}{c(s)}$$

Ray and dynamic equations are integrated in BELLHOP using a two-step polygon method. For a given output option the model writes ray coordinates for every launching angle, or calculates progressively travel time data, acoustic pressure or intensity on the grid specified by the user. The standard measure in underwater acoustics of the change in signal strength with range is *TL* defined as the ratio in decibels between the acoustic intensity $I_{(r,\,z)}$ at a field point and the intensity I_0 at 1 m distance from the source, i.e.

$$TL = -10\log\left(\frac{I(r,z)}{I_0}\right) = -20\log\left(\frac{|P(r,z)|}{|P_0|}\right)\ (dBre\ \ 1m)$$

COMPARISON OF TL WITH MODEL PREDICTIONS

Comparisons between the measured TL and the ray theory (BELLHOP) based propagation model predictions for three receiver depths 15,45and 75m for frequencies 1000 and 1600Hz are shown in Figure 4 respectively. In this figure continuous lines with markers are represents measured values and dotted are BELLHOP model predictions. In addition to ray theory based propagation model predictions, spherical spreading law with attenuation was also compared with measured TL. From the results it is observed that the TL increases with range for all frequencies and receiver depths. The measured TL data reasonable agreement with the model predictions based on ray theory. High TL was noticed for 1600 Hz compared to 1000 Hz which may be due to attenuation both in model and measurements.

High TL was noticed for receiver depths 45 and 75 m, since these were placed below the sonic layer whereas the source was within the sonic layer. However minimum TL was noticed at two ranges (15 and 28 km) due to low grazing bottom bounces. In addition to bottom bounces two shadow zones were also noticed: one near to source region and other one at 23 km range (Figure 3). All these phenomena are well reflected in BELLHOP model predictions. The measured TL matched with the geometric spreading (*17*log* *(range))* law with attenuation for the receiver depth 15m only which is due to the receiver in the sonic layer. Whereas for other receiver depths 45 and 75m, measured values were much higher than the geometric spreading law.

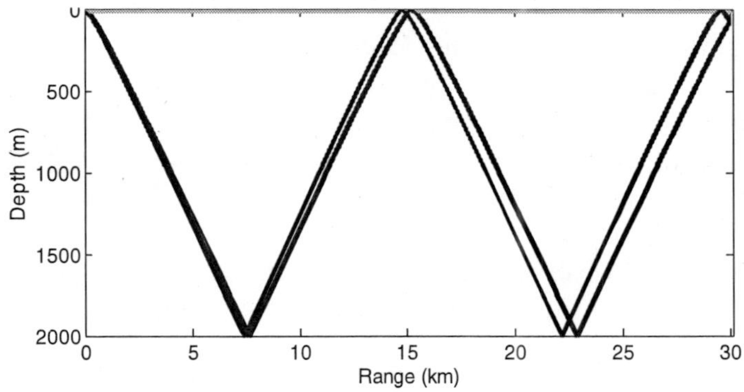

Fig. 3: Eigen Rays for Receiver Depth 75 m and Source Depth 9 m

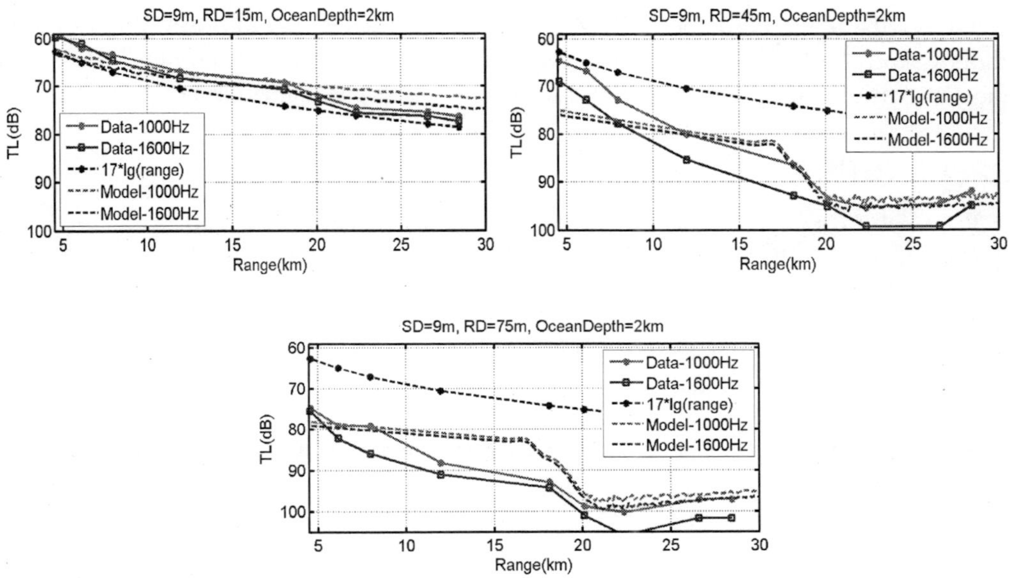

Fig. 4: Measured and Modeled Data Comparison of TL for Three Receiver Depths (15, 45, and 75 m) for Two Frequencies Respectively

SUMMARY AND CONCLUSIONS

Acoustic data was collected at a deepwater site off west coast of India to measure transmission loss (TL). The water depth was around 2000m. The field measured TLs were compared with TL obtained using geometric spreading law with attenuation and also compared with ray theory (BELLHOP) based propagation model predictions. Details of the BELLHOP model were also presented. Relevant environmental data were also collected during the experiment. Sound speed profiles of the water column indicate the presence of shallow sonic layer (~25 m).

High TL was noticed for receiver depths 45 and 75 m, since these were placed below the sonic layer whereas the source was within the sonic layer. However minimum TL was noticed at two ranges (15 and 28 km) due to low grazing bottom bounces. In addition two low TL zones were also noticed: one near to source region and other one at 23 km range. All these phenomena are well reflected in BELLHOP model predictions. The measured TL matched with the geometric spreading *(17*log (range))* law with attenuation for the receiver depth 15 m only which is due to the receiver in the sonic layer. Whereas for other receiver depths 45 and 75 m, measured values were much higher than the geometric spreading law.

ACKNOWLEDGEMENTS

The authors thank the Director, NPOL, for the encouragement to carry out this work. The authors acknowledge the support and guidance from Group Head, Ocean Science. The authors also acknowledge the support provided by the ship personnel during the experiment.

REFERENCES

[1] R.J. Urick, *Principles of Underwater Sound*, 3rd ed., McGraw-Hill: New York.

[2] Qihu Li (2011) Digital Sonar Design in Underwater Acoustics: Principles and applications, Zhejiang university press, China.

[3] M.B. Porter and H.P. Bucker, "Gaussian beam tracing for computing ocean acoustic fields," *J. Acoust. Soc. Am.* 82, 1349–1359 (1987).

[4] M.B. Porter, Ocean Acoustics Library, BELLHOP http://oalib.hlsresearch.com/, viewed April 10, 2012.

[5] F.B. Jensen, W.A. Kuperman, M.B. Porter and H. Schmidt, *Computational Ocean Acoustics* (AIP, New York, 1994).

Acoustic Cloaks for Plates and Cylinders

Beena Mary John and D.D. Ebenezer[1]

Rajagiri School of Engineering and Technology, Kakkanad, Kochi

[1]Naval Physical and Oceanographic Laboratory, Thrikkakara, Kochi

ABSTRACT: *A method is presented to determine the properties of thin, lossy, homogeneous coatings that can acoustically cloak plates and cylinders. Flat panels and cylinders are often used as basic building blocks in mathematical models of submarines and are, therefore, of interest. The frequency-dependent sound speed in the coating required to reduce the broadband backscattered pressure to zero is determined by using analytical models. Numerical results indicate that materials with the required sound speed can be developed. The sensitivity of the reflection coefficient to the complex sound speed is also presented to illustrate the effect of variations in sound speed that will occur during production.*

Keywords: Echo, Backscatter, Multilayer, Internal Loss.

INTRODUCTION

Rendering a submarine invisible to interrogating acoustic waves would constitute a gigantic leap in stealth technology. An ideal acoustic cloak conceals a submarine by causing the total pressure field to be the same as that due to the incident field alone; without echoes or shadows [1]. Cai [2] presents multilayer elastic cloaks for cylinders that, unlike the cloak in Ref. 1, do not have any material singularity, but perform very well in the low frequency region where there are no resonances. Cheng et al [3] present a cloak that is effective against waves with different wave-front shapes.

In this paper, a method is presented to design broadband acoustic cloaks for underwater plates and cylinders such that the reflected and backscattered pressures, respectively, are zero. Flat panels and cylinders are often used as basic building blocks in mathematical models of submarines and are, therefore, of interest. The method is used to determine the frequency-dependent properties of a homogeneous cloak for a flat plate - and it is likely that such materials can be developed. The effect of a small difference between the actual and ideal material properties on the effectiveness of the cloak is also studied. A model of a cylinder coated with a thin lossy layer is used to show that coatings without losses can amplify echoes whereas designed ones can reduce the backscattered pressure to zero.

ACOUSTIC CLOAKS FOR PLATES

Consider an infinite flat metal plate with a lossy homogeneous coating immersed in infinite fluid as shown in Figure 1. The objective is to determine the speed of sound in the coating that would make the reflected pressure zero when a broadband wave is normally incident. The thickness and density of the metal plate and the speed of sound in it, and the thickness and density of the coating are known. Losses in the coating are modelled using complex Lame's coefficients and the speed of sound is therefore complex.

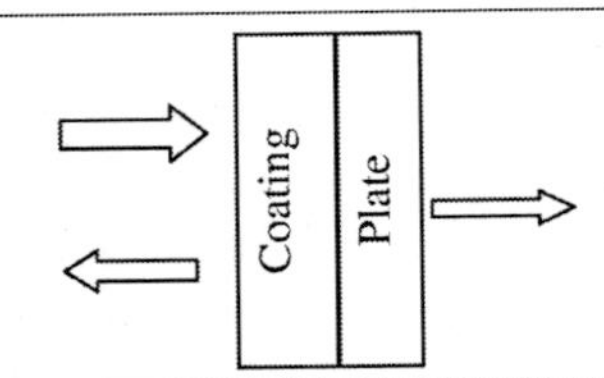

Fig.1. An infinite coated plate; and incident, reflected, and transmitted plane acoustic waves.

The pressures due to the incident, reflected, and transmitted waves are expressed as $P_i = A_1 e^{j(\omega t - k_1 x)}$, $P_r = B_1 e^{j(\omega t + k_1 x)}$, and $P_t = A_4 e^{j(\omega t - k_1 x)}$ respectively, where $k_1 = \omega / c_1$ is the acoustic wave number in water, ω is the angular frequency, c_1 is the speed of sound in water, and t denotes time. The normal stresses in the coating and the metal plate are expressed as $T_2 = A_2 e^{j(\omega t - k_2 x)} + B_2 e^{j(\omega t + k_2 x)}$ and $T_3 = A_3 e^{j(\omega t - k_3 x)} + B_3 e^{j(\omega t + k_3 x)}$, respectively. The continuity conditions for stress and particle velocity at the first, second, and third interfaces are used to determine the coefficients A_2, A_3, B_2, B_3, and A_4 when A_1 is known. Plates with several coatings are analyzed using a transfer matrix approach [4].

The reflection coefficient $R = B_1 / A_1$ is a function of the density and the sound speed in each layer and the thicknesses of the metal and coating layers. It is assumed that all these properties except the speed of sound in the coating, c_2, are known. The complex value of c_2 at which R becomes zero is determined. The method used to find c_2 is presented in a later section.

ACOUSTIC CLOAKS FOR CYLINDERS

Consider the solid infinite circular cylinder shown in Figure 2. It has a thin coating. A plane acoustic wave travelling along the direction perpendicular to the axis of the cylinder is incident on it. The backscattered acoustic wave is of interest and is determined using a model of a multilayer cylinder [5]. The effect of the material properties of the coating on the backscattered pressure is studied.

The analysis can be extended to study coated thin cylindrical shells.

The pressure in the incident wave is expressed in cylindrical coordinates using Bessel functions of the first kind [6]. The solutions to the equations of motion of the solid cylinder and the coating are determined using the method of separation of variables [7]. In the azimuthal direction, a Fourier series expansion with orthogonal terms is used. In the radial direction, Bessel functions are used. The scattered acoustic pressure is also expressed in terms of a Fourier-Bessel series. Each orthogonal component has two coefficients in the solid cylinder, four coefficients in the coating, and one coefficient in water.

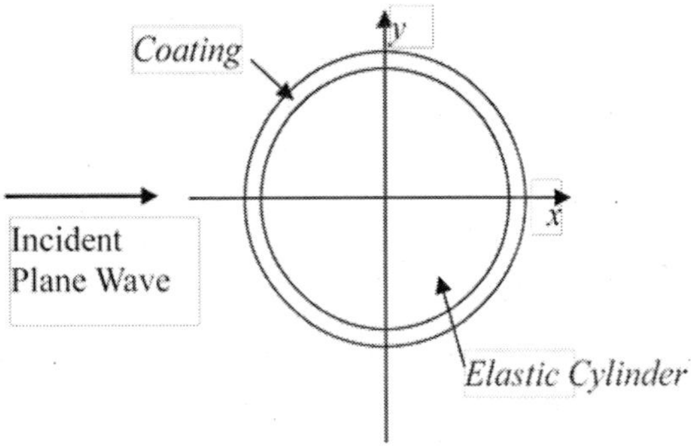

Fig. 2: A Solid Elastic Cylinder with a Thin Acoustic Coating. A Plane Wave Travelling along the x Axis is Incident on it

The seven coefficients in the solutions are determined using seven continuity conditions at the interfaces: at the water-coating interface, the normal stress is equal to the pressure, the shear stress on the surface of the coating is zero, and the radial displacement is continuous; at the coating-cylinder interface, the normal and shear stresses, and radial and tangential displacements are continuous. The coefficients can be used to determine the pressure in any direction. However, here, only the back-scattered pressure is computed because this is of interest when a monostatic sonar interrogates the object. Losses in the coating are modelled using complex Lame's coefficients in this case also.

RESULTS AND DISCUSSION

Numerical results are presented to illustrate cloaks that render plates and cylinders invisible to monostatic sonars. Results are presented for a steel plate of thickness 6 mm, density 7700 kg/m^3, Young's modulus 195 GPa, and Poisson ratio 0.3. The thickness of the coating is 40 mm for the plate and the density of the coating is 1200 kg/m^3.

The reflection coefficient, at 10 kHz, for a coated flat plate is shown in Figure 3 as a function of the real and imaginary parts of the speed of sound in the coating. It is seen that R has several local minima that are not nearly zero. However, it is seen from Figure

3 that R is approximately zero at the global minimum where $c_2 = 710+j100$ m/s. R does not change very rapidly in this neighborhood indicating that small deviations from the ideal c_2 will not result in a large change in R . The change is much more rapid at local minima. After visually finding the approximate global minimum using the 3D plot, the search for the ideal sound speed is continued by using 2D plots. This is easier than a fully automatic search.

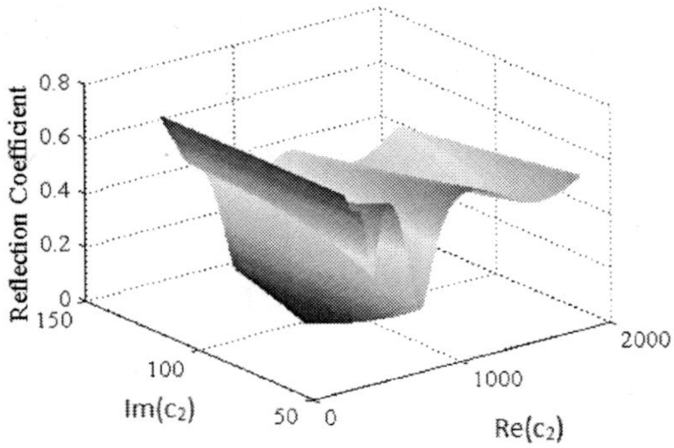

Fig. 3: R as a Function of Real and Imaginary Parts of c at 10 kHz

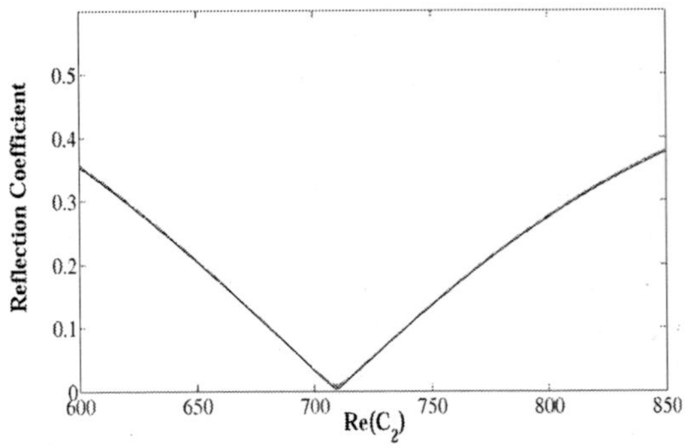

Fig. 4: Reflection Coefficient at 10 kHz. Five Lines at Integer Values of Im(c) Ranging from 97 to 101 are shown

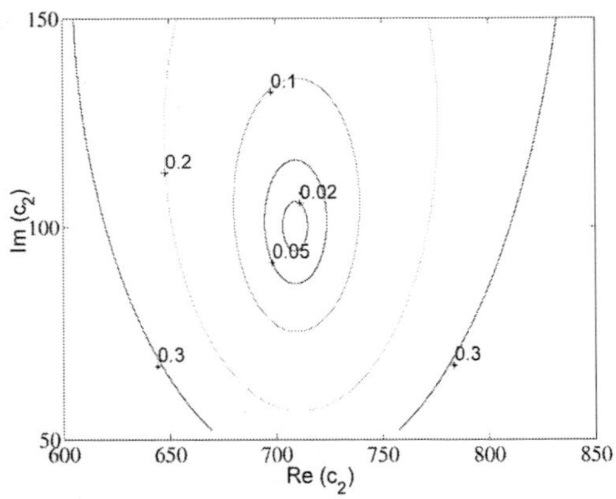

Fig. 5: Contour Plot of the Amplitude of the Reflection Coefficient at 10 kHz. Contour Lines are Shown at $R = 0.02, 0.05, 0.1, 0.2$, and 0.3

Table 1: Ideal Sound Speed in an Acoustic Cloak for a Steel Plate

Frequency (kHz)	Ideal sound speed (m/s)		Young's Modulus (MPa)
	Re (c)	Im (c)	
2	158	0.75	1.75 +0.01j
4	305.5	8.7	6.53+0.37j
6	450	28.7	14.14+1.81j
8	584.5	60	23.7+4.92j
10	710	99	34.65+9.85j

2D graphs are shown in Figure 4. Each line is for a constant integer value of $\text{Im}(c_2)$ ranging from 97 to 101. The five lines are nearly the same and cannot be distinguished. The ideal speed of sound for the cloak is the one corresponding to $R = 0$. If such a sound speed does not exist, the sound speed at which $|R|$ is minimum is the ideal sound speed.

The sensitivity of $|R|$ to the sound speed, in the neighborhood of the ideal sound speed, is illustrated using a contour plot in Figure 5. Contour lines are shown for $|R|$ = 0.02, 0.05, 0.1, 0.2, and 0.3. It is seen that $|R| < 0.1$ (–20 dB) in a region that is approximately elliptical. This indicates that the cloak will be effective even if the actual sound speed is a little different from the ideal sound speed either due to inability to achieve the ideal sound speed or due to variation during production.

The feasibility of using an acoustic cloak to conceal a cylinder at high normalized frequencies is illustrated by presenting results for rigid and Titanium cylinders of radius 1 m.

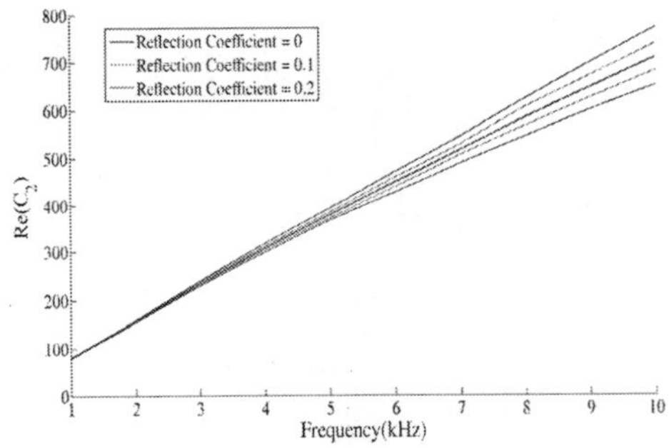

Fig. 6: Re(c_2) Corresponding to R = 0, 0.1 and 0.2. Im(c_2) = Ideal Value at Each Frequency

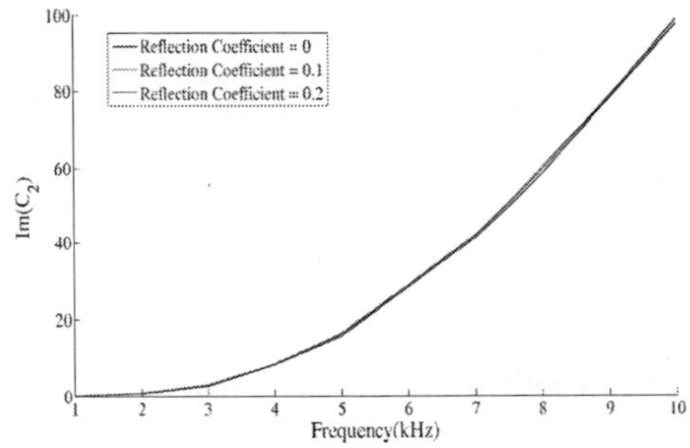

Fig. 7: Im(c_2) Corresponding to R = 0, 0.1 and 0.2. Re(c_2) = Ideal Value at Each Frequency

The thickness of the coating is 50 mm. The Young's modulus, Poisson's ratio, and density of Ti are 110 GPa, 0.3, and 4533 kg/m^3, respectively. The Young's modulus, Poisson's ratio, and density of the rubber coating are 0.11(1–j0.1) GN/m^2, 0.48(1–j0.1), and 910 kg/m^3, respectively.

The pressure backscattered by a rigid cylinder, at a distance of 500 m from the cylinder, is shown in Figure 8 for three cases: without coating, coated with lossless layer, and coated with lossy layer. At large distances from the cylinder, the pressure varies inversely with square root of radial distance. It is seen that the backscattered pressure is nearly independent of frequency if frequency > 1 kHz. Using a coating without losses causes rapid oscillations and higher backscattered pressure in several narrow frequency bands. However, there are no oscillations in the 3.5–8.5 kHz band and the backscattered pressure is very nearly the same as that from the rigid cylinder. When a layer with loss is used and frequency > 500 Hz, the backscattered pressure is less than that when there is no coating. There are no oscillations at frequency > 3 kHz. There is a minimum at about 7.2 kHz and the coating is a cloak at this frequency.

The backscattered pressure is shown in Figure 9 for a Ti cylinder. At high frequencies, the pressure oscillates about the pressure scattered by a rigid cylinder. In contrast to the rigid cylinder case, there are several oscillations even at high frequencies. A thin coating without loss does not significantly affect the backscattered pressure. However, when there is loss in the coating, oscillations are still present but the pressure is considerably reduced. The coating is nearly an ideal cloak at approximately 7 kHz.

It is feasible to compute the frequency-dependent characteristics of the ideal coating by using the method used for plates.

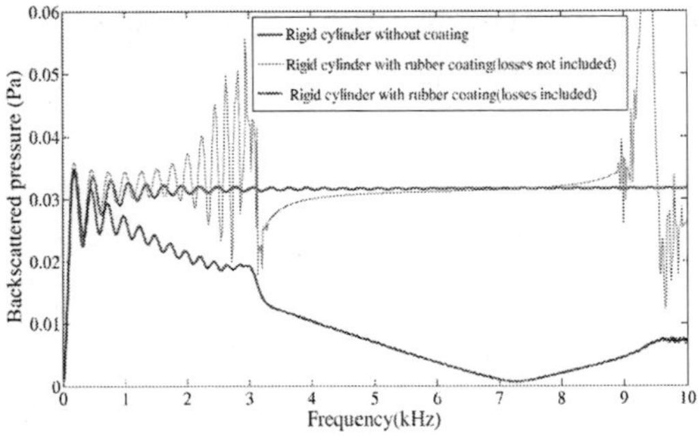

Fig. 8: Magnitude of Pressure Backscattered by a Rigid Cylinder of Radius 1 m

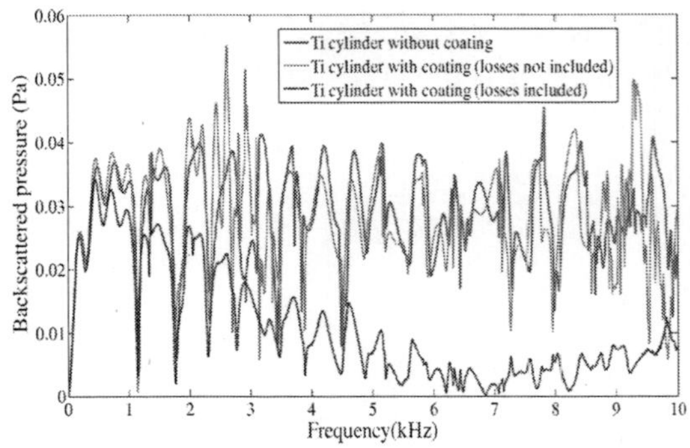

Fig. 9: Magnitude of Pressure Backscattered by a Titanium Cylinder of Radius 1 m

CONCLUSIONS

Analytical models are used to show that it is feasible to acoustically cloak plates and cylinders such that the backscattered pressure is zero. Models of multilayered flat plates and cylinders with lossy coatings, based on exact equations of motion, are used to compute the backscattered pressure at high normalized frequencies. It is shown that the frequency dependent complex sound speed in the coating can be suitably designed to achieve zero reflection. The effect of a small difference between the ideal and actual sound speed on the reflected pressure is also presented to assist in setting production tolerance.

ACKNOWLEDGEMENT

Facilities to do the work provided by Director, NPOL are gratefully acknowledged.

REFERENCES

[1] S.A. Cummer and D. Schurig, "One path to acoustic cloaking," New Journal of Physics 9, pp. 1–8 (2007).

[2] L.-W. Cai, "Optimizing imperfect cloaks to perfection," J. Acoustical Soc. America, 132, 2923– 2931 (2012).

[3] Y. Cheng, F. Yang, J.Y. Xu and X.J. Liu, "A multi layered structured acoustic cloak with homogeneous isotropic materials," Applied Phy. Let. 92, 151913 (2008).

[4] D.D. Ebenezer and Pushpa Abraham, "Effect of multilayer baffles and domes on hydrophone response," J. Acoustical Soc. America, 99, 1883–1893 (1996).

[5] J.S. Sastry and M.L. Munjal, "Response of a multi-layered infinite cylinder to two-dimensional pressure excitation by means of transfer matrices," J. Sound and Vibration, 209, 123–142 (1998).

[6] P.M. Morse and K.U. Ingard, Theoretical Acoustics (McGraw-Hill, New York, 1968).

[7] M.C. Junger and D. Feit, Sound, Structures, and their Interaction, (MIT Press, Cambridge, MA, 1972).

[8] L.E. Kinsler, A.R. Frey, A.B. Coppens and J.V. Sanders, Fundamentals of Acoustics (John Willey & Sons,1972), 4th ed.

Design Curves for Free-Flooded Deep-Submergence Ring Transducers in Annular Fill-Fluids

M.R. Subash Chandra Bose and D.D. Ebenezer

Naval Physical and Oceanographic Laboratory, Thrikkakara, Kochi, Kerala

ABSTRACT: *Free-flooded ring transducers are preferred for high-bandwidth, deep-submergence applications that require both transmission and reception of acoustic signals. A variant of the basic transducer has a radially polarized tube in an annular, acoustically transparent boot. Oil is used to fill the boot and it couples the transducer to the exterior fluid. Design curves are presented for free-flooded radially polarized piezoceramic tube in an oil-filled boot. ATILA—a software package for the analysis of sonar transducers—is used to study the effect of various dimensions of the piezoceramic tube and different types of fill-fluid on the transmission and reception characteristics of the transducer. Based on the study, dimensions are chosen to optimise the transmission and reception characteristics of a transducer.*

INTRODUCTION

A free-flooded cylindrical transducer is usually made up of a radially polarized piezoceramic ring and is encapsulated with rubber or polyurethane. There are two fundamental resonances associated with these transducers: the radial resonance of the shell, and cavity resonance of the enclosed water column. McMahon [1] has carried out detailed theoretical and experimental study on cavity resonance of open, radially polarized ceramic rings. The frequency of cavity resonance decreases with the increase of height, *h*, to radius, *a*, ratio. The radial resonance frequency of the ring is dependent on diameter and its size becomes large as frequency decreases. So, it can typically be used for frequencies above 4 kHz [2]. Problems related to the manufacture of a solid ceramic ring of large diameter for low frequencies can be solved by the use of segmented ring cylinders [3]. Metal straps over the boot [4] or fiber glass winding over the cylinder can be used to pre-stress the segmented cylinders [5].

Renna [6], in his US patent, has reported oil filled, fiber glass wound segmented ring transducer that enhances the bandwidth and efficiency by close coupling the cavity and hoop mode resonances. He also provides the nominal size of the transducer and suggests that the cavity resonance should be below the operating band, a minimum ratio of radial to cavity resonance, and minimum impedance variation to keep the mechanical *Q* low. Low ratios of element wall thickness, *t*, to radius, and height to radius will give a low mechanical *Q*.

The resonance frequency of the free flooded ring can be reduced by replacing alternate piezoceramic segments by more compliant metallic or non-metallic segments like Lucite. Butler [7] studied the segmented ring transducer with inactive segments and derived an equation of motion and equivalent circuit. Harris [8] proposed wedge shaped non-piezoceramic spacing strips and piezoceramic stacks that are cuboids, instead of the wedge shaped ceramics. This concept reduces much of the problem associated in making the segmented transducer by gluing a number of wedge shaped ceramic strips.

A free-flooded ring transducer is a useful directional source even with a size smaller than the acoustic wavelength. It has a toroidal directivity. The transducer can be made more directional by stacking more rings along the axis and forming an array. When a number of cylinders are to be stacked one over the other for large depth operation, decoupling members between the cylinders can be used to isolate them to improve its transmitting power [9]. Sreejith and Ebenezer [10] modeled and measured characteristics of free flooded radially polarized broadband transducers. Transmitting Voltage Response (TVR) and directivity are computed using the FEM software ATILA. The effect of average diameter, length, and wall thickness of the piezoceramic shell and wall thickness of the encapsulant on the TVR is studied by varying one parameter at a time.

DESIGN OBJECTIVE

In the current study, a radially polarized piezoceramic tube in an annular, acoustically transparent boot filled with oil is considered (Figure1). The oil in the boot couples the transducer to the fluid in which it is immersed. The cavity at the centre of the transducer is free-flooded. The objective is to generate design curves for transducers operating in the frequency band of 4–20 kHz with more than two octave band width and transmitting voltage response in excess of 130 dB, capable of operating at deep waters. An FEM package, ATILA, is a used to study the effect of various dimensions of the piezoceramic tube and different types of fill-fluid on the transmission and reception characteristics of the transducer. Based on the study, dimensions are chosen to optimise the transmission and reception characteristics of a transducer.

DESCRIPTION OF THE TRANSDUCERS STUDIED AND THE MESH

The construction details of the transducer studied in this report are shown in Figure 1. It consists of a radially polarized PZT-4 ring supported on PVC bushes inside a rubber boot and filled with oil. This type of construction helps in overcoming the problems related to rubber moulding over the ceramic ring and better coupling of the transducer to the transmitting medium through the fill fluid. The transducer and the excitation are axi-symmetric and the model is therefore axi-symmetric. The 2D axi-symmetric model of the transducer is shown in Figure 2. Eight noded quadrilateral elements are used to model the piezoelectric, elastic, fill fluid, and water surrounding the transducer. Effect of outside diameter (OD), length to radius ratio, wall thickness of PZT ring, and different fill fluids on Transmitting Voltage Response (TVR) and Receiving Sensitivity (RS) are considered in the study.

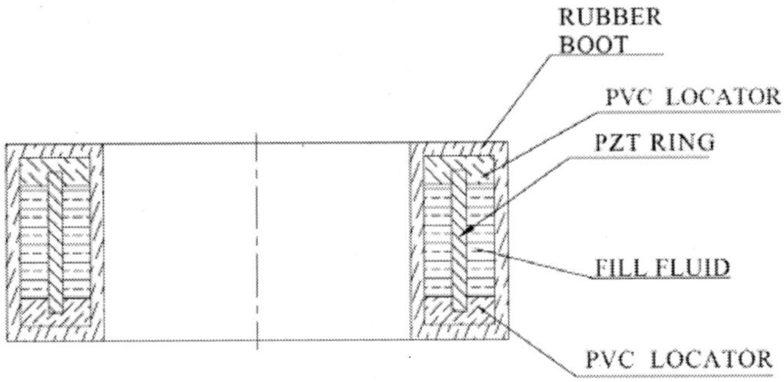

Fig. 1: Schematic of Transducer Cross-Section

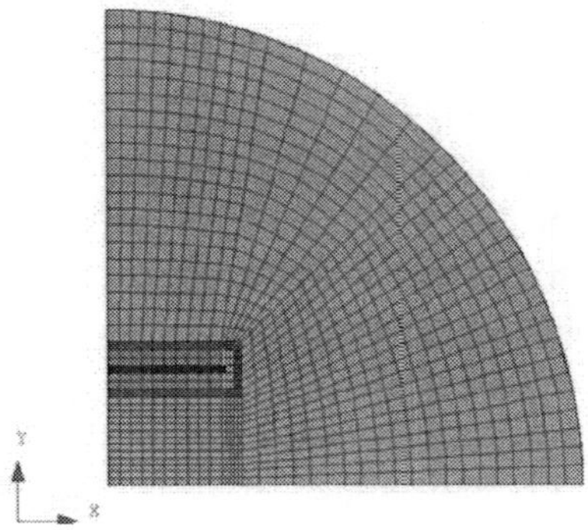

Fig. 2: 2D Axi-Symmetric Model of the Transducer

TRANSDUCER MODELS AND RESULTS

The dimensions of the base model of transducer considered are 150 mm OD, 140 mm ID (i.e. 5 mm wall thickness) and 150 mm length. So, h/a is 2. The effect of housing material on TVR is studied by modeling a bare PZT ring, then the PZT ring inside a steel housing, and then within a rubber boot. Finally the ring is modeled with a fill fluid around the ring, inside the rubber boot. TVR of the four cases are shown in Figure 3. From the graph it is clear that the response of the transducer with fill fluid, compared with other cases, is higher over the band of 3 to 11 kHz with a useful band width of close to two octaves.

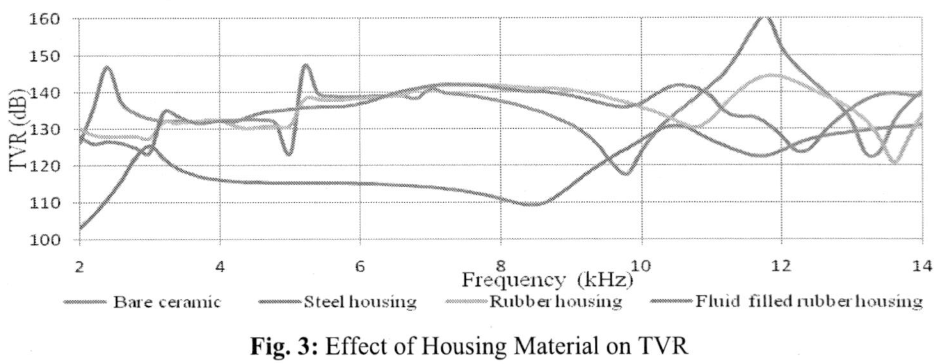

Fig. 3: Effect of Housing Material on TVR

In the second stage of modeling, the effect of *h/a* ratio on the transducers Transmitting and Receiving response characteristics are studied. The *h/a* ratio from 0.66 to 2.66 i.e. 50/75 to 200/75 are modeled and plotted in Figure 4 and 5. As *h/a* increases, the frequency at which first resonance occurs decreases and the TVR increases. For *h/a* equal to one, the band width of the transducer is flat from 4 to 22 kHz.

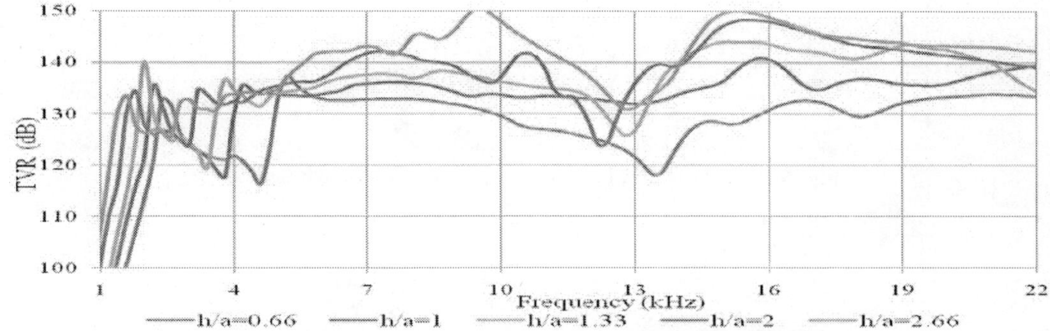

Fig. 4: Effect of *h/a* Ratio on Transmitting Voltage Response

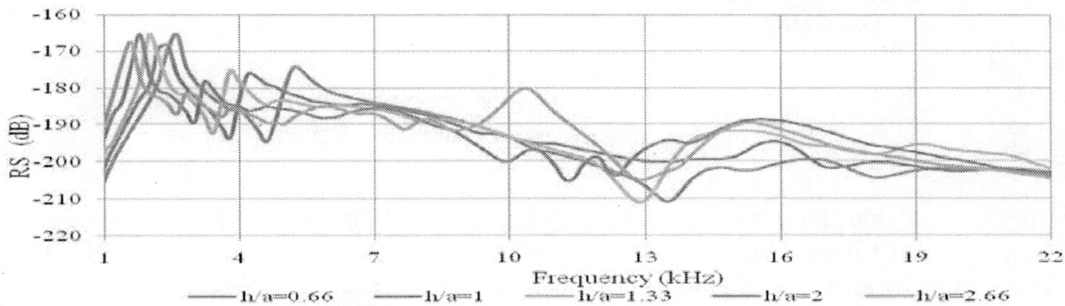

Fig. 5: Effect of *h/a* Ratio on Receiving Sensitivity

The effect of fill fluid on the TVR is studied by considering different fill fluids. The major fill fluids used in fluid filled transducer applications for deep submergence are Silicone oil, Caster oil, Isopar-L and Transformer oil. Water is also considered as a fill fluid for modeling sake. The results on the effects of fill fluid on the transducer TVR is shown in Figure 6. Here the transducer considered is with *h/a* ratio equal to one because it showed a flatter response in Figure 4. The result shows that there is no significant influence of the fill fluid on TVR except in the band of 13 to 16 kHz frequency.

Effect of thickness of the ceramic ring on TVR is studied by using wall thickness of 3, 5, 7, and 10 mm with ceramic ring length and outside diameter of 75 mm. The results are shown in Figure 7. The results show that TVR increases with reduction in wall thickness.

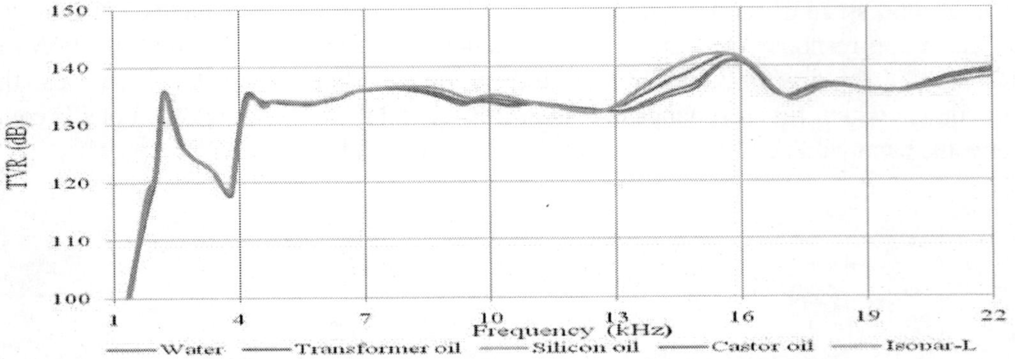

Fig. 6: Effect of Fill Fluid on Transmitting Voltage Response

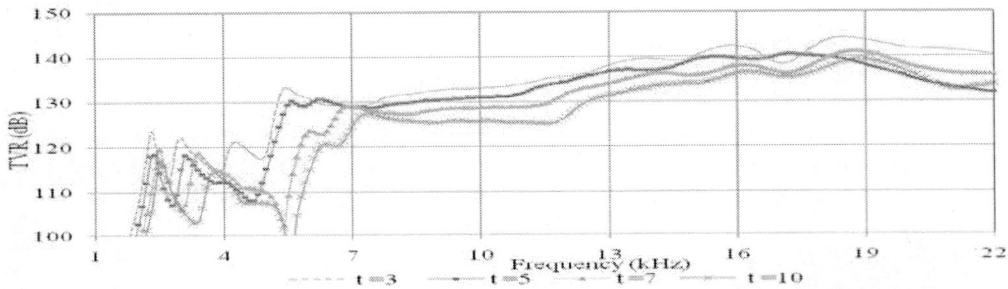

Fig. 7: Effect of Ceramic Thickness on Transmitting Voltage Response

Effect of outside diameter on TVR is studied by varying the outside diameter to 75, 100, 125, and 150 mm with ceramic ring length of 75 mm and wall thickness of 5 mm. The results are shown in Figure 8. It indicates that TVR increases as the outside diameter increases when other parameters are maintained same.

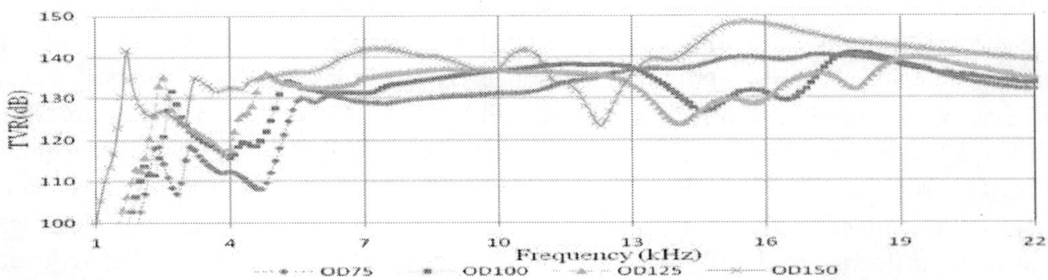

Fig. 8: Effect of Ceramic Outer Diameter on Transmitting Voltage Response

Based on the modeling results an optimized transducer is modeled with OD of 150 mm, 3 mm wall thickness and length of 75 mm that gives good bandwidth for TVR and RS over a band of 3.5 to 22 kHz as shown in Figures 9 and 10.

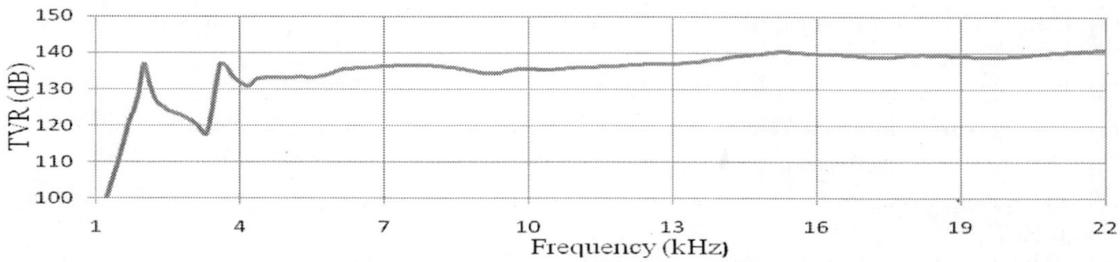

Fig. 9: Optimised Transducer Transmitting Voltage Response

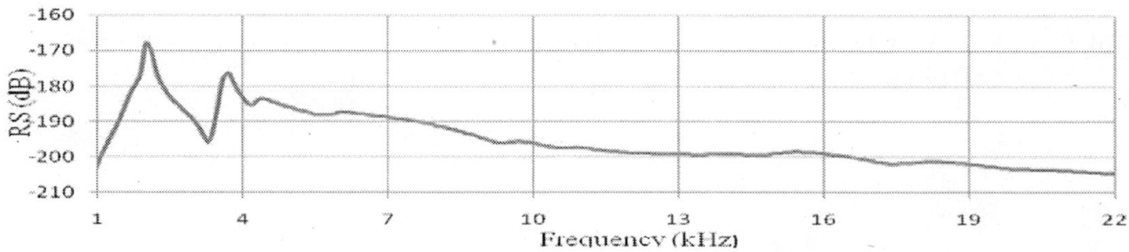

Fig. 10: Optimised Transducer Receiving Sensitivity

CONCLUSIONS

The studies reveal that the free flooded ring transducer made with RPC, housed in a fluid filled rubber boot has better coupling to the transmitting medium through the fill fluid compared to directly rubber moulded or metal-housed transducers. Low ratios of element height to radius and thickness will give a broader bandwidth. The results show that there is no significant influence of the fill fluid on TVR of the transducer studied. The optimized transducer has an *h/a* ratio of one, ceramic thickness of 3 mm and OD of 150 mm. It has high TVR and RS from 3.5 to 22 kHz and can be developed for deep submergence applications. The transducers receiving and transmitting characteristics are also suitable for underwater communication.

ACKNOWLEDGEMENT

Permission from Director, NPOL to publish the paper is gratefully acknowledged.

REFERENCES

[1] G.W. McMahon, Performance of open ferroelectric ceramic cylinders in underwater transducers, J. Acous. Soc. Am., 36, pp. 528–533 (1964).

[2] W.J. Hughes, Transducers, Underwater Acoustics, Encyclopedia of Applied Physics, Vol. 22, pp. 67–84 (1998).

[3] C.E. Green, Mosaic construction for electro-acoustical cylindrical transducers, US Patent No. 3177382, Apr 1965.

[4] W. L. Clearwaters, Electrostrictive transducer, US Patent 3043967, July 1962

[5] D.E. Parker, Reinforced Ceramic Cylinder Transducer, US Patent No. 3230505, Jan 1966.

[6] N. Renna, Jr., Underwater acoustic projector. US Patent 3706967, Dec. 1972

[7] J.L. Butler, Model for a ring transducer , J. Acous. Soc. Am. Vol. 59, No. 2, pp. 480–481, February 1976.

[8] W.T. Harris, Ring shaped transducers, US Patent 3142035, July, 1964.

[9] E. Sernit and B. Fromont, Underwater acoustic transmitter for large submersion, US Patent No. 5784341, 21, Jul. 1998.

[10] S. Sreejith Pillai, D.D. Ebenezer, and Subash Chander, Free-flooded, piezoelectric ceramic, broadband transducers, pp 149–157 SYMPOL (Kochi, 2007).

Optimization of a Class IV Flextensional Transducer

Laly Joseph, P. Prashant Satyanarayan, R. Krishna Kumar, P. Annadurai, K.P.B. Moosad, M.R. Subash Chandra Bose and D.D. Ebenezer

Naval Physical and Oceanographic Laboratory, Thrikkakara, Kochi, Kerala

ABSTRACT: *Class IV Flextensional Transducers (FTs) are extensively used for the generation of high-power, low-frequency underwater sound. The transducer is comprised of an end- capped elliptic cylindrical shell excited by a piezoceramic stack fitted along the major axis. The resonance frequency of a FT depends primarily on the dimensions, elastic properties, and density of the shell and weakly on the properties of the stack. In this paper, the effect of various parts of the transducer is investigated using a finite element software package ATILA with a view to optimizing the transmitting characteristics. It is shown that internal losses in the rubber used for the O-ring has a strong effect on the transmitting characteristics. Other parameters investigated are the size and volume of the piezoceramic and the shape of the inner profile of the shell.*

Keywords: Internal Losses, Rubber O-ring, ATILA.

INTRODUCTION

Flextensional Transducers (FTs) are extensively used for the generation of high intensity, low frequency underwater sound for research purposes as well as for Naval requirements. In contrast to the conventional Tonpilz transducers, that involve translatory motion of metallic masses, FTs make use of Flexural vibrations of shells, excited by the extensional vibrations of a driver stack. Depending on the type of shells used, FTs are classified into different Classes. In Class IV FTs, the shell is in the shape of an elliptical cylinder and the driver is, in most cases, a piezoceramic stack fitted along the major axis of the ellipse. The transducer is made water-worthy by means of end-plates at the top and bottom, attached to the shell using O-rings and tie-rods. A typical scheme is shown in Figure 1 [1]. When the shell vibrates, the O-rings undergo shear and a large amount of energy is dissipated in the O-rings. It is reported in Ref [2] that the temperature rise in the piezoceramic stack in a high power FT corresponds to a very small percentage of the energy dissipated in the FT. Since the typical value of electro-acoustic efficiency for FTs is around 75% [3], a good amount of energy must be getting dissipated in the O-rings. This calls for a careful design of the properties of the

O-ring material. In this paper, it is shown by Finite Element Modeling that the properties of the O-ring significantly affect the performance of the FT.

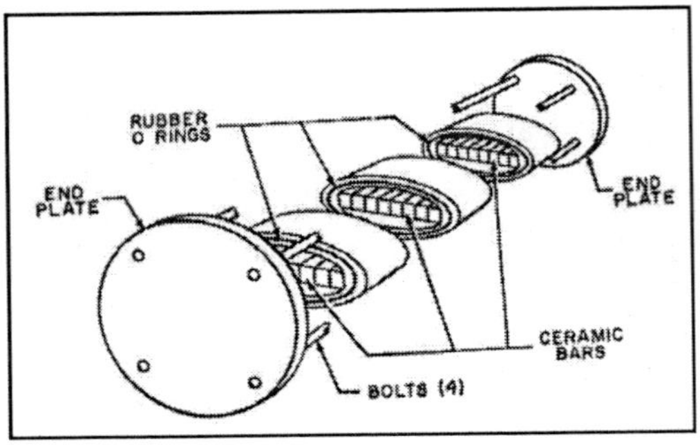

Fig. 1: Use of Rubber O-rings in Class IV FT Assemblies—A Typical Case (from Ref [1])

O-rings affecting the performance of FTs was discussed in one of the early papers on FT development. It is mentioned in Ref [4]: "In order for the transducer to work efficiently the radiation losses from the end-plates must be minimized by decoupling them from the vibrating shell. This has always been a difficult engineering problem as a very flexible seal is required capable of withstanding movements in operation of 1mm, but at the same time it has to withstand high static pressure differentials. The traditional solution has been to use either an O-ring seal here or a rubber boot has been moulded across the shell and end plate while maintaining a small gap between the two. Neither method has been found satisfactory for long term operation". However, O-rings are used to couple the shells to the end-plates in several FTs [1]. Moosad et al. [5] have examined the effect of O-ring rubber properties in the case of Class VII Flextensional Transducer. It was seen that the resonance frequency of the transducer increases with the hardness of the rubber. Also, the Transmitting Voltage Response (TVR) was found to be decrease significantly when the Young's modulus of the rubber changed from 1 MPa to 593 MPa. However, for commonly used values of Young's modulus (1 MPa to 5 MPa) the change in TVR was not much. The rubber properties used in modeling the ARE transducer [6] were the following: Young's modulus = 4 MPa, Poisson ratio = 0.45 and density = 1000 kg/m^3. The effect of losses in the O-ring on the characteristics of the transducer does not appear to have been presented earlier.

In this paper, the effect of various parts of the transducer is investigated using a finite element software package ATILA with a view to optimizing the transmitting characteristics. It is shown that the properties of the rubber used for the O-ring has a strong effect on the transmitting characteristics. Another important parameter affecting the transmitting characteristics is the total volume of piezocermic material inside. Maximizing the volume would some time mean a decrease in the total number of slabs in the stack and therefore will not yield the expected increase in TVR. Therefore the size of the stack is to be optimized for performance. This will have implications on the stack-to-shell interface also and therefore on the inner profile of the shell, which are also discussed.

MODEL DETAILS

3D modeling of a low frequency FT is carried out using the commercial FEM package ATILA. Extrusion is used to create the 3D model. The 3D model is shown in Figure 2. 5 mm thick piezoceramic slabs are used for making the driver stack that is fitted along the major axis of the elliptical shell. The shell is closed by means of steel end-plates that are coupled to the shell through a rubber gasket. (Instead of O-ring, gasket is used, the shape of which is close to that of the O-ring under compression).

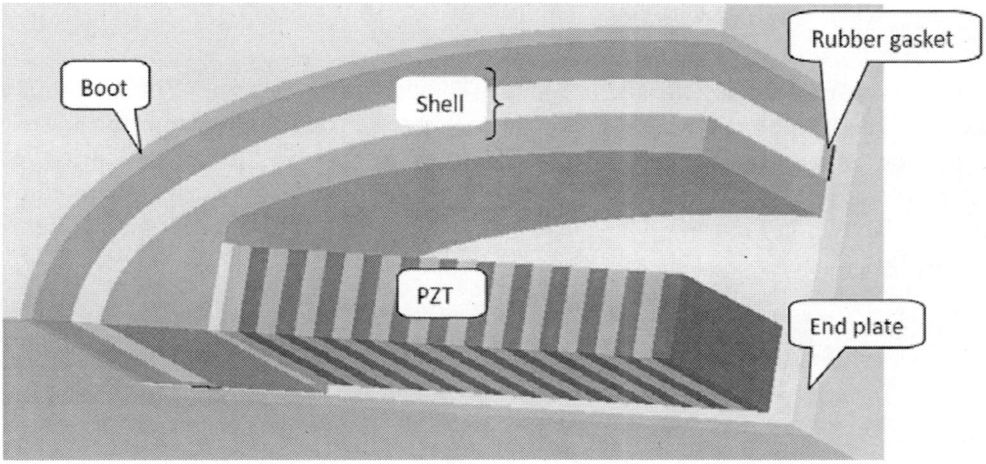

Fig. 2: 3D Model Showing the Class IV Flextensional Transducer

RESULTS AND DISCUSSIONS

TVR plots generated by ATILA are shown in Figure 3. As the Young's modulus of the O-ring is changed there is a change in the frequency at which the TVR is maximum. The peak TVR values are not affected much. The values of material constants used for the modeling are shown in Table 1.

Table 1: Properties of the Rubber Modeled

Type of Rubber	Young's modulus (MPa)	Poisson ratio	Density kg/m^3
1	1	0.45	1000
2	50	0.45	1000
3	1000	0.45	1000

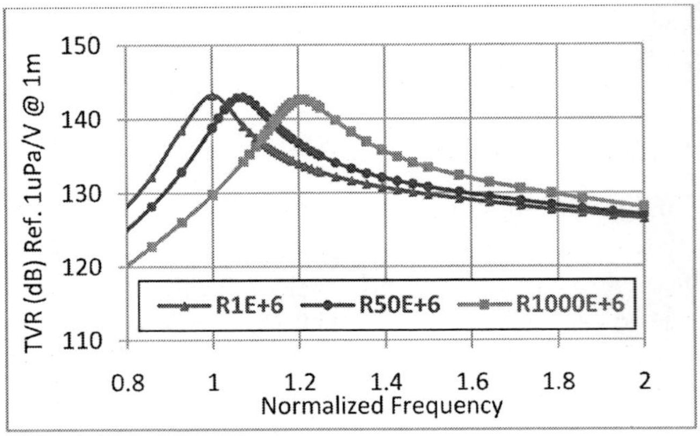

Fig. 3: Effect of the Young's Modulus of the O-ring Rubber on the TVR

All the above rubbers are loss-less rubbers, in the sense that the Young's modulus values are real. Rubber being a visco-elastic material, it is more appropriate to use the complex value of the Young's modulus. When Y = 50 (1+ j) MPa is used instead of 50 MPa, both the peak value of the TVR and the frequency at which it occurs showed changes as shown in Figure 4. This clearly shows that the O-rings for FTs are to be made from low loss rubbers.

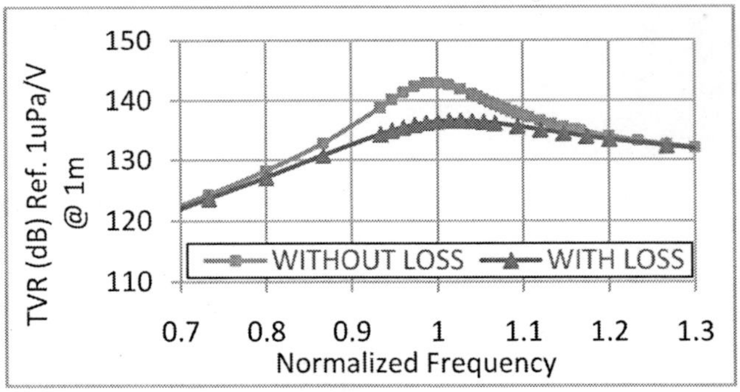

Fig. 4: Effect of Losses in the Rubber Gasket

Table 2: Properties of Two Different Rubber Formulations Used for the Experiments

Rubber formulation	Young's modulus MPa	Density kg/m^3	Poisson ratio	Loss (Tan δ)
1	1.69	1000	0.45	0.63
2	3.44	1000	0.45	0.92

To prove this experimentally, two different formulations were made having the properties shown in Table 2. These values are used in the ATILA models and TVR plots are generated. O-rings are made from these rubbers and FTs are assembled and TVR is experimentally measured. Figure 5 shows the predicted plot with experimental plot. The match is very good and shows that a 2 dB increase can be practically achieved by using an O-ring made from soft, low-loss rubber.

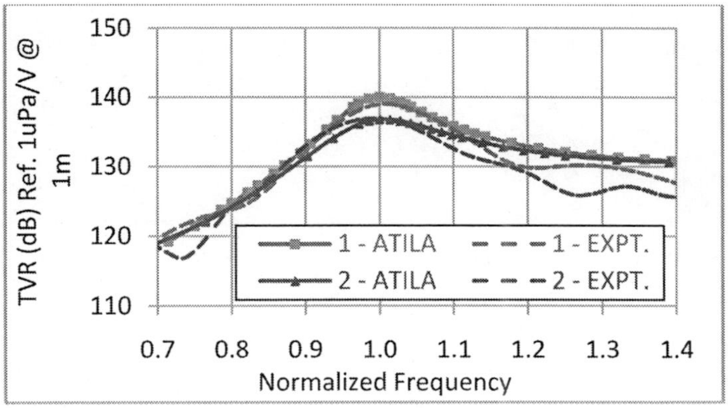

Fig. 5: Predicted and Experimental TVRs of FTs with Different O-rings

The effects of increasing the size of the piezoceramic plates used for the stack are then investigated. Figure 6 shows the TVR plot for various sizes of the piezoceramic plates used for making the stack. 100 × 50 × 5 mm plates are seen to be showing the best performance.

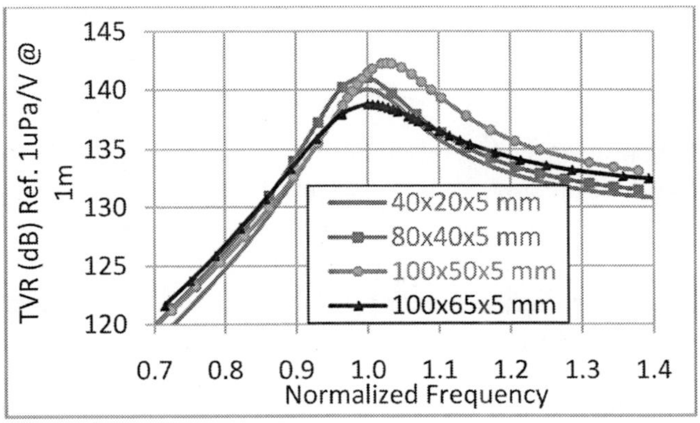

Fig. 6: Effect of the Size of the Piezoceramic Plates on the Resonance Frequency and the TVR

Figure 7 shows two different ways of mating the end of the piezoceramic stack to the curved surface of the shell. In one case, the inner profile of the shell is simple and straight (Case 1) whereas in the other (Case 2), it has a protrusion towards inside. These two cases were modeled and TVR plots were generated, as shown in Figure 8. There is not much of a difference in TVR values, but the resonance frequency is slightly higher for Case 1.

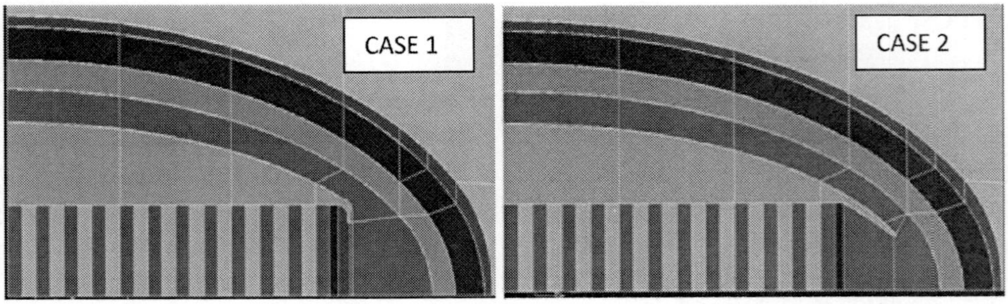

Fig. 7: Two Different Inner Profiles for the Shell Studied by Modeling

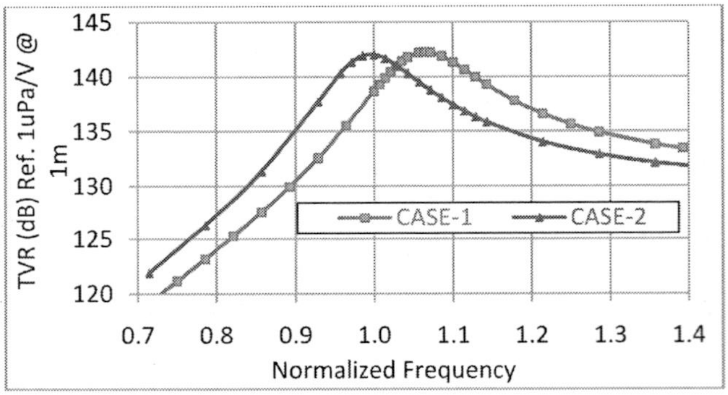

Fig. 8: TVR for the Two Inner Profiles Shown in Figure 7

CONCLUSIONS

Optimization of transmitting characteristics of a Class IV Flextensional transducer was carried out by FE modeling using ATILA. Effect of rubber properties used for O-ring was investigated. It is shown using model and experiment that loss in the rubber lowers the TVR of the FT. The size of the piezoceramic stack is optimized for best transmission characteristics and the inner profile of the shell is also modified accordingly. The optimized transducer is being used.

REFERENCES

[1] G.A. Brigham, "Analysis of the Class IV flextensional transducer by use of wave mechanics," J. Acoust. Soc. Am., Vol.56, No. 1, 1974, pp. 31–39.

[2] R. Rajesh, O.B. Satheesh Kumar, R. Krishna Kumar, Poturaju Prashant S., K.P.B. Moosad and D.D. Ebenezer, "Temperature rise in a high power underwater flextensional transducer", paper presented in NSA2010, Rishikesh, 11–13 Nov 2010.

[3] D.F. Jones and J.F. Lindberg, "Recent transduction developments in Canada and the United States," Proc. Inst. Acoustics Vol. 17 pp. 15–33 (1995).

[4] G.H. Ash and G. Bromfield, "The 350 Hz ARE Flextensional Transducer", Proc. UDT conf. 1988, pp. 109–114.

[5] K.P.B. Moosad, Pushpa Abraham and R.M.R. Vishnubhatla, "Design of a low frequency Class VII Flextensional Transducer", Proc. of 3rd Indian National Conference on Harbour and Ocean Engineering, NIO, Goa, Dec 7–9, 2004, pp. 785–791.

[6] R.J. Brind, "Finite Element Modelling of the ARE low frequency flextensional transducer", Proc. Inst. Acoustics, Vol. 10, Pt 9, 1988, pp. 105–128.

Detection of Onset of Acoustic Cavitation in Underwater Transducers

R. Krishnakumar, M. Ajesh Kumar and Kurian Isac

Naval Physical and Oceanographic Laboratory, Thrikkakara, Kochi, Kerala

ABSTRACT: *At low operating depths, cavitation limits the performance of underwater active acoustic transducers. Cavitation may occur when the local pressure in the water reaches a level below the vapour pressure of the liquid, and this happens during the negative half of pressure cycle. When cavitation occurs, bubbles are generated near the projector and they may implode during the positive half of the cycle. In this paper, three experimental approaches are presented to detect the onset of cavitation. The results obtained using the three approaches are nearly the same and are in agreement with theoretical estimates.*

Keywords: Acoustic Transducers, SONAR, Source Level, Total Harmonic Distortion (THD).

INTRODUCTION

High power electro-acoustic transducers are generally used to insonify the medium, and facilitate detection using active sonar systems. It is desirable to generate maximum amount of acoustic power to achieve maximum detection range. In attempting to achieve maximum power, one peculiar limitation is cavitation, a property of fluid medium into which the acoustic power is radiated. In general, cavitation occurs when the local pressure in the medium reaches a level below the vapour pressure of the liquid.

Cavitation can occur in high power transducers when rarefaction occurs. If the total pressure in the water is very low, tiny bubbles may form in the water and medium starts cavitates. The generated bubble implodes during the positive pressure cycle and results in the generation of shock waves. Cavitation in transducers gives rise to various undesirable effects like erosion of the projector surface, distortion in the output signal, reduction in the output acoustic power, and deviation from linear behavior of the transducer.

Cavitation threshold is the inception point of cavitation and depends upon the depth of operation, frequency and duration of the signal, impurities, and the air content of the medium. In the ocean, hydrostatic pressure is approximately proportional to depth and the cavitation threshold therefore increases with depth. As frequency increases cavitation threshold also increases and this increase in threshold is small up to 10 kHz, but may be up to a factor of about 30 (in power level) as frequency increased from 10 to 100 kHz [1]. An increase in threshold may also be observed if the pulse length is increased, this is in agreement with the nucleation process [1].

The available tools for monitoring acoustic cavitation are - direct observation of cavitation bubble field using high speed photography, analysis of the diffraction pattern obtained from a laser light propagated through a bubbly liquid, quantifying the effect of physical erosion of test object, determination of chemical species produced, and monitoring of light emission during cavitation [2].

In this paper, three methods are used to detect the onset of cavitation at various depths. A class IV Flextensional transducer is used for the experiments whereas it appears that Tonpilz transducers are used by earlier investigators. The radiating face of the class IV Flextensional transducer is an elliptical cylinder.

In general, some parts of a Flextensional transducer will vibrate and cause rarefaction while other parts cause compression at the same time. This is in contrast to Tonpilz transducers where all parts of the radiating face either cause compression or rarefaction at a given instant. The results of the experiments are used to obtain expressions for the depthlimited cavitation. The results are used in sonar systems to determine the maximum power that can be transmitted at a particular depth.

EXPERIMENTAL SET UP

The experimental set up used to measure the cavitation-limited electrical power of a Flextensional transducer is shown in Figure (1).

Measurements are carried out in an 18 m deep fresh water tank facility at NPOL. All measurements are done at the resonance frequency of the transducer which is only a few kHz.

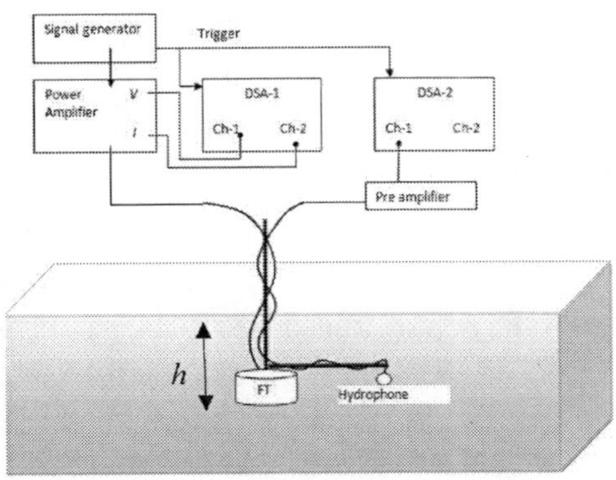

Fig. 1: Measurement Setup

First, the transducer is lowered to a depth of *h* meters, and a hydrophone, to record the acoustic characteristics of FT, is aligned along the major axis of the elliptical transducer. B&K-8105 hydrophone is used and kept at a distance of 0.6 meters away from the transducer. Signal generated from HP 3314-A function generator is amplified using L-10 power amplifier and directly fed into FT. Signal picked up by the hydrophone is amplified using B&K NEXUS pre-amplifier before analysing using DSA. Agilent 35670A Dynamic Signal Analyser (DSA I) is used for measuring input parameters to transducer like voltage, current, impedance, phase and electrical power. A similar signal analyser (DSA II) is used for measuring the output from the hydrophone.

METHODS FOR DETECTION OF ONSET OF ACOUSTIC CAVITATION

At low depths, cavitation is often the first factor that limits the performance of any high power electro-acoustic active transducers. Piezoelectric transducer behaves linearly at low power of operations. In the linear operation range, as voltage to the transducer increases, the acoustic pressure in the medium also increases proportionately. When cavitation occurs because of high amplitude of vibration, the transducer is no longer linear.

Three parameters are studied to determine their suitability for detecting onset of cavitation. One: Ratio of square of the radiated pressure to input power. Two: Total Harmonic Distortion (THD) in the hydrophone output. Three: Phase of the electrical impedance. At onset of cavitation, the dependence of these parameters on the input power is found to change as described below.

RESULTS AND DISCUSSIONS

Source Level (SL) of the transducer is the measure of far field pressure, capable of generating on its axis of maximum response. In linear operation range, SL proportionally increases with input power level (power in log). At the inception point of cavitation, transducer is no longer linear, and further increase in input power level does not yield a corresponding increase in SL. The SL of FT is measured atvarious depths for different power level to identify the inception point of cavitation.

The measured SL is shown in Figure 2. At nine meters depth, SL varies linearly up to 562 W and the respective SL is 194 dB. Again, the SL measurements are repeated for 11, 13 and 15 meters depth. For 11 m depth, transducer is linear up to 196 dB, and the power required to generate this SL is 732 W. Similarly, for 13 and 15 meters depth of operation, the respective power a level up to which transducer behaves linearly is 850 and 1011 watts. The cavitation limited power for each depth is marked in Figure 2 using dashed vertical lines.

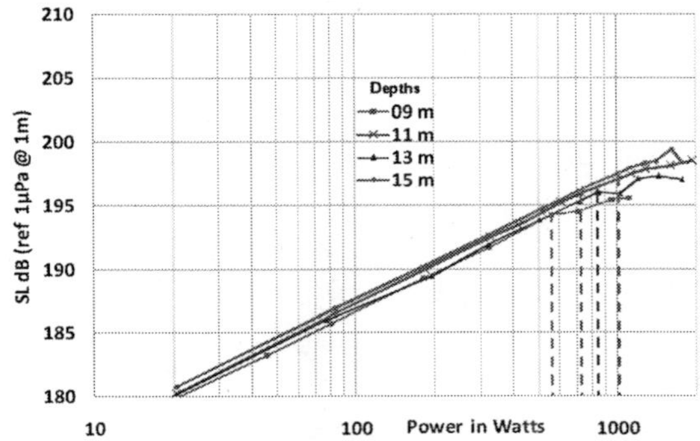

Fig.2. Measured SL for different Power

The phase of the electrical impedance of transducer is the characteristic property of piezoceramic transducer and the radiation reactance of the medium in which transducer radiates. The radiation reactance represents a mass, added to the transducer when it accelerates the adjacent medium and transfers kinetic energy to it. Phase characteristics of transducer vary abruptly when the medium starts cavitates. The phase response of FT for different power levels at various depths is shown in Figure 3. At nine meters depth, phase started varying suddenly at 562 W power. Similarly, for 11, 13 and 15 meters depth, the measured cavitation limited powers are 733, 850 and 1011 watts. The measured cavitation limited power using phase measurements are matching with the response shown in Figure 2.

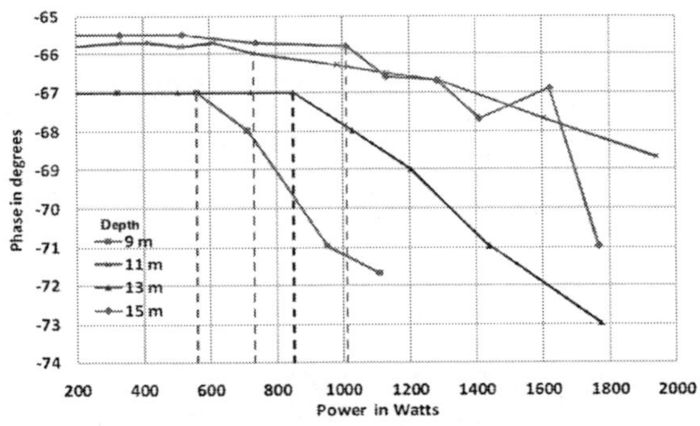

Fig. 3: Phase Response for Different Power

Total harmonic distortion (THD) of a signal is a measure of the harmonic distortions, and is defined as the ratio of the sum of the powers of all the harmonic components to the power of fundamental frequency. It is generally expressed in percentage. Cavitation creates bubbles in the medium and the generated bubbles starts to oscillate by absorbing energy from the acoustic waves. These oscillations are at the harmonics of the acoustic wave frequency, depending on the size of the bubble. As the acoustic power increases, higher order harmonics

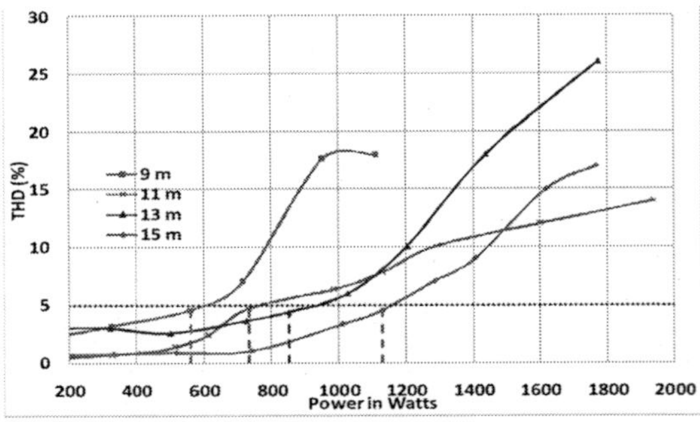

Fig. 4: THD of Hydrophone Output Levels

will be excited and that will leads to large THD. When THD goes above 5%, visible deviation from the sinusoidal pattern can be observed, which is consider as inception point of cavitation. The measured THD of hydrophone output (voltage) at various depths is shown in Figure 4. The power at which THD crosses 5% for various depths is marked in Figure 4.

Table 1: Measured Cavitation Limited Electrical Power of Low Frequency FT at Different Depths

Sl. No.	Depth of Operation (h)		Cavitation limited Electrical power (Wc in Watts)
	Meters	Feet	
1	9	29.5	562
2	11	36	732
3	13	42.6	850
4	15	49.2	1011

The measured cavitation limited electrical power using three methods are nearly the same, and this method can be adopted to any transducers to measure the cavitation limited performance. The measured cavitation limited power of FT at various depths is shown in Table 1.

CAVITATION LIMIT

The expression for cavitation threshold power varies significantly for different types of transducers. Established equations are available for piston type transducers with circular radiating heads. For low frequency (<10 kHz) high pulse length (>5 mS) acoustic transducers, the inception of cavitation is primarily decided by the hydrostatic pressure at the operating depth. Hydrostatic pressure linearly grows with operating depth, h. Each ten meters of depth in ocean represents an increased pressure of 1 atm.

In general, cavitation limited acoustic intensity, I_c, at h feet depth can be written as [3];

$$I_c = 0.3\gamma \left(x+\frac{h}{33}\right)^2 \quad Watt/cm^2 \qquad \dots (1)$$

For circular pistons, plane and hemispherical, the coefficient γ ranges from 0.3 to 0.6 and x is unity. The factor γ is introduced in the Eq. (1) to express the near field effect on the cavitation threshold. Using Eq. (1), the cavitation limited acoustic intensity at any depths can be estimated for piston (Tonpilz) transducers.

In FT, modes of vibrations are significantly different from piston transducers, hence, the coefficients used in the Eq. (1) cannot be used to predict the cavitation limited performance of FT. But by knowing the cavitation limited power at two depths, these coefficients can be estimated for any transducers. In the normal operation condition, cavitation limited acoustic intensity for any transducers at h feet depth is expressed as;

$$I_c = \gamma \, (x + \frac{h}{33})^2 \quad Watts\,/\,cm^2 \qquad \text{... (2)}$$

where x+h/33 is the hydrostatic pressure term and coefficient γ is the characteristics of the transducer.

In linear operation range, electric power to the transducer is proportional to acoustic intensity and using Eq. (2), cavitation limited electric power (W_c) of FT can be written as;

$$W_c = \gamma \, A \, (x + \frac{h}{33})^2 \quad Watts \qquad \text{... (3)}$$

where A is the radiating area of Flextensional Transducer in cm^2.

The coefficients γ and x are calculated using the two measured cavitation limited power values, shown in table 1. Using the measured power values at 9 and 13 meters, calculated γ and x coefficients are 0.2 and 0.84. The radiating area of this typical FT is 919 cm^2. Incorporating the coefficients γ and x, the expression for W_c is;

$$W_c = 0.2 \, A \, (0.84 + \frac{h}{33})^2 \quad Watts \qquad \text{... (4)}$$

Using Eq. (4), cavitation limited power can be estimated for any depths. The estimated power for various depths is shown in Table 2.

Table 2: The Estimated Cavitation Limited Power at Different Depths

Sl. No.	Depth of Operation (h)		Cavitation limited Electrical power (W_c in Watts)
	Meters	Feet	
1	9	29.5	552
2	11	36	685
3	13	42.6	835
4	15	49.2	999

The predicted power using Eq. (4) is matching with measured results shown in Table 1. This equation can be used in sonar systems to determine the maximum power, this FT can be transmitted at a particular depth.

CONCLUSION

Three different methods are discussed to detect onset of cavitation in the acoustic transducers at various depths. The expression for the onset of cavitation power for low frequency FT is derived and the predicted results are in agreement with measured results.

ACKNOWLEDGEMENT

Several discussions with Dr. D.D. Ebenezer, Head, Transducer group and Dr. K.P.B. Moosad, Head, Transducer Science Division and the facilities, encouragement, and permission given by Director, NPOL, are gratefully acknowledged

REFERENCES

[1] D. Stansfield, Underwater Electroacoustic transducers, Bath university, Bath, UK and Institute of Acoustics, St. Albans, UK, 1991.

[2] J. Frohly S. Labouret, C. Bruneel, I. Looten-Baquet and R. Torguet 'Ultrasonic cavitation monitoring by acoustic noise power measurement', J. Acoust. Soc. Am., Vol. 108, 2012–2020 (Nov 2000).

[3] R.J. Urick, Principles of Underwater Sound for Engineers, McGraw-Hill, USA, 1967.

Metal-Ceramic Segmented Cylinder Transducer

Antony Jose and K.P.B. Moosad

Naval Physical and Oceanographic Laboratory, Thrikkakara, Kochi, Kerala

ABSTRACT: *Piezocermic cylinders are often used as projectors for underwater applications. Monolithic cylinders with electrodes on the inner and outer surfaces and polarized along the radial direction (Radially Polarised Cylinders - RPCs) are used in certain applications. Cylinders formed by adhesively joining several segments, which are then polarized tangentially, are also used in some cases. Construction of such cylinders is an expensive process since it involves machining of the piezoceramic segments to close tolerances. An alternate method is to use piezocermic material in the regular geometry (rectangular prism) and join them using shaped metal pieces in between. Instead of a cylinder, the final geometry will be a several-sided polygon, but with a comparable performance. This is shown by means of Finite Element Modeling (FEM) of the all-ceramic and metal-ceramic segmented cylinders, using a commercial FE package ATILA.*

Keywords: Segmented Cylinder, Radially Polarized Cylinder, Transmitting Voltage Response, ATILA.

INTRODUCTION

Segmented cylinders as well as radially polarized monolithic cylinders of piezoceramic materials are used as underwater sound projectors for various applications [1,2]. While vibrating in the hoop mode, the performance parameters of the segmented cylinder are governed by the piezoceramic coefficient d_{33}, whereas for a radially polarized cylinder (monolithic), these are governed by the d_{31} coefficient. For piezoceramic materials d_{33} is inherently higher and therefore the performance parameters of a segmented cylinder are expected to be higher when compared with a monolithic cylinder of same dimensions. Although segmented cylinders are widely used in sonar industry, nothing much has been published regarding the design or properties of segmented cylinders. Butler, in a 1976 paper [3] had brought out an equivalent circuit model for segmented rings with inactive segments and discussed how the coupling coefficient reduces when number of inactive segments increase. K P B Moosad et al. [4] have presented a FE model of a segmented cylinder. Antony Jose [5] has earlier compared the performances of segmented cylinder vs monolithic cylinder and has shown that the segmented cylinder shows a much larger coupling coefficient.

The realization of a segmented cylinder involves precise machining of piezoceramic materials in the shape of segments. This incurs a lot of cost and therefore the segmented cylinder becomes prohibitively expensive. A simpler way of doing this is to make a multi-sided polygon instead of a cylinder. The polygon can be formed using relatively inexpensive piezoceramic material in regular geometries (plates or slabs), interspersed with wedge-shaped metallic pieces. In this work, it is attempted to compare the performance of a metal-ceramic segmented cylinder and an all-ceramic segmented cylinder of similar dimensions, by modeling them in ATILA, a commercial Finite Element Modelling (FEM) package.

MODEL DETAILS

To compare the performance of cylinder with that of a polygon, the problem is modeled in 2D, applying plane strain conditions. At first, the performance of an all-ceramic cylinder is compared with that of a cylinder with some segments replaced by metal segments. The geometries are shown in Figure 1 and 2. The metal used in modeling is brass. Figure 3 shows the mesh used for the modeling. Quadrilateral elements were used.

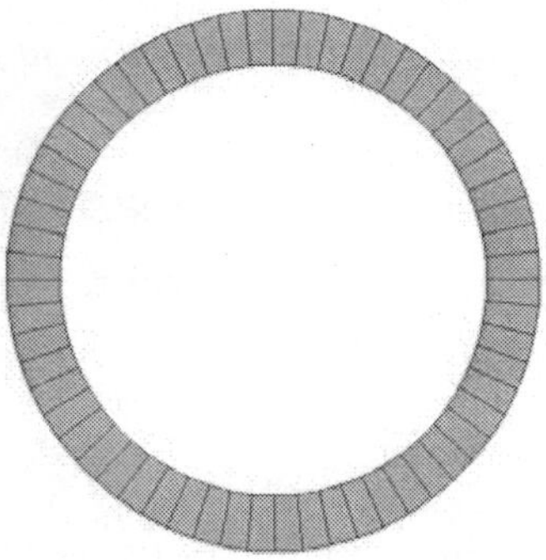

Fig. 1: All-ceramic Segmented Cylinder

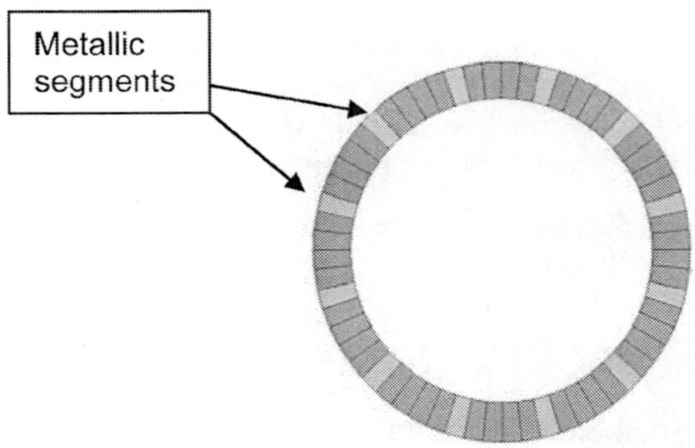

Fig. 2: Metal-ceramic Segmented Cylinder

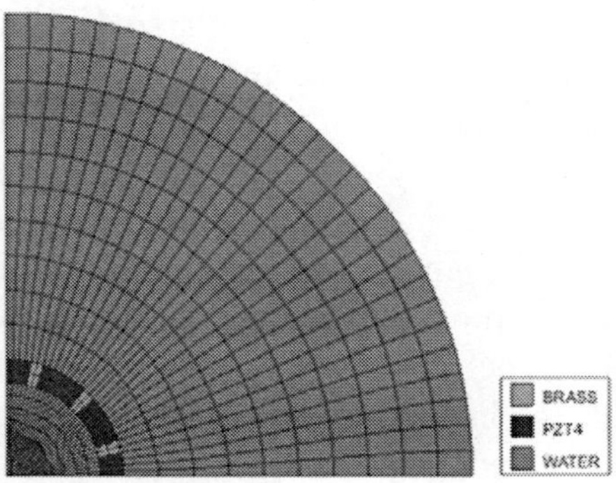

Fig. 3: Mesh Used for the 2D Models

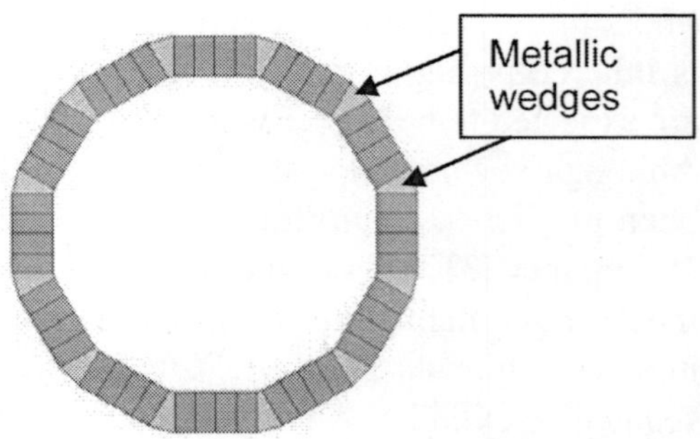

Fig. 4: Construction of Metal-Ceramic Segmented Cylinder Using Slabs of Regular Geometry

When these two are modeled in ATILA, the Transmitting Voltage Response (TVR) plots are found to be quite similar in shape and the difference in TVR values in the interested range of frequencies are found to be within 2 dB. It is also to be noted that the volume of piezocermic is also less by 20% in metal-ceramic case. To reduce the cost of manufacturing, the segments are to be replaced by regular slabs. This makes the cylinder a multi-sided polygon, as shown in Figure

4. In this, several rectangular slabs are adhesively joined together and are formed into a multi- sided polygon, by means of additional metallic wedges. To compare the performance of this assembly with that of the all-ceramic segmented cylinder, TVR plots are generated by ATILA, equating the inner and outer diameters of the ring in both the cases. The various TVR plots are compared in Figure 5.

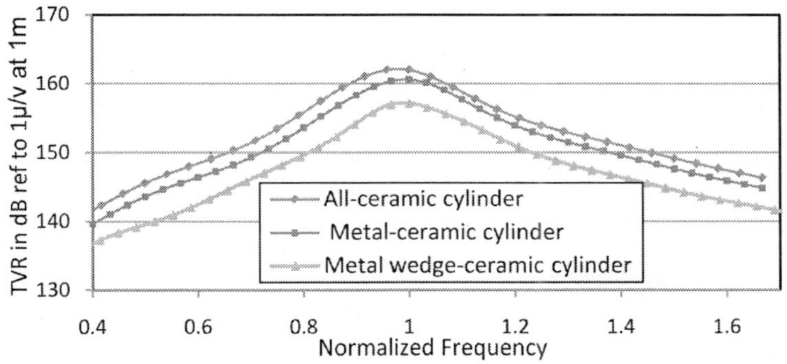

Fig. 5: Comparison of the TVR Plots for the Three Cases

Figure 5 shows that the performance of metal-ceramic segmented cylinder constructed using regular slabs of piezoceramic material is comparable with that of an all-ceramic segmented cylinder. In applications where cost is a major consideration, the latter one is a better option. To increase the TVR, and to bring the geometry closer to a cylinder, the number of wedges can be doubled and the thickness of the piezoceramic plates can be halved. The resulting geometries are shown in Figure 6 and their respective TVR plots are shown in Figure 7.

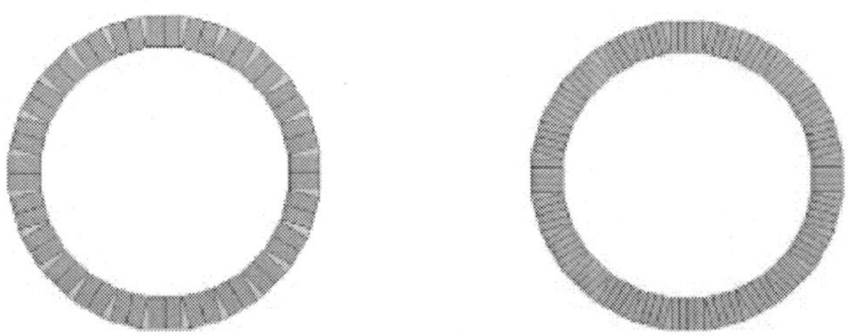

Fig. 6: Improved Constructions of Metal-Ceramic Segmented Cylinder Using Slabs of Regular Geometry

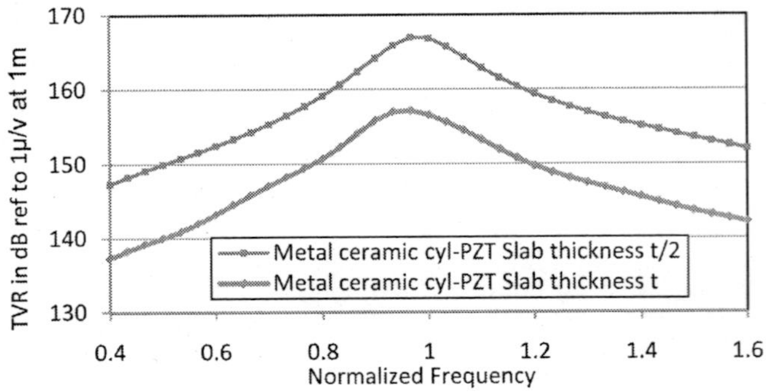

Fig. 7: TVR Plots for the Improved Geometries

The 2D models are useful in studying the comparative performance of various constructions. For a more accurate prediction of the performance figures, it is required to model the cylinder in 3D, in order to include the effects due to the definite length of the cylinder. 3D model of one of the cases in Figure 6 is shown in Figure 8. The TVR plot obtained from the model is shown in Figure 9. As explained in Ref [4], the first resonance is generally not observed in experiments. It can also be seen that the TVR values are comparable to those recorded experimentally from an all-ceramic segmented cylinder.

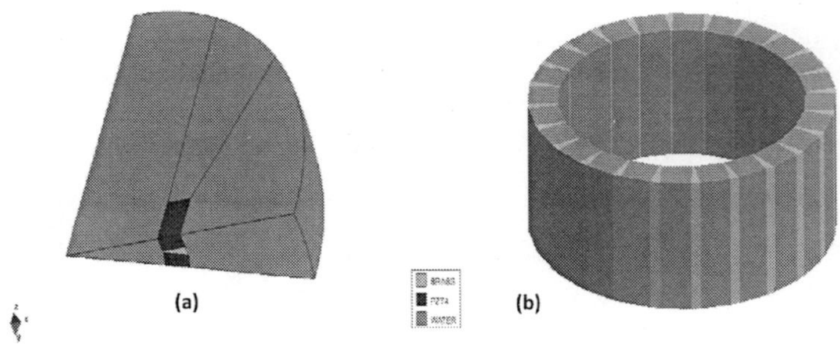

Fig. 8: 3D Model of Metal-Ceramic Segmented Ring Cylinder
(a) Model of a 30-degree Segment (b) Model of the Full Structure

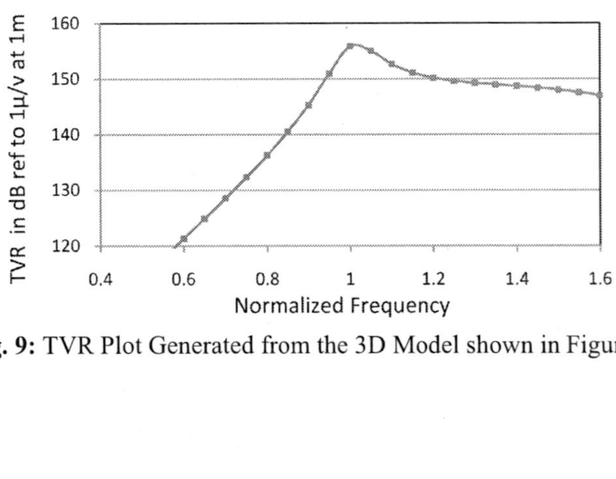

Fig. 9: TVR Plot Generated from the 3D Model shown in Figure 8

CONCLUSIONS

Performance of two types of piezoceramic segmented cylinder constructions are compared by modeling them in 2D, using a commercial Finite Element Modeling package ATILA. One is a proper cylinder, constructed by precise machining of piezoceramic material in the form of segments and the other one is constructed by using piezoceramic slabs of regular geometry, interspersed with shaped metallic pieces. Results show that both the constructions would give comparable performance. Performance improvement of the metal-ceramic segmented cylinder is attempted by increasing the number of metal wedges and decreasing the thickness of the piezoceramic plates. 3D model of the final geometry is also done and it is shown that the performance is comparable with the experimental values published earlier for an all-ceramic segmented cylinder.

REFERENCES

[1] D. Stanesfield, Underwater electroacoustic transducers, Bath University Press, Claverton Town, 1990, pp. 363–364, pp. 49–50; Charles H Sherman, John L. Butler, Transducers and Arrays for Underwater Sound, Springer, 2007, pp. 81–88.

[2] O.A. Godin, D. R. Palmer, In History of Russian Underwater Acoustics, ed. A Godin Oleg, World Scientific, Singapore, 2008, pp. 966–967.

[3] J.L. Butler, Model for a ring transducer with inactive segments, J. Acoustical Society of America, 1976, 59(2), 480–482.

[4] K.P.B. Moosad, Rahel Sebastian, R.M.R. Vishnubhatla, Study of an all-ceramic segmented ring transducer, Proceedings of SYMPOL 2003, pp. 185–190, Mc Graw Hill, New Delhi. [5] Antony Jose, Piezoceramic Segmented Cylinder and Monolithic Cylinder—a comparison, SEATECH - NPOL Technical Journal, Vol. 9, Issue 2, May 2012, pp. 43–46.

Performance Improvement of Proactive Condition Monitoring Systems for Marine Machinery

Suraj Kamal, Shameer K. Mohammed, P.R. SaseendranPillai and M.H. Supriya

Department of Electronics, Cochin University of Science and Technology, Kochi, Kerala

ABSTRACT: *Machine condition monitoring systems are becoming ubiquitous around the industrial space of mechanical systems due to the ever increasing demand for improved reliability and fail proof operation. Much of the downtime can be obviated with proactive maintenance by measuring vital machine parameters to discover imminent failures. The condition of the ship engine is monitored using the characteristic signatures present in the Acoustic Emanations (AE) generated by the vibrating components of the reciprocating engines along with their associated parts. The feeble noise of the malfunctioning components will not be detectable at the beginning stages as they are often buried in the noise floor and this creates a major challenge in prognostic reporting of the events that could lead to a catastrophic failure. In this paper an attempt is made to lift and isolate the fault signatures that are hidden behind the noise floor by utilizing Blind Source separation (BSS) technique. In the prototype system developed, Independent Component Analysis (ICA) algorithm are used for BSS and the classification and decision making is carried out by the Probabilistic Neural Networks (PNN).*

Keywords: Acoustic Emanations, Proactive Monitoring, PNN Classifier, ICA.

INTRODUCTION

Machine condition monitoring systems have been a thrust area of research during the past few decades and are becoming ubiquitous around the industrial space of mechanical systems due to the ever increasing demand for improved reliability and fail- proof operation. Merchant vessels, carrier or ocean liners, often travel months through the open ocean, keeping their engines at full throttle round the clock, without any ground assistance or maintenance support. With their engines and electronic control systems running throughout, they are extremely vulnerable to unexpected failures and the occasional downtimes pose immense challenges to the onboard maintenance crew. Hence, predictive monitoring is an essential requirement for taking preventive measures to ensure that these systems run to their maximum extent or at least to the next shore yard. All the mechanical systems including the ship engines often degenerate over time, before it eventually fails, through a sequence of states which can be detected and classified for proactive maintenance.

Proactive condition monitoring systems produce diagnostic data on the engine and its associated components thereby helping to detect problems before they lead to a catastrophic failure. A proactive monitoring system generally comprises of data acquisition, processing and prognostic report generation. The parameters to be monitored can be chosen according to the key indicators that hold the traces of eventual failures that are likely to occur, such as combustion emissions, temperature profiles and vibration data.

If the process is automated by incorporating adaptive signal processing and interpretation systems, the dependency on manual interpretation can be eliminated, thereby obtaining effective and consistent results. With the use of sophisticated online signal processing and machine learning algorithms, multilevel hierarchical alert points can be generated including multiple types of failure modes which in turn create opportunities to schedule maintenance tasks.

In this paper, the acoustic vibration data is being utilized for automating the condition monitoring process. The condition of the ship engine is monitored using the characteristic signatures present in the acoustic emanations, generated by the vibrating components of the reciprocating engines along with their associated parts. The real challenge with acoustic coupling is the detection of important signatures from the superposition of all the trivial and vital signals generated by all the vibrating components. Earliest detection of the signs of defects is utmost important as these defects can lead to unscheduled outages. However, the early signs are very feeble, often hidden behind the noise floor generated by the reciprocating components and gear trains associated with the engine. For effective detection of the imminent faults, even the feeble fault signals must be detected in the presence of a heavy noise background. Most of the state of the art systems rely upon the detailed knowledge of the source characteristics; however they can't detect these weak noise remnants that are buried below the combustion noise. In spite of taking a priori knowledge into account, an alternative approach has been followed for isolating the signals of interest from the composite signal.

METHODOLOGY

The composite signal captured by the sensors is decomposed into its component factors by using Blind Source Separation (BSS) methods that are originally inspired by the human ability to listen and focus to a particular source of interest among a mixture of distracting noise surroundings. In the prototype system, Independent Component Analysis (ICA), a well-known technique for doing BSS [1–3] with instantaneous mixing assumptions, has been incorporated for unmixing the vibration signals.

ICA reveals the latent signals from the observed mixture, thereby enabling the signals originating from different spatial points of the engine to be studied separately. The decomposed component signals corresponding to the various sources include noise signatures from malfunctioning bearings, gear teeth, combustion, valves, piston slaps and other rotating components of the ship engine. The ICA process thus reveals the otherwise undetectable vibration signatures that are hidden behind the noise floor rendered by the collective vibrations from all the moving mechanical components. Now the task of detection/classification is much easier as most of the interfering signals are removed or separated. The signals are continuously monitored for any possible defect signatures that may occur during the operation of the machinery. Artificial Neural Network based classifiers are widely used for this [4] and a Block description of the prototype system is outlined in Figure 1.

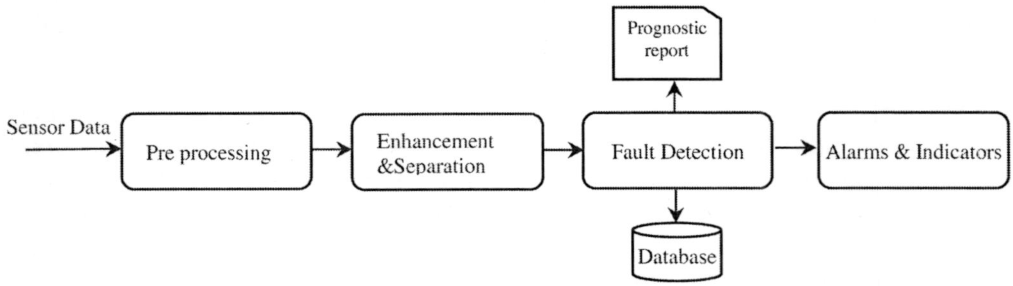

Fig. 1: Block Diagram of Prototype

INDEPENDENT COMPONENT ANALYSIS

Independent Component Analysis (ICA) is a multivariate statistical technique for decomposing a complex dataset into independent sub-components by assuming the statistical independence of the observed multivariate time series. ICA, an extension to PCA can separate the sources from a Gaussian mixture, thereby providing the solution to cocktail party problem and BSS. ICA decomposition of statistically independent data makes use of linear transformations to co-ordinates for extracting the machine modeling signals below the noise floor. ICA method works on certain assumptions that the source components are independent and have non-Gaussian distributions.

ICA decomposes the simultaneous signals from various sensors attached to the engine and power train assembly of the ship. The process of decomposition of these simultaneous independent signals starts with a signal pre-whitening stage followed by an optimization stage, which retrieves the independent source signals.[1],[5–7] ICA isolates and extracts the vital signals buried under the noise floor. The decomposed independent signals are feature extracted and then recognized with a PNN classifier.

FAULT DETECTION AND CLASSIFICATION

Classifiers are used to identify the defect signatures emanating from various components of the machinery.

MFCC Feature Extraction

Feature extraction is a form of dimensionality reduction to input independent acoustic signals that are decomposed by ICA, for reducing the computational complexity and memory consumption. Among the set of spectral features generally used for acoustic feature extraction, Mel Frequency Cepstral Coefficients (MFCC) has a special position due to its simplicity and robustness, which represents the signal as a sequence of compact Mel Frequency Spectral Coefficients or feature vectors. The MFCC feature estimation process divides the incoming signal stream into finite width frames and a windowing function (Hamming window) is used to remove the effect of frame discontinuities. The frames are converted into frequency domain using a DFT function. The DFT of the input signal $x(n)$ is given by $X(k)$ is multiplied by a sequence of triangular gain filters and the Mel-frequency filter bank [8–9] comprises of p filters computes the energy in each band m_j $(j = 1, 2, \ldots, p)$. The Mel-frequency cepstrum is then the discrete cosine transform [10] of the p filter outputs and is represented as

$$c_i = \sqrt{\frac{2}{N}} \sum_{j=1}^{p} m_j \cos\left(\frac{\pi i j}{p}\right) \qquad (3)$$

where c_i is the ith MFCC coefficient. By taking the inverse DFT of the converted scale, the signal is restored again into the time domain.

B. CLASSIFIER

Classifier predicts the membership or categorizes the input feature space to a corresponding class. Artificial neural networks and statistical methods can be used as the background technique for classification. ANN is a multilayer perceptron consisting of an input layer, output layer and intermediate layers which perform activation function and weight adjustment for training the network for a particular output. Probabilistic neural networks (PNNs) are a type of ANN with a hierarchical parallel structure which is widely accepted as a classifier [4],[11]. PNN is proposed for the classification due to the advantage of faster training process compared to back propagation algorithms and its flexibility to add or remove training samples without extensive retraining [12].

a) Probabilistic Neural Network

PNN is an implementation of a statistical algorithm called kernel discriminant analysis in which the operations are organized into a multilayered feedforward network with four layers such as Input layer,

Pattern layer, Summation layer and Output layer. The input layer unit simply distributes the input to the neurons in the pattern layer without performing any computation. On receiving a pattern x from the input layer, the neuron x_{ij} of the pattern layer computes its output [3]. The summation layer neurons manipulate the maximum likelihood of pattern vector x through summing and averaging operations.

$$p_i(x) = \frac{1}{(2\pi)^{d/2}\sigma^d} \frac{1}{N_i} \sum_{j=1}^{N_i} exp\left[-\frac{(x - x_{ij})^T (x - x_{ij})}{2\sigma^2}\right] \quad (5)$$

where N_i represents the number of samples in a particular class C_i. The patterns are classified into their corresponding class C_i with the help of the decision layer by making use of the Bayesian decision rule and the decision layer inputs.

$$\tilde{C}(x) = \arg\max \{p_i(x)\} \quad i = 1,2,3\ldots\ldots K \quad (6)$$

where K is the number of classes and $\tilde{C}(x)$ is the corresponding class of the pattern x.

IMPLEMENTATION

The prototype system has been implemented in Matlab in a modular structure. FastICA algorithm has been used for separating the signatures generated due to malfunctioning and that are hidden below the noise floor. All the extracted signals from the ICA stage are processed for feature extraction using MFCC method and the extracted features are given as the input to the classifier for detecting the faulty states.

The anomalous conditions have been detected with Probabilistic Neural Networks (PNNs) which is a realization of the Bayesian classifier with non-parametric kernel estimation and multilayer feed forward architecture. The network has been trained for detecting malfunctions at different states that sequentially lead to a catastrophic failure. Multiple alarm points have also been generated according to the severity of the observed damage and based on this, proactive maintenance actions can be scheduled. The PNN identifies the malfunction signatures present in the observed signals for which it is trained.

RESULTS AND DISCUSSIONS

The normal and malfunctioning conditions of the ship engine and auxiliaries were simulated in Matlab. A typical scenario of a bearing malfunction and its evolution through different states has been selected for evaluating the system performance. Vibration signals from various components of a ship engine have been synthesized. Because of the noise from the reciprocating components of the engine, the feeble noise of the malfunctioning bearing will not be detectable at the beginning of the fault sequence. The ICA algorithms are able to separate these weak signals that are hidden below the noise floor. From the observed mixture of the signals, the independent components are estimated using FastICA algorithm. The component signals are assumed to have different origins and *a priori* knowledge is needed to identify the components to which these signals correspond. Supervised learning classifiers have the ability to incorporate *a priori* knowledge. Because of its ability to recognize subtle patterns in the input data even in presence of noise, PNN classifiers trained with exemplar data have been used to detect and categorize the fault signatures. The prognostic fault stages are enlisted as 0 for Normal Running, 1 for Minor Defect,2 for Need Attention, and 3for Critical, Immediate shutdown

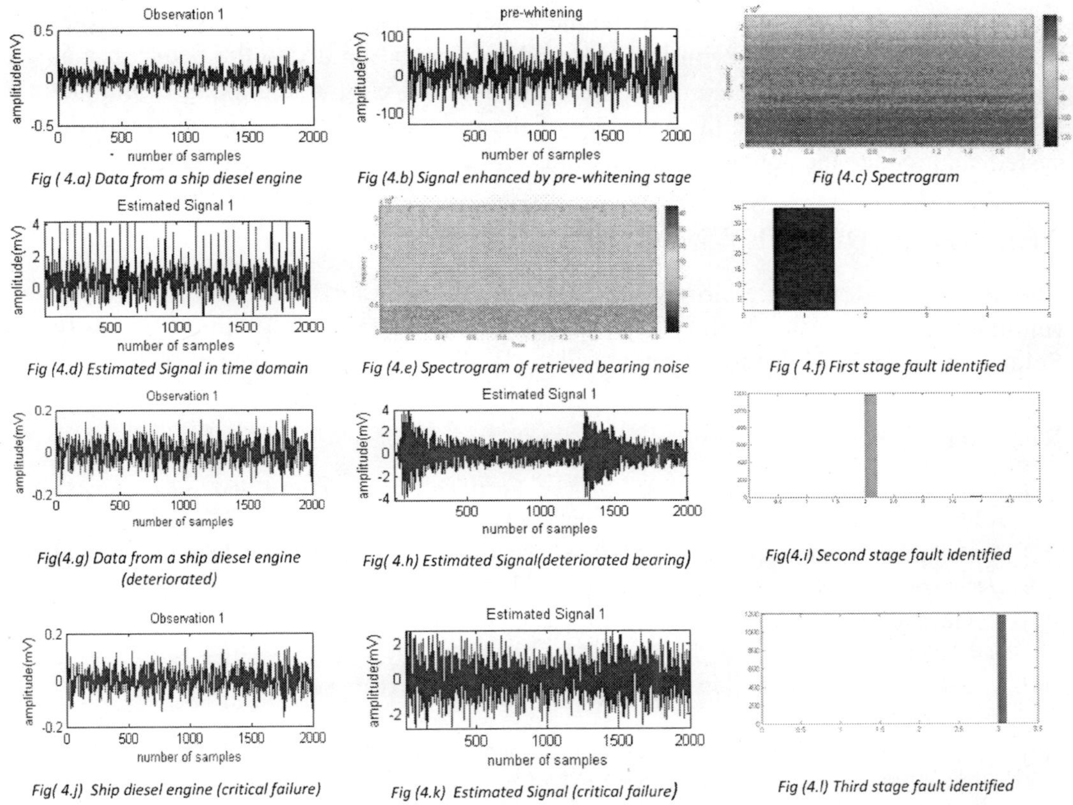

Fig (4.a) Data from a ship diesel engine *Fig (4.b) Signal enhanced by pre-whitening stage* *Fig (4.c) Spectrogram*

Fig (4.d) Estimated Signal in time domain *Fig (4.e) Spectrogram of retrieved bearing noise* *Fig (4.f) First stage fault identified*

Fig(4.g) Data from a ship diesel engine (deteriorated) *Fig(4.h) Estimated Signal(deteriorated bearing)* *Fig(4.i) Second stage fault identified*

Fig(4.j) Ship diesel engine (critical failure) *Fig (4.k) Estimated Signal (critical failure)* *Fig (4.l) Third stage fault identified*

Fig. 4

Figure 4.a shows the measurement data from a ship diesel engine with a faint signature tone of a bearing shell crack but severely masked by the collective noise from other sources. Figure 4.b shows the signals enhanced by a pre-whitening stage. Now the ringing tone of the bearing is slightly visible below the noise background. Figure 4.c shows the corresponding spectrogram. In the third stage, ICA estimates the independent signal components corresponding to different sources, including the bearing noise. Figure 4.d shows the signal in time domain and Figure 4.e shows the spectrogram of the retrieved bearing noise. The PNN classifier identifies the problem as stage 1, i.e. a minor defect.

The subsequent figures show the observations and results as the problem builds up eventually to critical stage.

CONCLUSIONS

The prototype system demonstrates the effectiveness of the BSS preprocessing method in improving the performance of the machine condition diagnosis. The PNN classifier has been found to be much faster in learning and converges rapidly to the decision making classification problem. The combination of both the ICA signal unmixing technique and PNN classifier can improve the accuracy of the marine machinery condition monitoring system and reduce the downtimes significantly.

ACKNOWLEDGEMENT

The authors gratefully acknowledge The Naval Research Board, New Delhi for the financial assistance and the Department of Electronics, Cochin University of Science and Technology for extending all the facilities for carrying out this work.

REFERENCES

[1] S. Amari, A. Cichocki and H. Yang. "A New Learning Algorithm for Blind Signal Separation". Advances in Neural Information Processing Systems 8, 1996.

[2] A. Bell and T. Sejnowski. "An Information Maximization Approach to Blind Separation and Blind Deconvolution". Neural Computation, 7:1129–1159, July 1995.

[3] C. Jutten and P. Comon. "Handbook of Blind Source Separation: Independent Component Analysis and Applications". Elsevier Ltd.

[4] Mao, K. Z. *et al.*, "Probabilistic Neural-Network Structure Determination for Pattern Classification," IEEE Transaction.

[5] A. Hyvärinen and Erkki Oja, "Independent Component Analysis: A Tutorial" (http://cis.legacy.ics.tkk.fi/aapo/papers/IJCNN99_tutorialweb).

[6] A. Hyvärinen and Erkki Oja. "Independent component analysis: algorithms and applications", Neural Networks Volume 13, Issues 4–5, June 2000.

[7] A. Hyvärinen. 'Fast and Robust Fixed-Point Algorithms for Independent Component Analysis'. IEEE Transactions on Neural Networks 10(3): 626– 634, 1999.

[8] Sirko Molau, Michael Pitz, Ralf Schluter and Hermann Ney, "Computing Mel-Frequency Cepstral Coefficients on the Power Spectrum", Proceedings of the IEEE International Conference on Acoustics, Speech and Signal Processing, 1, 2001, pp: 73–76.

[9] Mark D. Skowronski and John G. Harris, "Improving the Filter Bank of a Classic Speech Feature Extraction Algorithm",Proceedings of the IEEE International Symposium on Circuits and Systems, 4, 2003, pp. 281–284.

[10] Mubarak, O.M..; Ambikairajah, E. and Epps, J.; "Novel Features for Effective Speech and Music Discrimination", Proceedings of the IEEE International Conference on Engineering of Intelligent Systems, 2006, pp. 1–5.

[11] H.M. Khattab, Abdelaziz, S.F Mekhamer, M.A.L. Badr, "Static Security Assessment Using a Probabilistic Neural Network Based Classifier". In online journal on Electronics and Electrical Engineering (OJEE) Vol. 3, No. 4.

[12] Donald, F. Specht, "Probabilistic Neural-Networks orginal contribution on Neural Networks", Vol. 3, pp. 109–118, 1990.

Numerical Investigation of Non-Cavitating Underwater Propeller Noise Using FW-H Method

V. Rama Krishna, P.V.S. Ganesh Kumar, P. BangaruBabu[1] and S. Rama Krishna[2]

Naval Science and Technological Laboratory, Visakhapatnam, Andhra Pradesh

[1]*National Institute of Technology (NIT), Warangal, Andhra Pradesh*

[2]*GVP College of Engineering, Madhurawada, Visakhapatnam, Andhra Pradesh*

E-mail: ramki_40@rediffmail.com

ABSTRACT: *The Propeller is a vital component which is subtle for the safe operation of a ship at sea. The purpose of the propeller is to produce a thrust causing to move the ship in forward direction. But it is identified as major noise source. The noise produced by a propeller is very much importance to warship designers and military strategists for many years. A ship /Submarine noise signature of a few tens of watts could be sufficient to give enemy valuable information at a considerable range. Using noise signature analysis at remote locations, it is possible not only to determine which class of vessel has been located, but also if character of signature is sufficiently known to identify the particular ship. Clearly the ultimate goal of a warship designer must be to make the ship's signature vanish into the background noise of the sea which comprises contributions from the weather, marine life and also other shipping from a wide geographical area. This paper presents a numerical study on the non-cavitating noises of the underwater propeller using Computational Fluid Dynamics (CFD) Fluent software.Propeller behavior is investigated for its hydrodynamic parameters and then Computational Acoustic Analysis (CAA) is carried out using viscous model of large eddy simulation of Ffowcs Williams–Hawkings formulation(FW-H). Here solver is chosen as pressure based, unsteady formulation of second order implicit. Noise is predicted using time-domain acoustic analogy and finite volume method. Sound pressure levels are predicted at different receiver positions.*

Keywords: Propeller, Non-cavitating Noise, Computational Fluid Dynamics (CFD), Computational Acoustic Analysis (CAA), Large Eddy Simulation, Ffowcs Willams-Hawkings Formulation (FW-H).

INTRODUCTION

Sound generated by a propeller is critical in underwater detection, and is often related to the survivability of the vessels especially for military purposes. Marine propeller noise can be classified into cavitating and non-cavitating noise. Cavitation of the marine propeller is the most prevalent source of underwater sound in oceans and is often the dominant noise source of a single marine vehicle.However, submarines and torpedoes are usually operated deep enough under the sea to avoid cavitation. Salvatore *et al.* [1] presented propeller-induced noise emission is studied through a general hydro acoustics formulation based on the Ffowcs Williams-Hawkings (FW-H) equation that allows describing the acoustic pressure field generated by lifting bodies in arbitrary motion through a fluid. Chang [2] applied a finite volume CFD method in conjuction with the standard k-ε turbulence model to calculate the flow pattern and performance parameters of a DTNSRDC P4119 marine propeller in a uniform flow.

The essential step taken by Lighthill [3] was to incorporate the non-linear features of aerodynamic sound generation into a linear acoustic model. There are various ways to evaluate Ffowcs Williams and Hawkings [4] equation and the three types of noise source terms (monopole, dipole, and quadrupole) were proposed. Farrasat and Myers [5] proposed a time-domain formulation that can predict noise from an arbitrarily shaped object in motion without the numerical differentiation of the observer time. Through these studies, the dominant noise source of marine propellers is analyzed, which will provide basis for proper noise control strategies under non-cavitating conditions. Jin-ming *et al.* [6] analyzed the blade frequency noise of non-cavitation propeller in a uniform flow in time domain. The unsteady loading (dipole source) on the blade surface is calculated by a potential-based surface panel method. Then the time-dependent pressure data is used as the input for FW-H formulation to predict the acoustic pressure.Yang *et al.* [7] presented a new method to measure the propeller noise level in the ship engineering by coupling with Large Eddy Simulation and boundary element numerical acoustics methods in the frequency domain to predict the underwater non cavitation noise of ship.

Compared with the extensive amount of literature concerning cavitation noise of propellers, works concerning the non-cavitation noise of propellers are hard to find and also many authors used Reynolds-averaged Navier- Stokes (RANS) for turbulence modelling. However RANS have limitation of capturing details of such transient events like unsteady force and pressure fluctuation of the blade with an acceptable accuracy. In that regard, Large Eddy Simulation (LES) has been considered having merits to resolve the least large scale turbulent structures, which is further justifiable for the unsteady flow around propellers operated at the stern of ships.

So in this paper an attempt is made to predict non-cavitation noise of propeller using FW-H equation coupled with LES computer code based on cell-centered finite volume method (FVM) on unstructured meshes for viscous flow field around propeller.

FFOWCS WILLIAMS-HAWKINGS EQUATION

The theory of hydrodynamic sound was developed by Lighthill [3], who rewrote the Navier-Stokes equations into an exact, inhomogeneous wave equation whose source terms are important only within the turbulent region. Lighthill acoustic analogy equation is given below,

$$\left(\frac{1}{c^2}\frac{\partial^2}{\partial t^2} \nabla^2\right)\left[c^2(\rho-\rho_0)\right] = \frac{\partial^2\ T_{ij}}{\partial_{xi}\partial_{xj}} \quad \ldots (1)$$

Ffowcs Williams and Hawkings [4] extended the acoustic analogy developed by Lighthill [3], by taking into account the displacement of a rigid surface placed in the flow. The essential step taken by Lighthill was to incorporate the non-linear features of aerodynamic sound generation into a linear acoustic model. The acoustic analogy generalized by Ffowcs Williams and Hawkings is often applied in the prediction of the noise emission generated by the rotors of helicopters, axial fans, propellers etc. Ffowcs Williams and Hawkings equation is written in the following form:

$$\frac{1\partial^2 p}{c^2\partial t^2} - \Delta^2 p = \frac{\partial q}{\partial t} - \frac{\partial F_i}{\partial x_i} + \frac{\partial^2 T_{IJ}}{\partial x_i \partial x_j} \quad \ldots (2)$$

and shows three types of sources on the right hand side:

1. Monopolar, which results from introducing a mass (per unit volume) into the considered area.
2. Dipolar, that takes into account the hydrodynamic forces (F_i).
3. Quadrupolar, due to the turbulence and represented by the Lighthill's tensor.

The noise due to the hydrodynamic forces applied to the surface of the blades is dominant compared to the other sources of noise in the case of subsonic flow fans. The Ffowcs Williams—Hawkings equation is an exact rearrangement of the continuity and the momentum into the form of an inhomogeneous wave equation. This equation has two surface source terms, known as thickness and loading sources, and a volume source term, known as the quadruple source derived from the original Lighthill theory.

PROBLEM DESCRIPTION AND METHODOLOGY

In the present paper, propeller consisting of seven blades was considered for study. The diameter of propeller is 3.1 m and hub to propeller diameter ratio is 0.2. In the present simulations for prediction of non-cavitation noise of propeller at rotating speed of propeller is taken as 242 rpm and the speed of ship is 9.26 m/sec.

Modeling of the propeller is done using CATIA V5R16. In order to model the blade, it is necessary to have sections of the propeller at various radii. These sections are drawn and rotated through their respective pitch angles. Then all rotated sections are projected onto right circular cylinders of respective radii as shown in Figure 1. Now by using multi section surface option, the blade is modelled. After enclosing the entire surfaces solid is created as shown in Figure 2.

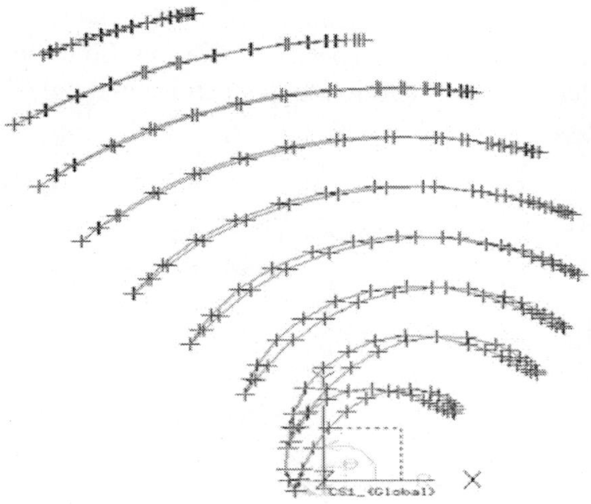

Fig. 1: Points on Surface of the Blade

Fig. 2: Solid Model of Propeller

The objective of the present CFD and CAA analyses is to find overall SPL of a propeller at receiver positions. Three dimensional structural tetrahedral grids are generated to descretize the domain. In order to obtain better results tetrahedral element has been used by taking advantage of mesh consistency. The domain is discretized by minimum 10 linear elements per source wavelength to increase the accuracy of the acoustic propagation. Commercially available grid generation code GAMBIT is used to mesh the entire domain of propeller. Convergence is checked with various element sizes and it is observed that close results are observed with element size 10 and beyond. Hence finally element size 10 is used for mesh generation. After convergence there are 436445 and 680134 tetrahedral elements in the inner volume (propeller volume) and outer volume (far field volume) respectively. Figure 3 shows the meshed model of propeller in inner volume.

The numerical simulations have been carried out with a finite volume code method using FLUENT. The turbulent nature of the flow is incorporated through the LES. LES is chosen as viscous model because, it needs time dependent solution for hydrodynamic solution and it is not highly dependent on geometrical conditions. For steady-state simulations, the Multiple Reference Frames (MRF) model is used, and for unsteady simulations, the sliding mesh model is adopted. Surfaces that rotate relatively are defined as "moving wall". Moreover, as they are dependent on the fluid around them and as they rotate, they are defined as "relative to adjacent cell zone" and "rotational motion".

Cylinder walls are defined as "stationary wall" and the inlet and outlet are defined as "velocity inlet (9.26 m/sec)" and "outflow". Fluid zone in the inner volume is defined as "moving mesh" and 242 rpm in x-direction. However, fluid in the outer volume is defined as "velocity inlet (9.26m/sec)". Figure 4 shows the boundary conditions on entire domain of propeller. In order to give sliding mesh property, the same surfaces in the inner and outer volume families are defined as "interface", and then merged equations resolved iteratively at time increment of 6.88×10^{-4} s for angular speed of 242 rpm. Solution is stopped when changes in solution variables from one iteration to the next is negligible.

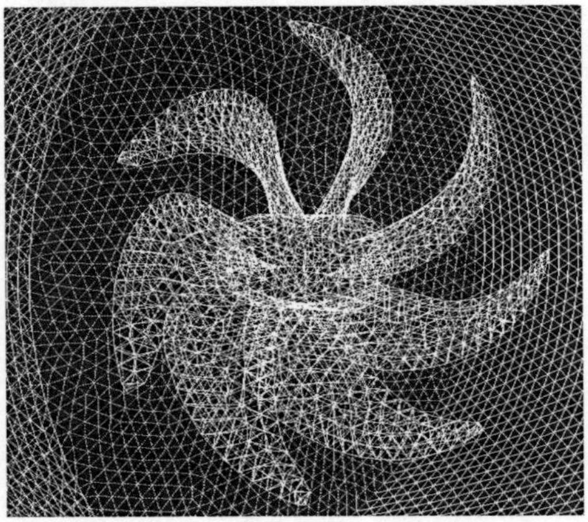

Fig. 3: Meshed Model of Propeller

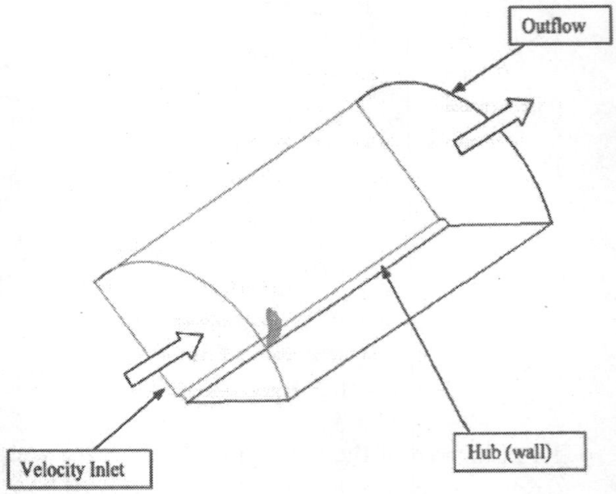

Fig. 4: Boundary Conditions on Propeller Domain

RESULTS AND DISCUSSION

Basic flow fields of propeller are obtained using a LES solver around the 7-bladed propeller in uniform flow and noise prediction results are presented for non-cavitating propeller at different receiver positions. The density and speed of sound in the sea water are considered as 1026 kg/m^3 and 1500 m/s, respectively for the present simulations. The receiver positions considered are 10m and 15m in radial direction to the propeller. Figure 5 and 6 show the noise prediction graphs at 10m and 15m corrected to 1m from propeller. Figures 5 and 6 clearly show that peak SPL values are observed at shaft rotational frequency. The basic noise source in propeller is dipolar noise source as generated due to hydrodynamic pressure fluctuation.

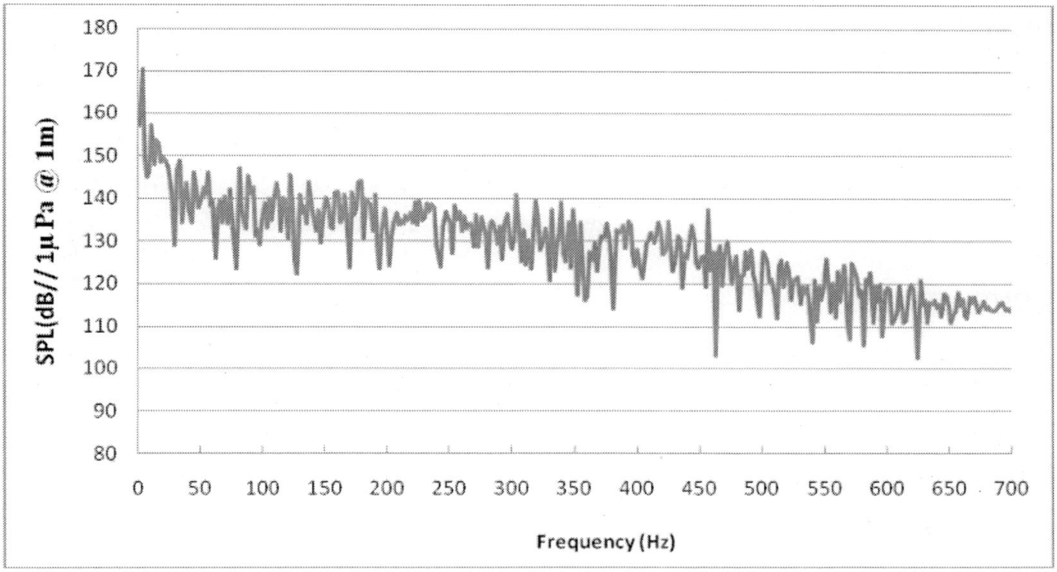

Fig. 5: Noise Prediction Graph for Receiver at 10 m Corrected to 1 m Distance from Propeller

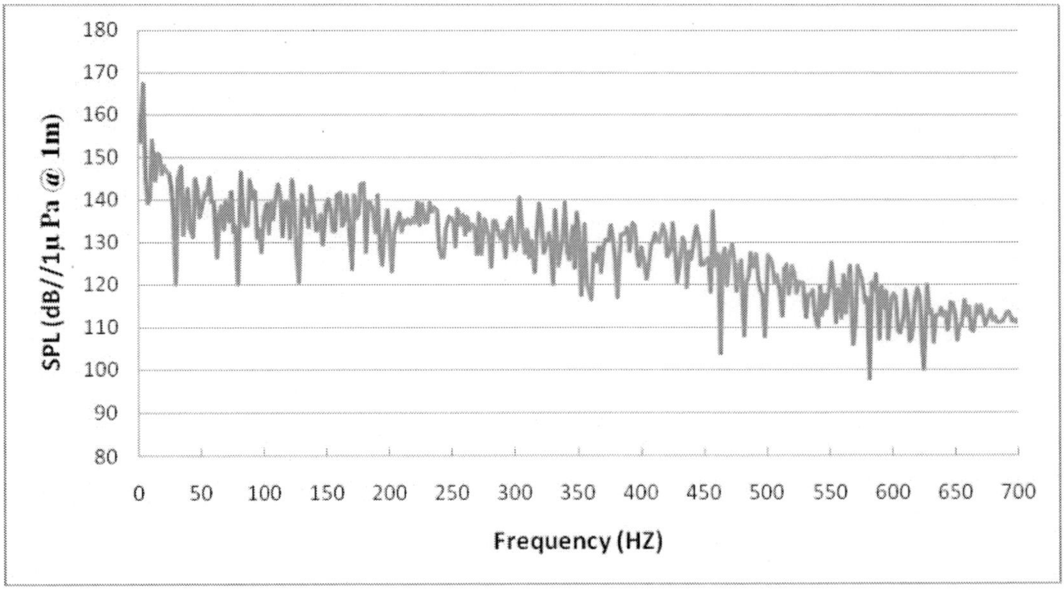

Fig. 6: Noise Prediction Graph for Receiver at 15 m Corrected to 1 m Distance from Propeller

CONCLUSIONS

Non-cavitation noise generated by underwater propeller has been analysed numericallyin this study using FW-H equation.Sound pressure levels are predicted and plotted at different receiver positions. From this study, noise prediction methodology has been developed with LES and FW-H equation.

ACKNOWLEDGMENTS

Authors expresses their sincere thanks to Sri S.V. Rangarajan, Sc'H', Director NSTL, Visakhapatnam for his continuous support for publishing this paper.

REFERENCES

[1] F. Salvatore, C. Testa, S. Ianniello and F. Pereira, 2006, Theoretical and modelling of unsteady cavitation and induced noise, INSEAN, Italian Ship Model Basin, Rome, Italy. Sixth International Symposium on Cavitation, CAV2006, Wageningen, The Netherlands.

[2] B. Chang, 1998, Application of CFD to P4119 Propeller, 22nd ITTC Propeller RANS/Panel Method Workshop, France.

[3] M.J. Lighthill, 1952, On sound generated aerodynamically: I. General theory, Proc. Royal Society (London), A 211, pp. 564–587, 1952.

[4] J.E. Ffowcs Williams and D.L. Hawkings, 1969, Sound generated by turbulence and surfaces in arbitrary motion, Philosophical transactions of Royal Society Series, A 264, No. 1151, pp. 321–342.

[5] F. Farassat and M.K. Myers, 1988, Extensions of Kirchhoffs formula to radiation from moving surfaces, Journal of Sound and vibration, 123(3), pp. 451–460.

[6] Jin-ming Ye, Ying Xiong, Chang-run Xiao, Yi Bi, 2012, Prediction of non-cavitation propeller noise in time domain, China Ocean Engineering, 25(3), pp. 531–538

[7] Q. Yang, Y. Wang and M. Zhang, 2011, Calculation of propeller's load noise using LES and BEM numerical acoustics coupling methods, Transactions of the Wessex Institute, Paper DOI: 10.2495/BE110081.

SECTION–V

Ultrasonics

Study of Molecular Interactions and Ultrasonic Velocity on Binary Mixture of Dehpa [di-(2-ethylhexyl) Phosphoric acid] + Ethanol at 303 K–318 K

Sujata Mishra and Rita Paikaray

Department of Physics, Ravenshaw University, Cuttack

E-mail: sujataananya@yahoo.co.in; r_ paikaray@rediffmail.com

ABSTRACT: *Ultrasonic speed and density have been measured at 303 K–318 K in binary system (di-(2-ethyl hexyl) phosphoric acid) of dehpa with ethanol over entire composition range. The acoustical parameters such as adiabatic compressibility (K_s), free length (L_f), acoustic impedance (Z), Relative association (R_A), wada's constant (W), Available volume (V_a), interaction parameter (χ), surface tension (σ) and also excess parameters such as excess adiabatic compressibility (K^E_s), Excess velocity (C^E), Excess free length L^E_f, Excess molar volume (V^E) and Excess acoustic impedance(Z^E) are computed for the binary mixture at different temperatures. The results are interpreted in terms of molecular interaction between the components of the mixture. The ultrasonic speeds calculated using Nomoato relation, impedance dependence relation Rao's specific sound velocity relation and free length theory are compared with those obtained experimentally to reflect the molecular interaction existing in the present system.*

Keywords: Binary Mixtures, Acoustic Parameters, Intermolecular Interaction, Adiabatic Compressibility.

INTRODUCTION

Intermolecular interactions and thermodynamic properties of the binary liquid mixtures can be estimated more precisely using ultrasonic technique [1,2]. The study of solution properties of liquid mixtures consisting of polar as well as non-polar components finds applications in industrial, chemical and biological processes, speed of sound, itself is highly sensitive to the structure and interactions present in the liquid system as it is fundamentally related to the binding force between the constituents of the medium and also sensitive to temperature. The measurement of speed of sound in liquid mixtures used to determine some useful Acoustic and thermodynamic parameters and their excess values, that are very sensitive to molecular interactions [3,4].Dehpa [$C_8H_{17}O)_2$ PO_2H] is a organophosphorus acid used for extraction of uranium, cobalt, nickel, rare earths and vanadium. It is characterized by the presence of two functional groups in their Molecules, hydroxyl (OH^-) and phosphonate ($P = O$). The simultaneous participation of both groups affects the resulting extraction ability. The extraction efficiency of the extractants improves with the addition of suitable organic diluents and modifier [5,6]. The organic solvents such as alcohols, ketones, ethers etc contain lone pair of electrons on the oxygen atom and can therefore directly solvate protons and metal ions and bring about their extraction. It is interesting to note that alcohols solvate better than ketones and ethers. Ethanol is a good solvent that can dissolve both polar and non-polar components. Ethanol may be the effective diluents or modifier, which increases the extraction efficiency of dehpa. The present investigation is carried out to study the nature and extent of molecular interaction in binary liquid mixture of dehpa and Ethanol in the temperature range 303K–318K, covering entire composition. Theoretical evaluation of ultrasonic velocity in liquid mixtures and its comparison with experimental values reflects the molecular interactions in these mixtures. Ultrasonic velocities evaluated using Nomato's relation (NR) [8]. Free length theory (FLT) [9], Impedance dependence relation (IDR), and Rao's specific sound velocity relation[10] have been compared with those obtained from the experimental values.

EXPERIMENTAL DETAILS

Chemicals used in this study were of Merck grade chemicals (99.9% purity) and used without further purification. The ultrasonic velocities in liquid mixtures have been measured using an ultrasonic interferometer supplied by Mittal Enterprises, New Delhi working at IMHZ frequency with a high degree of accuracy. The temperatures were controlled by circulating water around the liquid cell from a thermostatically controlled water bath. The density of the pure liquid and liquid mixture was determined with a specific gravity bottle with a 25ml capacity.

THEORETICAL CONSIDERATIONS:

The experimentally measured ultrasonic speed (C) and density (ρ) are used to calculate the derived parameters like isentropic compressibility (K_s), Intermolecular free length (L_f), molar volume (V) acoustic impedance (Z), Relative association(R_A), Interaction parameter (χ), Wada's constant (W) and excess parameters by using the following expressions:

$$K_s = \frac{1}{\rho C^2} \quad \ldots (1)$$

$$= k\,(K_s)^{1/2} \quad \ldots (2)$$

$$V = (x_1 M_1 + x_2 M_2)/\, \rho_{mix} \quad \ldots (3)$$

$$R_A = (\rho/\rho_o)\ (C_o/C)^{1/3} \qquad \ldots (4)$$

$$Z = \rho C \qquad \ldots (5)$$

$$\chi = (C^2_{exp}/C^2_{ideal})-1 \qquad \ldots (6)$$

$$W = V\ (K_s)^{-1/7} \qquad \ldots (7)$$

$$\sigma = 6.3 \times 10^{-4}\ \rho\ C^{3/2} \qquad \ldots (8)$$

where $C_{ideal} = x_1C_1 + x_2C_2$ is ideal mixing velocity.

$$A^E = A_{exp} - (x_1A_1 + x_2\ A_2) \qquad \ldots (9)$$

where x_1 and x_2 are mole fractions of DEHPA and Ethanol, ρ_0 = density of DEHPA Co = Ultrasonic velocity in pure DEHPA at particular temperature, ρ_{mix} = density of mixture. M_1 and M_2 are molecular weight of DEHPA and Ethanol. k (Jacobson's constant) [value, 93.875 + 0.375T) × 10^{-8}, T is absolute temperature, A^E stands for excess property of parameters (C^E, $K^E{}_s$, Z^E). The theoretical values of sound speeds are evaluated using free length theory (FLT) [9], Nomato's relation (NR) [8], impedance dependence relation and Rao's specific sound velocity relation (R) (IDR) [10].

RESULTS AND DISCUSSIONS:

The experimental values of ρ and C were used to calculate various acoustic parameters and their excess values. The values of ρ, C, Ks, L_f, Z, χ, R_A, w, Va, σ, as functions of molefraction of dehpa at temperature range 303 K–318 K are listed in Table 1. Representative graphs of excess velocity (C^E), excess isentropic compressibility ($K_S{}^E$), excess acoustic impedance (Z^E) and wada's constant (w) as a function of concentration and temperature are presented in Figures 1–4. Table2 compares experimental and theoretical values of sound speed. It is observed (Table1) that the values of ρ, C, R_A, W, χ, Z, σ, increase while those of K_S and L_f decrease with increase of mole fraction of dehpa at all temperatures taken for study. The variation of C through the mixture depend upon increase or decrease of L_f. In general C and L_f have been reported to vary inverse of each after with composition of the mixture [11,12], as in the present system. The non-linear decrease of K_S and L_f, and also an increase in R_A, Z and σ with the mole fraction of dehpa at all temperatures indicates the presence of molecular interaction between unlike components. Considering the effect of temperature, both L_f and K_S increases with rise in temperature at all concentrations (Table 1). These results are supported by pandey *et al.* [12] view that an increase in temperature results increase of inter molecular distance in liquids, then by increasing the distance between surfaces of two molecules. The increase in values of L_f and K_S and decrease of R_A, χ, σ, with increase in temperature clearly reveals that the interactions becomes weaker at high temperatures. Wada's constant (W) increase linearly with concentrations of dehpa and remain independent of temperature. It indicates strong solute solvent interaction without complex formation.

The molecules of dehpa are polar and exist as dimmers in pure form due to hydrogen bonding. The molecules of Ethanol are polar and associated through hydrogen bond in pure state. The hydrophilic-OH group of ethanol can dissolve the polar whereas the short hydrophobic alkyl group can dissolve the non-polar (7). The mixing of dehpa with Ethanol would favour formations of new hydrogen bonds and dipole-dipole interaction between unlike molecules. The dipoles in dehpa and Ethanol arise due to electro negativity difference between oxygen, hydrogen and phosphorus. They are in order O> H >P. In order to highlight the presence of interactions between the molecules, it is essential to study the excess parameters.

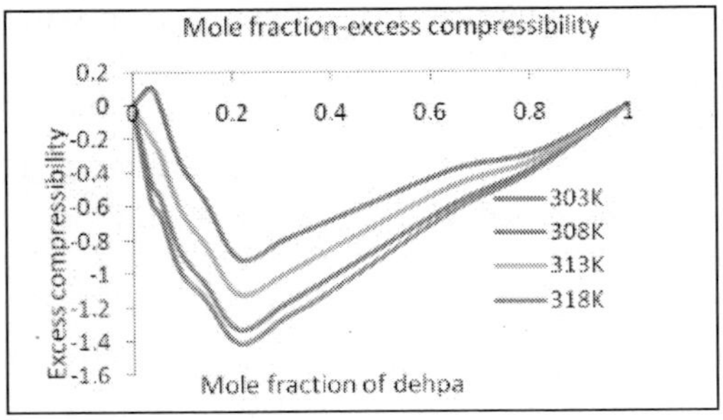

Fig. 1

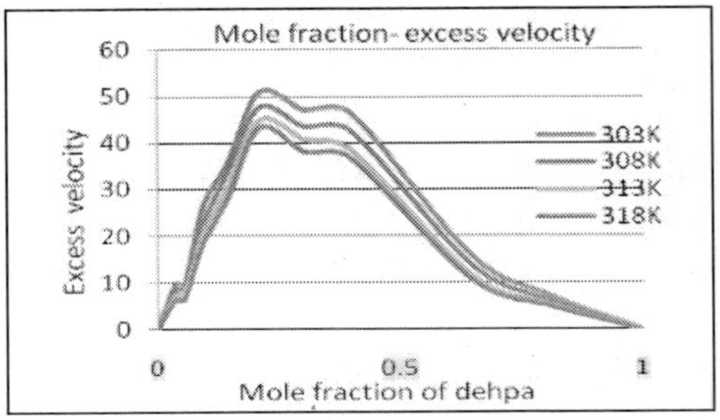

Fig. 2

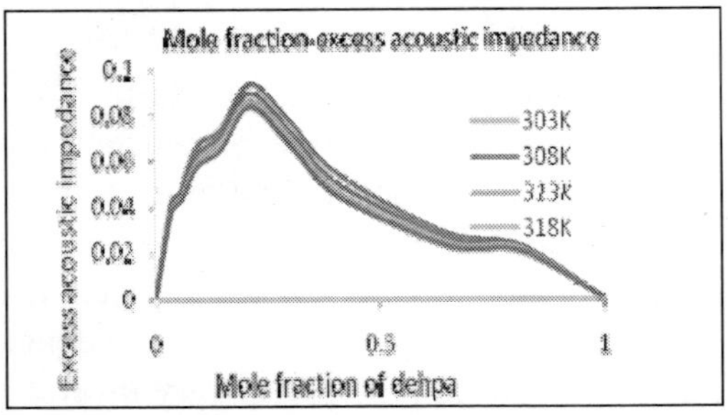

Fig. 3

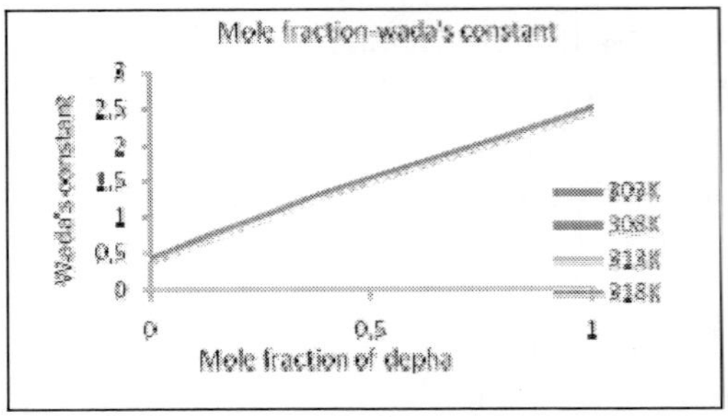

Fig. 4

Figures 1 and 2 show that K_S^E is negative (except in 0.6 mole fraction of dehpa at 318K) and C^E is positive at all concentration of dehpa. K_S^E and C^E can be considered as arising from two types of interactions between the component molecules. (i) Physical interaction consisting mainly of dispersion forces making a positive deviation in compressibility and negative deviation in sound speed; and (ii) chemical or specific interaction which include strong dipole-dipole interactions, forming of Hydrogen bonds and other complex forming interactions resulting in negative deviation in compressibility and positive deviation in sound speed.

Table 1: Computed Values of Acoustical Parameters at Different Temperatures for Dehpa + Ethanol

(X_1)	C m/s	ρ $kg\ m^{-3}$	Ks $10^{-10}\ N^{-1}m^2$	Z $10^{-6}kgm^{-2}s^{-1}$	L_f $10^{-11}\ m$	χ	R_A	W $m^3\ mol^{-1}$ (Nm^{-2})	V_a $10^{-5}\ m^3\ mol^{-1}$	σ N/m
At 303K										
0.0000	1138.0	785.1	9.8353	0.8934	6.5074	0.0000	–	0.4234	16.943	18.986
0.0557	1158.0	832.9	8.9534	0.9644	6.2088	0.0174	0.8622	0.5395	20.381	20.675
0.0938	1181.5	847.4	8.4608	1.0012	6.0356	0.0460	0.8714	0.6263	22.221	21.680
0.1458	1200.6	858.7	8.0790	1.0309	5.8978	0.0625	0.8782	0.7463	25.105	22.503
0.2119	1228.0	879.3	7.5416	1.0797	5.6983	0.0887	0.8926	0.8917	27.665	23.837
0.3050	1241.0	888.9	7.3047	1.1031	5.6081	0.0804	0.8991	1.1040	32.905	24.481
0.4023	1258.0	894.4	7.0649	1.1251	5.5153	0.0777	0.9006	1.3300	37.580	25.140
0.6451	1272.0	947.9	6.5202	1.2057	5.2984	0.0251	0.9510	1.8109	48.518	27.090
0.8123	1294.0	984.3	6.0674	1.2736	5.1111	0.0107	0.9819	2.1250	53.600	28.863
1.0000	1321.6	1009.4	5.6720	1.3340	4.9418	0.0000	–	2.4931	55.580	30.549
At 308K										
0.0000	1126.5	777.9	10.0130	0.8763	6.6253	0.00000	–	0.4262	17.526	18.528
0.0557	1145.0	826.0	9.2344	0.9457	6.3625	0.0157	0.8547	0.5417	21.158	20.160
0.0938	1165.5	840.8	8.7555	0.9799	6.1953	0.0399	0.8650	0.6281	23.278	21.075
0.1458	1186.0	852.4	8.3404	1.0109	6.0466	0.9597	0.8719	0.7485	26.215	21.933
0.2119	1211.5	873.4	7.8008	1.0581	5.8478	0.0838	0.8870	0.8934	29.125	23.202
0.3050	1223.5	883.6	7.5602	1.0810	5.7569	0.0750	0.8944	1.1052	34.762	23.823
0.4023	1239.4	889.7	7.3170	1.1026	5.6635	0.0719	0.8967	1.3383	39.833	24.454
0.6451	1253.5	944.7	6.7368	1.1841	5.4344	0.0223	0.9485	1.8085	51.428	26.412
0.8123	1275.0	982.3	6.2623	1.2524	5.2395	0.0091	0.9808	2.1198	55.947	28.172
1.0000	1302.2	1008.5	5.8474	1.3132	5.0629	0.0000	–	2.4847	59.506	29.855
At 313K										
0.0000	1115.2	770.1	10.0441	0.8588	6.6950	0.0000	–	0.4303	18.126	18.066
0.0557	1132.0	818.4	9.5354	0.9264	6.5232	0.0135	0.8493	0.5442	21.965	19.636
0.0938	1151.2	833.3	9.0551	0.9592	6.3568	0.0362	0.8599	0.6307	24.233	20.534
0.1458	1170.5	845.1	8.6367	0.9891	6.2082	0.0547	0.8673	0.7511	27.432	21.318
0.2119	1195.8	866.4	8.0716	1.0360	6.0017	0.0797	0.8829	0.8962	30.508	22.569
0.3050	1207.2	877.0	7.8242	1.0587	5.9090	0.0709	0.8908	1.1081	36.491	23.172
0.4023	1221.2	883.5	7.5896	1.0789	5.8197	0.0659	0.8939	1.3326	42.137	23.751
0.6451	1235.0	939.4	6.9793	1.1601	5.5808	0.0188	0.9469	1.8097	54.480	25.684
0.8123	1257.7	977.6	6.4667	1.2295	5.3720	0.0094	0.9796	2.1202	59.205	27.468
1.0000	1283.4	1004.6	6.0434	1.2893	5.1932	0.0000	–	2.4828	63.508	29.098

The actual values depend upon the relative strengths of two opposing effects. The observed negative values of K_S^E and positive values of C^E for the mixture imply that the specific interactions dominate over the dispersive interactions between unlike molecules. At about 0.2 mole fraction of dehpa, both K_S^E and C^E clearly indicates that the hydroxyl group at Ethanol has more electronegative centre, i.e. oxygen which form strong inter molecular interaction with positive centre of dehpa.

There is a possibility of H- bonding between oxygen atom of dehpa (with its lone pair of electron) and the hydrogen atom of –OH group of Ethanol. But the less magnitude of negative K_S^E and positive C^E reveals that the two bulky alkyl groups in dehpa create steric hindrance for the formation of H- bonds between –OH groups of Ethanol and dehpa and existence of strong dipole-dipole interaction between unlike molecules. This fact is supported by large difference in molecular sizes of dehpa (molar volume = 31.9427×10^{-5} m^3 mol^{-1}) and Ethanol (molar volume = 5.8680×10^{-5} m^3 mol^{-1}). It seems that the component molecules pack well into each other's structure. This results in decrease of free length and compressibility and hence negative K_S^E and positive C^E values. This conclusion is fortified by the positive values of Z^E (Figure 3) and linear variation of wada's constant (W) fig4 with increase of concentration of dehpa. Similar observations are found by others [13,14,15] when temperature increases the negative values of K_S^E and positive values of C^E and Z^E decreases. It indicates the weakening of interactions at higher temperature, as the greater thermal motion overcomes the mutual orientating effects of the dipoles [18]. This fact is supported by the decreases of interaction parameter (χ) and relative association (R_A) with increase of temperature.

The experimental values of sound speed along with the theoretical values and percentage of deviation for free length theory (FLT), Nomoto's relation (NOM), Impedance dependence relation (IDR), Rao's specific sound velocity relation (R) are presented in Table 2 at temperature 303 K. The negative deviations of experimental values from values calculated using Rao's relation and Nomoto relation imply non-additivity of Rao's specific sound velocity and molar sound velocity in liquid mixtures.

It indicates absence of specific molecular interactions leading to complex formation. The positive deviations observed in Impedance relation implies that acoustic impedance in liquid mixture is greater than the ideal values [10]. This indicates that component molecules in liquid mixture have a compact arrangement than in component liquids forming the mixture [16,17]. The velocity values calculated using free length theory also confirms above predictions. Among the four theories taken up for the prediction of sound velocity, (FLT) is best suited for the evaluation of ultrasonic velocity.

Table 2: Application of Various Theoretical Approaches and Percentage of Deviation for Dehpa + Ethanol System at 303 K

|ΔC/C| %

Mole Fraction of DEHPA (X_1)	C^{EXP} ms^{-1}	C^{NOM} ms^{-1}	C^{IDR} ms^{-1}	C^{FLT} ms^{-1}	C^{R} ms^{-1}	C^{NOM}	C^{IDR}	C^{FLT}	C^{R}
0.0000	1138.0	1138.0	1138.0	1138.0	1138.0	0.0000	0.0000	0.0000	0.0000
.0557	1158.0	1180.9	1150.8	1158.0	1317.3	1.9775	0.6217	0.0000	13.756
.0938	1181.5	1202.0	1159.5	1181.6	1358.6	1.7350	1.8620	0.0084	14.989
.1458	1200.6	1224.1	1170.9	1200.8	1373.2	1.9573	2.4737	0.0166	14.376
.2119	1228.0	1244.8	1185.0	1228.5	1420.3	1.3680	3.5016	0.0471	15.659
.3050	1241.0	1265.5	1204.1	1241.0	1390.9	1.9742	2.9734	0.0000	12.078
.4023	1258.0	1280.6	1223.1	1258.3	1338.5	1.7965	2.7742	0..238	6.399
.6451	1272.0	1303.9	1266.5	1272.2	1375.1	2.5078	0.4323	0.0157	8.1053
.8123	1294.0	1313.7	1293.6	1294.4	1385.9	1.5224	0.0391	0.0309	7.1020
.1.0000	1321.6	1321.6	1321.6	1321.6	1321.6	0.0000	0.0000	0.0000	0.0000

CONCLUSION

The variation of derived parameters with concentrations of dehpa and temperature indicate the presence of molecular interaction between component molecules. Excess values reveals that strong dipole- dipole interactions are existing between unlike molecules. The temperature variation indicates that the strength of inter molecular interaction decreases with rise of temperature. The positive deviations of experimental values from values calculated using impedance relation shows existence of strong intermolecular interactions in liquid mixtures. Free length theory is best suited for the evaluation of ultrasonic velocity in the present system.

REFERENCES

[1] S. Jayakumar, N. Karunanithi and V.Kamappan, *Indian J Pure Appl Phy*. 34 (1996)761.

[2] S.K. Dash, V. Chakravorthy and B.B. Swain, Accoust Leto 19(1996) 142 .

[3] S. Mishra and R. Paikaray, J. Acoustic Soc-India 37 (2010)20.

[4] S. Mishra and R.Paikaray, J. Acoustic Soc-India No.3 (2011)144.

[5] S. Mishra and R. Paikaray, proceedings of National Symposium Acoustic, 2009.

[6] S. Mishra and R. Paikaray, proceedings of National Symposium Acoustics, 2011.

[7] I.G. Wade, Jr. M.K. Singh, Organic Chemistry, 6th edn (Pearson Education),2008, ISBN:81-7758–739–0.

[8] O. Nomoto, *J. Phys. Soc. Jpn* 13(1958)1528.
[9] B. Jacobsons, *J. Chem.Phys* 20 (1952)927.
[10] D. Saravan Kumar and D. Krishna Rao, *Ind. J Pure appl. Phy.* 45 (2007)210.
[11] H. Eyring and JF Kincaid, *J. Chem Phys*, 6(1938) 620.
[12] A. Ali, A.K. Nair and M Kamil, thermochim. Acta 274(1996)209.
[13] R. Thiyagrajan and L Palaniappan, *Indian J Pure Appl Phys* 46 (2008)852.
[14] J.D. Pandey, R. Dey and D.K Dwivedi, *Pramana-J. Phys* 52(1999) 187.
[15] T. Savitha Jyotsna and N. Satyanarayana, *India J. Pure Appl Phy* 43(2005)591.
[16] R. Mehra and M. Pancholi, *Indian J Phys.* 80 (2006) 253.
[17] R. Mehra and M. Pancholi, *Indian J Pure Appl. Phy.* 45(2007)580.

Ultrasonic Study of Molecular Interactions in Binary Mixture of Di-n-Butyl Pathalate with Phenol at Different Frequencies

N. Mohanty and R. Paikaray

PG Department of Physics, Ravenshaw University, Cuttack, Orissa

E-mail: mohantyninima@gmail.com; r_paikaray@rediffmail.com

ABSTRACT: The Ultrasonic Velocity and density of Binary mixture of DBP with phenol at frequencies 1MHz, 3 MHz, 5 MHz and 7 MHz are measured over entire composition range at 318K. The acoustical parameters such as adiabatic compressibility (Ks), Intermolecular free length (Lf), acoustic impedance (Z),Molecular interaction parameter (χ) and also excess parameters such as excess velocity (UE),excess free length (LEf), and excess compressibility (KEs) are computed for the binary mixture at different frequencies. The variation of these parameters with composition of the mixture helps us in understanding the nature and extent of interaction between unlike molecules in the mixture. Further theoretical values of ultrasonic velocity were evaluated using theories and empirical relations. Merits of these theories were discussed.

Keywords: Binary Mixture, Ultrasonic Velocity, Isentropic Compressibility, Intermolecular Free Length and Acoustic Impedance.

INTRODUCTION

Ultrasonic Investigation in binary mixture has an extensive application to study the physicochemical behaviour of liquid mixture [1–2]. Different acoustical parameters and their excess values basing on ultrasonic velocity measurement provide qualitative information on molecular interaction in liquid mixture [3–4]. These acoustical properties have a wide use in engineering, textile industries, pharmaceutical industries etc.

Survey of literature shows that no effort has been made to acoustical parameters for the binary mixture of DBP and phenol at different frequencies. So the present study is undertaken to have a deeper understanding of acoustical parameters between the components of the above binary mixtures at different frequencies.

EXPERIMENTAL

Ultrasonic velocity was measured by using a single crystal variable path interferometer at different frequencies of 1 MHz, 3 MHz, 5 MHz and 7 MHz at 318 K. Density was determined with a pyknometer of 25 cm capacity by standard procedure.

THEORY

Experimental measured values of ultrasonic velocity (U) and density (ρ) are used to calculate the parameters such as isentropic compressibility (β), intermolecular free length (L_f) and their excess values basing on the following relations[5,6].

$$\beta = 1/\rho U^2,\ L_f = K\,\rho^{1/2}$$

where K is Jacobson's constant. This is temperature dependent.

T the excess values are calculated as

$$\beta^E = \beta - (X_A\,\beta_A + X_B\,\beta_B)\ L_f^E = L_f\text{-}(X_A L_{fA} + X_B L_{fB})$$

where X_A and X_B are mole fraction of DBP and toluene respectively.

The obtained experimental values of ultrasonic velocity are compared with the following relation[7]:

1. *Nomoto's relation(NR)[8]:*

$$U_{NR} = [(X_A R_A + X_B R_B)/(X_A V_A + X_B V_B)]^3 \quad \dots (1)$$

where U_{NR}, R_A, R_B, V_A and V_B are ultrasonic velocity as per Nomoto relation, molar sound velocities and molar volume of components of mixture respectively.

2. *Jacobson's Free length theory(FLT):*

$$U_{FLT} = [K/L_f\,\rho^{1/2}] \quad \dots (2)$$

3. *Impedance Dependence Relation (IDR)*

$$U_{IDR} = (X_A Z_A + X_B Z_B)/(X_A\rho_A + X_A\rho_B) \quad \dots (3)$$

Z_A, Z_B – Acoustic impedance of components of mixture.

4. *Ideal Mixture Relation (IMR)/Van Dael and Vangeal Relation[9]*

$$U_{IMR} = [1/(X_A M_A + X_B M_B)]^{1/2}\ [X_A/M_A U^2{}_A + X_B/M_B U^2{}_B]^{-1/2} \quad \dots (4)$$

5. *The modulus of percentage deviation in ultrasonic velocity between experimental and theoretical values are calculated as*

$$(du/u)\ \% = ((U_{exp} - U_{thep})/U_{exp}) \times 100 \quad \dots (5)$$

6. *Interaction parameter* (χ) *to measure strength of interaction is given by,*

$$\chi = (U_{exp}/U_{ideal})^2 - 1 \qquad \ldots (6)$$

RESULT AND DISSCUSSION

Measured values of ultrasonic velocity U and density ρ of the binary mixture DBP with Phenol were used to calculate β, L_f, Z and excess values of ultrasonic velocity U^E, β^E, L^E_f, excess molar values using standard relation with accuracy up to four decimal places. The ultrasonic velocity studies carried out in the present investigation reveal that the velocity varies with concentration due to interaction of mixture through molecular association. The existing particle- particle resistance initiates some more interactions and this is supported by the measured parameters.

The variation of ultrasonic velocity with mole fraction of DBP at different frequencies is shown in Figure 1. It is observed from the graph that ultrasonic velocity decreases gradually with increases in mole fraction DBP over entire range of composition. However, the decrease in velocity is not uniform, which indicates different type of molecular association exist in the binary mixture at different concentration of DBP. At X_b = 0.93 there is a drop of ultrasonic velocity. It may be explained in terms of intermolecular hydrogen bonding between solvent-solvent due to presence of free H^+ ion from phenol. Due to solvent-solvent association through hydrogen bonding ultrasonic velocity drops at certain point but phenolet (C_6 HO^-) ion which is a strong and stable ion enhances ion-solvent interaction between free DBP and the phenolet ion. Similar trend is observed for all frequencies. Enhancement of frequency decreases the ultrasonic velocity due to less interaction.

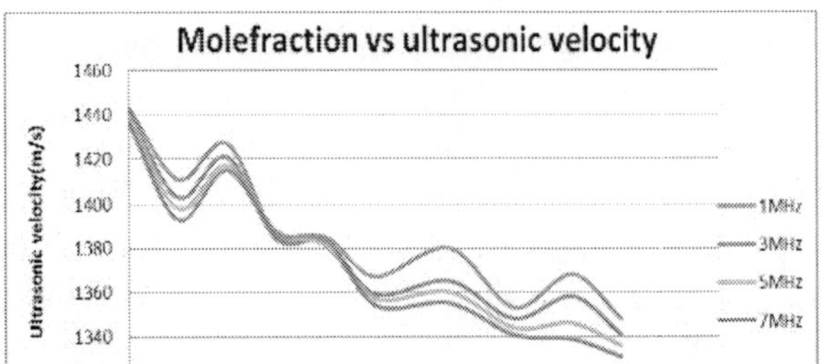

Fig. 1

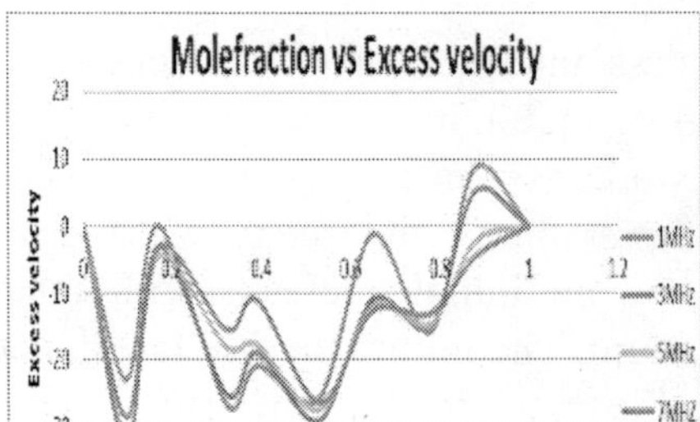

Fig. 2

Excess thermodynamics parameters play the measure role in understanding the nature of molecular interaction in liquid mixture and are studied by several workers [5–10]. Figure 2 shows the negative deviation of excess velocity in all range of composition except higher concentration of DBP (0.88 mole fraction) at low frequencies of 1 MHz and 3 MHz and (0.7 mole fraction of DBP at 1 MHz frequency). Negative contribution is due to breaking of self-association of DBP with Phenol.

The excess adiabatic compressibility values are negative in the entire range of composition at different frequencies. The negative peak value of excess compressibility is around 0.4 to 0.7 mole fraction of DBP are shown in the Figure 3. The negative excess value is due to breaking of hydrogen bond in DBP and Phenol which leads to an increase in free space in the mixture contributing to negative deviation in sound velocity and positive deviation in compressibility. Negative value of β^E in the binary mixture indicates the presence of molecular interactions in liquid mixtures forming dipole –dipole interaction as well as interstitial accommodation of molecules making the structure more compact [11]. According to Fort Moore [12] an increase in negative β^E indicates grater interaction between components. From the graph the trend of excess compressibility is nearly equal but at peak region there is slightly higher deviation for higher frequency.

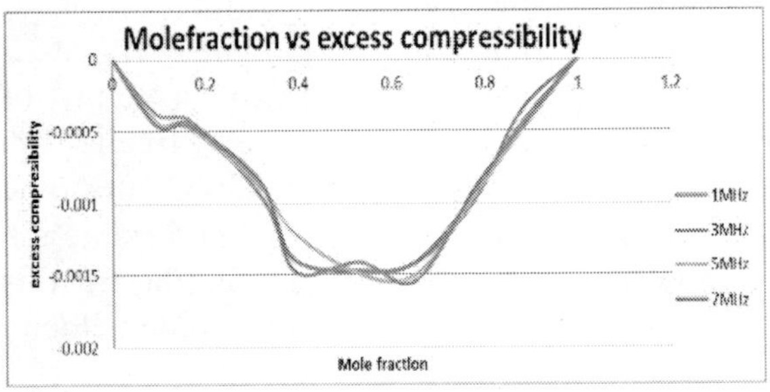

Fig. 3

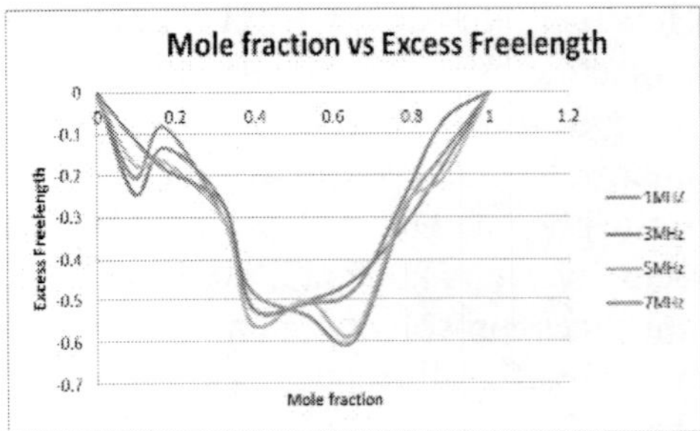

Fig. 4

Excess intermolecular free length is negative for the entire range of composition as shown in Figure 4 which suggests structural changes in the mixture. When the liquid mixtures assume a less compressible phase of decreased flow as it is evident by deviation in compressibility, the free length might be affected by same structural changes[13]. Figure 4 shows an initial decrease with increase in the composition DBP and phenol in the mixture. The minimum free length occurs around 0.6 and this indicates that the structural readjustment in the liquid mixtures are in the direction of less compressible phase or closer of molecules. Hence association of molecules takes place either through physical forces like dipole-dipole interaction of hetero molecules or chemical forces such as H-bonding between homogeneous molecules leading to contraction in volume [14–16].

The experimental values, theoretical values of ultrasonic velocity and modulus of percentage of deviation in velocities and interaction parameter at 318K are calculated by using equation (1–6) and are shown in Table 1.

It is observed from the Table 1, FLT gives the best agreement with the experimental values of ultrasonic velocity. Nomoto's relation and IDR value closely follows the experimental value in the mixture. An overall comparison of sound velocity basing on the present study at 318K of mixture is as:

$U_{FLT} > U_{NR} > U_{IR} > U_{IMR}$

There is a deviation of theoretical values from experimental values due to limitations and approximations of the theories [17]. In the Nomoto's theory, interaction between the components of liquid mixtures has not been taken into account as the volume is supposed to be not change on mixing. In case of ideal mixing, the ratio of specific heats of ideal mixtures and volumes are equal by not taking into the consideration of molecular interactions. Various types of forces such as dipole-dipole, dipole induced dipole, hydrogen bonding, ion-ion, ion-dipole and dispersive forces are operative due to interactions when two liquids are mixed. Interaction parameters are found to be negative indicating weak molecular interaction. Hence free length theory is found to be in good agreement with mixture having weak interactions. The deviation of intermolecular parameter of DBP and Phenol at various mole fractions is a direct measure of non-ideality of the system, which is a consequence of association or other type of interactions.

Table 1: Values of Experimental, Theoretical Ultrasonic Velocity, the Modulus of Percentage Deviation in Velocity and Molecular Interaction Parameter in (IP) at 318 K at 1 MHz

Mole Fraction	U_{exp}	*Density*	*U*				*(dU/U)%*				*IP* (χ)
			U_{NR}	U_{FLT}	U_{IDR}	U_{IMR}	% U_{NR}	% U_{FLT}	% U_{IDR}	% U_{IMR}	
X_A	$m\ s^{-1}$	$g\ cm^{-3}$	$m\ s^{-1}$								
0	1443	1.0545	1443	1443	1443	1443	0	0	0	0	0
0.095	1411	1.0514	1419.55	1411	1434.22	1365.35	0.6059	0.00001	1.6457	3.2351	–0.0318
0.17	1427	1.0489	1405.94	1426.85	1427.32	1320.56	1.4756	0.0107	0.0227	7.4589	0.0002
0.32	1386	1.0440	1386.47	1385.53	1413.20	1261.88	0.0339	0.0339	1.9694	8.9553	–0.0373
0.39	1385	1.0418	1379.7	1385.07	1406.59	1245.53	0.3824	0.0050	1.5587	10.0702	–0.0296
0.53	1367	1.0372	1369.01	1367.29	1393.36	1230.11	0.1471	0.0214	1.9279	10.0140	–0.0365
0.65	1380	1.0333	1361.97	1380.23	1381.86	1233.88	1.3064	0.0169	0.1348	10.5885	–0.0018
0.78	1353	1.0290	1355.87	1353.33	1369.41	1256.11	0.2121	0.0244	1.2129	7.1580	–0.0231
0.88	1368	1.0257	1351.98	1368.40	1359.75	1288.06	1.1714	0.0290	0.6030	5.8438	0.0127
1	1348	1.0218	1348	1348	1348	1348	0	0.082	0	0	0

CONCLUSION

It is studied that acoustical parameters and the excess values of the binary mixture of DBP and phenol shows the molecular association at different concentrations of DBP. The interactions seem to be similar in all frequencies. But increase in frequency decreases the ultrasonic velocity which shows less interaction. This indicates DBP shows non-ideal behavior when mixed with polar diluents. Further, the binary mixture of DBP with Phenol is of non-ideal system as the intermolecular parameters are negative and FLT model is the most applicable theoretical model for this binary mixture.

REFERENCES

[1] G. Douheret, M.I. Davis, I.J. Fjallngar and H. Hailand, J. Chem. Soc., Faraday Trans, 93, 1943 (1997).

[2] K. Tamura, T Sonoda and S. Murakami, J. Solution Chem 28, 777 (1999).

[3] A. Ali and A.K. Nain, Pramana, Journal of Physics, Vol. 58, No. 4, April 2002, pp. 695–701.

[4] Jacobson. B., Acta Chem. Scnd. 5 1214 (1951), 6 1485(1952), J Chem. Physics 20.927 (1952).

[5] G. Nath, S. Sahu and R. Paikaray Indian Journal of Physics, 83(11) 1567 (2009).

[6] R. Paikaray and N. Mohanty, Journal of Acoustical Society of India, 37(2) 70–73 (2010).

[7] N. Mohanty and R. Paikaray, Orissa Journal of Physics), 19(1) 117–120 (2012).

[8] O. Nomoto J. Physics, Soc. Jpn 13.1528 (1958).

[9] W. Vandil and E Vangeal, poroc. International conference on Caloriemetry thermodynamics Warsaw (1969).

[10] Prakash, *et al.*, Acoustic leters, 12(2), 1988, 35.

[11] R.D. Rai, *et al.*, Chem themodymanics, 21, 2125.

[12] R.J. Fort and W.R. Moore, Trans Faraday Soc 61(1965), 2102, 18.

[13] G.V. Rama Rao, A.V. Sharma and C. Rambabu, IJP and Appl Physics Vol. 42, November 2004, pp. 820–826.

[14] Morakova L., *et al.*, I J Chem, Thermodynamics, 37(2005)810.

[15] K.S. Howard and Pike, Chem. engg. data 4 (1959)3331.

[16] J.P.E. Grolier and Benson, J Chem. engg data. 20(1975) 243.

[17] Palaniappan P.L. RM, Pichaimuthu A. and Kamappan A.N., 1998, Ind. J Phys., Vol. 72B (2), p. 175.

Ultrasonic Study of Chemical Mixture Used for Preparation of Fiber from Date Palm Leaves

G. Nath, A. Sarangi and R. Paikaraay[1]
Department of Physics, VSS University of Technology, Burla, Sambalpur, Odisha
[1]Department of Physics, Ravenshaw University, Cuttack, Odisha
E-mail: ganeswar.nath@gmail.com

ABSTRACT: *Propagation of ultrasonic waves in chemical mixture have greater importance in study of different acoustic parameter which are provides information about the strength of different composite materials prepared from different agricultural wastes.Measurement of ultrasonic velocity and density at different concentration of solvent mixture has been carried out at different temperature using multi frequency ultrasonic interferometer. The data obtained are used to evaluate various acoustic parameters in view of identification of the presence of different interaction under prevailing conditions. The variation of ultrasonic velocity, density and other acoustic parameters with concentration under different physical conditions provides information about the mechanical strength of the fiber.*

Keywords: Ultrasonic Velocity, Date Palm Leaves, Solvent Mixture, Compressibility, Acoustic Impedance, Intermolecular Free Length.

INTRODUCTION

High intensity ultrasonic waves are generally employed in the areas of cleaning, plastic welding, machining etc.In addition to these applications, it is observed that if a liquid mixture is subjected to irradiation of high intensity ultrasonic waves, alternate compression and expansion modes takes place which leads to change in some thermodynamic properties of the liquid mixture. As a result of which the variation in speed of ultrasonic waves takes place which affects some of the mechanical properties of the natural fiber. Ultrasonic methods are directly sensitive to these changes and can be used to assess the integrity of the composite structure. The absorption of ultrasound in polymer composite systems is governed by local modes of motion and cooperative because of the existence of strong intermolecular interaction within the liquid mixture. The manner in which the propagation of the ultrasonic wave is affected by structure of the liquid results in parameters that can lead to the binding between the natural fibers. Variation in thermal and acoustic properties with composition provides added information regarding the intermolecular interactions present in a system. The sign and magnitude of the nonlinear deviations from ideality as a function of composition and frequency may be ascribed to the presence of weak or strong type of interaction between unlike molecules. The excess acoustic parameters of binary mixtures have been satisfactorily used in explaining the extent of interactions between mixing components [1–3].The present investigation deals with two important liquids namely acetone and methanol. Both the liquids under investigation are very useful chemicals and of industrial significance. The mixture of these liquids is used in the preparation of date palm leaf fiber. This treatment removes a certain amount of lignin, wax and oils covering the external surface of the fiber cell wall, depolymerizes cellulose and exposes the short length crystallites. This treatment has two effects on the fiber i.e. it increases surface roughness resulting in better mechanical interlocking, and it increases the amount of cellulose exposed on the fiber surface, thus increasing the number of possible reaction sites. Date palm leaves as agricultural waste generated from regular pruning of older leaves have low economic utilization values. Normally, farmers leave this waste outdoors which could provide good environment for certain pests like rodents and insects. Therefore, from both environmental and economic points of view, simultaneous utilization of plastic and date palm leaves wastes for producing a plastic-palm leaves fiber composite wood with good physical, mechanical and chemical properties can be as alternative to natural wood.

EXPERIMENTAL DETAILS

In the present study the chemicals used are of analytical grade purified by standard procedure [4–6] and redistilled before use. Density was determined with a Pyknometer of 25 cm^3 capacity, calibrated with de-ionized double distilled water. Ultrasonic speed was measured by a single crystal variable path ultrasonic interferometer model MX-3, Mittal Enterprises New Delhi, India operating at different frequencies of 1 MHz, 3 MHz and 5 MHz. The temperature stability is maintained within 0.1 K by circulating thermo stated water around the interferometer cell that contains the liquid, with circulating pump. Binary mixtures of acetone were prepared with methanol and kept in an enclosed container properly.

THEORY

The experimental measured values of ultrasonic speed and computed values of density are used to compute acoustic parameters such as intermolecular free length (Lf), isentropic compressibility (β), acoustic impedance (Z) and bulk modulus (K) and their excess values. The above acoustic parameters are determined with the help of the following relationship [7].

Isentropic compressibility,

$$\beta = (\rho C^2)^{-1} \quad \ldots (1)$$

Intermolecular free length,

$$L_f = K\beta^{1/2} \quad \ldots (2)$$

Acoustic impedance,

$$Z = \rho C \quad \ldots (3)$$

Bulk modulus,

$$K = \rho C^2 \quad \ldots (4)$$

and their excess values are calculated as,

$$\beta^E = \beta_{mix} - (X_A\beta_A + X_B\beta_B) \quad \ldots (5)$$

$$L_f^{\ E} = L_{fmix} - (X_A L_{fA} + X_B L_{fB}) \quad \ldots (6)$$

$$Z^E = Z_{mix} - (X_A Z_A + X_B Z_B) \quad \ldots (7)$$

$$K^E = K_{mix} - (X_A K_A + X_B K_B) \quad \ldots (8)$$

where X_A, X_B, β_A, β_B, β_{mix}, L_{fA}, L_{fB}, L_{fmix} Z_A, Z_B, Z_{mix} B_A, B_B and B_{mix} are mole fraction, isentropic compressibility,inter molecular free length, acoustic impedance bulk modulus of methanol, acetone and mixture respectively. The constant K is temperature dependent which is given as $[93.875 + (0.375T)] \times 10^{-8}$ [8].

RESULT AND DISCUSSION

The ultrasonic speed in methanol increases smoothly with mole fractions of acetone as shown in Figure 1. Acetone contains carbonyl functional group, which is polar and hence it can interact with methanol like polar molecules through dipole- dipole interaction. In pure acetone there is dipole-dipole as well as the dispersive interactions.

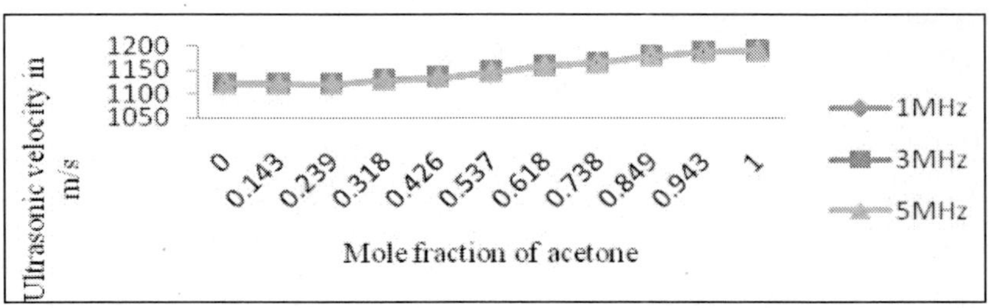

Fig. 1: Variation of Ultrasonic Velocity in Methanol with Acetone

As the concentration of acetone gradually increases the different type of interaction comes into existence due to tautomerisim. In acetone molecules the hydrogen bonding comes into existence between the like molecules of aldol at –OH end. In addition with hydrogen bonding, keto and enol form of tautomerisim comes into exisistence with maximum percentage of keto form for which the dipole-dipole interaction takes vital role for increasing the ultrasonic speed in high concentration of acetone .As seen from the profile, the frequency has no significant contribution in variation of ultrasonic speed in the binary mixture though the frequency varies from 1 MHz-5 MHz. Figures 2 and 3 demonstrates the variation of β^E and $L_f{}^E$ for binary mixture of acetone and methanol which are found to be negative over the entire composition range.

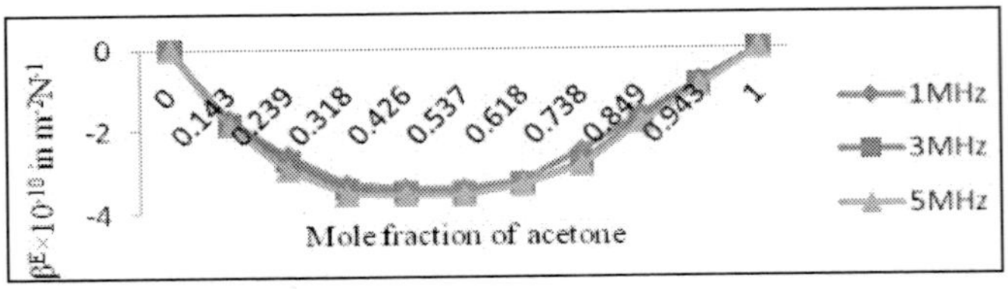

Fig. 2: Variation of Excess Isentropic Compressibility Methanol with Acetone

The negative values of excess isentropic compressibility indicates that the liquid mixture is less compressible than the pure liquids forming the solution and molecules in the mixture are more tightly bound than in pure liquids. Thus negative values of excess isentropic compressibility indicate strong specific interactions between component molecules and interstitial accommodation of smaller molecules in the void created by bigger molecules. The negative excess isentropic compressibility results reduction of volume of mixture favoring the fitting of component molecules in to each other. Greater the negative value of β^E stronger is the attractive interaction between the component molecules such as hydrogen bonding; dipole-dipole interaction and other specific interactions between unlike molecules are operative in the system [9]. This indicates that the interaction is stronger in acetone-methanol mixture .This supports the facts that acetone—methanol mixture is more compatible in bleaching of the surface date palm leaves due to interaction between the composition of date palm leaves and chemical mixture. Again, the variation of frequency is also not given any significant result in excess values of isentropic compressibility except in equimolar region due to strong interaction there is slight separation in profiles. The variation of excess free length is presented in Figure 3.

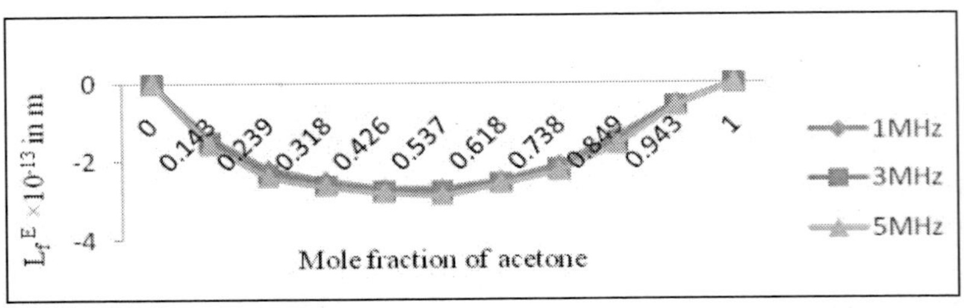

Fig. 3: Variation of Excess Intermolecular Free Length in Methanol with Acetone

The excess free length is negative over the entire range of composition investigated exhibiting a broad minimum around equimolar region of acetone and methanol. Considering the effect of frequency, it was observed that no significant results also obtained in this case like the above situations. The variation of excess free length further supports the variation of other excess parameters like excess acoustic impedance and excess bulk modulus of the mixtures.

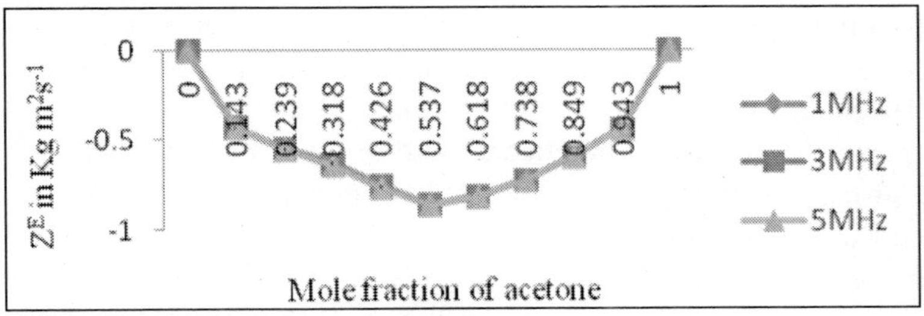

Fig. 4: Variation of Excess Acoustic Impedance in Methanol with Acetone

The excess acoustic impedance in the mixtures was found to be negative for the entire composition range of acetone. From the Figure 4 it is instructive that the variation of acoustic impedance is very smooth in acetone-methanol mixture. This due to fact that the occurrences of different intermolecular interactions like dipole-dipole interactions, dipole-induced dipole interactions, Vander Waals interactions in the liquid mixtures. The liquid mixture acetone-methanol has a clear minima in indicating the fact that more structural changes takes place in this mole fraction of chemical mixture which may favors the interaction of the liquid mixture with the chemical composition of the date palm leaves. As the bulk modulus is the reciprocal of compressibility so its variation is supported by variation of excess compressibility.

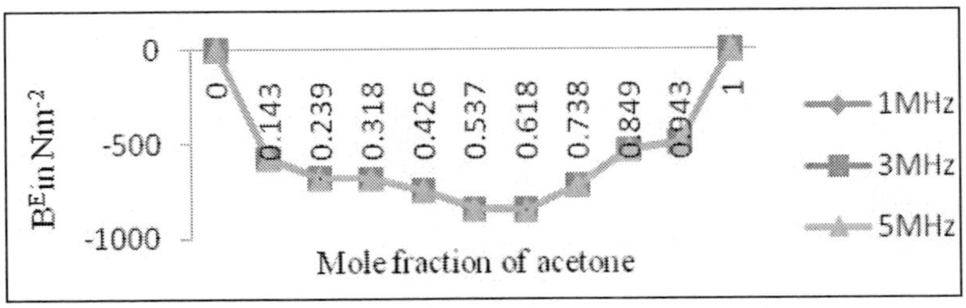

Fig. 5: Variation of Excess Bulk Modulus in Methanol with Acetone

From the profiles for excess bulk modulus as shown in Figure 5 it is informative that with decrease of volume of the mixture pressure increases. As a result of which the intermolecular interaction between the liquid molecules increases and the components of the mixture are tightly bound with each other and also with the chemical composition of the date palm leaves. This results in increase of surface roughness and interlocking in edges of the date palm leaves.

CONCLUSION

The behavior of different acoustic parameter clearly indicates that there is compatibility between the selected solvent mixtures. The compatibility is strongly dependent on different concentration of binary mixture rather than factor frequency of ultrasonic waves. This compatibility leads to presence of different intermolecular interactions like dispersions, H-bonding, dipole-dipole interaction and dipole-induced dipole interaction etc. Due to presence of such fundamental interactions, the treatment of this mixture increases the surface roughness removing certain amount of lignin wax and oils covering the external surface of the date palm leaf cell wall. As a result it is found that better mechanical interlocking takes place and increases the no. of possible reaction sites which improves the fiber matrix adhesion, increases the strength of the composites, decreases its water absorption and improves its thermal stability.

REFERENCES

[1] A. Ali and A.K. Nain, Indian J. Pure and Appl. physics, Vol. 19, 41(1997).
[2] J.D. Pandey, P. Jain and V. Vyas, Praman, J.Physics, 43, 361(1994).
[3] G. Nath, S. Acharaya and R. Paikaray, Journal of acoustic society of India, Vol. 34, No. 4, pp. 135–139 (2007).
[4] A.I. Vogel, Text Book of Practical Organic Chemistry, Third Edition (Longmans, London) (1937).
[5] J.A. Riddick and W.B. Bunger, Organic Solvents (Wiley-Inter-Science, New York), (1970).
[6] A. Weisberger, Techniques of organic chemistry, (Interscience, New York), Vol-III, (1955).
[7] K.P. Thakur and N K Gupta, Acustica, 53, 59 (1983).
[8] B. Jacobson, J. Chem. Phys. 20, 927 (1952).
[9] R.J. Fort and W.R. Moore, Trans Faradays, Soc., 61, 2102, 18 (1963).

Absorption Studies of Aniline and Esters

K. Sathi Reddy, D. Shanmukhi Jyothi[1] and D. Linga Reddy[2]

Department of Physics, College of Natural Science, Arba Minch University, Ethopia

[1]Department of Physics, Sardar Patel College, Secunderabad, A.P.

[2]Department of Physics, UCS, Osmania University, Hyderabad

E-mail: jyothi6physics@gmail.com

ABSTRACT: *Ultrasonic velocity(v),density(ρ),viscosity(η) for binary mixtures of Aniline with Methyl Acetate, Ethyl Acetate, Amyl Acetate have been measured at room temperature of 300.15 K over entire volume component percentage range. The ultrasonic velocity (v) is measured with Pulse Echo Overlap technique at a frequency of 2 MHz. The density (ρ) measurements have been carried out by 10ml specific gravity bottle. The viscosity (η) measurements have been carried out by using Canon-Fenske Viscometer with an accuracy of ±0.3% .Attenuation Coefficient (α) is measured using a Cathode Ray Oscilloscope for a transmitted pulse and n^{th} echo. Relaxation times (τ), Absorptivity (A) are computed using measured data of ultrasonic velocity (v), density (ρ), viscosity (η).*

INTRODUCTION

Ultrasonic is a useful tool to know the nature of molecular interaction in pure, binary and ternary liquids. But no sufficient data is available in the binary mixture of Aniline in Methyl Acetate, Ethyl Acetate, Amyl Acetate. Ultrasonic method found extensive application [1–2] in characterizing the various aspects of the physico-chemical behavior of liquid mixtures such as molecular interaction and association .Molecular association and complex formation in liquid mixtures through ultrasonic velocity measurements have been studied by Reddy and Naidu[3]. Ultrasonic velocity of a liquid is related to binding forces between atoms and molecules. The sound velocity in liquid mixtures is known to be a quantity which depends on concentration in a variety of manners according to the nature of each component liquid. Accurate measurements of ultrasonic velocity deliver the information about the physical and chemical behavior of solutions and liquid mixtures.

In a material scattering and absorption are produced by various dynamical interactions which typically arise from enharmonic force between the atoms, the presence of free charge carriers, cycling to localized spin states and the presence of impurities [4]. When acoustic waves are propagated through a liquid, dissipation of acoustic energy that is associated with changes in the molecular structure of the medium results from the finite time, which is required for these changes to take place. Ultrasonic velocity has been a subject of active interest during the recent years [5–7]. Nomoto and co researchers made successful attempt to evaluate the sound velocity in binary mixtures. In the present communication we focus on the ultrasonic behavior of liquid mixtures at 300.15K temperature with Aniline as the base and few esters Methyl Acetate, Ethyl Acetate and Amyl Acetate as mixtures.

EXPERIMENTAL

The samples are taken for the present study are of Analar reagent grade and procured from Sd.Fine Chemicals Limited,Bombay,India. The liquids are thoroughly distilled to remove the dissolved impurities using standard chemical procedures like steam distillation. Liquids practically are immiscible with water, volatile in steam and possessing a high boiling point.

By taking two liquids in separate burettes, and with Job's variation method of continuous variation has been used to prepare the mixtures of required proportions. The ultrasonic velocity measurements are made with the help of microprocessor based ultrasonic Pulse Echo Overlap technique (model UX4400 M) with built in time Velocity and thickness measuring feature with a frequency of 2 MHz. supplied by Roop Telesonic Ultrasonic Ltd., Bombay, India. The internal circuit of (ULTRASONIX4400 M) is designed with full solid state version. Further outstanding stability of operation is ensured by dynamic use of integrated circuit and PCB mounted gold switches. It has special memory features which enable one to enter testing procedures like angle and type of transducer for permanent storage and direct digital read out of Ultrasonic velocity up to an accuracy of ±1 m/s.

The Pulse Echo Overlap (PEO) method is very versatile and highly accurate technique for measuring the velocity of the ultrasonic waves in materials of liquids and solids structures. A highly absolute accuracy is obtained from this system, since the method is capable of accurately measuring from any cycle of one echo to the corresponding cycle of the next echo. The PEO method is able to handle diffraction (beam spreading) and phase correction properly, so that absolute accuracy of the PEO method may exceed the accuracy of most of the other methods. However, precision of some other methods is better than PEO method. The PEO method may operate either with transducer bonded to the specimen or directly with RF bursts. On the negative side, because of transducer effects, the leading cycles of successive echoes are attenuated and trailing cycles are burst up, so that great care has to be taken to obtain proper cycle for cycle match. In view of the above advantages and accuracy of the system the PEO technique has been adopted for measuring velocity and attenuation.

The densities of all the mixtures have been determined with 10 ml. specific gravity bottle and mass (m) of a given volume of the liquid is determined by using a single pan optoelectrical balance. The results of the densities are accurate to ±0.5%.

The internal circuit of pulse echo overlap system is designed with fully solid state version, which allows immediate calculation of the ultrasonic wave velocity as given in the following equation,

$$v = 2l / t$$

Where l is the liquid length and t is the time interval.

The long ultrasonic waves are passed through the binary mixture. They are reflected from the reflector face and received by the same transducer, which now acts as a receiver. The transmitted pulse and the received echoes are seen in the CRO screen. Once, the echo pattern is achieved on the screen of the CRO, the pattern is adjusted to a suitable height. The transmitted pulse height is measured in mv/cm and noted as 'a_0' on the screen. The height of first echo is noted as 'a_1' and second echo as 'a_2' and so on. The path length is noted using the micrometer attached to the reflector and is noted as 2l.

∴ The attenuation coefficient (α) is calculated as $\alpha = 1/2 \log_e (a_0/a_n)$ nepers/unit length.
Where $a_0 \rightarrow$ amplitude of transmitted pulse.

$a_n \rightarrow$ amplitude of nth echo.

The absorptivity is measured by dividing α by the square of the frequency (2MHZ) of the ultrasonic wave propagated into the liquid media, i.e.

$$\text{Absorptivity, } A = \alpha/f^2$$

where f is the frequency of the ultrasonic wave propagated.

The coefficient of absolute viscosity (η) has been determined using a Cannon-Fenske Viscometer. The results of viscosities are accurate to ± 0.3%.

RESULTS AND DISCUSSION

Ultrasonic velocity decrease in Aniline-Methyl Acetate, Aniline-Ethyl Acetate, Aniline-Amyl Acetate. Ultrasonic velocity is larger in Aniline-Amyl Acetate the reason being that Amyl Acetate is more alcoholic. Aniline is a basic component and possesses less number of H^+ ions which causes for freer movement of Ultrasonic waves through the medium. Therefore Ultrasonic velocity increases with respect to the increase in Aniline concentration .These are in tune with [4]. The density values decrease with respect to the increase in Ester concentration in all the systems. The density values are lesser in Aniline-Amyl Acetate system and larger in Aniline-Methyl Acetate system. The order of densities is Amyl Acetate < Ethyl Acetate < Methyl Acetate.

The attenuation coefficient is a quantity that characterizes how easily a material or medium can be penetrated by a beam of light, sound, particles, or other energy or matter [10–14]. A large attenuation coefficient means that the beam is quickly "attenuated"

Table 1: Ultrasonic Velocity (v), Density (ρ), Viscosity (η), Relaxation Time (τ), Attenuation Coefficient (α), Absorptivity (A) of Binary Mixture of Aniline in Methyl Acetate at 300.15 K

Volume Component Percentage		Ultrasonic Velocity (V)m/s	Density (ρ) 10^{-3} Kgm^{-3}	Viscosity (η) poise	Relaxation Time(τ) 10^{-8} sec	Attenuation Coefficient (α)	Absorptivity (A) 10^{-13}
Aniline	Methyl Acetate						
90	10	1502	1.0182	0.025	1.4511	0.5531	1.3827
80	20	1439	1.0068	0.0178	1.1383	0.6192	1.548
70	30	1407	1.0053	0.0151	1.011	0.6247	1.56175
60	40	1370	0.9962	0.0117	0.8343	0.6347	1.58675
50	50	1327	0.9820	0.0098	0.7556	0.6491	1.62275
40	60	1285	0.9784	0.0084	0.6932	0.6568	1.642
30	70	1240	0.9718	0.0072	0.7002	0.6642	1.6605
20	80	1200	0.9521	0.0066	0.6418	0.6715	1.67875
10	90	1150	0.945	0.0053	0.5654	0.6821	1.70525

Table2. Ultrasonic Velocity (v), Density (ρ), Viscosity (η), Relaxation Time (τ), Attenuation Coefficient (α), Absorptivity (A) of Binary Mixture of Aniline in Ethyl Acetate at 300.15 K

Volume Component Percentage		Ultrasonic Velocity (V)m/s	Density (ρ) 10^{-3} Kgm^{-3}	Viscosity (η) poise	Relaxation Time(τ) 10^{-8} sec	Attenuation Coefficient (α)	Absorptivity (A) 10^{-13}
Aniline	Ethyl Acetate						
90	10	1502	1.0095	0.0248	1.4519	0.5531	1.3827
80	20	1457	1.0017	0.0194	1.2164	0.6247	1.56175
70	30	1412	0.9905	0.0166	1.1207	0.6347	1.58675
60	40	1373	0.9879	0.0137	0.9808	0.6491	1.62275
50	50	1329	0.9853	0.0111	0.8504	0.6568	1.642
40	60	1288	0.9717	0.0095	0.7857	0.6642	1.6605
30	70	1244	0.9489	0.0081	0.7354	0.6715	1.67875
20	80	1203	0.9289	0.007	0.6942	0.6821	1.70525
10	90	1153	0.9243	0.0059	0.6402	0.6977	1.74425

Table 3: Ultrasonic Velocity (v), Density (ρ), Viscosity (η), Relaxation Time (τ), Attenuation Coefficient (α), Absorptivity (A) of Binary Mixture of Aniline in Amyl Acetate at 300.15 K

Volume Component Percentage		Ultrasonic Velocity (V)m/s	Density (ρ) 10^{-3} Kgm^{-3}	Viscosity (η) poise	Relaxation Time(τ) 10^{-8} sec	Attenuation Coefficient (α)	Absorptivity (A) 10^{-13}
Aniline	Amyl Acetate						
90	10	1504	1.0142	0.0249	1.4471	0.5531	1.3827
80	20	1460	1.0097	0.0186	1.1522	0.6491	1.62275
70	30	1417	0.9781	0.0160	1.0862	0.6568	1.642
60	40	1390	0.9641	0.0106	0.7587	0.6642	1.6605
50	50	1348	0.9520	0.0089	0.6859	0.6821	1.70525
40	60	1301	0.9423	0.0080	0.6687	0.6821	1.70525
30	70	1278	0.9208	0.0072	0.6383	0.6977	1.74425
20	80	1240	0.9023	0.0064	0.6150	0.7083	1.77075
10	90	1203	0.8906	0.0052	0.5379	0.7153	1.78825

(weakened) as it passes through the medium, and a small attenuation coefficient means that the medium is relatively transparent to the beam [15–16]. The ultrasonic waves are passed through the specimen under study. They are reflected from the opposite face and received by the same transducer, which now acts as a receiver. The Pulse Echo Wave train pattern is observed on the screen. Gate pulse was made to coincide with the bottom of the secondary echo pulse and amplitude is adjusted to 80% of its maximum value. The amplitude pulse noted as a_o and the amplitude of the nth echo pulse as a_n.

A perusal of attenuation coefficient (α) obtained for various binary mixtures of the Aniline with Esters reveals the following factors.

1. Attenuation coefficients (α) of binary mixture of Aniline with Methyl acetate lower compared to binary mixture of Ethyl acetate and Amyl acetate with Aniline. This means the chain length of the Esters is deciding the nature of binary mixture.

2. Attenuation coefficients (α) for Amyl acetate is higher than Methyl acetate and Ethyl acetate. Generally it can be said that, higher the chain length of the Esters, the higher will be the attenuation coefficient (α). In case of Methyl acetate, Ethyl acetate and Amyl acetate with Aniline binary mixtures is observed as Aniline-Amyl acetate > Aniline-Ethyl acetate > Aniline-Methyl acetate.

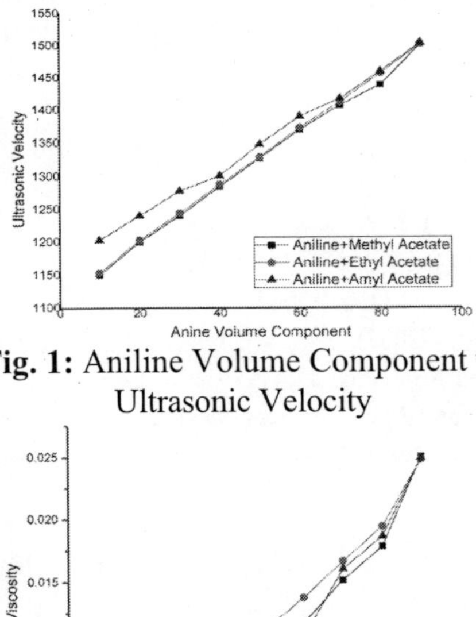

Fig. 1: Aniline Volume Component vs Ultrasonic Velocity

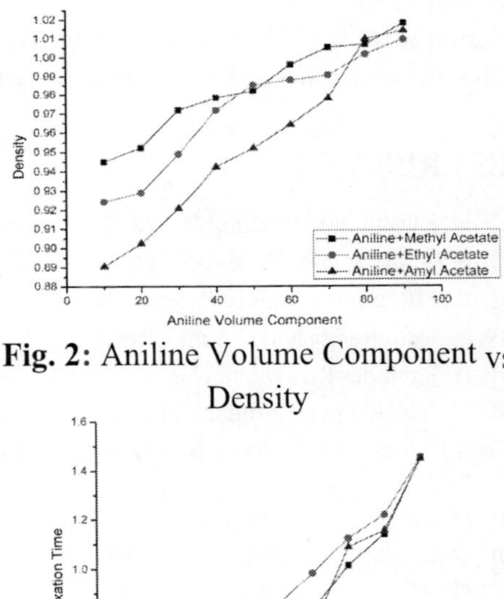

Fig. 2: Aniline Volume Component vs Density

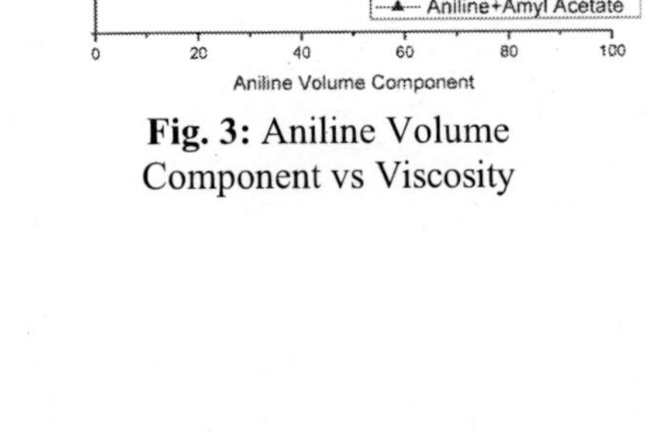

Fig. 3: Aniline Volume Component vs Viscosity

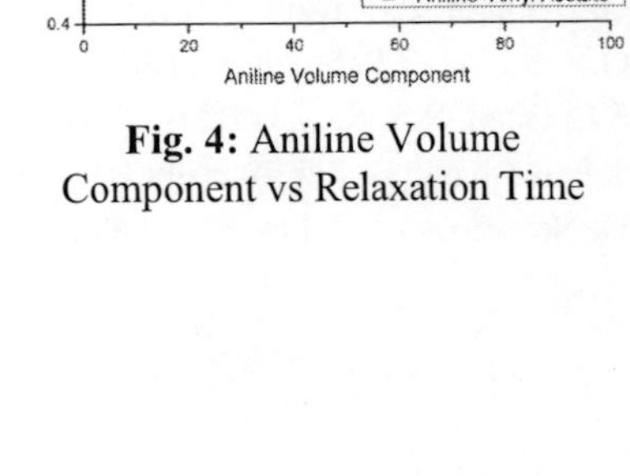

Fig. 4: Aniline Volume Component vs Relaxation Time

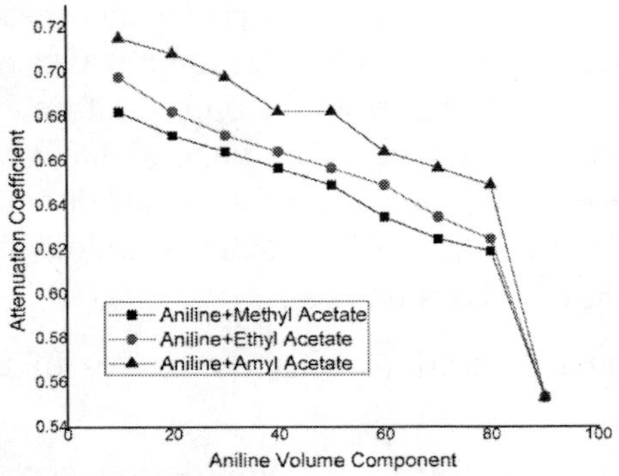

Fig. 5: Aniline Volume Component vs Attenuation Coefficient

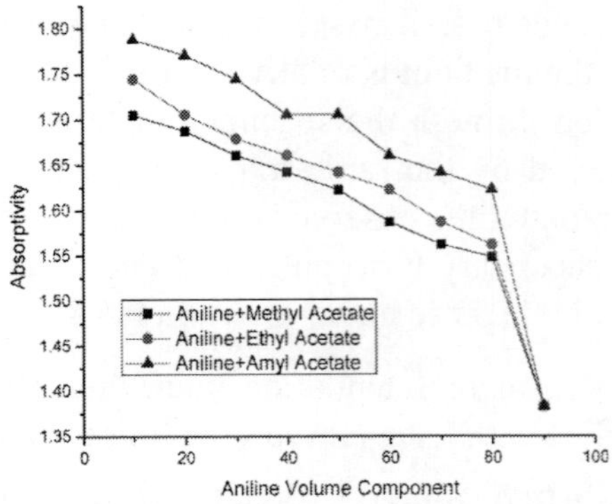

Fig. 6: Aniline Volume Component vs Absorptivity

CONCLUSION

Absorption studies are carried out apart from ultrasonic studies on Methyl Acetate, Ethyl Acetate, and Amyl Acetate with Aniline over entire volume component percentage range. It is found that Ultrasonic velocity increases with increase in Ester concentration. The density values decrease with respect to the increase in Ester concentration in all the systems. The Attenuation Coefficient and Absorptivity decrease as the concentration of the Esters increase.

REFERENCES

[1] H. Eyringand Hirschfelder, J.O., J. Phy. chem, 41 (1937), 249.
[2] R.J. Fort and W.R. Moore, Trans, Faraday Society 61 (1979), 687.
[3] K. Subramanyam Reddy and P.R. Naidu, Aust. J. Chem 32 (1979), 687.
[4] K. Sathi Reddy and D. Linga Reddy, Journal of Ultrasonics (2007).
[5] B. Ravinder Reddy and D. Linga Reddy, Ind. J. Pure and Appl. Phy. 37, (1999) 56.
[6] N. Jaya Kumar, Karunanidhi and V. Appan, Ind. J. Pure and Appl. Phy.36, (1999]76
[7] S.V. Ranganayakulu, C. Sreenivas Reddy and D. Linga Reddy, J. Acoustical Society of India 31,(2003) 295.
[8] H. Eyring, J.F. Kincaid, J. Chem. Phy.6,(1938) 67.
[9] V.K. Syal, Kumari, U. Chauhan, S. Chauhan, M.S. Sud, S.P. Singh, Indian J. Pure and Appl. Phy. (1992).
[10] S. Rajagopalan, Acoustica, 66, (1988) 63.
[11] K.S. Reddy, P.R. Naidu, J. Chem. Liq.15, (1985) 147.
[12] C. Bachem, Z. Physik 101, (1936) 565.
[13] G.V. Reddy, R.P. Singh, Acustica 23, (1970) 31.
[14] S.C. Bhatt, R.S. Rawat and B.S. Semwal, J. Acoust. Soc. India, 27 (1999) 297.
[15] J. Prakash and B. Rai, Acoustica, 64, (1987), 226.
[16] O. Nomoto *et al.*, J. Phy. Soc. Japan Vol. 8, 4 July–August 1953.

Acoustic Impedance, Intermolecular Free Length and Transport Behaviour of Peptides using Ultrasonic Velocity

R. Padmavathy, E. Jasmine Vasantharani, N. Radha, S. Sujatha[1] and R. Yogarani

Department of Physics, Seethalakshmi Ramaswamy College, Tiruchirarappalli, Tamil Nadu
[1]Department of Physics, Srimathi Indira Gandhi College, Tiruchirarappalli, Tamil Nadu

ABSTRACT: *Knowledge of the solute-solute interactions is very important to understand the fundamental phenomena like solubility of proteins, folding/unfolding processes, denaturation of proteins etc, in solutions. Peptides have been used in study of protein structure and function. In the present study non-aqueous solutions of Glycyl-L-Leucine have been taken with different molalities. The density, viscosity and ultrasonic velocities were measured in the temperature range from 5°C to 55°C. The main advantage of considering the internal pressure in calculating the transport properties lies in the fact that it is experimentally measurable and it depends on molar volume. By measuring the fundamental parameters, the specific acoustic impedance, Rao's constant, intermolecular free length, internal pressure and free volume were computed. An attempt will be made to identify the entry of solvate into the sample and the entry of biological molecules in the solvates.*

Keywords: Peptides, Solute-Solute Interaction, Ultrasonic Velocity.

INTRODUCTION

Amino acids belong to an important family of biomolecules which serve primarily as the basic building block of all proteins[1]. Ultrasonic studies of solutions yield valuable information about the molecular interactions, the nature and strength of the interactions. In the present study, the density, viscosity and ultrasonic velocity of glycyl-L-Leucine in formamide are measured at various molalities from 5°C to 55°C. From the above values, acoustic and thermodynamic parameters are computed.

EXPERIMENTAL TECHNIQUE

Density of the solutions is measured using 25 ml specific gravity bottle with accuracy of 0.0001 gm/cc. Cannon Fenske viscometer is used for the viscosity measurements, with an accuracy of ± 0.5%. Mittal's interferometer of frequency 2 MH_Z with an accuracy ±0.5 m/s is used for the measurement of ultrasonic velocity.

The following thermodynamic and acoustic parameters are computed using the formulae.

Internal pressure (π_i) = bRT $(K\eta/u)^{1/2}$ * $(\rho^{2/3}/M_{eff}^{7/6})$ atms
Free volume (v_f) = $[M_{eff} * u/K\eta)^{3/2}$ cc
Rao's constant (R) = $M_{eff}(U)^{1/3}$ ρ
Intermolecular free length (L_f) = $K/[u(\rho)^{1/2}]$A°
Specific acoustic impedance Z = ρu Rayl.

Table 1: Internal Pressure (atms)

Molality (m)	*5°C*	*15°C*	*25°C*	*35°C*	*45°C*	*55°C*
0.001	20181	17725	14434	13006	12006	11018
0.005	18869	16585	14136	12914	11887	10798
0.01	19860	16749	14608	13100	12124	10805
0.015	18619	16247	13940	12871	11910	10720
0.02	18924	15588	13683	12513	11661	10555

Table 2: Free Volume (cc)

Molality (m)	*5°C*	*15°C*	*25°C*	*35°C*	*45°C*	*55°C*
0.001	0.008	0.012	0.025	0.037	0.051	0.071
0.005	0.009	0.015	0.026	0.037	0.052	0.075
0.01	0.008	0.014	0.024	0.036	0.049	0.075
0.015	0.010	0.016	0.027	0.038	0.052	0.077
0.02	0.009	0.018	0.029	0.041	0.055	0.080

Table 3: Rao's Constant (R)

Molality (m)	*5°C*	*15°C*	*25°C*	*35°C*	*45°C*	*55°C*
0.001	2143.12	2149.96	2162.62	2174.91	2182.52	2195.45
0.005	2148.35	2159.40	2167.92	2181.34	2186.78	2200.56
0.01	2153.43	2164.43	2173.47	2184.17	2195.62	2204.88
0.015	2157.25	2164.40	2176.12	2191.09	2200.23	2211.62
0.02	2166.42	2174.42	2183.15	2192.63	2205.40	2214.61

Table 4: Intermolecular Free Length (A°)

Molality (m)	*5°C*	*15°C*	*25°C*	*35°C*	*45°C*	*55°C*
0.001	0.1073	0.1107	0.1148	0.1187	0.1217	0.1243
0.005	0.1069	0.1104	0.1144	0.1182	0.1211	0.1237
0.01	0.1065	0.1100	0.1136	0.1171	0.1199	0.1224
0.015	0.1059	0.1094	0.1134	0.1167	0.1194	0.1221
0.02	0.1055	0.1088	0.1127	0.1163	0.1188	0.1215

Table 5: Specific Acoustic Impedance (Rayl)

Molality (m)	*5°C*	*15°C*	*25°C*	*35°C*	*45°C*	*55°C*
0.001	18.83	18.65	18.38	18.00	17.77	17.59
0.005	18.91	18.67	18.43	18.07	17.86	17.67
0.01	18.97	18.73	18.57	18.25	18.04	17.87
0.015	19.08	18.86	18.60	18.31	18.12	17.90
0.02	19.14	18.95	18.71	18.38	18.21	18.01

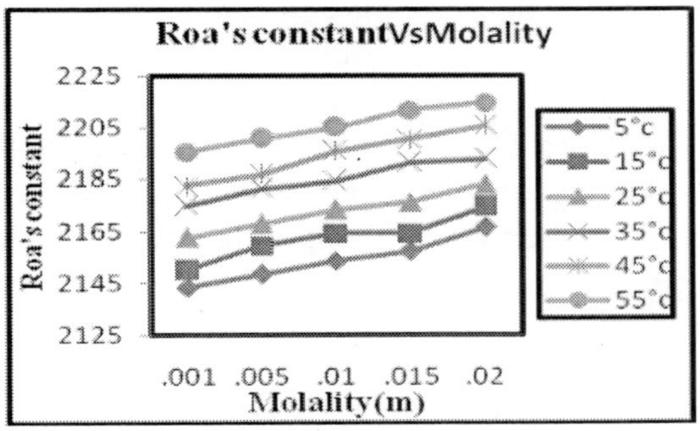

Fig. 1: Intermolecular Free Length (A°)

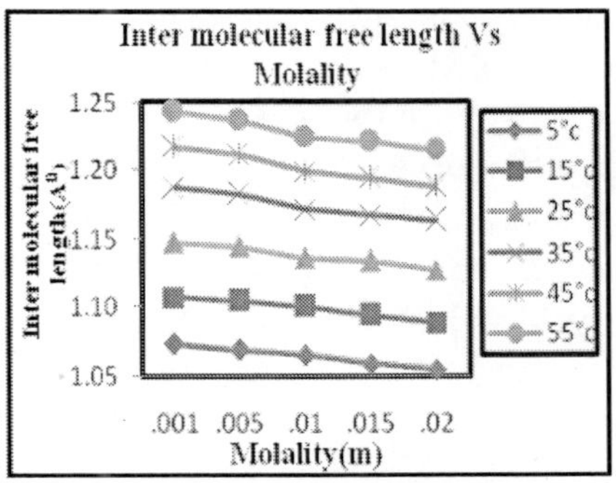

Fig. 2: Rao's Constant (R)

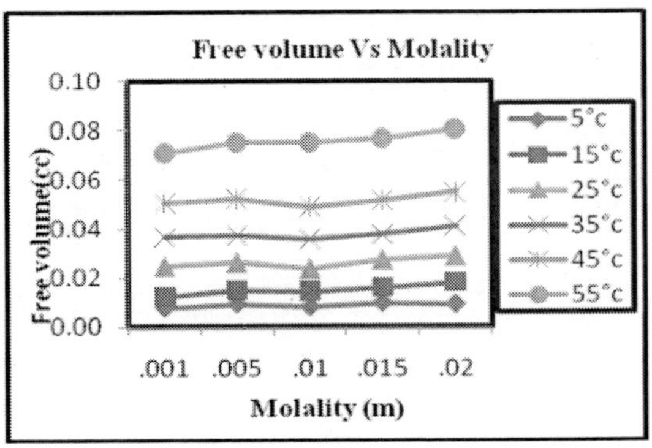

Fig. 3: Free Volume (cc)

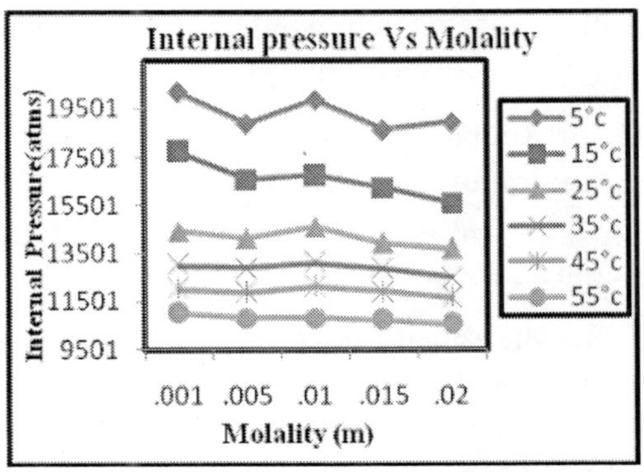

Fig. 4: Internal Pressure (atms)

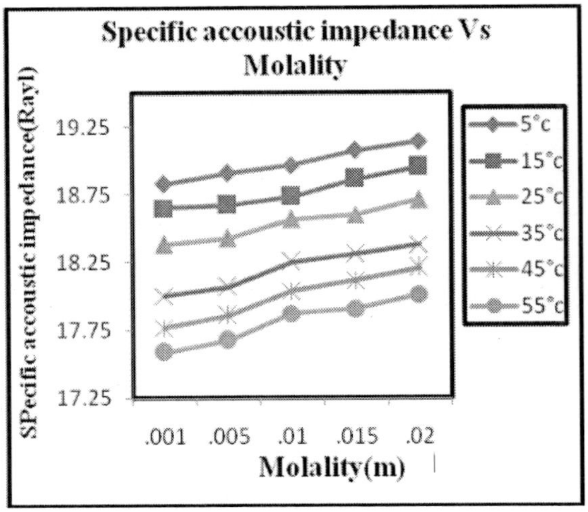

Fig. 5: Specific Acoustic Impedance (Rayl)

RESULTS AND DISCUSSION

Suryanarayana and venkatesan [2] recognized that internal pressure of the medium was closely related to degree of dissociation, ionic mobility, ion pairing, dielectric constant and viscosity. In the present investigation, the internal pressure decreases with the rise in temperature because when the temperature is increased, there is a tendency for the ions to move away from each other, reducing the possibility of interaction, which may further reduce the cohesive forces and ultimately leads to a decrease in the internal pressure [3]. In Glycyl-L-Lecucine, π_i is found to increase sharply at 0.01 molality at all temperatures which indicates the strong solute-solvent interaction existing in the solution. This sharp rise may be attributed due to zwitterionic behavior of Glycyl salts.

Free volume is one of the significant factors in explaining the variations of the physico-chemical properties of liquids, solutions and liquid mixtures. The free volume increases with increasing temperature [4] as well as with increasing concentration. Thus due to structural stability, Glycy1–1-Leucine, behaves as a structure maker.

Molar sound velocity or Rao's constant is found to be increasing with increasing concentration and also with increasing temperature for Glycy1-L-Leucine. This behavior suggests that there is a molecular association taking place between the molecules of the solute and solvent.

When the salt is added to a solvent, the strength of the interaction increases and hence intermolecular free length decreases with molality, [5] increases with rise in temperature. These changes reveal the fact that there is a strong solute-solvent interaction existing between the solute and the solvent [6].

The specific acoustic impedance is governed by the internal and elastic properties of the medium. In the present investigation, Z increases with increasing concentration and decreases with increasing temperature, which indicates the absence of complex formation and the existence of solute-solvent interaction. The results are tabulated in Tables (1–5) and shown in graphs (1–5).

CONCLUSION

The field of protein folding and stability has been a critically important area of research for years and remains today as one of the greatest unsolved mysteries. In the present work, a detailed analysis of acoustic and thermodynamic properties was made. The study reveals the fact that the addition solute in the solvent enhances the solute-solvent interaction.

This behavior of glycy1-L-Leucine in formamide emphasizes that it act as a structure maker due to its structural stability and Zwitterionic nature.

REFERENCES

[1] G. Ravichandran, S. Arumugam and T.K. Nambi Narayanan, J.Pure and Appl. Ultrasonics.34 (2012) pp. 17–21.

[2] Suryanarayana, C.V. and Venkatesan.V.K., Acta. chem. Hung, 19, (1959), 441–551.

[3] Kalyanasundaram, S., Manuel Stephen, H., Gopalan. A., Polym., J. and Mater, R. (1995) 323–333.

[4] Geetha, G., *et al.*, Proceedings of the National Sysposium on Acoustits, Vol. (2005) 385–390.

[5] Agnihotri, S.C. and Omprakash., J. of. Acous. Soc. of Ind. Vol. XIII, No. 4 (1985) 186.

[6] Tewari, K., Patra, C., Padhy, S. and Chakravarthy, V., Phys.Chem. Lig (32) (1996) 149.

Ultrasonic Investigation of Internal Pressure Free Volume Andacoustical Parameters of Sulphanilamide in Non-Aqueous Medium

R. Padmavathy, E. Jasmine Vasantharani, S. Santhakumari[1] and A. Thamarai Selvi

Department of Physics, Seethalakshmi Ramaswamy College, Thiruchirappalli, Tamil Nadu

[1]Department of Physics, Srimathi Indira Gandhi College, Thiruchirappalli, Tamil Nadu

ABSTRACT: *Ultrasonic technique is the vital probe in evaluating the thermo-dynamical parameters such as internal pressure and free volume. Sulphanilamide is the parent compound of all the sulfa-drugs which is important in urinary tract infections and meningococcal meningitis profilaxes. In the present work, non-aqueous solutions of Sulphanilamide have been prepared with different concentrations and the experiments were carried out from a low temperature of 5°C to a high temperature of 55°C. The ultrasonic velocity data combined with density and viscosity provides the standard means for determining the internal pressure, free volume and acoustic parameters. The various interactions occurring in the solutions are interpreted in terms of ion-ion and ion-solvent interactions.*

Keywords: Ion-Solvent Interactions, Thermo-Acoustic Parameters.

INTRODUCTION

Sulfonamide compounds identified as chemotherapeutic agents, possess broad spectrum of biological properties[1]. The passage of ultrasonic waves through solutions and liquids disturb the equilibrium between solute and solvent molecules. The velocity of such a wave is a thermodynamic quantity and is related to the physico-chemical properties of the medium[2].

Internal pressure in liquids provides wealth of information about the state of liquids. The free volume is generalized aspect of the idea that each molecule is enclosed by its neighbors in a cell. Specific acoustic impendence in solutions can be used in estimating the strength of intermolecular attraction[3]. The present work deals with the ultrasonic investigations study of sulphanilamide in formamide. Internal pressure, free volume and acoustical parameters such as adiabatic compressibility (β), Intermolecular free length (L_f), Specific acoustic impedance (z). Infrared spectrum is an important record which gives sufficient information about the structure of a compound.

EXPERIMENTAL ANALYSIS

The density of the solution is measured with 25 ml of specific gravity bottle with an accuracy of 0.0001 gm/cc. Cannon Fenske viscometer is used for the viscosity measurements, with an accuracy of ± 0.5%. Mittal's interferometer of frequency 2 MH_Z, with an accuracy of ±0.5 m/s is used for the measurement of ultrasonic velocity. Internal pressure (π_i) and free volume (v_f) and acoustic parameters are evaluated by the formulae,

1. Internal pressure (π_i) = $bRT\,[k\eta/u]^{1/2}\rho^{2/3}/M^{7/6}$ atms
2. Free volume (v_f) = $[M_{eff}\,u/k\eta]^{3/2}$cc
3. Adiabatic compressibility (β) = $[1/u^2\rho]$ cm^2/dynes
4. Specific acoustic impedance (Z) = ρu (Rayl)
5. Intermolecular free length (L_f) = $K_T\,(\beta_{ad}{}^{1/2})$A°

RESULTS AND DISCUSSION

In the present study internal pressure, exhibits an increasing trend with respect to increasing molalities. These variations may be due to the structure making nature of the solute in the solvent [4]. It implies that there is a strong solute- solvent interaction taking place in the solution.

The internal pressure values decreases with rise in temperature of the solution. Since when the temperature is increased there is a tendency for molecules to move for away which causes a reduction in the cohesive forces and ultimately leads to a decrease in internal pressure. The variations of V_f with molalities also confirms the structure making nature of the solute in the solvent.

The adiabatic compressibility increases with decrease in concentration and rise in temperature. [5]

The linear increase in specific acoustic impendence with concentration at a given temperature is observed in solution. This may be due to the effect of solute- solvent interactions [6] occurring in the solutions.

On the model of sound propagation given by Eying and Kin card [7] the increase in intermolecular free length is due to the process of mixing results in the lowing of ultrasonic velocity. This indicates that L_f is predominant factor in determining the nature of interaction between the components of the solution.

The L_f of the given sample is found to decrease with molality and increase with temperature. This indicates that there is a strong interaction between the solute and solvent and strength of interaction increases as L_f decreases with molality.

Table 1: Internal Pressure (atms)

Molality (m)	*5°C*	*15°C*	*25°C*	*35°C*	*45°C*	*55°C*
0.001	19448	15828	14076	12895	12087	10933
0.005	19843	16591	14080	12983	12239	10992
0.01	19658	16774	13943	12867	12090	10703
0.015	19779	16493	14002	12934	12117	10786
0.02	19376	16535	13916	12842	11875	10582

Table 2: Free Volume (cc)

Molality (m)	*5°C*	*15°C*	*25°C*	*35°C*	*45°C*	*55°C*
0.001	0.008	0.017	0.027	0.037	0.049	0.072
0.005	0.008	0.015	0.026	0.037	0.048	0.071
0.01	0.008	0.014	0.027	0.038	0.049	0.077
0.015	0.008	0.015	0.027	0.037	0.049	0.075
0.02	0.008	0.015	0.027	0.038	0.052	0.080

Table 3: Adiabatic Compressibility × 10^{-11} (cm^2/dynes)

Molality (m)	*5°C*	*15°C*	*25°C*	*35°C*	*45°C*	*55°C*
0.001	3.23	3.28	3.41	3.52	3.62	3.69
0.005	3.19	3.27	3.35	3.45	3.59	3.66
0.01	3.16	3.24	3.32	3.39	3.55	3.61
0.015	3.14	3.22	3.28	3.36	3.51	3.58
0.02	3.13	3.20	3.28	3.33	3.44	3.54

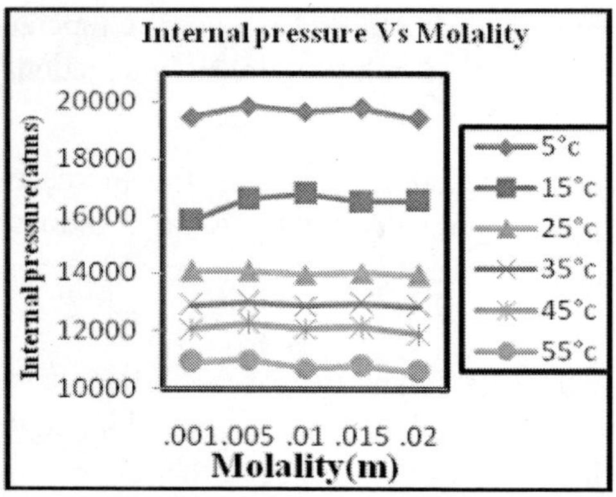

Fig. 1: Internal Pressure (atms)

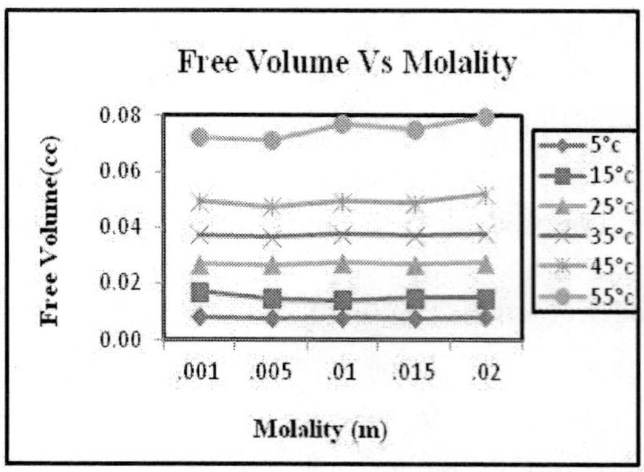

Fig. 2: Free Volume (cc)

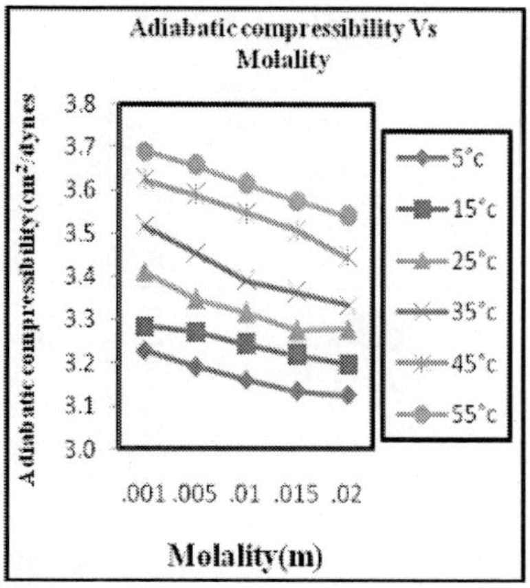

Fig. 3: Adiabatic Compressibility (cm^2/dyne)

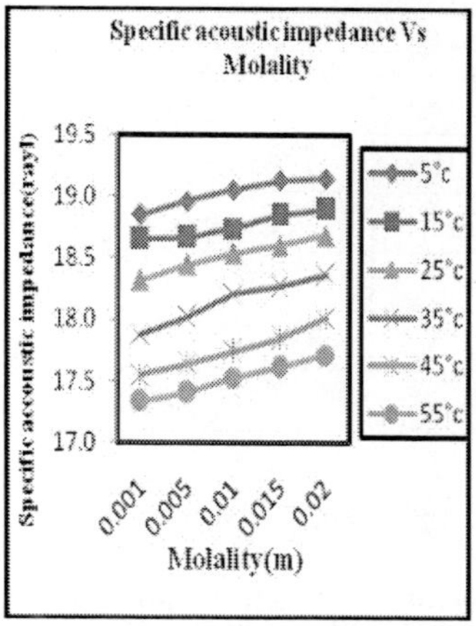

Fig. 4: Specific Acoustic Impedance (Rayl)

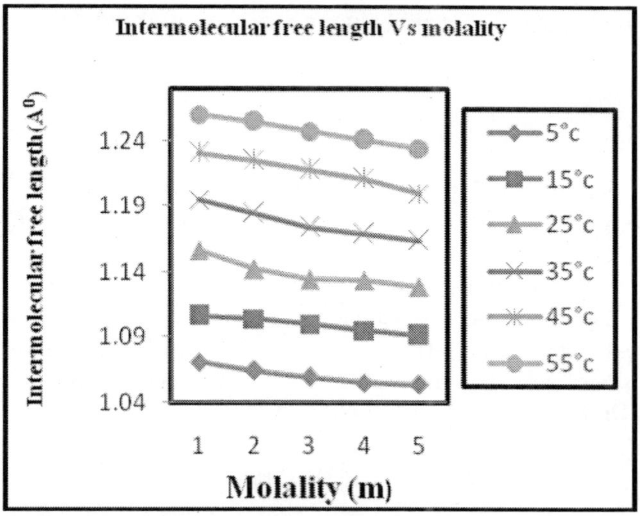

Fig. 5: Intermolecular Free Length (A^0)

Table 4: Specific Acoustic Impedance (Rayl)

Molality (m)	*5°C*	*15°C*	*25°C*	*35°C*	*45°C*	*55°C*
0.001	18.85	18.65	18.22	17.87	17.55	17.34
0.005	18.96	18.66	18.45	18.02	17.63	17.40
0.01	19.05	18.74	18.54	18.21	17.74	17.53
0.015	19.13	18.85	18.60	18.27	17.84	17.61
0.02	19.15	18.90	18.68	18.36	18.01	17.71

Table 5: Intermolecular Free Length (A^0)

Molality (m)	*5°C*	*15°C*	*25°C*	*35°C*	*45°C*	*55°C*
0.001	1.071	1.106	1.156	1.195	1.231	1.260
0.005	1.065	1.104	1.142	1.184	1.225	1.255
0.01	1.060	1.100	1.134	1.173	1.218	1.247
0.015	1.055	1.095	1.133	1.169	1.211	1.241
.02	1.054	1.092	1.128	1.164	1.200	1.234

CONCLUSION

Sulfanilamide is a potent antibiotic because this drug undergoes metabolic alterations in tissues. The detailed study of acoustic parameters suggests that there is a molecular association existing in the solution. The addition of solute into the solvent brings about a strong solute-solvent interaction in the solution. This behavior is confirmed by the infrared spectral analysis and that reveals the structure making nature of the solute in the solvent.

REFERENCES

[1] R.A. Herole, V.S. Velingkar, Int. Journal of pharmaceutical chemistry (2005).

[2] R. Palani, G. Srinivasan and B. Geeta lakshmi, Int. Journal of Chem. Tech, Vol. 3, No. 1 (2011), 284–289.

[3] Shinde B.R., et al., Journal of Engg.Research and Studies, Vol. 1, (2010) 128–137.

[4] Thirumaran S., Karthikeyan N., Int. Journal of Chemical Research, Vol. 3, (2011) 83–98.

[5] Jasmine Vasantha Rani E., Ph.D., thesis entitled Internal pressure, free volume, solvation number,thermochemical and electrochemical analysis of quaternary ammonium salts submitted to Bharathidasan University, Tamil Nadu, India (1997).

[6] Syal V.K. *et al.*, J. Acous. Soc. Ind., 30 (2002).

[7] Erying and Kin card J.F., J. Chem. Phys 41 (1937) 249.

Study of Molecular Interactions in Binary Mixtures through Ultrasonic Measurements

S. Ponnancy and Beulah J.M. Rajkumar
PG and Research Department of Physics, Lady Doak College, Madurai
E-mail: Physics.dept@gmail.com

ABSTRACT:*The density (ρ), viscosity (η) and ultrasonic velocity (U) have been measured for the binary mixtures of Dimethylformamide (DMF) with Isopropanol, DMF with Carbon tetrachloride and Carbon tetrachloride with Isopropanol over the entire range of composition (0:20 to 20:0 in steps of 2 ml). The measured data are used to compute thermodynamic acoustical parameters such as acoustic impedance (z), adiabatic compressibility (β), free length (L_f), free volume (V_f). Excess values of the above said parameters were plotted and interpreted in terms of intermolecular interactions.*

Keywords: Binary Mixtures, Acoustic Impedance, Adiabatic Compressibility.

INTRODUCTION

Interaction between solute-solvent molecules brings out tremendous change in nature. Physicochemical properties of two different liquids change when they mix together. These changes lead to various applications in industry, medicine and in chemical engineering processes [1]. The interaction between liquid mixtures may be weak or strong depending on the whether the liquid is polar, non-polar, protic or aprotic. Ultrasonic investigation is a powerful tool to probe these interactions [2]. Many research works have been carried out in analyzing the structural changes of binary mixtures with ultrasonic and thermo acoustical parameters. This paper focuses on the analysis of structural changes of like and unlike molecules through ultrasonic parameters.

METHODOLOGY

Binary mixtures of ten different proportions from 0:20 to 20:0 were prepared for DMF+ 1-ol, DMF + CCl_4 and CCl_4 + 1-ol. The density of the mixture has been found by the relative measurement method. Viscosity of the liquid mixture was measured by Oswald viscometer and the ultrasonic velocity was measured by using the ultrasonic interferometer B1 Microsoft model –83 with high degree of accuracy operating at 3 kHz.

The thermo acoustical parameters can be calculated from the measured data.

Acoustical impedance (z) $z = U\rho$

Adiabatic compressibility (β) $\beta = \dfrac{1}{U^2\rho}$

Free length (L_f) $L_f = K\sqrt{\beta}$

Free volume (V_f) $V_f = \left(\dfrac{M_W U}{k\eta_l}\right)^{3/2}$

where, U – Ultrasonic velocity of binary mixture (ms^{-1})

ρ – Density of binary mixture ($Kg\ m^{-3}$)

K – Temperature dependent Jacob constant $(93.875+ 0.3757)10^{-8}$ [3]

M_W – Molecular weight (Kg)

k – Temperature independent constant (4.28×10^9)

η_l – Viscosity of binary mixture ($Ns\ m^{-2}$)

The excess values of above said parameters can be found by the expression

$$E = P - \sum_{i=0}^{i=1} m_i X_i$$

where, P is the experimental value of the binary mixture, m_i is the theoretical value of each component and X_i is the mole fraction.

The degree of molecular interaction can be found from the expression $a = \dfrac{U_{exp}^2}{U_{IMR}^2} - 1$

where, U_{IMR} is ideal mixture relation[4] of ultrasonic velocity and is given as

$$\frac{1}{(X_1M_1 + X_2M_2)U_{IMR}^2} = \frac{X_1}{M_1U_1^2} + \frac{X_2}{M_2U_2^2}$$

RESULT AND DISCUSSION

DMF + Isopropanol

Figure 1 gives the variation of the different excess values with increasing DMF concentration. Figure 2 represents the degree of molecular interaction (α) as calculated from theoretical considerations using the ideal mixture relation. The density of the mixture increases linearly with mole fraction of DMF. As DMF is the denser component, the resultant increase in density causes sound energy to be quickly transferred in the medium which increases ultrasonic velocity. The increase in ultrasonic velocity also indicates a large association among the molecules of the mixture. The random non-linear variation in viscosity with increasing molar concentration of DMF indicates the presence of intermolecular interaction [5]. The decrease in adiabatic compressibility indicates the formation of a large number of tightly bonded systems leading to a decrease of inter molecular free length was observed in Table 1.

Table 1: Thermo Acoustic Parameters of DMF + Isopropanol

Mole Fraction of DMF	*Density* ρ 10^3 Kgm^{-3}	*Viscosity* η 10^{-3} Nsm^{-2}	*Ultrasonic Velocity U* ms^{-1}	*Acoustic Impedance* Z $10^3\ Kgm^2s^{-1}$	*Adiabatic Compress-ability* $\beta\ 10^{-10}$ m^2N^{-1}	*Free Length* L_f $10^{-11}\ m$	*Free Volume* V_f $10^{-7}\ m^3$
0.0000	0.7646	1.4360	1118.067	0854.8155	10.4631	6.7362	0.3616
0.0912	0.7683	1.2531	1149.273	0883.0340	09.8537	6.5371	0.4760
0.2399	0.8428	1.2314	1187.437	1000.7480	08.4152	6.0411	0.5376
0.2835	0.7756	1.0243	1213.090	0940.8742	08.7614	6.1641	0.7416
0.3996	0.8186	1.0383	1251.963	1024.7898	07.7942	5.8140	0.7890
0.4863	0.8004	1.0313	1279.876	1024.3535	07.6275	5.7514	0.8452
0.5998	0.8540	1.0312	1309.593	1118.3921	06.8276	5.4415	0.9042
0.7073	0.8686	1.0120	1346.230	1169.2951	06.3527	5.2488	0.9994
0.8116	0.8825	0.9881	1378.559	1216.6277	05.9623	5.0850	1.1051
0.9103	0.9139	1.0300	1405.655	1284.6256	05.5379	4.9007	1.0983
1.0000	0.9581	1.0575	1441.508	1381.1492	05.0228	4.6672	1.1232

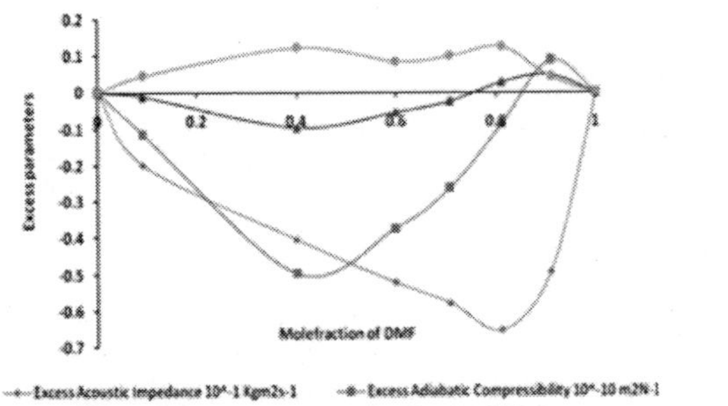

Fig. 1: Excess Parameters of DMF + IPA

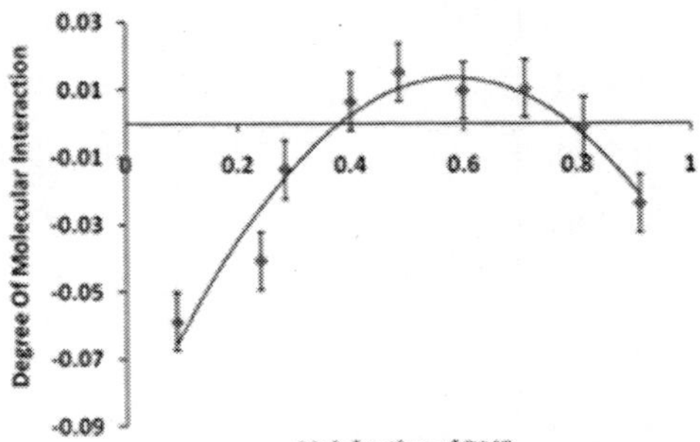

Fig. 2: Molecular Interaction of DMF + IPA

The positive excess volume reaches its maximum when the concentration of both species in the binary mixture is almost the same, indicating that the interaction of both species is responsible for the increase in free volume. Dipolar interaction between the two molecules can mutually form dipoles and this dipolar interaction supports the existence of more space between the components. Because of this structure formation, the actual acoustic impedance is smaller than the expected value leading to a negative excess value.

Negative deviation in excess adiabatic compressibility may be attributed to the presence of dipole-dipole interaction between the compound molecules [1]. Negative excess adiabatic impedance suggest that the rupture of the hydrogen bond chain of isopropanol dominates over the hydrogen bond formed between the unlike molecules.

At lower concentrations of DMF there is a high molecular association between the IPA molecules due to hydrogen bonding. This leads to a low degree of molecular interaction between the two compounds. As the concentration of DMF is increased, these bonds break and DMF may occupy the interstitial position of IPA. With a further increase of DMF, the degree of molecular interaction decreases due to repulsive force between the DMF molecules.

Table 2: Thermo Acoustic Parameters of DMF + CCl_4

Mole Fraction of DMF	*Density* ρ 10^3 Kgm^{-3}	*Viscosity* η 10^{-3} Nsm^{-2}	*Ultrasonic Velocity U* ms^{-1}	*Acoustic Impedance* Z 10^3 Kgm^2s^{-1}	*Adiabatic Compress-ibility* $\beta\ 10^{-10}$ m^2N^{-1}	*Free Length* L_f 10^{-11} m	*Free Volume* V_f 10^{-7} m^3
0.0000	1.5610	1.4907	894.3550	1396.0852	8.0090	5.8935	1.0012
0.1115	1.4841	1.4363	938.1960	1392.4142	7.6549	5.7617	1.0390
0.2899	1.4387	1.4884	990.1569	1424.5247	7.0897	5.5450	0.9126
0.3299	1.3288	1.4046	1037.0050	1377.9921	6.9980	5.5090	1.0276
0.4561	1.2849	1.4255	1092.3020	1403.5062	6.5229	5.3187	0.9587
0.5391	1.1990	1.3733	1149.2730	1377.9221	6.3147	5.2331	1.0015
0.6468	1.1852	1.3679	1230.4250	1458.2897	5.5732	4.9163	0.9868
0.7524	1.0969	1.2137	1285.9850	1410.6534	5.5124	4.8894	1.1061
0.8400	1.0380	1.1518	1327.3850	1377.8827	5.4675	4.8695	1.1145
0.9277	0.9841	1.1198	1398.8960	1376.6178	5.1928	4.7455	1.1057
1.0000	0.9581	1.1186	1441.5080	1381.1492	5.0228	4.6672	1.0323

DMF + CCl_4

Figues 3 and 4 give the variation of the different excess values and the calculated degree of molecular interaction respectively with increase in the DMF concentration The binary mixture of DMF and CCl_4 involves a zero dipole moment liquid CCl_4. Like in the case of

DMF+IPA, the viscosity decreases in a non-linear fashion indicating interaction between the molecules. The acoustic impedance changes randomly with increasing mole fraction of DMF although the velocity increases uniformly. This further justifies the possibility of molecular interactions between the solute and the solvent molecules. Since DMF is a polar solvent, it induces a dipole in CCl_4 which orients itself such that there is repulsion between molecules thereby increasing the free volume. Therefore weak dipole-induced dipole interactions exist between them.

The negative deviation in excess adiabatic compressibility and a small negative deviation in excess free length were observed between 0.6 and 0.8. This indicates weak interaction involving dispersion forces. Decrease in free length between the component molecules indicates that the system is more resistant to the sound propagation, which is supported by the positive values of excess acoustic impedance. The random positive excess acoustic impedance observed indicates structural changes due to hydrogen bonding between the solute molecules. Figure 2.2 indicates that the formation of complex structure due to dipole –induced dipole interaction.

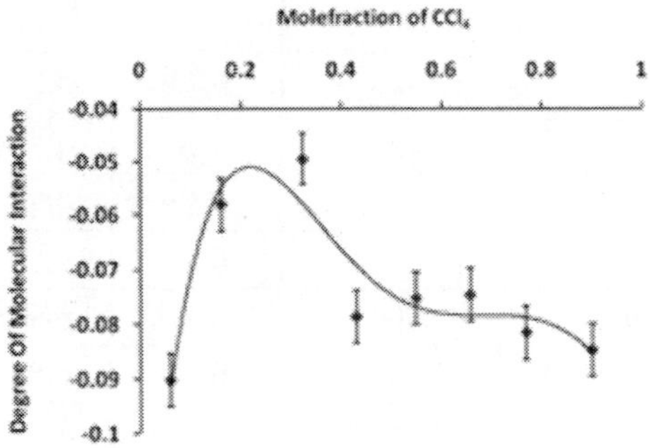

Fig. 3: Excess Parameters of DMF + CCl_4

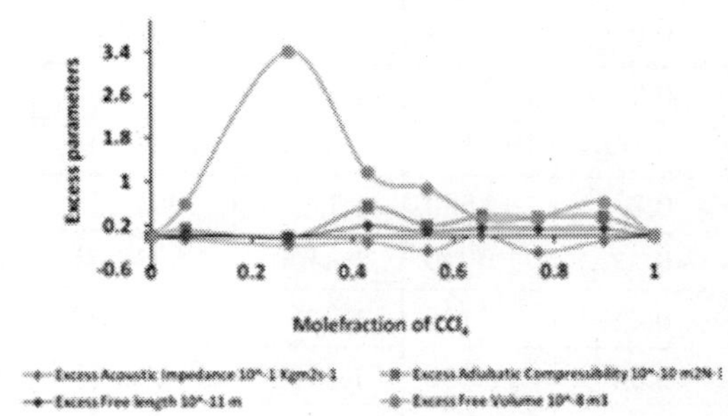

Fig. 4: Molecular Interaction of DMF + CCl_4

CCl_4 + Isopropanol

Figure 5 and 6 depict excess value variations and a negative degree of molecular interaction with increasing CCl_4 concentration. The ultrasonic velocity of the mixture correspondingly decreases while the viscosity and density increases. The behavior of viscosity and velocity shows that the interaction between the CCl_4 and IPA is very weak[6]. Table 3 shows a random decrease in the free length. This is because the addition of non-polar molecules in the mixture causes the molecular structure to be loosely packed contributing to a decrease in velocity. This is further supported by the increasing values of acoustic impedance and decreasing values of adiabatic compressibility. It is also clear that the value of free volume increases with increasing mole fraction of CCl_4.

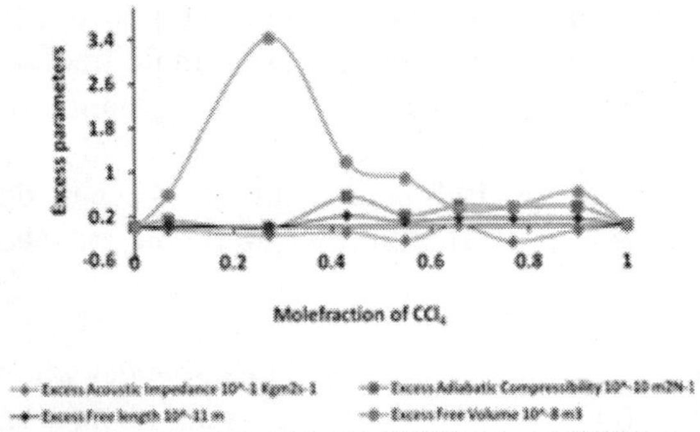

Fig. 5: Excess Parameters of CCl_4 + IPA

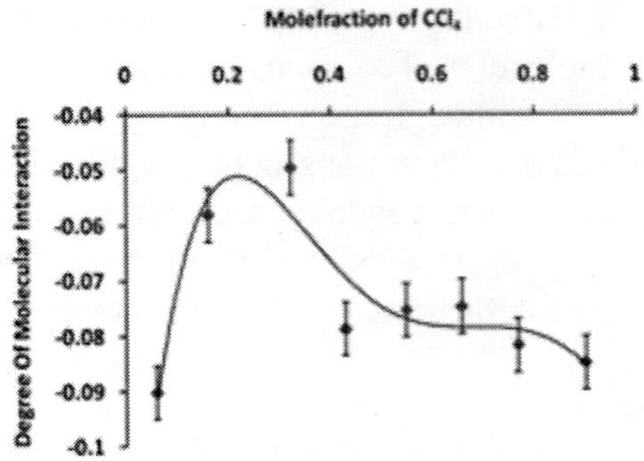

Fig. 6: Molecular Interaction of CCl_4 + IPA

Table 3: Thermo Acoustic Parameters of CCl_4 + Isopropanol

Mole fraction of CCl_4	*Density* ρ 10^3 Kgm^{-3}	*Viscosity* η 10^{-3} Nsm^{-2}	*Ultrasonic velocity U* ms^{-1}	*Acoustic Impedance* Z 10^3 Kgm^2s^{-1}	*Adiabatic Compressibility* $\beta\, 10^{-10}\, m^2N^{-1}$	*Free length* L_f 10^{-11} m	*Free volume* V_f 10^{-8} m^3
0.0000	0.7645	1.4569	1118.067	854.8155	10.4631	6.7362	3.5385
0.0674	0.8177	1.3583	1079.860	883.0287	10.4872	6.7439	4.3342
0.1605	0.9174	1.3620	1048.206	961.6186	9.9209	6.5593	4.9672
0.2721	0.8580	1.1726	1081.355	927.8027	9.9672	6.5747	7.9216
0.3224	1.0999	1.3700	993.7779	1093.038	9.2061	6.3186	5.9889
0.4293	1.1128	1.3259	950.669	1057.918	9.9430	6.5667	6.8920
0.5482	1.2430	1.3988	929.215	1155.018	9.3174	6.3567	7.1975
0.6561	1.2991	1.5018	914.210	1187.672	9.2100	6.3200	7.1919
0.7681	1.3843	1.5246	899.884	1245.731	8.9205	6.2198	7.7744
0.9001	1.4654	1.5290	891.420	1306.299	8.5877	6.1027	8.7294
1.0000	1.5610	1.6462	894.355	1396.085	8.0090	5.8935	8.6280

The positive trend in excess adiabatic compressibility and excess free volume is due to rupture of hydrogen bonded chains of IPA which predominate over the attractive force between unlike molecules [7]. This is also confirmed by negative deviation in acoustic impedance. The positive contribution of intermolecular free length arises from the breaking of hydrogen bonds of the self associated structure of alcohols. Degree of molecular interaction is negative which indicates that there is no significant interaction between the components.

CONCLUSION

From the thermo acoustical studies, it can be concluded that molecular association between DMF and Isopropanol arise from the hydrogen bonding between the solute and solvent molecules, structure breaking tendency because of repulsive force between the dipole and induced dipole and solute-solute interaction occurs between the polar and non-polar mixtures, DMF and CCl_4 and there is no significant interaction between CCl_4 and Isopropanol.

REFERENCES

[1] Mchaweh, A., Alsaygh, A. and Moshfeghian, M.A., *Fluid Phase Equilibria*, 224, 157–167 (2004).

[2] V. Tiwari and J.D. Pandey, *Indian J. Pure Appl. Phys.,* 18, 51 (1980).

[3] S. Prabakar and K. Rajagopal, *J. Pure Appl. Ultrason.*, 27, 41–48 (2005).

[4] J.D. Pandey, Gyan Prakash Dubey, B.P. Shukla and S.N. Dubey, *Pramana- J. Phys.*, 37(6), 497–503 (1991).

[5] P. Vasantharani, V. Pandiyan and A.N. Kannappan, *Asian J. Appl. Sci.*, 2(2), 169–176 (2009).

[6] Madhu Rastogi, Aashees Awasthi, Manisha Guptha and J.P.Shukla, *Ind. J. Pure Appl. Phys.*, 40, 256–263 (2002).

[7] R. Vadamalar, P. Mani, R. Balakrishnan and V. Arumugam, *Invertis J. Sci. Tec.*, 2(4), 252–260 (2008).

A Study of Solvation Effect of Imidazolium Bromide in Non-Aqueous Medium Using Ultrasonic Velocity

E. Jasmine Vasantha Rani, Seethalakshmi, R. Padmavathy, S. Shalini and K. Karthika

Department of Physics, Seethalakshmi Ramaswami College, Trichy, Tamil Nadu
E-mail: jasstephen @yahoo.com; seethalakshmiramasamy@yahoo.in

ABSTRACT: *Ultrasonic velocity of liquids gives valuable information regarding the nature and strength of various interactions and the formation of hydrogen bonding. Compressibility measurements are highly accurate and yield interesting information. The Ultrasonic velocity data combined with density constitute the standard means of determining the compressibility of the system. The sign and magnitude of compressibility indicate the solvation effects. When an electrolyte is added to a solvent, the volume of the solvent changes and hence there is a change in compressibility. These changes have been determined quantitatively as solvation number. In the present study, the solvation of BMIMBr in non-aqueous medium is taken for interpreting the various interactions occuring between the solute and solvent and the effect of temperature and concentration on solvation is also analyzed.*

Keywords: Ionic Liquids, Solvation Number, Solute-Solvent Interaction.

INTRODUCTION

Ionic liquids (ILs) have been investigated as an ion conductive matrix due to their unique physical properties, high ionic conductivity, a wide voltage window and non-volatility. Therefore, ILs have attracted much attention as a promising alternative to the conventional organic electrolyte and the organic solvent systems.[1–2] The liquid structure of ILs seems to be heterogeneous, unlike that of molecular solvents.[3–6] In the present investigation, solvation effects of BMIMBr are studied in non-aqueous medium. By measuring the fundamental quantities such as density and ultrasonic velocity of BMIMBr in formamide from 5°C to 55°C at different concentration, the solvation number is computed.

EXPERIMENTAL

Here, 1-butyl-3-methyl imidazolium bromide and formamide are purchased from Sigma Aldrich chemicals, USA with high purity (99%) and hence used without any further purification. A solution of BMIMBr in formamide is prepared at five different molalities (0.001 m, 0.005 m, 0.01 m, 0.015 m, 0.02 m). Density of the solution is measured using 25 ml specific gravity bottle with an accuracy of 0.0001gm/cc. The ultrasonic velocity is measured using Mittal's interferrometer, with an accuracy of $\pm$0.5 m/s.

COMPUTATION

The following thermodynamical parameters are computed using the standard formulae[7]

(i) Adiabatic compressibility:

$$\beta = \frac{1}{U^2\rho}\frac{cm^2}{dyne}.$$

(ii) Solvation Number:

$$n_h = \left(\frac{n_s}{n_i}\right)\left[1 - \frac{\beta}{\beta_0}\right]$$

where U – ultrasonic velocity in cm/sec.

ρ – the density of the solution in gm/cc.

M_{eff} – effective molecular weight of the solution in gm

β_0 & β – the compressibility of the solvent and solution in cm^2/dynes.

n_h, n_s and n_i are the primary solvation number, moles of solvent and moles of ions respectively.

RESULTS AND DISCUSSION

Adiabatic Compressibility

The rise and fall is observed in adiabatic compressibility of BMIMBr. The values are reported in Table 1 and the graphical representation is shown in Figure 1. The adiabatic compressibility decreases at 0.01 m showing a weakening of interactions at specific value and then it increases at higher molalities. These variations suggest that there is a strong solute-solvent interaction. Compressibility method is used to study the solvation effect of the solution.

Table 1: Adiabatic Compressibility × 10^{-11} (cm^2/dyne)

Molality (m)	*5°C*	*15°C*	*25°C*	*35°C*	*45°C*	*55°C*
0.001	3.19	3.37	3.45	3.54	3.61	3.75
0.005	3.2	3.32	3.45	3.58	3.64	3.76
0.01	3.16	3.27	3.43	3.52	3.64	3.66
0.015	3.2	3.4	3.41	3.54	3.64	3.73
0.02	3.24	3.45	3.44	3.55	3.64	3.8

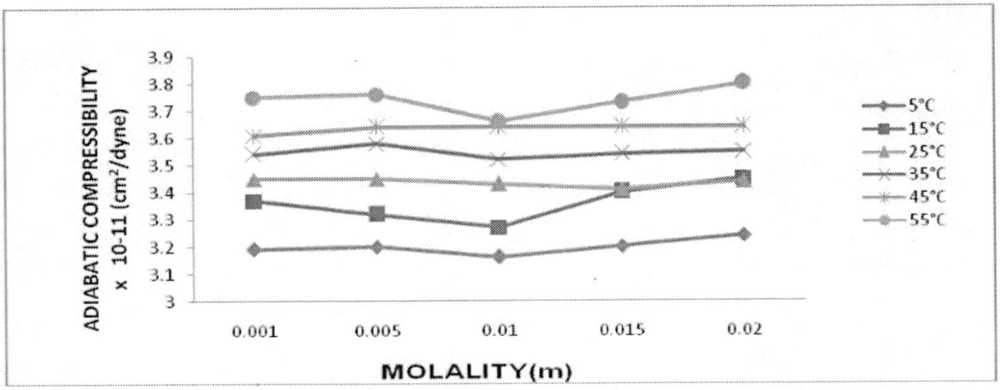

Fig. 1: Adiabatic Compressibility vs Molality

Solvation Number

Solvation Number of BMIMBr are reported in Table 2 and the variations are shown in Figure 2. In the present study, the solvation number is computed from compressibility measurements. Solute–solvent inter action, or solvation of solute ions or molecules, play a crucial role in interactions, such as dissolution of salts, metal-ion complexation, etc. in solution, because solute ions and molecules are solvated [8]. Here, electron-pair donating and accepting abilities of solvent play a significant role in determining the strength of solvation of cation and anion, respectively[9]. For (BMIMBr), above the room temperature (45°C and 55°c) the positive solvation number is observed. The variation in solvation number is observed only at lower concentrations. As the temperature increased from 5°C to 55°C, the value of solvation number is decreased parabolically from the positive high value and it increases above the room temperature. The low value of solvation number is found to be at higher concentration. Negative solvation number is observed below room temperature. It emphasize that the solutions are more compressible than solvent.Above room temperature (35°C, 45°C, 55°C), positive solvation is occurring. These temperature effects observed in this solution support the strong solute-solvent interaction.[10] It is also confirmed from IR spectral studies.

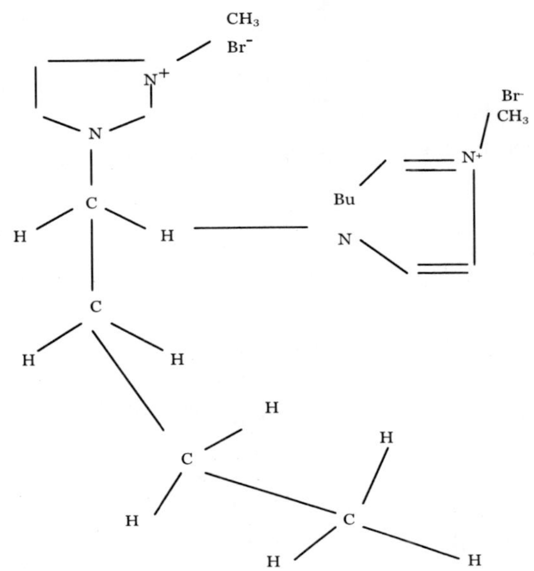

Table 2: Solvation Number

Molality (m)	*5°C*	*15°C*	*25°C*	*35°C*	*45°C*	*55°C*
0.001	352.	–190	–289	84	543	426
0.005	51	17	–51	–29	76	74
0.01	–11	44	–10	25	36	95
0.015	18	–29	–1	10	25	36
0.02	3	–36	–8	3	19	6

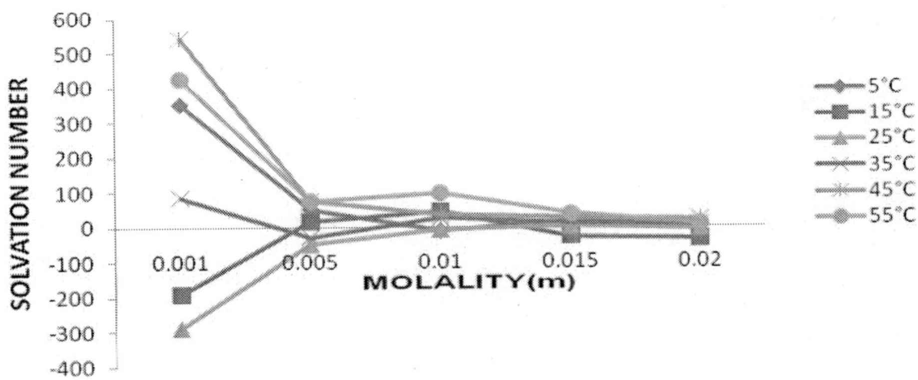

Fig. 2: Solvation Number vs Molality

CONCLUSION

In the present work, the acoustic studies are carried out to analyze the various molecular interactions occurring in the solution of BMIMBr. These results suggest that the solute-solvent binding is large at tertiary nitrogen of butyl group.

REFERENCES

[1] Patrick Ngoy Tshibangu, Silindile Nomathemba Ndwanwe, Ezekiel Dixon Dikio Int. J. Electrochem. Sci., 6(2011) 2201–2213.
[2] Bull. Korean Chem. Soc. 2006, Vol. 27, No. 6, 847 Synthesis and Properties of Ionic Liquids.
[3] J.N. Canongia Lopes and A.A.H. Padua, *J. Phys. Chem.*, B 110, 3330 (2006).
[4] J.N. Canongia Lopes, M.F.C. Gomes and A.A.H. Padua, J. Phys. Chem. B 110, 16816 (2006).
[5] A. Triolo, O. Russina, H.-J. Bluif, E. D. Cola. J. Phys. Chem. B 111, 4641 (2007).
[6] R. Atkin, G.G. Warr. J. Phys. Chem. B 112, 4164 (2008).
[7] Suryanarayana C.V., J. Acou .Soc. India 7(1979) 131–136
[8] J. Burgess, Ions in Solution: Basic Principles of Chemical Interactions, Harwood, Chichester (1999).
[9] C. Reichart, Solvents And Solvent Effects In Organic Chemistry, Vch, New York (1988).
[10] Seethalakshmi K. *et al.,* Elixir Ultrasonics 51 (2012), 10889–10891 .

Ultrasonic and Hydration Study of Aqueous Solutions of Ammonium Bromide

J.H. Rakini Chanderasekaran, A. Beulah Mary[1], V. Divya[1] and R. Shobana[1]

Department of Physics, Arignar Anna Government Arts College, Musiri

[1]Department of Physics, Bishop Heber College, Tiruchirappalli

ABSTRACT: *Ultrasonic study of liquids and solutions is a simpler but accurate method of analyzing the interactions between the molecules. The present work focus the study of acoustic parameters, hydration number and its related parameters of aqueous solutions of Ammonium Bromide. For studying these parameters, density and ultrasonic velocity through the aqueous solutions of various concentrations are measured at different temperatures ranging from 35°C to 55°C at 5°C. The apparent molal volume, apparent molal compressibility and molar solvated volume are all determined using the hydration number and traditional formulae. The variation of all these parameters with respect to concentration and temperature are analysed and also molecular interactions between solute and solvent are studied.*

INTRODUCTION

Ammonium Bromide is a wood preservative, soluble in water and used in pharmaceutical preparations. The aim of this paper is to study the physical properties of Ammonium Bromide.

The ultrasonic velocity in a liquid solution helps us to study and probe the properties of liquids. The physical and chemical behaviour of solutions and their molecular interactions are studied by measuring the density, viscosity and ultrasonic velocity. Measurements of these parameters with different concentration and temperature will help us to determine the acoustic parameters such as Adiabatic compressibility, Intermolecular free length, Specific Acoustic Impedance, Rao's constant and Wada's constant.

The interaction of ion with water molecules is explained in terms of hydration number. The hydration number is the number of water molecules that get attached with each ion during the process of interaction. The hydration number has been studied by various investigators. The compressibility of the solution is [1,2] assumed to be mainly due to the contribution of free solvent molecules. Kalyanasundaram *et al.* [3,4] has reported the methodology of computing hydration number and its related parameters such as the apparent molal volume, apparent molal compressibility and molar solvated volume using the hydration number and traditional formulae of the electrolyte solution. The molar hydration number is calculated using Barnartt's relation whereas molal hydration number is determined using Passynsky's equation. In this paper, the hydration number is determined for aqueous solutions of Ammonium Bromide for various concentrations in the temperature range 35°C to 55°C at 5°C intervals. From the hydration number, the variation of all these parameters with respect to concentration and temperature are analysed and their values are also compared with the traditional equations.

EXPERIMENTAL WORK

The aqueous solutions of various concentrations of Ammonium Bromide are prepared using AR grade salt and double distilled water. The density measurements are carried out using 10 ml specific gravity bottle and single pan digital balance, viscosity using Ostwald's viscometer and sound velocity in the solution using Mittal type ultrasonic interferometer with crystal frequency of 2 MHz with an accuracy of ±2 m/s. The measurements are carried out in the temperature range of 35° to 55°C. The temperature is maintained by a thermostat which can maintain temperature within ±0.1°C.

COMPUTATION

Formulae used for computation of different parameters:

Adiabatic compressibility

$$\beta = 1/(u^2\rho) \qquad \dots (1)$$

Intermolecular free length

$$L_f = K/(u^2\rho)^{1/2} \qquad \dots (2)$$

Specific acoustic impedance

$$Z = u\,\rho \qquad \dots (3)$$

Rao's constant

$$R = (M_{eff}/\rho)\, u^{1/3} \qquad \dots (4)$$

Wada's constant

$$W = (M_{eff}/\rho)\, \beta^{-1/7} \qquad \dots (5)$$

where the terms carry the usual meaning.

The **molar hydration number** n_h is calculated using Barnartt's [5] relation,

$$n_h = \frac{1}{n_2}\left[n_1 - \frac{\beta}{\beta_0}N\right] \qquad \dots (6)$$

where β_0 is the compressibility of water, n_l is the number of moles of the solvent present in the 1000 cc of the solution of molar concentration n_2 and N is the number of moles of the solvent in 1000 cc of the solvent.

The **molal hydration number** $n_h^{'}$ has been computed through Passynsky's [6] equation

$$n_h^{'} = \frac{N'}{n_2^{'}}\left[1-\frac{\beta}{\beta_0}\right] \qquad \dots (7)$$

where β_0 is the compressibility of water, n_l is the number of moles of the solvent present in the 1000 gm of the solution of molar concentration $n_2^{'}$ and N is the number of moles of the solvent in 1000 gm of the solvent.

The **apparent molal volume** of the solution is calculated using the relation,

$$\phi_v = \left[\frac{N}{n_1}n_h^{'} - n_h\right]V_m \qquad \dots (8)$$

It is also computed by the traditional method using the equation,

$$\phi_v = \frac{1000}{n_2\,\rho_0}(\rho_0 - \rho) + \frac{M_2}{\rho_0} \qquad \dots (9)$$

The **apparent molal compressibility** has been calculated using the relation,

$$\phi_k = -V_m\beta_0 n_h \qquad \dots (10)$$

and the traditional equation,

$$\phi_k = \frac{1000}{n_2\,\rho_0}(\rho_0\,\beta - \beta_0\,\rho) + \frac{M_2}{\rho_0}\beta_0 \qquad \dots (11)$$

The molar solvated volume is calculated from the following equations,

$$\phi_s = \phi_v + n_h\,V_m \qquad \dots (12)$$

$$\phi_s = \frac{V\Delta\beta}{n_2\beta_0} \qquad \dots (13)$$

where $\Delta\beta = \beta - \beta_0$

RESULTS AND DISCUSSION

The acoustic parameters such as adiabatic compressibility, intermolecular free length, specific acoustic impedance, Rao's constant and Wada's constant of the aqueous solutions are computed and given in Table 1. The adiabatic compressibility decrease with

Table 1: Acoustic Parameters of Ammonium Bromide

Temp	*Molality*	*β(*E10)*	*Lf(*E11)*	*Z (*10⁴)*	*R(*10³)*	*W(*10³)*
°C	*M*	m^2/N	*m*	*Rayl*	$[m^3/M]/[N/m^2]^{1/3}$	$[m^3/M]/[N/m^2]^{1/7}$
35°C	0.0	4.3314	4.18	1515042	0.209	0.3943
	0.5	4.1991	4.12	1554941	0.2469	0.4681
	1.0	4.0666	4.05	1603584	0.2773	0.5279
	1.5	3.9670	4.00	1641961	0.3048	0.5821
	2.0	3.8769	3.96	1677650	0.3290	0.6302
	2.5	3.7859	3.91	1715335	0.3497	0.6717
	3.0	3.7104	3.87	1748173	0.3684	0.7095
	3.1	3.6886	3.86	1757080	0.3717	0.7161
40°C	0.0	4.2983	4.21	1518484	0.2100	0.3960
	0.5	4.1605	4.14	1560886	0.2477	0.4695
	1.0	4.0373	4.08	1606633	0.2787	0.5303
	1.5	3.9398	4.03	1644903	0.3063	0.5846
	2.0	3.8487	3.98	1681438	0.3305	0.6326
	2.5	3.7599	3.94	1718801	0.3513	0.6743
	3.0	3.6835	3.90	1752089	0.3701	0.7122
	3.1	3.6622	3.88	1760607	0.3735	0.7191
45°C	0.0	4.2707	4.22	1522541	0.2100	0.3968
	0.5	4.1319	4.15	1566802	0.2478	0.4696
	1.0	4.0155	4.09	1610359	0.2792	0.5311
	1.5	3.9149	4.04	1649362	0.3069	0.5856
	2.0	3.8254	3.99	1685709	0.3312	0.6338
	2.5	3.7411	3.95	1722046	0.3521	0.6757
	3.0	3.6643	3.91	1755723	0.3709	0.7135
	3.1	3.6469	3.90	1763301	0.3743	0.7204
50°C	0.0	4.2444	4.24	1525785	0.2110	0.3979
	0.5	4.0909	4.17	1575108	0.2481	0.4700
	1.0	3.9832	4.11	1616013	0.2800	0.5323
	1.5	3.8853	4.06	1654428	0.3078	0.5871
	2.0	3.8006	4.02	1689626	0.3323	0.6356
	2.5	3.7139	3.97	1726865	0.3532	0.6775
	3.0	3.6368	3.93	1760695	0.3722	0.7156
	3.1	3.6168	3.92	1769025	0.3756	0.7225
55°C	0.0	4.2328	4.28	1526021	0.2110	0.3990
	0.5	4.0839	4.19	1575324	0.2486	0.4708
	1.0	3.9759	4.14	1616160	0.2806	0.5333
	1.5	3.8783	4.09	1654508	0.3086	0.5883
	2.0	3.7894	4.04	1690804	0.3331	0.6368
	2.5	3.704	3.99	1727631	0.3541	0.679
	3.0	3.6289	3.95	1761190	0.373	0.717
	3.1	3.6156	3.95	1767161	0.3767	0.7243

increase in concentration at all temperatures indicating that there are ion-solvent interactions [7]. Intermolecular free length is also found to decrease with concentration and this indicates that there is enhanced molecular association. Specific acoustic impedance, Rao's constant and Wada's constant increase with concentration as expected at all temperatures. Acoustic parameters values are found to increase with increase in temperature except adiabatic compressibility.

Table 2: Hydration Number of Ammonium Bromide

Temp	*Molality*	n_2	n_1	n_h	n_h'	ϕ_v (*10^6) m^3/mole		ϕ_k (*10^{14}) m^5/N/mole		ϕ_s (*10^6) m^3/mole	
35°C	0.5	0.49110	54.518	3.7	5	24.780	24.574	–2.874	–2.875	90.950	90.950
	1.0	0.96462	53.543	2.5	4.1	31.210	30.838	–1.942	–1.942	75.665	75.665
	1.5	1.42308	52.66	2.1	3.7	32.211	32.143	–1.645	–1.645	70.113	70.112
	2.0	1.86089	51.645	1.7	3.4	34.538	34.461	–1.352	–1.352	65.685	65.685
	2.5	2.28220	50.671	1.5	3.2	36.071	35.840	–1.166	–1.166	62.748	62.748
	3.0	2.67951	49.576	1.2	2.9	37.291	37.925	–0.934	–0.934	59.49	59.490
	3.1	2.75658	49.357	1.2	2.9	37.974	38.305	–0.908	–0.908	59.269	59.268
40°C	0.5	0.48891	54.275	3.4	4.8	27.292	27.414	–2.640	–2.640	88.832	88.832
	1.0	0.95943	53.255	2.2	3.9	33.069	33.302	–1.739	–1.739	73.751	73.750
	1.5	1.41236	52.263	1.8	3.5	35.147	35.381	–1.377	–1.377	67.415	67.415
	2.0	1.84283	51.144	1.4	3.2	37.992	38.155	–1.061	–1.061	62.836	62.836
	2.5	2.25951	50.167	1.2	3	38.623	38.982	–0.919	–0.919	60.356	60.356
	3.0	2.65354	49.096	0.9	2.8	40.025	40.529	–0.74	–0.740	57.748	57.748
	3.1	2.72971	48.876	0.9	2.8	40.595	40.862	–0.722	–0.722	57.663	57.663
45°C	0.5	0.48824	54.201	3.2	4.7	28.252	27.967	–2.472	–2.472	85.841	85.842
	1.0	0.95661	53.098	2	3.8	35.595	35.257	–1.522	–1.522	70.887	70.887
	1.5	1.40687	52.060	1.6	3.4	37.182	37.399	–1.208	–1.208	65.689	65.689
	2.0	1.83446	50.912	1.2	3.1	39.844	40.067	–0.904	–0.904	61.231	61.231
	2.5	2.25168	49.993	1	3	40.638	40.073	–0.816	–0.816	59.174	59.174
	3.0	2.64218	48.886	0.8	2.8	42.295	41.776	–0.632	–0.632	56.572	56.572
	3.1	2.72020	48.706	0.8	2.7	41.036	41.781	–0.626	–0.626	56.441	56.441
50°C	0.5	0.48657	54.016	3	4.6	30.915	31.105	–2.317	–2.317	85.686	85.686
	1.0	0.9526	52.875	1.8	3.7	37.732	37.713	–1.368	–1.368	69.94	69.940
	1.5	1.40098	51.842	1.4	3.4	39.376	39.090	–1.104	–1.104	65.107	65.107
	2.0	1.82426	50.629	1	3	41.671	42.145	–0.755	–0.755	59.929	59.929
	2.5	2.23220	49.560	0.8	2.8	42.489	43.172	–0.604	–0.604	57.405	57.405
	3.0	2.63035	48.667	0.7	2.7	42.109	42.830	–0.577	–0.577	56.428	56.428
	3.1	2.70712	48.472	0.7	2.7	42.383	42.931	–0.573	–0.573	56.422	56.422
55°C	0.5	0.48538	53.883	2.8	4.4	31.159	31.223	–2.141	–2.142	81.824	81.825
	1.0	0.94951	52.703	1.6	3.6	39.008	38.673	–1.229	–1.229	67.719	67.719
	1.5	1.39627	51.668	1.3	3.3	40.004	39.854	–1.006	–1.006	63.630	63.630
	2.0	1.81841	50.467	0.9	3	42.713	42.677	–0.708	–0.708	59.396	59.396
	2.5	2.22336	49.364	0.7	2.8	43.953	43.967	–0.541	–0.541	56.739	56.738
	3.0	2.60925	48.277	0.5	2.6	44.486	45.083	–0.402	–0.402	54.579	54.580
	3.1	2.68500	48.076	0.5	2.6	44.989	45.178	–0.387	–0.387	54.31	54.310

The computed values of hydration number, apparent molal volume, apparent molal compressibility and molar solvated volume of Ammonium Bromide solutions are given in Table 2. At all temperatures, the molal and molar hydration numbers are found to decrease with concentration. There is a slight decrease in the value of hydration number with temperature.

The values $\phi_{v,}$ ϕ_k and ϕ_s using traditional formula are found to be in good agreement [8] with the values computed with hydration number.

The apparent molal volume ϕ_v and the apparent molal compressibility ϕ_k increase with increase in concentration at all temperatures. The value of apparent molal compressibility ϕ_k is found to increase with temperatures whereas that of ϕ_v has a decreasing trend. The molar solvated volume ϕ_s is found to decrease with increase in concentration as in the case of hydration number. This shows that hydration number plays an important role [9] in deciding the value of ϕ_s.

CONCLUSION

The variations in the acoustical parameters with respect to concentrations and temperations reveal the existance of molecular association between solute and solvent molecules.Also the results of thermochemical parameters show the molecular interactions taking place in the solution. The acoustical and thermochemical results are found to be in good agreement with structural promoting behaviour of the solutes.

REFERENCES

[1] S. Kalyanasundaram and R. Bhagwath, 1990. *Trans. SAEST*, 25, 99.

[2] S. Kalyanasundaram and S. Saraswathi, 1994. *Indian J. Pure and Appl. Phys.*, 32, 320.

[3] R. Varadarajan, Piathomas and S. Kalyanasundaram, 1995. *J. Acoust. Soc.*, XXIII, 58–61.

[4] K. Maria Eugenie Pia and K. Ganthimathi, 2000. *J. Acoust. Soc. Ind.*,28(1–4),313–315.

[5] S. Barnartt, 1952. *J. Chem.Phys.*, 20, 278.

[6] A. Passynsky, 1930. *Acta Physiochem*, 8, 385.

[7] A. Beulah Mary, G. Bharathi. K. Ganthimathi, J.H. Rakini Chanderasekaran, *J. Acoust. Soc. Ind.*, Vol. 38, No. 2, (ISSN 0973–3302).

[8] Pia Thomas and K. Ganthimathi, 1999. *J. Acoust. Soc. Ind.*, 27 (1–4), 331–333.

[9] A. Dhanalakshmi, G. Rani and T. Arivudainambi, 2000. *J. Acoust. Soc. Ind.*, 28 (4), 385–388.

Intermolecular Interaction Studies of N-Benzoyl-L-Arginine in Non-Aqueous Solution

E. Jasmine Vasantharani, K. Kannagi[1] and R. Padmavathy
Department of Physics, Seethalakshmi Ramasamy College, Trichy
[1]Department of Physics, Bharathidasan University Constituent College Lagudi, Trichy
E-mail: jasstephen@yahoo.com; kanagi.natraj@gmail.com

ABSTRACT: *The ultrasonic studies of biological materials can provide knowledge of the acid-base properties of proteins. The thermo acoustic analysis of amino-acids in non-aqueous solution would be of great interest and it also gives the structural behavior of the sample. Ultrasonic velocity (v), density (ρ), viscosity (η) of N. benzoyl L. Arginine in non aqueous medium at 278 k to 328 k have been measured. Using these experimental data, the acoustical parameters such as adiabatic compressibility (β) apparent molal compressibility (φ_k), apparent molal volume (φ_v) were calculated. These experimental values have been analyzed and eventually emphasizing the possible molecular interactions in terms of structure promoting nature of the sample in the solvent.*

INTRODUCTION

N-Benzoyl L-Arginine (L_NBA) is an essential component of important biological molecules [1]. The synthesis of L-∝-amino acids remains of continuing interest for obtaining a particular amino acids- via protein hydrolysis which implies its separation from other amino acids and their possible wastage, particularly on large scales [2]. Major application of ∝-amino acids are to study of peptide science and protein engineering places significant demands [3]. Acoustical and thermodynamical study contributes a major part, in order to understand the physico-chemical properties and molecular interaction in various liquids [4,5]. Investigation of amino acids through ultrasonic techniques has been stimulated scientific interest of researchers.

EXPERIMENTAL TECHNIQUE

Solution of samples of different concentrations is prepared with AR grade salts. These reagents were used without further purification. The density of the solutions is determined using 25 ml specific gravity bottle, using the thermostatic bath with a compressor unit. A

Canan Fenske viscometer was used for the viscosity measurements. Variable path interferometer having a frequency of 2 MHz (Mittal Enterprises, New Delhi, model: F-81) with an overall accuracy of 0.1% was used for velocity measurements. A constant temperature path (digitalized) was used to circulate water through the double – walled measuring cell of steel containing the experimental solution at the desired temperature (5°C to 55°C).

The following formulae are used for the computation of adiabatic compressibility, apparent molal volume, and apparent molal compressibility.

FORMULAE

1. Adiabatic compressibility (β) is given by

$$\beta = \left[\frac{1}{u^2 \rho}\right] cm^2 / dyne.$$

u – ultrasonic velocity in cm/sec.,
ρ – density of the solution in gm/cc.

2. Apparent Molal Volume

$$\varphi_v = \frac{1000}{m_1 \rho_0}(\rho_0 - \rho) + \frac{M}{\rho_0} ml / mol$$

m_1 – molal concentration of the solute
M – molecular mass of the solute
ρ – density of the solution
ρ_0 – density of the solvent

3. Apparent molal compressibility (ψ_k) is calculations by

$$\varphi_k = \frac{1000}{m\rho_0}(\rho_0 \beta - \rho \beta_0) + \left[\frac{\beta_o M}{\rho_0}\right] ml / mol / cm^2 / dyne$$

ρ_0 – Density of the Solvent
β_0 – Compressibility of the Solvent
ρ – Density of the Solution
β – Compressibility of the Solution
M – Molecular Mass of the Solute
m – Molal concentration of the Solute

RESULTS AND DISCUSSIONS

Adiabatic Compressibility

The structural changes of molecules in the solution takes place due to the existence of electrostatic field between the interacting molecules [6]. The arrangement of molecules results in the effect of adiabatic compressibility.

The adiabatic compressibility (β) with molal concentration of L-NBA in formamide solution. The values of β in the system shows a increasing trend. Since the temperature increases, the intra-molecular bonds also tends to increase due to stretching and hence the compressibility increases [6]. This indicates that the molecular association is greater in this system. This is shown in Table 1 and Figure 1.

Apparent Molal Compressibility

The positive values of apparent molal compressibility supports the structure making nature of the solute in formamide [11]. At low molality (0.005) and at high temperatures (45°C and 55°C) L-NBA shows a negative value. This result supports that there is a strong ion-ion interaction occurring in the solution [12]. It is represented in Table 3 and Figure 3.

Apparent Molal Volume

At higher temperatures apparent-molal volume increases as concentration increases, from 5°C to 45°C the apparent molal compressibility of L-NBA is greater. This behavior suggests that there is a strong ion-ion interaction occurring in the solution [10]. It is shown in Table 3 and Figure 3.

From the IR spectroscopic analysis the red shift of amide vibration from 1604 cm^{-1} in the solid to 1672 cm^{-1} in formamide is also a strong evidence for the solute and solvent interaction. It confirms structure making nature of L-NBA.

Nα Benzoyl –L-Arginine

Table 1: Adiabatic Compressibility

Molality (m)	*5°C*	*15°C*	*25°C*	*35°C*	*45°C*	*55°C*
0.001	32.08	33.38	33.93	35.10	35.94	37.36
0.005	32.77	34.25	34.86	35.93	36.85	38.06
0.01	33.18	34.65	35.25	36.95	37.63	38.42
0.015	33.45	34.97	35.81	37.35	37.79	38.66
0.02	34.60	35.84	36.70	37.82	38.06	38.79

Table 2: Apparent Molal Compressibility

Molality(m)	*5°C*	*15°C*	*25°C*	*35°C*	*45°C*	*55°C*
0.001	–58.68	–26.34	9.37	–67.00	–136.61	–113.14
0.005	4.93	17.73	16.05	8.30	–5.52	–4.82
0.01	7.92	14.28	12.39	15.00	6.50	1.51
0.015	7.63	11.65	12.56	13.43	5.57	3.32
0.02	12.81	14.25	14.91	12.87	5.97	3.36

Table 3: Apparent Molal Volume

Molality(m)	*5°C*	*15°C*	*25°C*	*35°C*	*45°C*	*55°C*
0.001	–3851	–6979	–4129	–2001	–2372	–2517
0.005	–494	–378	–36	486	–49	129
0.01	–38	59	–55	69	131	34
0.015	65	–47	–10	162	100	171
0.02	197	119	115	159	154	158

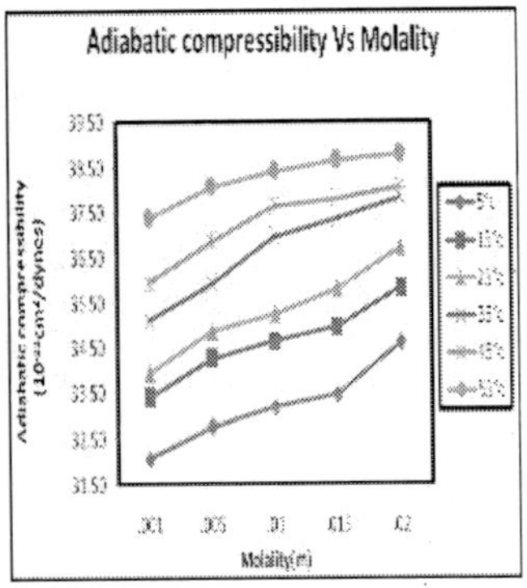

Fig. 1

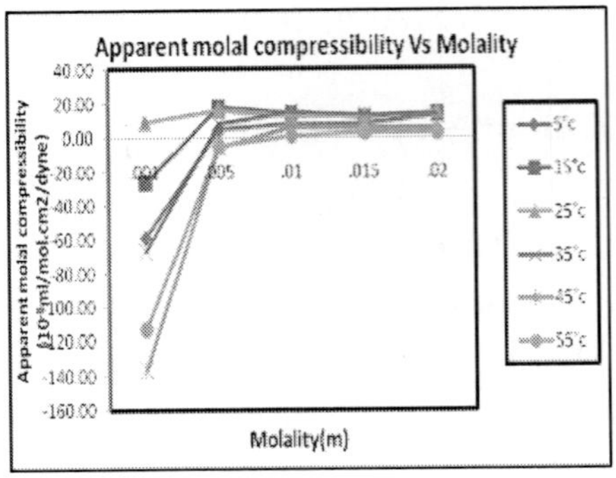

Fig. 2

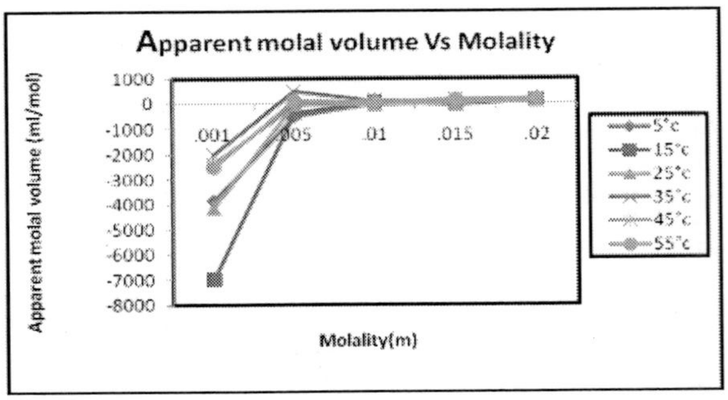

Fig. 3

CONCLUSION

The study of structural elucidation of the solute in the solvent shows that N-∝-Benzoyl L-Arginine in formamide acts as a structure maker. The increase in adiabatic compressibility indicates that there is a significant interaction taking place between N-∝-Benzoyl L-Arginine and formamide molecules. Thermochemical parameters also support the structure making nature of the solute in the solvent.

REFERENCES

[1] M. Aydro, M.H. Baskar, S. Yakan. S.F. Ulak, M. Aydmol, B. Aydmol, B. Aydowl and M. Buyum, Radiat. Eff. Defect solids 1163, 41 (2008).
[2] Hardy P.M. in comprehensive organic Chemistry, Berton, D.H.R; other W.D. Haslam E, Eds; Pergamon, Oxford, 1979, Vol. 5. pp. 187–214.
[3] EPR Study o M. Aydin, Rom Journal Phys. Vol. 56, Nos. 9–10, pp. 1156–1161, Bucharest, 2011.
[4] M. Aydin; Braz. J. Phys. 40, 499 (2010).
[5] Thiverny M. Philouze, C.; Charant, P.Y. Blandia V. Org. Biomol. Chem. 2010, 8, 1399–1405.
[6] Beilstein, J. Org. Chem, 2012, 8, 1393–1399.
[7] Y. Zhou and W.H. Nelson, J. Phys. Chem B, 114, 5567 (2010).
[8] J. Sinclair and Codella, J. Chem, Phys, 59, 1569(1973).
[9] A.N. Kannappan, and V. Rajendran, Indian J. Pure and Appl. Phys, 1992, 30, 176.
[10] Islam S. Waris B.N, Thermo Chemical, Acta; 424 (2004),165–174.
[11] Sathyarathi A.V., Acoustica, 28 (1973) 177.
[12] Vishnu Wadhwani R and Yasmin Akhtar, Ind. J. Chem., 34A (1995), 954.

Thermodynamic and Spectroscopic Study on Non-Aqueous Solution of Imidazolium Halide

E. Jasmine Vasantha Rani, K. Seethalakshmi, R. Padmavathy, N. Radha[1] and J. Bhuvaneshwari

Department of Physics, Seethalakshmi Ramaswami College, Tiruchirapalli

[1]Department of Chemistry, Seethalakshmi Ramaswami College, Tiruchirapalli

E-mail: seethalakshmiramasamy@yahoo.in; jasstephen @yahoo.com

ABSTRACT: *The state of any system can be defined completely by four observable properties, namely composition, pressure, volume and temperature. Study of thermodynamical properties of liquids and liquid mixtures play an important role in elucidation of the nature of molecular interactions. The liquid molecules may be held together because of the cohesive forces due to the presence of ions.The measurement of cohesive force is provided by the internal pressure. Internal pressure (π_i) and free volume (v_f) are the thermodynamical parameters, which are useful in understanding the intra and inter molecular interactions. In the present work, the ultrasonic velocity, density and viscosity measurements have been made for various temperatures and at different concentrations of imidazolium halide in non-aqueous solution. The variation of internal pressure and free volume are studied. The results are also compared with FT-IR spectrum.*

Keywords: Ultrasonic Velocity, Internal Pressure, Free Volume.

INTRODUCTION

Acoustical and thermodynamic parameters have been used to understand various types of intermolecular interaction and their strengths, molecular association in solution. Like other methods, the measurement of ultrasonic velocity in the solution has also been found to be an important tool to study the physico-chemical behavior.[1] To meet the needs of applications, ultrasonic velocity measurement are generally carried out at different temperatures. The present investigation is carried out to study internal pressure, free volume, specific acoustic impedance and intermolecular free length[2]. In this case, 1-butyl-3-methyl imidazolium chloride (BMIMCl) is taken as a solute and formamide as solvent.

EXPERIMENT

Here, 1-butyl-3-methyl imidazolium chloride and formamide are purchased from Sigma Aldrich chemicals, USA with high purity (99%) and hence used without any further purification. The solutions are prepared in the saturation molality by mass using electronic balance precise to 0.0001 gm. Ultrasonic velocity of (BMIMCl) in formamide solution is measured at the frequency of 2 MHz with a variable path interferometer with an accuracy of $\pm$0.05 m|s.

Computation of Internal pressure (π_i) and free volume (v_f) requires measurements of ultrasonic velocity (U), density(ρ) and viscosity(η) are made with usual procedures are determined using the following formulae[4].

Internal Pressure (π_i)

$$\Pi_i = bRT(k\eta/U)^{1/2} * (\rho^{2/3} / M_{eff}^{7/6}) \text{ atms};$$

Free Volume (v_f)

$$V_f = (M_{eff} * U/k\eta)^{3/2} \text{cc}$$

Where η = viscosity of the solution in poise
U = ultrasonic velocity in cm/sec
ρ = density of the solution in gm/cc
b = cubic constant (2)
R = gas constant (8.314×10^7)
T = temperature in Kelvin
k = constant equal to 4.28×10^9
M_{eff} = effective molecular weight of the solution in gm

The FTIR spectrum of this solution is recorded between 4000–400 cm^{-1} by using (PERKIN ELMER) model spectrum RXI FTIR spectrometer. The spectrum resolution is 1.0 cm^{-1}.

RESULTS AND DISCUSSION

Internal Pressure

From the graph, the internal pressure shows a general increase with increase in concentration (Table 1, Figure 1). This trend suggests that there is a strong solute-solvent interaction between the solute and the solvent molecules due to high cohesive energy. The dependence of this parameter on temperature has also been used to explain the strength of interactions with change in temperatures.[3][4] This may be due to the H-bonding. This emphasizes that the addition of salt in the solvent behaves as a structure maker.

Table 1: Internal Pressure π_i (atms)

Molality (m)	*5°C*	*15°C*	*25°C*	*35°C*	*45°C*	*55°C*
0.001	18944.6	15710.1	13366.8	12595.0	11358.5	10508.4
0.005	19149.6	15845.0	13500.0	12709.4	11789.9	10638.8
0.01	19277.5	16046.1	13565.9	12801.9	11776.8	10835.1
0.05	19339.6	16265.6	13763.5	13128.7	11785.3	10945.1
0.1	19703.2	16286.6	13764.1	13002.3	12079.4	10927.9

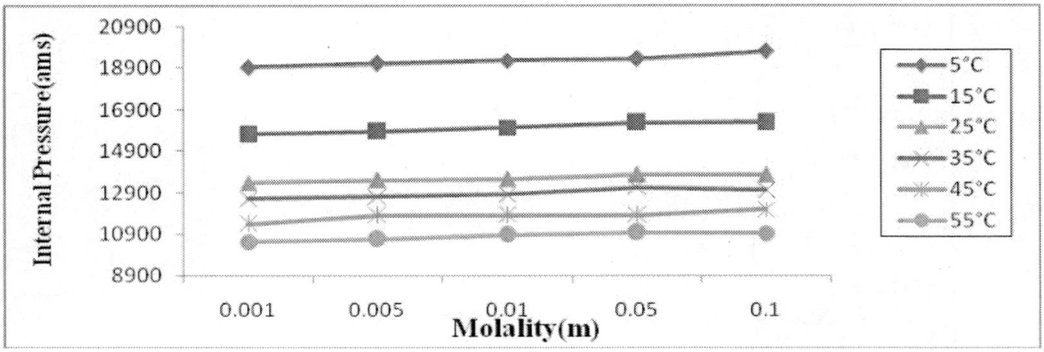

Fig. 1: Internal Pressure v_s Molality

Free Volume

The decreasing values of free volume with molalities and the increase in free volume in its decreasing trend is in agreement with the structure making behavior of the solute. The continuous decrease in free volume indicates that there is close packing of molecules with increase in solute concentration[5] (Table 2, Figure 2).

Table 2: Free Volume V_f (cc)

Molality (m)	*5°C*	*15°C*	*25°C*	*35°C*	*45°C*	*55°C*
0.001	0.00902	0.017407	0.030921	0.04023	0.05941	0.08098
0.005	0.00872	0.016924	0.029954	0.03898	0.05317	0.07652
0.01	0.00855	0.016295	0.02952	0.03816	0.05304	0.07388
0.05	0.00835	0.015492	0.02798	0.03487	0.05253	0.07140
0.1	0.00778	0.015196	0.027532	0.03517	0.04800	0.07007

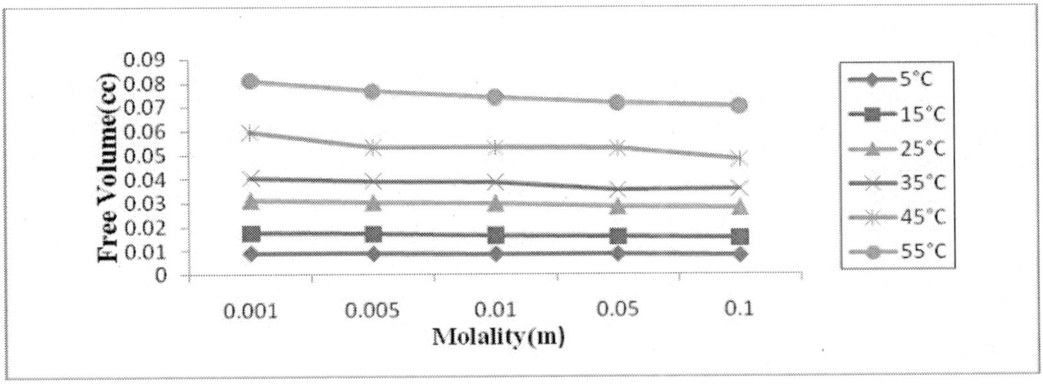

Fig. 2: Free Volume v_s Molality

IR Spectral Analysis

FTIR Spectrum of Formamide

At 3409 cm^{-1} ν_{NH} stretching vibration is observed (Figure 3, Table 3). At 2886cm^{-1}, a sharp peak of CH stretching vibration is found. At 2771 cm^{-1} and 2689 cm^{-1}, the stretching vibrations are assigned to the CH of the enolic form. A small amount of enolic form is identified by the $\nu_{C=N}$ at 2395 cm^{-1} and 2198 cm^{-1} vibrations.[6] The CO group appears as a sharp peak at 1688 cm^{-1}. The amide II bending modes are observed at1390 cm^{-1} and 1308 cm^{-1}, ν_{C-O} vibration is observed at 1054 cm^{-1}.

FTIR Spectrum of 1-butyl - 3 Methyl Imidazolium Chloride

1-butyl-3-methyl imidazolium chloride in the solid state (Figure 4) exhibits the characterization of a >N^+CH_3 group, the ammonium salt vibration at 2886 cm^{-1}. A series of peaks are observed in this region. At 2469 cm^{-1}, a broad absorption due to C=N is found. At 1624 cm^{-1}, C=C stretching vibration and CN bending vibration appeared as broad band. Further, at 1298 cm^{-1}, a sharp peak is observed due to CH bending vibration. At 1094 cm^{-1} and 898 cm^{-1} two bands of equal intensities are found[7]. This may be due to the resonating forms of the salt .At 549 cm^{-1}, the CCl vibrations are observed.

FTIR Spectrum of Solution of 1-Butyl-3-Methyl Imidazolium Chloride and Formamide

The ν_{NH} vibration is observed at 3409 cm^{-1} in pure formamide and in the case of solution, the ν_{NH} vibration is changing from 3409 cm^{-1} to 3414 cm^{-1}(Figure 5). A red shift of 5 cm^{-1} is observed in this solvent peak. The ν_{CH} vibration at 2887 cm^{-1} has not undergone much shift. Hence solvation has occurred at the N-H bond in formamide. The important peak corresponding to amide-I vibration is noticed at 1689 cm^{-1}($\nu_{C=O}$) has shifted only by 2 cm^{-1}.[8] The characteristic change in absorption frequency ($\nu_{C\text{-}O}$) is found at 1059 cm^{-1} in the solution and in the pure solvent, it is found at 1054 cm^{-1}. The solution spectrum is thus indication of solvation at NH_2 group and the enolate position of formamide (Figure 6). This is further supported by the C=N vibration of formamide at 2395 cm^{-1} which is blue shifted (11 cm^{-1}) to 2384 cm^{-1} in the solution.

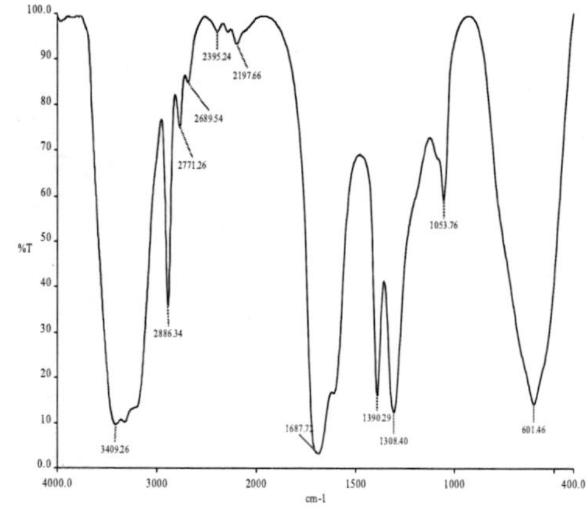

Fig. 3: FT – IR Spectrum of Formamide

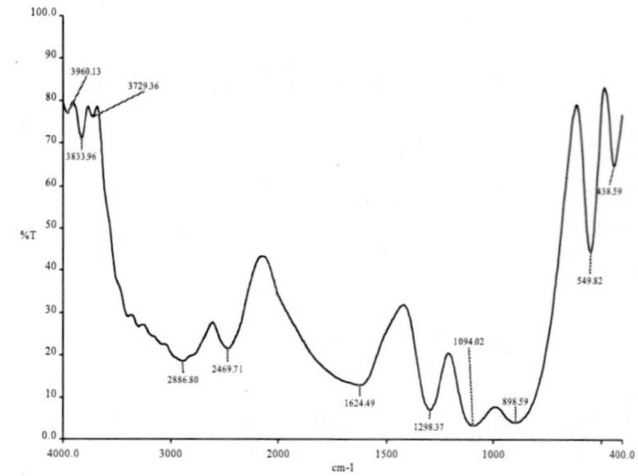

Fig. 4: FT – IR Spectra of 1– Butyl – 3 – Methyl Imidazolium Chloride

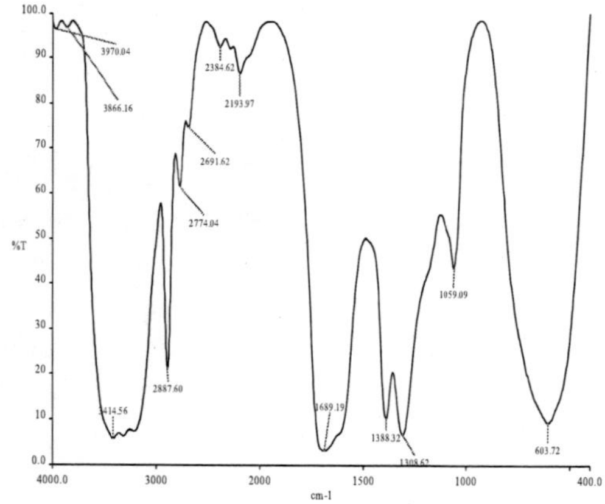

Fig. 5: FT – IR Spectra of 1– Butyl –3 Methyl

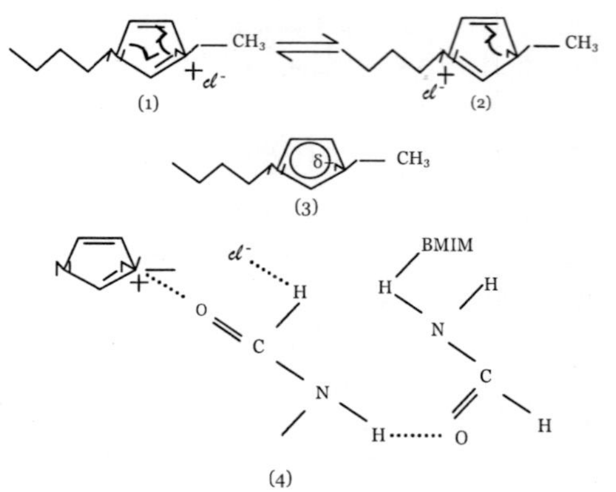

Fig. 6: Imidazolium Chloride with Formamide

Table 3: FT-IR Spectral Assignments

Band Assignment	*νNH cm^{-1}*	*νCH cm^{-1}*	*νC=N+cm^{-1}*	*νC=Ocm^{-1}*	*νC-Ocm^{-1}*	*νC=Ccm^{-1}*	*νC=Ncm^{-1}*	*δC–N cm^{-1}*	*νCl– cm^{-1}*
Solvent	3409	2886	2771 2689 2395 2198	1688	1054	–	–	–	–
Solid	2886	1298	2469	–	–	1624	–	1094 898	549
Solution	3414	2887	2384	1689	1059	–	–	–	–

CONCLUSION

In the present work, the acoustic and thermodynamic study suggests that the intermolecular H- bonding exists between the solute and solvent molecule. The variations in the internal pressure with different concentrations have been used to explain the ion-solvent interactions and hence study the structural changes in the solution. The results support that 1-butyl-3-methyl imidazolium chloride solution behaves as structure maker in formamide. It is also confirmed from the IR spectrum of solvent, solid and saturated solution.

REFERENCES

[1] Patrick Ngoy Tshibangu, Silindilev Nomathemba Ndwandwe, Ezekiel Dixon Dikio. Int. J. Electrochem. Sci., 6 (2011), 2201–2213.

[2] A.V. Paulin Edwidge Mary, K.M.E. Pia and A. Inigo, Proceedings of National Symposium on Acoustics, Nov. 2011.

[3] Jasmine Vasantha Rani E., Padmavathy R., Kannagi K., Acoustic waves, 11(2011) (152–162) ISBN : 978-81-8329-441-6.

[4] Jasmine Vasantha Rani, E, Geetha, K., Solachi, V., Padmavathy, R. and Kavitha, J.J. Acou .Soc. India , vol.31,1–4(2003), 323–326.

[5] Suryanarayana C.V. J. Acou .Soc. India 7(1979) 131–136.

[6] Mu, Z.G.; Wang, X.X.; Zhang, S.X.; Liang, Y.M.; Bao, M.; Liu, W.M., Investigation of tribological behavior of Al-Si alloy against steel lubricated with ionic liquids of 1-diethylphosphonyl-n-propyl-3-alkylimidazolium tetrafluoroborate. J. Tribol.- Trans. ASME 2008, 130, Art. No. 034501.

[7] Suzuki, A.; Shinka, Y.; Masuko, M., Tribological characteristics of imidazolium-based room temperature ionic liquids under high vacuum. Tribol. Lett. 2007, 27, 307–313, Molecules 2009, 14, 2906.

[8] Mazyar, O.A.; Jennings, G.K.; McCabe, C. Frictional dynamics of alkylsilane monolayers on SiO2: effect of 1-n-butyl-3-methylimidazolium nitrate as a lubricant. Langmuir 2009, 25, 5103–5110.

Excess Thermodynamical Studies in the Binary Mixtures of Methyl and Ethyl Acetate in 2-Methoxyethanol at Different Temperatures

G. Ravichandran and D. Gopinath

PG & Research Department of Physics, Aringar Anna Government Arts College, Villupuram, Tamil Nadu

ABSTRACT: *Density and ultrasonic velocity have been measured for binary liquid mixtures of methyl acetate and ethyl acetate in 2-methoxyethanol at the temperature ranges from 303.15 K to 323.15 K. The measured data are used to compute the acoustical excess thermodynamical parameters namely excess adiabatic compressibility (β_s^E), excess intermolecular free length (L_f^E) and excess molar volume (V^E). The excess values of the above said parameters were plotted against the mole fraction of methyl acetate and ethyl acetate over the entire composition range, indicates that the strong and weak interaction of hydrogen bonding between the molecules of the liquid mixtures.*

INTRODUCTION

The concentration and temperature dependence of acoustic properties has proved to be a significant observation of intermolecular interactions in liquids, liquid mixtures and solutions[1–6]. Since liquid mixtures have found applications in medicine, engineering, agriculture and other industrial applications, the study and understanding of thermodynamic and transport properties are more essential.[7–10] Intermolecular interactions in liquid mixtures modify the structural arrangements and hence change the shape of the molecule. Measurements of ultrasonic velocity and density have been used to calculate various parameters related to different types of molecular interactions in liquid mixtures. The compressibility behaviour of solutes, which is the second derivative of the Gibbs energy, is a very sensitive indicator of molecular interactions and can provide useful information about these phenomena.[11–14] The non-rectilinear behaviour of ultrasonic velocity, compressibility and other thermodynamical parameters of liquid mixtures also reveal the strength of interactions. To get additional information about the nature and strength of molecular interactions, other related acoustical parameters such as free length, adiabatic compressibility, free volume and their excess parameters have been calculated in the liquid mixtures.[15, 16] The excess thermodynamic functions are sensitive to the intermolecular forces as well as to the size of the molecules. In order to study all these molecular-kinetic properties of liquids and liquid mixtures, low amplitude ultrasonic wave is very valuable.

MATERIALS AND METHODS

The chemicals used in the present work are spectroscopic (SR) grade with a minimum assay of 99.9%. These chemicals were purchased from SD Fine chemicals, India and are used as such without further purification. In all systems, the various concentrations of the binary liquid mixtures were prepared in terms of mole fraction.

The density of pure liquids and their liquid mixtures are measured using a 10 ml specific gravity bottle with the experimental liquid is immersed in a temperature controlled water bath by relative measurement method with a reproducibility of ±0.01 Kgm^{-3}.

The ultrasonic velocity of the liquid mixtures has been measured using an Ultrasonic Interferometer (Model F81) supplied by Mittal Enterprises, New Delhi at an RF frequency of 2 MHz with an accuracy of ±0.1 ms^{-1} at different temperatures. The temperature was controlled by circulating water around the liquid cell from a thermostatically controlled water bath (accuracy ±0.1 °C).

Using the measured values of ultrasonic velocity and density, various excess thermodynamical parameters such as excess adiabatic compressibility (βs^E), excess free length (L_f^E) and excess molar volume (V^E) have been calculated using the standard equations.

RESULTS AND DISCUSSION

The ultrasonic velocity measurements are carried out in the binary mixtures of methyl acetate–2methoxyethanol (MA-2ME) and ethyl acetate-2methoxyethanol (EA-2ME) at different temperatures. The experiment was carried out in the composition range of X_1 = 0 to 1 mole fraction of methyl and ethyl acetates.

Using the measured values of ultrasonic velocity and density, variation of excess adiabatic compressibility (βs^E), excess freelength (L_f^E) and excess molar volume (V^E) with increasing concentration of methyl and ethyl acetates (X_1 = 0 to 1 mole fraction) are shown in Figures 1–6.

Methyl Acetate—2 Methoxyethanol System (MA-2ME)

The variation of excess adiabatic compressibility with increasing mole fraction of MA at 303.15, 308.15, 313.15, 318.15 and 323.15 K is shown in Fig1. Figure 1 shows that the excess adiabatic compressibility at 303.15 K has a negative deviation for the entire concentration range of methyl acetate. The magnitude of negative deviation reaches a maximum at $X_1 = 0.4983$ mole fraction of MA and then becoming less and less negative with further increase in concentration of MA in 2ME. The excess free length and excess molar volume also exhibits a similar behaviour as that of excess adiabatic compressibility at 303.15 K. The observed negative deviation of βs^E, L_f^E and V^E at 303.15 K from the ideal behaviour can be explained as follows.

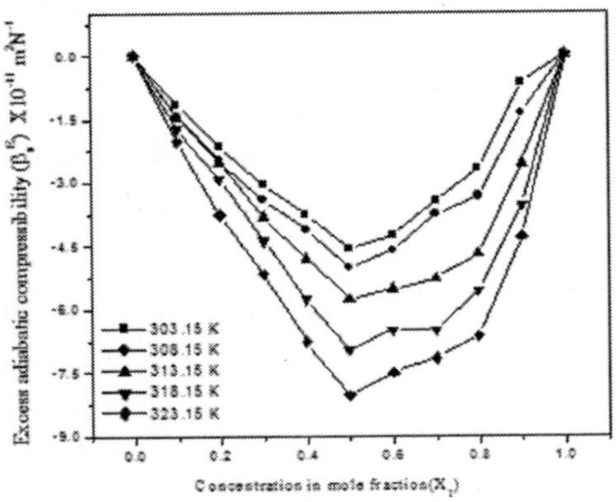

Fig. 1: Variation of Excess Adiabatic Compressibility (β_s^E) with Concentration in Mole Fraction (x_1) for MA-2ME System at Different Temperatures

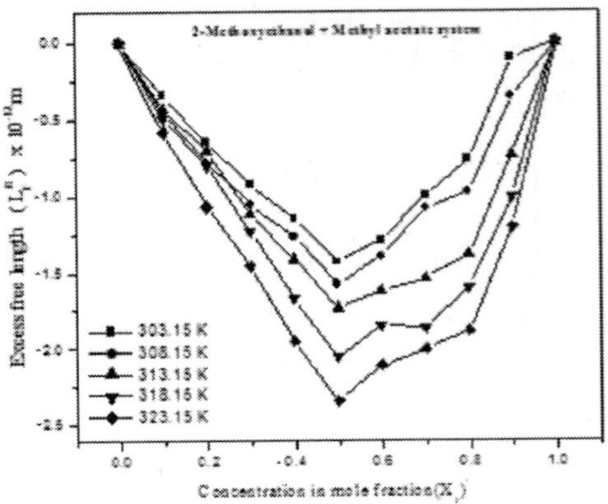

Fig. 2: Variation of Excess Free Length (L_f^E) with Concentration in Mole Fraction (x_1) for MA-2ME System at Different Temperatures

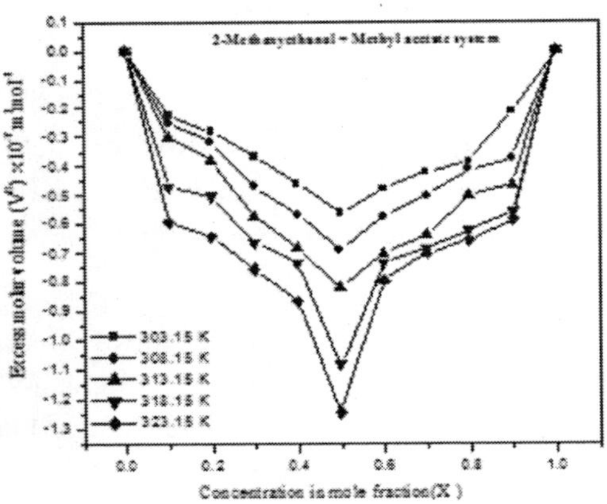

Fig. 3: Variation of Excess Molar Volume (V^E) with Concentration in Mole Fraction (x_1) for MA-2ME System at Different Temperatures

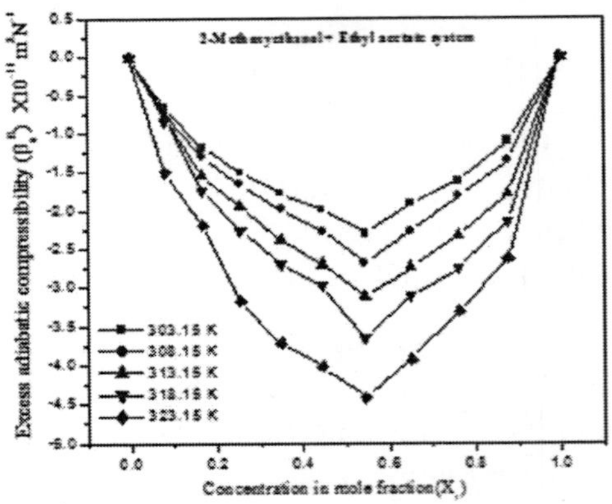

Fig. 4: Variation of Excess Adiabatic Compressibility (β_s^E) with Concentration in Mole Fraction (x_1) for EA-2ME System at Different Temperatures

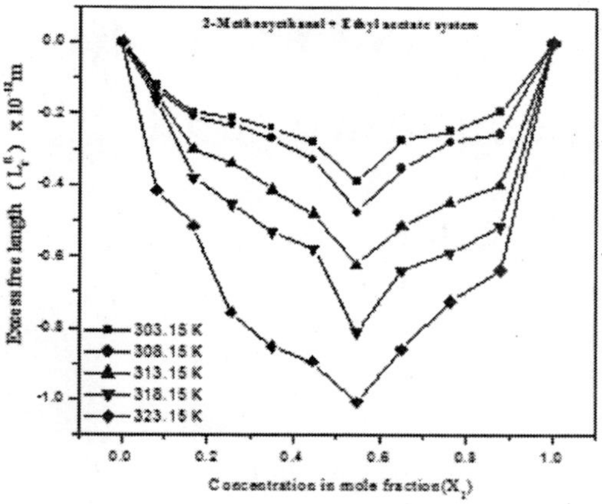

Fig. 5: Variation of Excess Free Length (L_f^E) with Concentration in Mole Fraction (x_1) for EA-2ME System at Different Temperatures

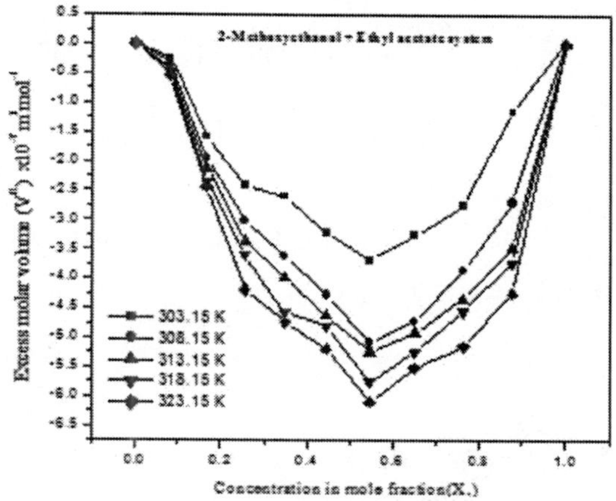

Fig. 6: Variation of Excess Molar Volume (V^E) with Concentration in Mole Fraction (x_1) for EA-2ME System at Different Temperatures

REFERENCES

[1] K. Tamura, T. Sonada and S. Murakami, J. Solution Chem. 28 (1999) 777.

[2] B. Garcia, R. Alcalde, J.M. Leal, J.S.Matos, J. Chem. Soc., Faraday Trans. 93 (1997)1115.

[3] S.L. Oswal, P. Oswal, R.D. Phalak, J. Solution Chem. 27 (1998) 507.

[4] Tiwari K, Patra C and Chakravortty V, *Acoustic Letters*, 19 (1995) 53–59.

[5] Awwad Akl M, *Acoustic Letters*, 12 (1989) 145–147.

[6] M. Jorg, A. Ghoneim, G. Turky, M. Stockhausen, Phys. Chem. Liq. 29 (1995) 26

[7] S.K. Chauhan, V.R. Singh, Indian J. Pure Appl. Phys. 31 (1993) 635.

[8] Babu P, Chandrasekhar G and Prabhakara Rao, *Indian J Pure and Appl Phys*, 38 (2000) 88–95.

[9] Ramanjappa T, Bhavani J A, Rajagopal E and Manaohara Murthy N, *Indian J Pure and Appl Phys,* 38 (2000) 301–305.

[10] R.L. Blokhra, A. Nag, Indian J. Pure Appl. Phys. 29 (1991) 756.

[11] B. Das, P.K. Muhuri, D.K. Hazra, Acoust. Lett. 18 (1994) 69.

[12] P.K. Muhuri, B. Das, D.K. Hazra, Indian J. Chem. 35A (1996) 288.

[13] P.J. Victor, P.K. Muhuri, B. Das, D.K. Hazra, J. Phys. Chem. B 103 (1999) 11227.

[14] P.J. Victor, B. Das, D.K. Hazra, J. Solution Chem. 30 (2001) 435.

[15] M. Gupta, J.P. Shukla, Indian J. Pure Appl. Phys. 34 (1996) 769.

[16] A.N. Kannappan, R. Palani, Indian J. Phys. 70B (1) (1996) 59.

Acoustical Studies on the Effect of Electrolyte on the Micelle of Aqueous Surfactant

G. Ravichandran and G. Lakshiminarayanan

PG & Research Department of Physics, A.A. Government Arts College, Villupuram, Tamil Nadu

ABSTRACT: *In the present studies, the ultrasonic velocity and absorption are carried out in aqueous solutions of sodium cholate in the concentration range of 0.006 molar to 0.017 molar by adding 0.1 M, 0.3 M and 0.5 M KCl, and LiCl using digital ultrasonic pulse echo velocity meter (model VCT 70 A) at a fixed frequency of 2 MHz and at a fixed temperature of 303.15 k.*

From the measured values of ultrasonic velocity (C) and observed absorption $(\alpha/f^2)_{obs,}$ *varies other parameters namely adiabatic compressibility* (β_s)*, free length* (L_f)*,excess absorption* $(\alpha/f^2)_{exe}$ *and relaxation time* (τ) *are calculated and reported. The addition of KCl and LiCl in aqueous solution of sodium cholate, the CMC value of aqueous solution of sodium cholate shift towards higher concentration side. The results are discussed in terms of micelle formation and electrostatic attractions of different cation in these solutions.*

INTRODUCTION

The bile salts are derivative of bile acid, which are naturally occurring detergents and form micellar aggregates in aqueous solutions [1].Bile salts helps to solubilize and disperse dietary lipids in small intestine. Bile salts have also been studied as alternative to conventional detergents for chemical analysis. Specific area of application include chemical separation [2] and luminescence analysis [3]. The bile salts micelle are smaller and more rigid than those of conventional detergents resulting in unique aggregation behaviour with respects to self-association as well as solubilization of hydrophobic molecules [4–6]. From the recent literature surveys it can be seen that most of ultrasonic studies in surfactants solutions are confined only to aqueous surfactants [7,8], surfactant dissolved in alcoholic [9,10], organic mixed solvents [11] and binary mixture of aprotic solvents in water [12–14] indicate that aprotic solvents are effective breaking power in water structure. Our earlier ultrasonic velocity studies in aqueous solution of oleate with aprotic solvents [15] and aqueous solution of sodium taurocholate [16] gives interesting results. But, Acoustical studies on electrolytes in aqueous solution of bile salts are scandy. So the present ultrasonic studies have been undertaken in aqueous solutions of sodium cholate (SC) by adding KCl, and LiCl. Hence, these studies has been undertaken to throw more light on the effect of KCl, and LiCl on the micelles formed by sodium cholate.

MATERIAL AND METHODS

The surfactant (sodium cholate) and electrolytes (KCl, and LiCl) used here are AR/BDH grade purchased from SD-fine chemical ltd, India and they are used as such without further purification. The surfactants solutions having different concentrations are prepared by dissolving the required amounts of surfactants and KCl, LiCl in double distilled water. The ultrasonic velocity and absorption are measured using a digital ultrasonic pulse echo velocity meter (model VCT-70A, Vi-microsystems, Pvt. Ltd., Chennai, India) at a fixed frequency 2MHz and at a fixed temperature of 303.15 K by circulating water from a thermostatically controlled water bath and temperature being maintained to an accuracy of ± 0.1°C. Accuracy in velocity measurements is ± 2 part in 10^5. Shear velocity and density of aqueous solutions of sodium cholate by adding KCl and LiCl are determined using an Oswald's viscometer and a graduated dilatometer respectively.

From the measured values of ultrasonic velocity(C), density (ρ), viscosity (η) and observed absorption $(\alpha/f^2)_{obs}$, various other parameter such as adiabatic compressibility (β_s), intermolecular free length (L_f) and classical absorption $(\alpha/f^2)_{cla}$ are calculated using standard formulae [17].

RESULT AND DISCUSSION

The ultrasonic velocity and observed absorption in aqueous solutions of sodium cholate by adding 0.1 M, 0.3 M and 0.5 M KCl, and LiCl increases with increasing concentration of sodium cholate in the entire concentration range studied exhibiting a break at CMC. The variation of ultrasonic velocity, adiabatic compressibility, intermolecular free length and observed absorption with concentration of sodium cholate studied is shown graphically in Figures 1–8.

Aqueous Solution of Sodium Cholate with KCl

With the addition of 0.1 M, 0.3 M and 0.5 M KCl to aqueous solution of sodium cholate, the ultrasonic velocity also increases. From Figure 1, it can be seen that the ultrasonic velocity for a particular concentration of sodium cholate is higher in the solution with 0.5 molar KCl than in the solutions with 0.3 molar and 0.1 molar KCl. Moreover, the velocity curves for the aqueous solutions of sodium cholate 0.1, 0.3 and 0.5 molar Kcl can be seen as the intersection of two straight lines at a particular concentration of sodium cholate that is the critical micelle concentration (CMC). The present study indicates that the CMC is 0.013 molar, 0.014 molar and 0.015 molar for aqueous solution of sodium cholate with 0.1, 0.3 and 0.5 molar KCl respectively.

The variation of ultrasonic velocity with increasing concentration of sodium cholate in 0.1 molar, 0.3 molar and 0.5 molar KCl water mixtures may be explained as follows:

In aqueous solution of sodium cholate with KCl, water molecules hydrogen bonded with hydroxyl group of sodium cholate. Na^+ ion obtained due to dissociation of sodium cholate in aqueous medium may also contributes towards increase of ultrasonic velocity by its water structure making property [18]. In addition, the electrostatic attraction between the K^+ ion (obtained by dissociation of KCl in water) and cholate ions.

The before the CMC value, all the above effect contributes the increase of ultrasonic velocity and decrease of compressibility by increasing concentration of sodium cholate with 0.1molar, 0.3molar and 0.5 molar. The increase of ultrasonic velocity when the concentration of sodium cholate are increased beyond the CMC may be due to the aggregation of sodium cholate molecule leading to micelle formation [19]. The above CMC, aggregation of molecules can occur by poly functional hydrogen bonding. Since sodium cholate molecules have both residual hydrogen bond donor and acceptors groups. Hence, poly functional hydrogen bonding may be possible and this may perhaps lead to the formation of higher aggregate [16, 17], the formation of higher aggregate as shown in Figure 1.

It is observed from the present study that the CMC's of aqueous solutions of sodium cholate with 0.1, 0.3 and 0.5 molar KCl is being shifted to 0.013 molar (a shift of 0.003 M), to 0.014 molar (a shift of 0.004 M) and 0.015 molar (a shift of 0.005 M) compared to 0.010 molar of aqueous sodium cholate without KCl. This may attributed to the electrostatic attraction between K^+ ion and cholate ion and increases of dielectric constant of the medium, still there is appreciable shift in CMC towards higher concentration

The ultrasonic absorption measurements undertaken in aqueous solutions of sodium cholate with 0.1 molar, 0.3 molar and 0.5 molar KCl shows a change of slope at CMC (as shown in Figure 4) thereby supporting the explanation offered for the velocity studies.

A similar effect can also occur in aqueous solutions of sodium cholate with LiCl. But the shifted CMC's value differ from aqueous solution of sodium cholate with KCl, these mechanism can be explained as follows.

Among the electrolytes, potassium have higher atomic radius compared to lithium. But all are having same charge. So, the electrostatic force between lithium and cholate is more, less with potassium. These leads to micellar property and ultrasonic velocity is found be in the order.

With lithium mixture › potassium mixture.

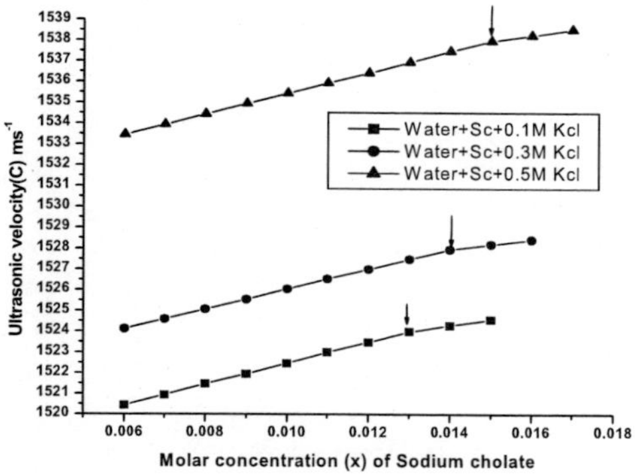

Fig. 1

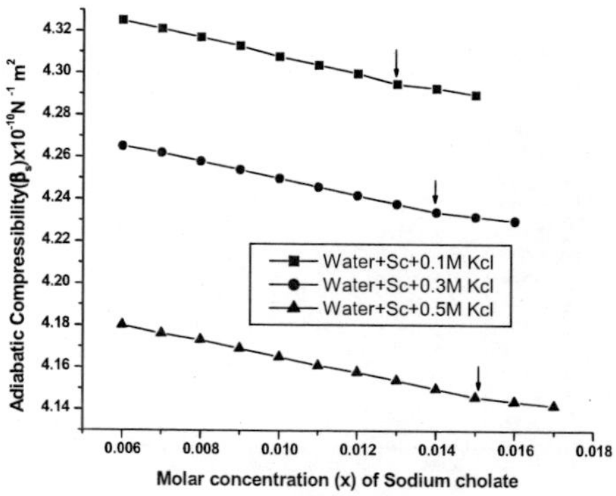

Fig. 2

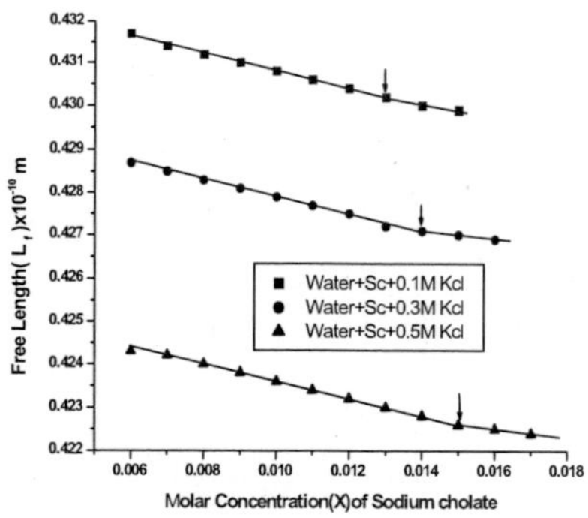

Fig. 3

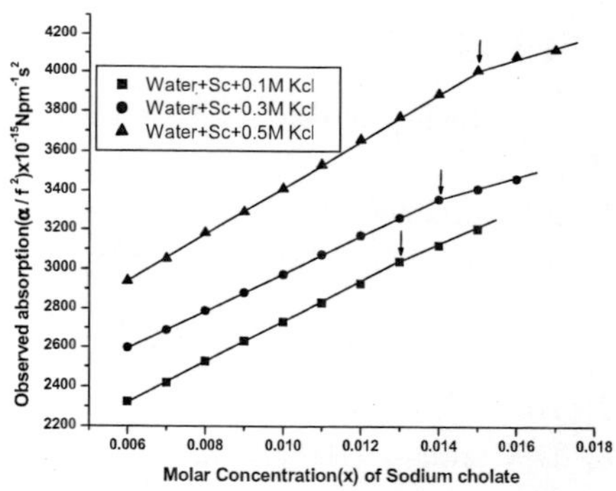

Fig. 4

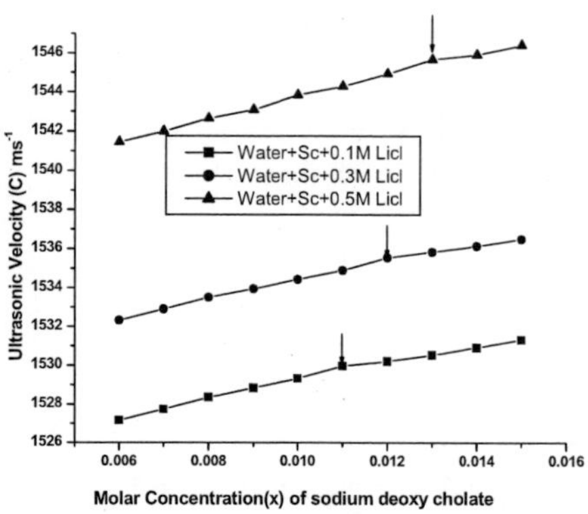

Fig. 5

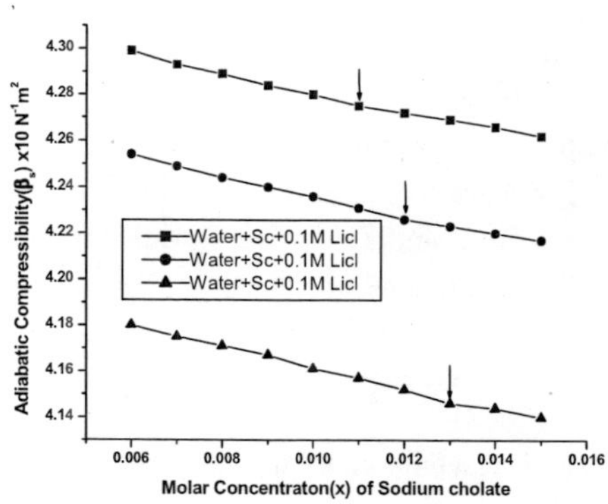

Fig. 6

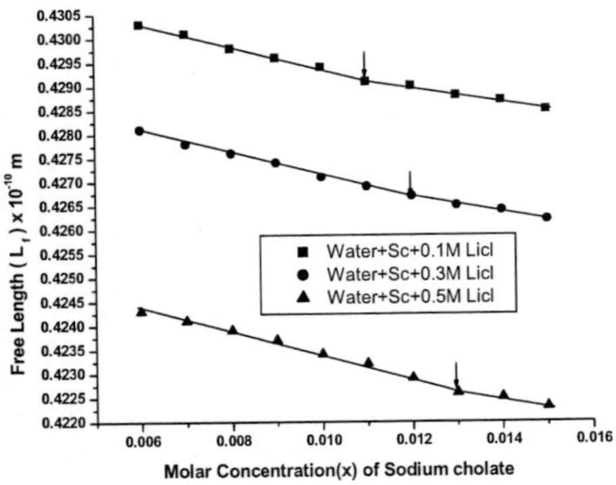

Fig. 7

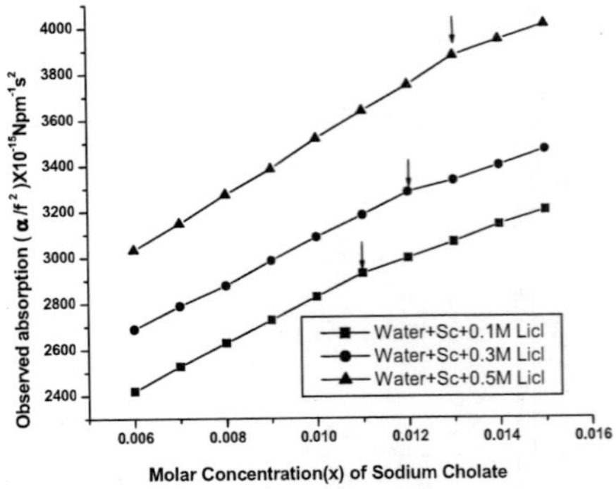

Fig. 8

REFERENCES

[1] D.M. Small "In Bile Acids" –Eds. P.P. Nair and D. Kritchevsky, Plenum press, New York (1971).

[2] J.J. Buckley, D.B. Wetlaufer Chromatoger., 464, 61 (1989).

[3] L.B. McGown, D.S. Kreiss Flourescence Detection II, Proc, SPIE-Int. Soc. Opt. Eng, 910,73 (1988).

[4] R. Zana, G.Guveli J. Phys, Chem., 89, 1687 (1985).

[5] P. Mukerjee, J. Morai, M. Murafa, A.Y.S. Yang Hepatology, 61S, 4 (1984).

[6] L. Fisher and D. Oakenfull Aust. J. Chem., 32, 31 (1979).

[7] Andrzej Balcerzak, Archives of Acoustics, 33, No 4, 4132–427 (2008).

[8] D. Geetha, T. Thilagavathi, P.S. Ramesh and C. Rakkappan. ARPN J. of Engineering and Applied Science, 4, No 6, (2009).

[9] Kim Gracie, Paul Wiseman, R. Paiepu, physics and chemistry of liquids 37. Issue 2, 107–123 (1999).

[10] Sadakatsu Nishikawa and Hua Huang., Buttetin of the chemical society of Japan, 75 No. 6, 1215–21 (2002).

[11] Cigdam Batigoe, Halide AKbas, Masut Boz .J of chemical thermodynamics, 43, No 9, 1349–1354 (2011).

[12] Mandeep Singh Bajshi, J. chem., Soc. Faraday Trans., 89, 24, 4323–26 (1993).

[13] Zelana. V. Zeitschrift fur Physikalische chemie Neue Folge, 138, 31–43 (1983).

[14] Mirti.P and Zelona.V., J. Molecular liquids, 37,281–87 (1988).

[15] G.Ravichandran, A.Periandy, A.Srinivasa Rao, K.Govindan, T.K. Nambinarayanan, Acoustic letters.18, No5, 84–88 (1994).

[16] G. Ravichandran, G. Rajarajan and T.K. Nambinarayanan J. Molecular liquids.102/1–3, 267–276 (2003).

[17] G. Ravichandran, Indian .J.Phys. 76 B. 3, 1–6 (2003).

[18] J. Kavanau, "Water and Solute-Water interactions" Holden Day Inc., Amsterdam 1964.

[19] D. Oakenfull and L.Fisher J. Phys.Chem., 81, 1839 (1977).

Ultrasonic Velocity and Absorption Studies on Some Amino Acids in Water-Aprotic Solvent Mixtures

G. Ravichandran and S. Arumugam

PG & Research Department of Physics, A.A. Government Arts College, Villupuram, Tamil Nadu

ABSTRACT: *Ultrasonic velocity and absorption studies have been carried out in the amino acids namely dl-norvaline and l-isoluecine in Water-dioxane, Water–tetrahydrofuran and Water–dimethylformamide mixed solvents. These studies are carried out in the amino acid concentration range 0.05 to 0.4 molar of dl-norvaline and 0.05 to 0.3 molar of l-isoluecine at a fixed frequency of 2 MHz and at a fixed temperature of 303.15 K. The ultrasonic velocity in aqueous solutions of amino acids containing 10% V/V of DN, THF and DMF increases non-linearly with increase of concentration of amino acids. The non-linear increase in ultrasonic velocity is explained on the basis of ionic hydration, hydrogen bonding and the structure breaking power of various aprotic solvents used. The excess absorption has been explained as due to ionic hydration, weakening of hydrogen bonding and proton transfer reactions occurring in these solutions.*

INTRODUCTION

Amino acids belong to an important family of biomolecules which serve primarily as the basic building blocks of all proteins. Proteins play may different biological roles in the living systems. They are the molecular instruments through which all genetic informations are expressed[1]. Several works on ultrasonic studies[2,3] of aqueous solutions of amino acids have been reported to understand their behavior in aqueous environment. Hence, in order to shed more light on the nature of molecular interactions of amino acids in water-aprotic solvent mixtures, ultrasonic velocity and absorption studies are carried out in aqueous solutions of dl-norvaline and l-isoluecine containing 10% v/v of DN, THF and DMF in the present work. It is more useful to extend the ultrasonic studies of amino acids to water-aprotic solvent mixtures because aqueous mixed solvents are extensively used in chemistry and other fields to control factors such as solubility, reactivity and stability of systems. In addition to this, the biological fluids are ultimately not pure water and hence this type of study would help us to understand the role of biological fluids in a much better way.

MATERIALS AND METHOD

The amino acids used in the present study are of AR/BDH grade purchased from SD-fine Chemical Ltd, India and they are used as such without further purification. The solvents used namely DN, THF and DMF are of spectroscopic grade. Triply distilled deionised water is used for preparing the solutions of amino acids. Ultrasonic velocity and absorption studies are carried out at a fixed frequency of 2 MHz in the concentration range 0.05 molar to 0.4 molar of dl-norvaline and 0.05 molar to 0.3 molar of l-isoluecine. Ultrasonic velocity studies are carried out using a Digital ultrasonic velocity meter (Model VCT-70A, Vi-Microsystems Pvt. Ltd., Chennai., India). The ultrasonic absorption studies are made using Pulsed Power Oscillator. Both ultrasonic velocity and attenuation studies are carried out at a fixed temperature of 303 K by circulating water from a thermostatically controlled water bath and the temperature being maintained to an accuracy of ±0.1°C. The accuracy of velocity measurement is ±2 parts in 10^5. Shear viscosity and density of aqueous solutions of dl-norvaline and l-isoluecine containing 10% v/v of DN, THF and DMF are determined using an Ostwald's viscometer and a graduated dilatometer respectively. From the measured values of velocity, density, viscosity and observed absorption, various other parameters such as adiabatic compressibility, intermolecular free length, classical absorption, excess absorption and volume viscosity are calculated using standard equations.

RESULTS AND DISCUSSION

The ultrasonic velocity measured in aqueous solutions of amino acids and in aqueous solutions of amino acids containing 10% v/v of DN, THF and DMF increases non-linearly with increasing concentration of amino acids. The variation of ultrasonic velocity with concentration of amino acids (dl-norvaline and l-isoluecine) in all the systems studied is shown graphically in Figures 1 and 3 respectively. The variation of adiabatic compressibility for aqueous amino acids and for aqueous amino acids with 10% v/v of DN, THF and DMF is shown in Figures 2 and 4. The excess ultrasonic absorption in all systems studied increases non-linearly with increasing concentration of amino acids as shown in Figures 5 and 6.

Ultrasonic Velocity Studies in Aqueous Solutions of Dl-Norvaline and L-isoluecine

Dl-norvaline when dissolved in water, the amino (NH_2) and carboxylic (COOH) groups present in dl-norvaline strongly interact and exist as dipolar ions (NH_3^+ and COO^-). Water monomers present in the medium surround the dipolar ions of dl-norvaline due to strong electrostatic forces leading to the formation of ionic hydration. This strengthens the cohesion among the water molecules in the medium. The increased cohesion in the medium brings about an increase in ultrasonic velocity and decrease of adiabatic compressibility in aqueous solutions of dl-norvaline as shown in Figures 1 and 2 respectively. On increasing the concentration of dl-norvaline, the ultrasonic velocity also increases. This may be due to the fact that the increase in concentration of dl-norvaline leads to increase of ionic hydration.[4] This in turn further strengthens the cohesion among the water molecules in the medium. This invariably results in increase of ultrasonic velocity. The variation of freelength with concentration also supports the explanation offered for the velocity and compressibility studies in aqueous solutions of dl-norvaline. The explanation given for the variation of ultrasonic velocity and adiabatic compressibility for aqueous solutions of dl-norvaline also holds good for aqueous solutions of l-isoluecine in the concentration range studied.

Ultrasonic Velocity Studies in Aqueous Solutions of Dl-norvaline and L-isoluecine with 10% v/v of DN, THF and DMF

When 10% v/v of each DN, THF and DMF is added to aqueous solutions of amino acids, a further increase in ultrasonic velocity is observed. This can be explained as follows using 'Flickering Cluster Model of Liquid Water'.

According to this model, water is supposed to consists of less dense hydrogen bonded clusters called 'Open packed structure' and high dense nonbonded water molecules called 'Close packed structure'. Water is a dynamic mixture of these two. The breakdown of clusters is a co-operative process, when one hydrogen bond in the cluster breaks down the whole cluster breaks down and increases the population of high dense close packed structure of water, i.e. dense monomers.[5]

When 10% v/v of DN is added to aqueous solutions of amino acids, DN breaks the hydrogen bonds in the open packed structure of water thereby releases high dense water monomers. The released water monomers form hydrogen bonding with two free oxygen atoms of DN molecule. Also, the released water monomers may surround the dipolar ions of the amino acids leading to the formation of ionic hydration sheath. Both these effects lead to strengthening of cohesion among the water molecules in the medium. This leads to a further decrease of adiabatic compressibility and increase of ultrasonic velocity in aqueous solutions of amino acids with 10% v/v of DN as shown in Figures 1, 3, 2 and 4. Thus, one can expect a higher value of ultrasonic velocity in aqueous solutions of amino acids containing 10% v/v of DN compared to aqueous solutions of amino acids. A similar effect also takes place in aqueous solutions of amino acids containing 10% v/v of THF and DMF respectively.

The ultrasonic velocity measured in aqueous solutions of amino acids with 10% v/v of DN is found to be slightly higher than the velocity measured in aqueous solutions of amino acids with 10% v/v of THF in the entire amino acid concentration range studied.

This may be due to the fact that each DN molecule has two free oxygen for the formation of hydrogen bonding with water whereas each THF molecule has only one free oxygen for hydrogen bonding. As a result, the cohesion in aqueous solutions of amino acids containing 10% v/v of DN must be slightly higher than the cohesion in aqueous solutions of amino acids with 10% v/v of THF.

The ultrasonic velocity measured in aqueous amino acids with 10% v/v of DMF is found to be higher than the ultrasonic velocity measured in other two systems namely 10% v/v of THF and 10% v/v of DN. This can be attributed to the higher values of donor number, dipole moment and dielectric constant of DMF compared to DN and THF. In general, the difference in the ultrasonic velocity measured in aqueous solutions of amino acids containing 10% v/v of DN, THF and DMF for any particular concentration of amino acid may be attributed to the difference in the structure breaking power, donor number and dipole moment of these aprotic solvents.

Ultrasonic Absorption Studies

From the Figures 5 and 6, it is clear that the excess absorption in aqueous solutions of amino acids with 10% v/v aprotic solvents increases non-linearly with increasing concentration of amino acids. This can be explained as follows:

When an amino acid is dissolved in water, it dissociates and exists in dipolar form called 'zwitterions'. Water molecules are attached strongly to these zwitterions by electrostatic forces which introduce a greater cohesion among the water molecules leading to ionic hydration. The cohesion due to ionic hydration in these solutions generally increases with increase of amino acid concentration. This increased cohesion will bring about an increase in the ultrasonic absorption and such an increase in absorption is observed in the present study in aqueous solutions of amino acids.

When 10% v/v of DN is added to aqueous solutions of dl-norvaline and l-isoluecine, the following effects takes place:

1. DN breaks the open packed structure of water and thereby releases water monomers. The released water monomers form hydrogen bonds with the two oxygen of DN molecule. When ultrasonic waves pass through the medium, part of its energy is used in weakening or breaking of hydrogen bonds already formed. So, it is very likely that during the compression cycle of ultrasonic wave, hydrogen bonds are pushed closer resulting in a partially irreversible weakening or breaking of hydrogen bonds due to the absorption of ultrasonic energy. As the pressure wave has passed through the medium completely, hydrogen bonds are formed again between DN and water molecules. Such type of making and breaking of intermolecular bonds as the pressure wave passes through the medium contributes to the observed increase in ultrasonic absorption.

2. The released water monomers may surround the dipolar ions of amino acids leading to the formation of ionic hydration. This increases the cohesion among the water molecules, thus leading to increased ultrasonic absorption.

Both of these effects taking place in aqueous solutions of dl-norvaline an l-isoluecine with 10% v/v of DN contribute significantly to the observed ultrasonic absorption. The above explanation offered for aqueous solutions of amino acids containing 10% v/v of DN is also equally applicable for aqueous solutions of amino acids containing 10% v/v of THF and 10% v/v of dimethyl formamide.

According to Applegate *et al.*,[6] perturbation of proton-transfer equilibrium

$$R\text{-}NH_3^+ + OH^- \rightarrow R\text{-}NH_2 + H_2O$$

is responsible for the ultrasonic absorption. This process also leads to increased cohesion between the water molecules and hence to an observed increase in ultrasonic absorption.

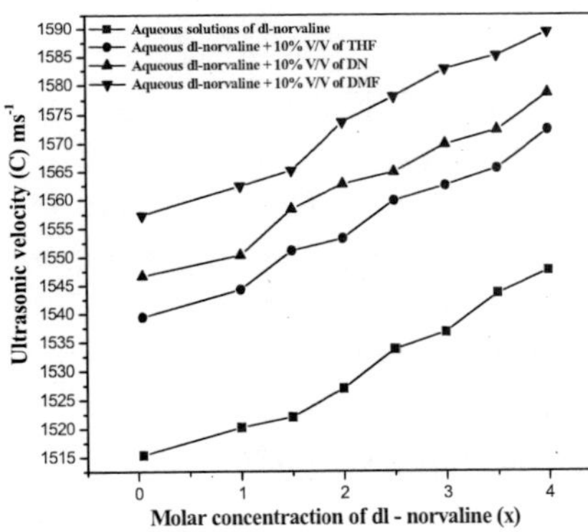

Fig. 1: Variation of Ultrasonic Velocity (C) with Molar Concentration (X) of Dl-norvaline

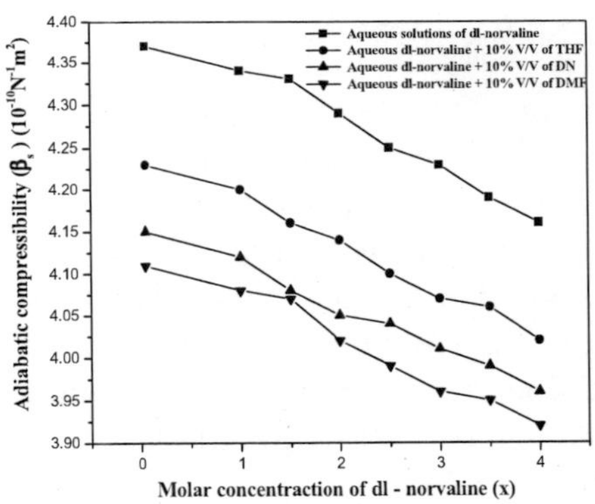

Fig. 2: Variation of Adiabatic Compressibility (β_s) with Molar Concentration (X) of Dl-norvaline

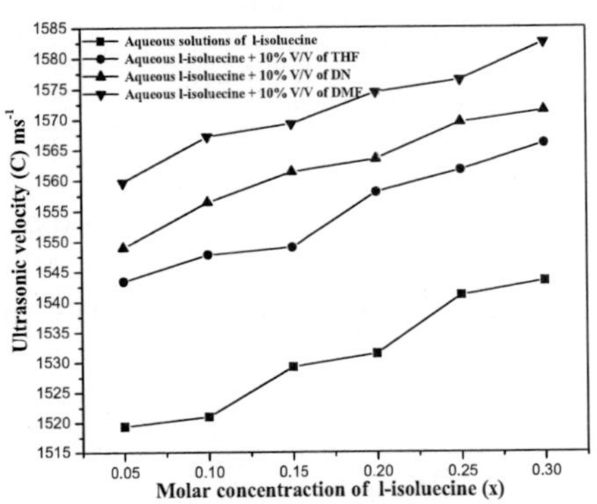

Fig. 3: Variation of Ultrasonic Velocity (C) with Molar Concentration (X) of L-isoluecine

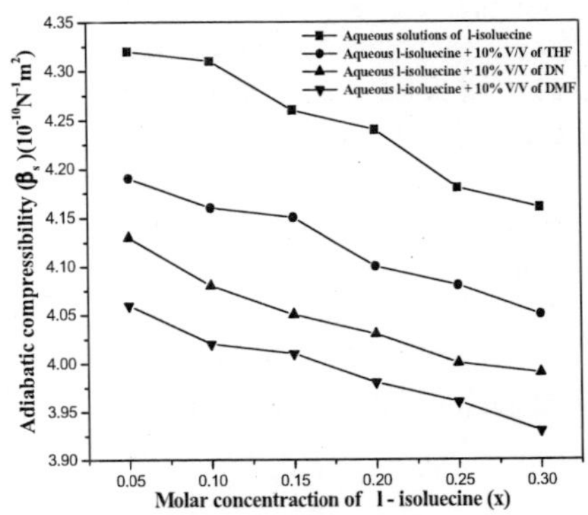

Fig. 4: Variation of Adiabatic Compressibility (β_s) with Molar Concentration (X) of L-isoluecine

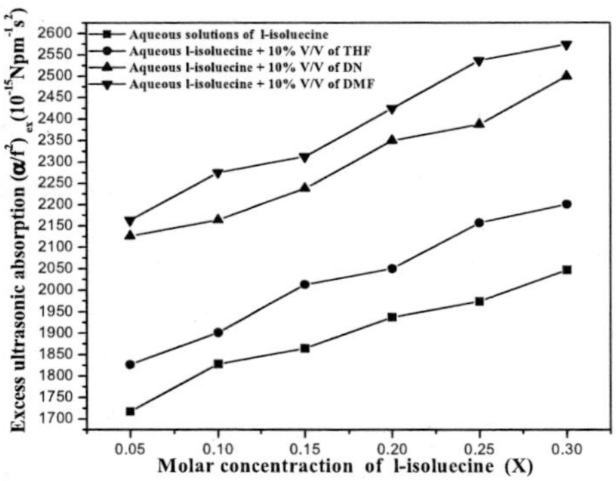

Fig. 5: Variation of Excess Ultrasonic Absorption $(\alpha/f^2)_{exc}$ with Molar Concentraction (X) of Dl-norvaline

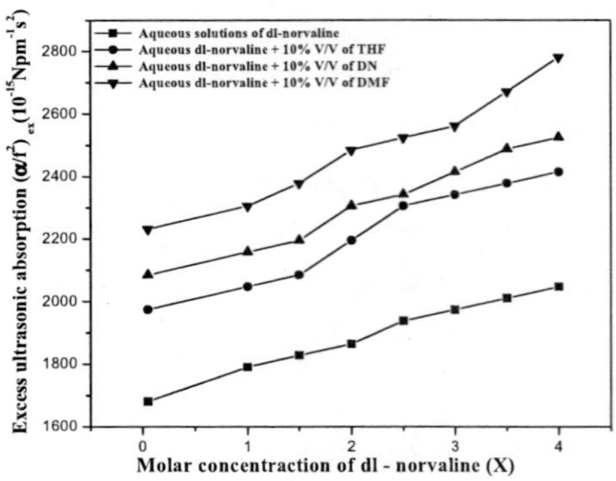

Fig. 6: Variation of Excess Ultrasonic Absorption $(\alpha/f^2)_{exc}$ with Molar Concentration (X) of L-isoluecine

CONCLUSION

The ultrasonic velocity measured in aqueous solutions of dl-norvaline and l-isoluecine increases non-linearly with increasing concentration of these amino acids due to ionic hydration taking place in the solutions. When 10% v/v of each DN, THF and DMF is added to aqueous solutions of amino acids, the non-linear increase in ultrasonic velocity is both due to ionic hydration and the formation of hydrogen bonds in the mixed solutions. The observed ultrasonic absorption in both the amino acids studied increases with increase of amino acid concentration due to ionic hydration, making and breaking of hydrogen bonds and proton transfer reactions taking place in the solutions.

REFERENCES

[1] Lehninger A., *Principles of Biochemistry* (Worth Publishers Inc., New York) 1984.

[2] Banipal, T.S. and Gagandeep Sehgal, Partial molal adiabatic compressibility of transfer of some amino acids and peptides from water to aqueous solutions of sodium chloride and aqueous glucose solutions, Thermochimica Acta, 262 (1995), 175–183.

[3] Banipal, T.S. and Singh, G., Thermodynamic study of some amino acids, diglycine and lysozyme in aqueous and mixed aqueous solutions,Thermochimica Acta, 412 (2004), 63–83.

[4] Nishikawa, S., Ohno, T., Huang, H., Yoshizuka, K. and Jordan, F., Ultrasonic absorption in aqueous solutions of amino acids at neutral pH, J. Acoust. Soc. Am, 113 (2003) 2884–2888.

[5] Amalendu Pal and Suresh Kumar, Volumetric and ultrasonic studies of some amino acids in binary aqueous solutions of $MgCl_2$ 6 H_2O at 298.15K, J. Mol. Liqs., 12 (2005) 148–155.

[6] Blandamer M.J., *Introduction to Chemical Ultrasonics* (Academic Press, London) 1973.

Acoustical Behaviour of Esters in Sesame Oil

Suhashini Ernest, P. Kavitha and S. Kanchana
Department of Physics, Urumu Dhanalakshmi College, Tiruchirappalli
E-mail: pkavithaa@yahoo.com

ABSTRACT: *The analysis of ultrasonic speed in binary liquid mixtures is a reliable property for the study of molecular interactions and structural determination. Several food and oil based industries need data on the physical properties of oil and solvent mixtures. In the present paper, measurement of ultrasonic velocity and density are made in binary liquid mixtures of three different esters namely ethyl acetate, isopropyl acetate and butyl acetate with sesame oil over the entire range of composition at 298.15K. From the obtained data, size dependent parameters namely molar volume, molecular radius, molar surface area and surface tension are evaluated and the intermolecular interactions are discussed in terms of size of the solvent molecules. Various excess acoustical parameters like excess ultrasonic velocity, excess density, excess molar surface area and excess surface tension are calculated and from the magnitude of the excess parameters, the bonding nature of solvent molecules with voluminous oil molecules are analysed.*

Keywords: Sesame Oil, Esters, Molar Volume, Excess Parameters.

INTRODUCTION

Sesame oil is a known dietary source having putative antioxidant property since a millennium. The lignans present in sesame oil are thought to be responsible for many of its unique chemical and physiological properties, including its antioxidant and antihypertensive properties[1–3]. In oil refineries and food technology, the properties of oil mixtures with organic solvents are becoming important in processing and product design. The acoustical study of liquid mixtures provides useful information about physical forces acting between the molecules of the same substance in pure liquids and molecules of different substances in liquid mixtures[3–5]. The sign and magnitude of the excess parameters is used to study the intermolecular interactions between the components in the liquid mixture[7–10]. The literature survey reveals that the studies on the oil and mixture of oils and organic solvents are scarce. Hence the study has been made to elucidate the physico-chemical properties of mixture of common edible sesame oil and three different esters.

MATERIALS AND METHODS

The solvents are purchased from Merck and sesame oil is provided by Bunge India Private limited, Tamil Nadu. The ultrasonic velocities are measured by an ultrasonic interferometer (Mittal) at a frequency of 2 MHz with an accuracy of 0.01 m/s. The density is measured by relative density method using 10 ml relative density bottle with an accuracy of 0.001 kg/m^3. The measurements are made at 298.15 K with the help of digitally controlled (Raagaa) temperature bath with an accuracy of 0.01 K.

RESULTS AND DISCUSSION

The speed of ultrasound (u) and density (ρ) of the binary mixtures are used to evaluate molar volume (V_m), molecular radius (r), molar surface area (Y), surface tension (S_T) and excess parameters of ultrasonic velocity (u^E), density (ρ^E), molar surface area (Y^E) and surface tension (S_T^E) are evaluated using standard formulae and their values are given in Tables 1, 2 and 3.

The graphical representation of molar volume (V_m) with mole fraction of different esters is shown in Fig. 1 and those magnitudes decreases almost linearly. From the Tables 1, 2 and 3 the values of molar volume at equimolar composition, say at 0.5 mole fraction of solvents for the three binary systems, is in the order of increasing chain length, indicating the expansive trend of molar volume with molecular weights of acetates. Thus the interaction is lesser in the esters of high molecular weight. From the plot of molecular radius (r), molar surface area (Y) and surface tension (S_T) with esters (Figures 2, 3 and 4), it is observed that the values decrease with increase in solvent concentration establishing more structured packing of molecules.

Table 1: Acoustical Parameters for the Mixtures of Ethyl Acetate (X_1) and Sesame oil (X_2)

X_1	V_m ($\times 10^{-3}$ m^3/mol)	r ($\times 10^{-9}$ m)	Y ($\times 10^{-8}$ m^2)	S_T ($\times 10^4$ N/m)	u^E (m/s)	ρ^E (Kg/m^3)	Y^E ($\times 10^{-8}$ m^2)	S_T^E ($\times 10^4$ N/m)
0.10	0.87	4.42	3.49	3.19	24.76	1.88	0.07	0.08
0.20	0.79	4.27	3.26	3.18	52.40	3.58	0.13	0.17
0.30	0.70	4.11	3.00	3.15	76.12	5.41	0.18	0.25
0.40	0.61	3.93	2.74	3.11	95.77	7.06	0.22	0.31
0.50	0.53	3.74	2.46	3.07	113.76	8.50	0.25	0.37
0.60	0.44	3.52	2.17	3.01	128.98	9.84	0.26	0.42
0.70	0.36	3.28	1.86	2.93	135.64	10.69	0.25	0.44
0.80	0.27	2.99	1.52	2.82	134.68	10.74	0.21	0.43
0.90	0.18	2.63	1.14	2.60	97.62	8.90	0.13	0.31
0.98	0.12	2.25	0.79	2.30	27.97	3.11	0.03	0.09

Table 2: Acoustical Parameters for the Mixtures of Isopropyl Acetate (X_1) and Sesame oil (X_2)

X_1	V_m ($\times 10^{-3}$ m^3/mol)	r ($\times 10^{-9}$ m)	Y ($\times 10^{-8}$ m^2)	S_T ($\times 10^4$ N/m)	u^E (m/s)	ρ^E (Kg/m^3)	Y^E ($\times 10^{-8}$ m^2)	S_T^E ($\times 10^4$ N/m)
0.10	0.87	4.42	3.49	3.18	27.06	4.51	0.06	0.09
0.20	0.79	4.27	3.25	3.15	54.65	8.76	0.12	0.18
0.30	0.70	4.11	3.00	3.12	82.34	12.80	0.17	0.28
0.40	0.62	3.95	2.75	3.08	105.96	16.52	0.20	0.36
0.50	0.54	3.76	2.47	3.03	126.11	19.99	0.23	0.42
0.60	0.45	3.55	2.18	2.95	139.38	22.80	0.24	0.47
0.70	0.37	3.32	1.88	2.86	148.94	24.62	0.23	0.49
0.80	0.29	3.04	1.55	2.71	142.94	24.35	0.19	0.47
0.90	0.20	2.71	1.19	2.46	104.76	19.72	0.12	0.34
0.98	0.13	2.37	0.85	2.06	8.92	6.74	0.02	0.04

Table 3: Acoustical Parameters for the Binary Mixtures of Butyl Acetate (X_1) and Sesame Oil (X_2)

X_1	V_m ($\times 10^{-3}$ m^3/mol)	r ($\times 10^{-9}$ m)	Y ($\times 10^{-8}$ m^2)	S_T ($\times 10^4$ N/m)	u^E (m/s)	ρ^E (Kg/m^3)	Y^E ($\times 10^{-8}$ m^2)	S_T^E ($\times 10^4$ N/m)
0.10	0.87	4.42	3.50	3.19	21.36	3.53	0.06	0.08
0.20	0.79	4.27	3.25	3.16	42.60	7.22	0.12	0.15
0.30	0.71	4.12	3.02	3.13	59.65	10.14	0.16	0.21
0.40	0.62	3.95	2.77	3.10	78.18	12.89	0.19	0.27
0.50	0.54	3.78	2.51	3.05	88.34	15.33	0.21	0.31
0.60	0.46	3.58	2.23	3.00	102.29	17.40	0.22	0.36
0.70	0.38	3.35	1.94	2.93	107.97	18.58	0.21	0.37
0.80	0.29	3.08	1.62	2.83	107.94	18.25	0.18	0.37
0.90	0.21	2.77	1.28	2.65	85.73	14.30	0.12	0.29
0.98	0.15	2.45	0.97	2.40	33.51	4.96	0.04	0.11

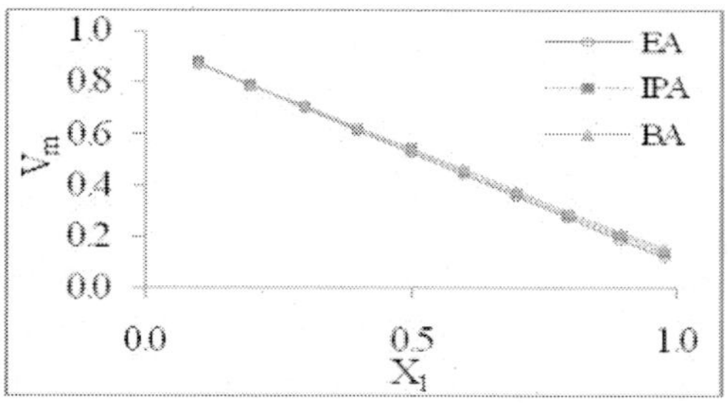

Fig. 1: Molar Volume vs. Mole Fraction of Esters

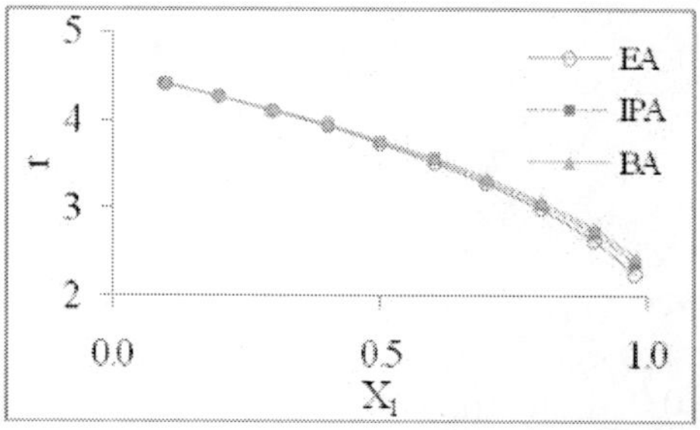

Fig. 2: Molecular Radius vs. Mole Fraction of Esters

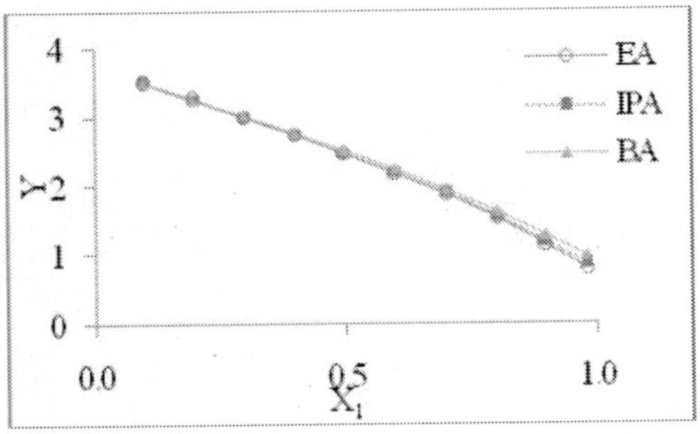

Fig. 3: Molar Surface Area vs. Mole Fraction of Esters

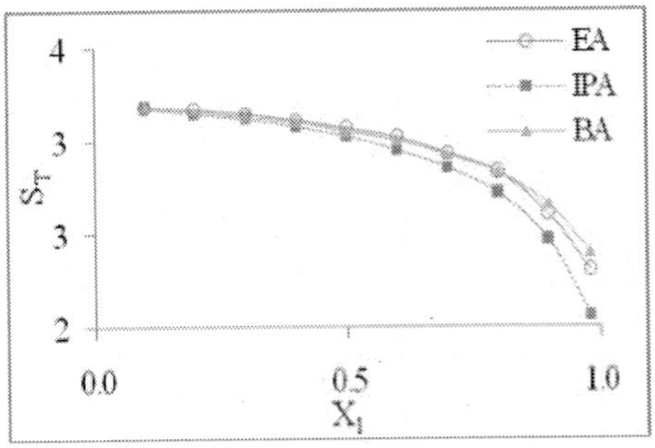

Fig. 4: Surface Tension vs. Mole Fraction of Esters

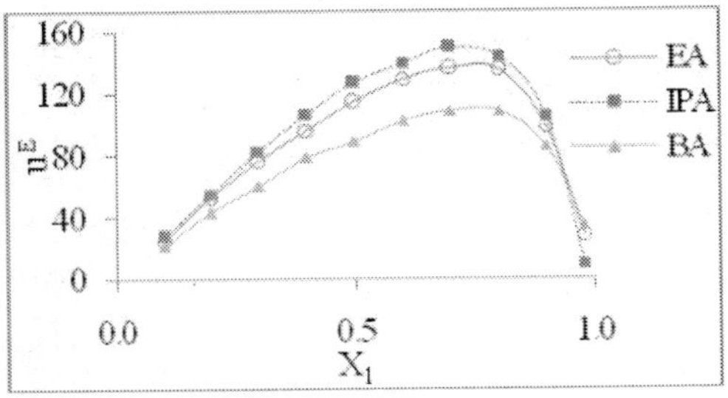

Fig. 5: Excess Velocity vs. Mole Fraction of Esters

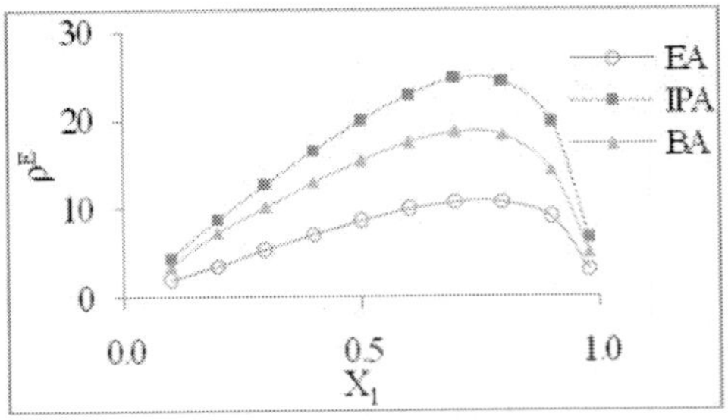

Fig. 6: Excess Density vs. Mole Fraction of Esters

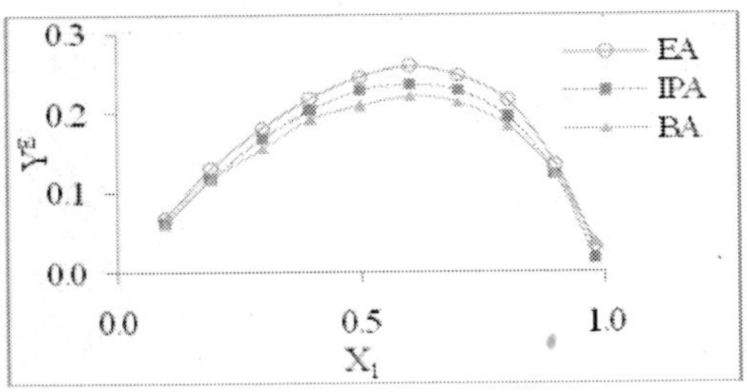

Fig. 7: Excess Surface Area vs. Mole Fraction of Esters

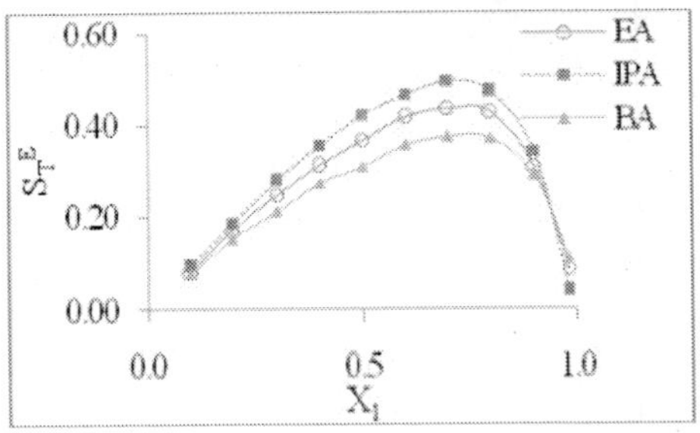

Fig. 8: Excess Surface Tension vs. Mole Fraction of Esters

The curves in Figures 5, 6, 7 and 8 show that the positive excess ultrasonic velocity (u^E), excess density (ρ^E), excess surface area and excess surface tension have maximum values at 0.7 mole fractions of esters, which may be attributed to the significant interaction between the unlike molecules in the binary mixture and interstitial accommodations of smaller molecules in the voids created by bigger oil molecules. The interaction decreases with increases in carbon atoms of esters in the following sequence:

Sesame oil+Ethyl acetate > Sesame oil+Isopropyl acetate > Sesame oil+Butyl acetate.

The variation of parameters may be attributed to the effects of dispersive and dipolar interactions between the components of mixtures.

CONCLUSION

The molecular association for binary mixtures formed with solvents of smaller molar mass is larger than for the mixtures formed with solvents of larger molar mass. This is attributed to the interstitial accommodation of smaller molecules into the voluminous sesame oil molecule. The interaction due to dispersive and dipolar forces is seen to be significant around the 0.7 mole fraction of solvents from the study of excess parameters.

REFERENCES

[1] Fukuda Y., *Nippon Shokuhin Kogyo Gakkaishi* (1990), 37, 484–492.

[2] Matsumura Y., Kita S., Morimoto S., Akimoto K., Furuya M., Oka N., Tanaka T., *Biol Pharm Bull.* 1995; 18: 1016–1019.

[3] Morris M.C., *J Cardiovasc Risk. (*1994).

[4] Pandey J.D., Dubey G.P. and Tripathi N., *Journal of Chemical Sciences*, 1999, 111(2), 361–367.

[5] Palaniappan L. and Karthikeyan V., *Indian J. Phys*, 2005, 79(2), 155.

[6] Nithyanantham and Palaniappan L, *Acta Ciencia Indica,* 2005, 36(4) 533–538.

[7] Fort R.J. and Moore W.R. *Trans. Faraday Soc.* 61 2102 (1965).

[8] Dash S.K., Chakaravorthy V. and Swain B.B., *Acoust. Lett.* 19 42 (1996).

[9] Gencia B., Alcalde R., Leal J. and Maton J.S., *J. Chem. Soc. Faraday Trans.* 93 1115 (1997).

[10] Mehra R. and Panchali M., *Indian J. Phy*, (2006) 80, 195.

Ultrasonic Study of Vinyl Acetate and Sesame Oil Mixture at Different Temperatures

P. Kavitha, Suhashini Ernest and S. Kanchana

Department of Physics, Urumu Dhanalakshmi College, Tiruchirappalli

E-mail: suha_ernest@yahoo.co.in

ABSTRACT: *Ultrasonic techniques have gained much attention for elucidating the physico- chemical properties of binary liquid mixtures. Food and oil industries rely on the research data on the properties of oil and solvent mixtures. The present study deals with the measurement of ultrasonic velocity and density in binary liquid mixtures of vinyl acetate with sesame oil over the entire range of composition at 298.15 K, 303.15 K and 308.15 K. From fundamental data, acoustical parameters namely isentropic compressibility and intermolecular free length are calculated along with their excess parameters and the intermolecular interactions existing between the components of mixtures are discussed.*

Keywords: Sesame Oil, Vinyl Acetate, Isentropic Compressibility, Intermolecular Free Length, Excess Parameters.

INTRODUCTION

The measurement of speed of sound in liquids and liquid mixtures plays important role in determining their physico-chemical behavior. The measurement of ultrasonic velocity gives more details about the bonding between the molecules of the components of the binary mixtures[1,2]. For the past decades, the acoustic and thermodynamic properties are studied for various liquid mixtures[3–7]. But the literature survey shows scarce results on the study of edible oils with organic solvents which are in great demand for the oil industry. In this work, for the binary mixture of sesame oil and vinyl acetate the ultrasonic velocity and density are measured at various temperatures and used to compute the thermo acoustical parameters namely isentropic compressibility, intermolecular free length and their excess parameters.

EXPERIMENTAL PROCEDURE

The densities of vinyl acetate and Sesame oil and the binary mixtures formed between them were measured using 25 ml relative density bottle with an accuracy of 0.001 kg/m^3. The ultrasonic velocities are measured by an ultrasonic interferometer (Mittal) at a frequency of 2 MHz with an accuracy of 0.01 m/s. The measurements are made at different temperature of 298.15 K, 303.15 K, and 308.15 K, with the help of digitally controlled temperature bath (Raagaa) with an accuracy of 0.01 K.

RESULTS AND DISCUSSION

Isentropic compressibility (K_s) has been calculated from the equation,

$$K_s = \frac{1}{\rho u^2} \qquad \dots (1)$$

where ρ and u are density and ultrasonic velocity in liquid mixture.

The intermolecular free length (L_f) is determined using the Jacobson's relation,

$$L_f = \frac{K}{up^{1/2}} \qquad \dots (2)$$

where K is the temperature Jacobson's constant given by the relation K = (93.875 + 0.375 T) × 10^{-8}.

The experimentally measured values of ultrasonic velocity along with the computed values of acoustical parameters at different temperatures are presented in Tables 1. The ultrasonic velocity is observed to decrease with increase in concentration of vinyl acetate, indicating the presence of interactions between the components of the mixtures. The ultrasonic velocity is found to decrease with increase with temperature also. The intermolecular free length and isentropic compressibility of the binary mixtures show the opposite trend to that of the velocity. Eyring and Kinkaid[8] proposed that the free length

Table 1: Acoustical Parameters for the Binary Mixture of Vinyl Acetate and Sesame Oil at 298.15 K

X1	*X2*	*U (m/s)*	K_s TPa^{-1}	L_f $(x10^{-11}\ m)$	K_s^E	L_f^E
0.1	0.9	1429	534.2	4.75	–22.1	–0.08
0.2	0.8	1426	536.6	4.76	–55.1	–0.21
0.3	0.7	1416	544.7	4.80	–82.5	–0.31
0.4	0.6	1401	556.5	4.85	–106.1	–0.40
0.5	0.5	1406	550.4	4.83	–147.7	–0.56
0.6	0.4	1382	572.8	4.92	–160.8	–0.61
0.7	0.3	1377	576.7	4.94	–192.3	–0.73
0.8	0.2	1338	608.8	5.07	–195.7	–0.74
0.9	0.1	1265	679.8	5.36	–160.1	–0.59
0.95	0.05	1208	743.4	5.61	–114.3	–0.41

and compressibility increases with decrease in velocity and vice versa. The increase in intermolecular free length leads to an increase in compressibility with rise in temperature (Pandey)[9]. An increase in the adiabatic compressibility shows a tendency towards ordering resulting in a decrease of ultrasonic velocity of the binary mixture (Tabhane and Patki)[10]. Thus the variations of intermolecular free length and adiabatic compressibility with concentration of solvent and temperature suggests that molecular interactions between the components of mixture are concentration and temperature dependent.

Table 2: Acoustical Parameters for the Binary Mixture of Vinyl Acetate and Sesame Oil at 303.15 K

X1	*X2*	*U* *(m/s)*	K_s TPa^{-1}	L_f $(x10^{-11}\ m)$	K_s^E	L_f^E
0.1	0.9	1422	541.4	4.83	–28.3	–0.10
0.2	0.8	1413	548.6	4.86	–59.4	–0.22
0.3	0.7	1391	566.4	4.94	–79.9	–0.29
0.4	0.6	1382	573.9	4.97	–110.8	–0.41
0.5	0.5	1390	566.4	4.94	–156.6	–0.59
0.6	0.4	1371	583.4	5.01	–177.9	–0.67
0.7	0.3	1360	593.7	5.06	–206.0	–0.77
0.8	0.2	1320	628.2	5.20	–209.8	–0.78
0.9	0.1	1250	699.1	5.49	–177.2	–0.64
0.95	0.05	1189	771.6	5.77	–123.8	–0.44

Table 3: Acoustical Parameters for the Binary Mixture of Vinyl Acetate and Sesame Oil at 308.15 K

X1	*X2*	*U* *(m/s)*	K_s TPa^{-1}	L_f $(x10^{-11}\ m)$	K_s^E	L_f^E
0.1	0.9	1415	548.4	4.90	–34.4	–0.13
0.2	0.8	1400	560.4	4.96	–63.8	–0.23
0.3	0.7	1368	587.2	5.07	–78.4	–0.28
0.4	0.6	1359	595.3	5.11	–111.7	–0.40
0.5	0.5	1383	574.2	5.02	–174.1	–0.65
0.6	0.4	1353	601.0	5.13	–188.7	–0.70
0.7	0.3	1352	601.9	5.14	–229.2	–0.85
0.8	0.2	1298	652.2	5.35	–220.3	–0.80
0.9	0.1	1248	704.8	5.56	–209.1	–0.75
0.95	0.05	1168	804.5	5.94	–130.2	–0.45

The observed negative values of excess isentropic compressibility (K_s^E) and the excess intermolecular free length (L_f^E) over the entire range of composition confirm that specific interactions are present between the molecules of oil and vinyl acetate. The negative values show that the liquid mixture is less compressible than the pure components forming the mixture. Oswal et al[11] have reported that the negative intermolecular free length indicate the structural readjustment in the liquid mixtures towards less compressible phase of fluid and closer packing of molecules.

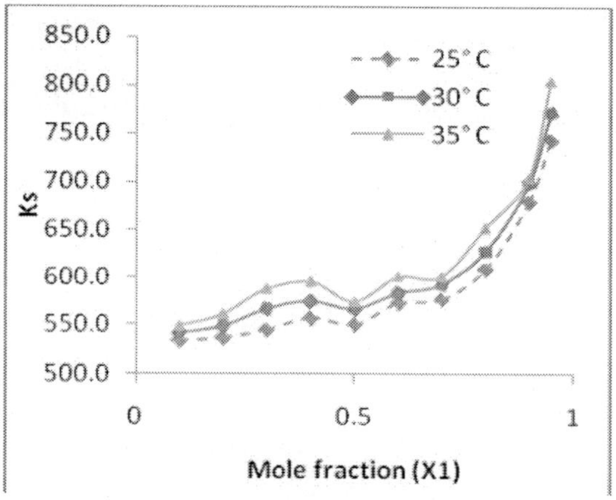

Fig. 1: Isentropic Compressibility vs. Mole Fraction of Solvent

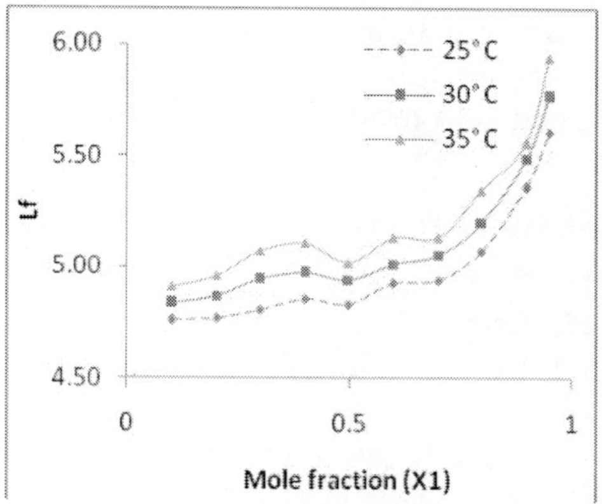

Fig. 2: Intermolecular Free Length vs. Mole Fraction of Solvent

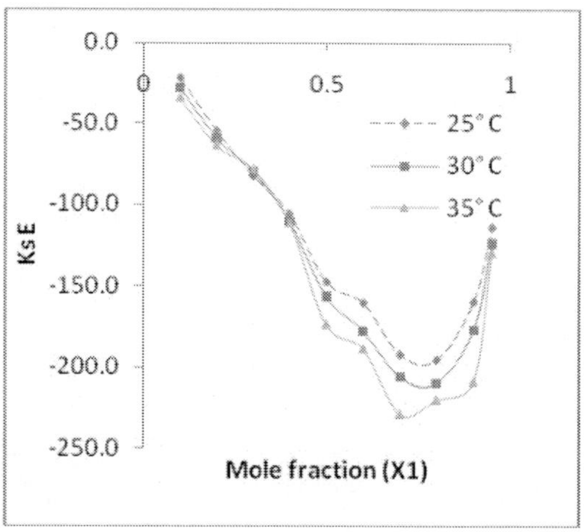

Fig. 3: Excess Isentropic Compressibility vs. Mole Fraction of Solvent

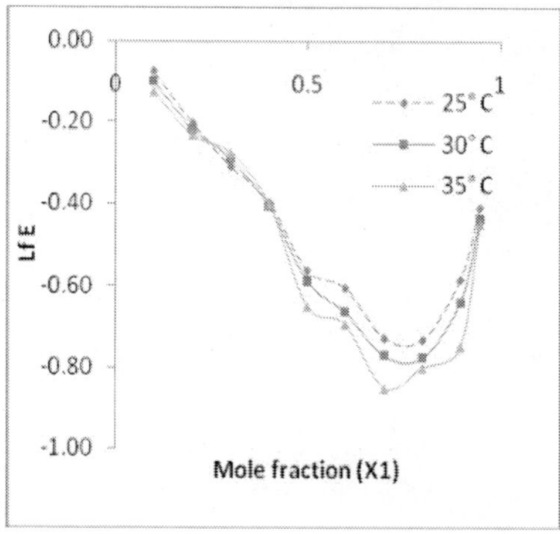

Fig. 4: Excess Intermolecular Free Length vs. Mole Fractions of Solvents

The interaction between the oil and acetate molecules may be attributed to the dispersive, dipolar forces and hydrogen bonding .A sharp deviation present around 0.7 mole fraction of solvent suggests that significant interactions are present at that composition. The increase of excess values of K_s^E and L_f^E above 0.7 mole fraction of vinyl acetate may be due to the self association of ester due to hydrogen bonding.

CONCLUSION

The ultrasonic velocity of binary liquid mixture sesame oil and vinyl acetate is found to decrease with increase in concentration of the solvent and increase in temperature. The binary mixture is more compressible than the components forming the mixture due to interactions resulting from dispersive and polar forces.

REFERENCES

[1] Y. Wada, *J. Phys. Soc. Japan.* 4, (1949) 280.
[2] Jacobson B., *Acta Chem. Scand. 6* (1952) 1485.
[3] N. Schaffs, *Acustica* 4 (1954) 636.
[4] O. Nomoto, *J. Chem. Soc. Jpn.* 11 (1956) 1146.
[5] Beyer R.T. and Letcher S.V., *Physical Ultrasonics,* Academic Press, N Y (1969).
[6] McClements, D.J. and M.J.W. Povey, *Ultrasonics* 30: 383–388 (1992).
[7] Coupland, J.N. and McClements, D.J., *J. Am. Oil Chem. Soc,* 74: 1559–1564 (1997).
[8] Eyring H. and Kincaid J.F., *J. Chem. Phys*. 6,(1938)620.
[9] Pandey, H.C., Jain R.P. and Pandey J.D., *Acustica* 34 (1975) 123.
[10] Tabhane V.A. and Patki B.A., *Acoustica.* 52, (1982) 44.
[11] Oswal, S.L. and Patel, N.B., J. Chem Engg. Data, 45 (2000) 11.

Study on Thermodynamical Properties of Binary Mixtures of Salicornia Biodiesel and Petroleum Diesel

K. Rajagopal, G. Johnson Newton[1], J. Selwin Rajadurai[2] and C. Adhikesavan[3]

Department of Physics, Government College of Engineering, Tirunelveli, Tamil Nadu

[1]M.S. University, Tirunelveli, Tamil Nadu

[2]Department of Mechanical Engineering, Government College of Engineering, Tirunelveli, Tamil Nadu

[3]Anna University of Technology, Tirunelveli, Tamil Nadu

E-mail: biofuel.newton@gmail.com

ABSTRACT: *This paper deals with the accurate measurement of density, viscosity and ultrasonic speeds and computations of several thermo dynamical parameters such as bulk modulus, surface tenison, internal pressure and intermolecular free length for the mixtures of salicornia biodiesels with petro diesel for various blends (B0, B5, B10, B15, B20 and B–100) at four different temperatures (303.15, 308.15,313.15 and 318.15) K.*

Keywords: Biodiesel, Transesterification, Thermodynamic.

INTRODUCTION

As petro diesel source is depleting at alarming rate, biodiesel is being recognized as the promising alternative fuel. Biodiesel is an environmentally friendly fuel, which can be used in diesel engines without any modification. In comparison to diesel, the higher cetane number of biodiesel results in shorter ignition delay and longer combustion duration lead to low particulate emission and reduction in smoke opacity. Biodiesel is a biodegradable, non toxic fuel and free from sulphur. Moreover the biodiesel possess physical properties like density, viscosity, calorific value, and flash point etc similar to petro diesel. Biodiesel can be prepared from both edible and non-edible sources. Many countries are exploring both edible and non edible seeds for the production of biodiesel. In India, because of the higher demand for edible oil, biodiesel has to be produced from non edible seeds. Biodiesel is produced by transesterification reaction of vegetable oil with low molecular weight alcohols namely methanol using sodium hydroxide as a catalyst. Biodiesel can be blended at any level with petroleum diesel and used in compression–ignition (diesel) engine.

Vicente *et al.*[1] discussed the properties combustion, advantages and production technologies involved in the use of methyl esters of sunflower oil fatty acids as an alternative to diesel.Baiju *et al.*[2] studied methyl esters of karanja oil can be used as fuel in compression engine without any engine modification.

Ramadhas *et al.*[3] studied biodiesel production from high free fatty acid rubber seed oil. They developed a two –step transesterification process to convert the high free fatty acid oils to its mono esters.The various factors affecting the yield of ester such as molar ratio, amount of catalyst,reaction temperature and reaction duration analyzed.

Malaya Naik *et al.* [4] worked on the production of biodiesel from high free fatty acid karanja (pongamia pinnata) oil. They conclude that the dual –step process of transesterification using acid –catalysd and followed by base-catalysed reaction used for the production appropriate quality of karanja biodiesel as per the ASTM specification.

From detailed literatures survey, it has been identified that no author had attempt to correlate the performance, emission and characteristics of diesel engine with the physical parameters. We are reporting for first time in literature, the effect of temperature on several thermo dynamical parameters such as bulk modulus, surface tension, internal pressure and intermolecular free length from the experimental values of density, viscosity and ultrasonic speed for the mixtures of petro diesel with salicornia biodiesel at four temperatures.

In this paper several thermodynamic parameters like bulk modulus, surface tension, internal pressure, intermolecular free length are calculated from the accurate measurement of density, viscosity and ultrasonic speed of petro diesel with salicornia biodiesel at different temperatures (303.15 K to 318.15 K) .The work on the study of these thermo dynamical parameters on diesel engine performance, emission and combustion characteristics are underway and will be reported elsewhere in detail.

EXPERIMENTAL PROCEDURE

Materials

Salicornia branchiata seeds are collected from the coastal belts of Tuticorin, Tamil Nadu. Analar grades of n-hexane, methanol, Conc.Sulphuric acid and Sodium hydroxide (E.merck, India) were used in transesterification process.

Extraction of Salicornia Oil

The conventional oil extraction is not suitable for salicornia branchiata seed.However the oil extraction from these seeds can be done by using soxhlet procedure, in which oil from seeds are extracted by repeating percolation with an organic solvents like hexane under reflux in a special glass ware.

Production of Biodiesel

The acid value of unrefined salicornia seed oil is 16.4 mg/KoH/g .Thus free fatty acid present in the salicornia seed oil is 9.1%. It is known that alkaline-catalysed transesterification does not occur when FFA content in the oil is more than 3%.Thus a two step transesterification process is developed to produce biodiesel from unrefined salicornia seed oil.

Biodiesel Characterization

The densities of the blends were measured using single pycnometer (pyrex glass) of bulb capacity of 12×10^{-3} dm^3 having graduated stem with 5×10^{-8} m^3 divisions. The mark of the stem of the pycnometer was calibrated by making use of doubly distilled water at different temperatures [5]. The reproducibility in the density measurement was within ± 0.1 kg m^{-3}.The ultrasonic velocity of the mixture was measured in a multifrequency interferometer (ModelM-84, Mittal enterprises' New Delhi) operated at 5 MHz .The reproducibility in the ultrasonic speed measurement was ±0.03% [6]. The viscosities of the mixtures were measured using ubbelohde type suspended level viscometer having a flow time of 204 at seconds at 303.15 K for triply distilled water. The flow time was measured using a digital stop watch (Racer, India) capable of registering time accurate to ±0.01 s. An average of four sets of flow times for each solution was taken for the purpose of calculation of the viscosity. The reproducibility in the viscosity measurements was within $\pm 2 \times 10^{-3}$ mpa.s.[7]. The mass measurements were carried out using a higher sensitive digital AND electronic balance (Model HR 300, Japan) with a precision of ±0.1 mg.

The calorific value measured by bomb calorimeter. The presence of acid in the biodiesel is determined by titration of the biodiesel versus potassium hydroxide base.

THEORY

Thermo Dynamic Relation [8]

The various thermodynamic and transport parameters, which are calculated from the measurement of ultrasonic speed, density and viscosity for the binary mixtures are as follows:

Bulk Modulus

$$\beta t = u^2 \rho \qquad \text{... (1.1)}$$

Surface tension

$$\sigma = 6.3 \times 10^{-3} \rho u^{3/2} \qquad \text{... (1.2)}$$

Intermolecular free length, L_f

$$L_f = K/u\ \rho^{1/2} \qquad \text{... (1.3)}$$

Internal pressure

$$Pi = \alpha T/\beta_T \qquad \text{... (1.4)}$$

where α and β thermal expansion coefficient and Isothermal compressibility. They are 100 evaluated using the following equations,

$$\beta = 1.71 \times 10^{-3}/T^{4/9}\, u^2\, \rho^{4/3};\ \alpha = (0.0191 \times \beta)^{1/4}$$

where K is Jacobson constant that depends on temperature and is given by $[K = (93.875 + 0.375\ T) \times 10^{-8}]$.

RESULTS AND DISCUSSION

The biodiesel is characterized by comparing some of its physical properties with a standard diesel fuel are which given in the Table 1. Thus density and viscosity are important properties of biodiesel in this values those values are within the standard range. Calorific value is another fuel property which measure the energy content in a fuel (ie) it has a direct correlation with the efficiency of the fuel.

Table 1: Physico-Chemical Properties of Salicornia Biodiesel

Properties	*Salicornia*	*Standard*	*Range*
Density (15°C)	887.5	ASTM D751	0.860–0.900 g/cm^3
Viscosity (40°C)	3.9865	ASTM D445	1.9–6.0 mm^2 s^{-1}
Acid value	0.1852	ASTM D664	0.50 mg KOH g^{-1} Max.
Calorific value	38.52	ASTM D240	NA
Specific gravity	0.8906	ASTM D4052	0.87–0.90

Generally, the fatty acid compositions have a significant correlation with the physical and chemical properties of biodiesel such as cetane number, density and viscosity.

The FFA content in the salicornia biodiesel was evlaluated at central salt and marine chemical research institute at Gujarat. The results are presented in Figure 1.

From the Table 2, it clearly shows that the values of density, viscosity and ultrasonic speed are increase with increasing concentration of biodiesel percentage and decrease with increase in temperatures. Some of the thermodynamical parameter such as bulkmodulus, surface tension, internal pressure and intermolecular free length are evaluated and given in Figures 2–5.

Table 2: The Values of Density and Ultrasonic Speed for the Mixtures of Petrodiesel with Salicornia Biodiesel at Different Temperatures

T (K)	ρ $(kg\ m^{-3})$	η $mm^2\ s^{-1}$	u $(m\ s^{-1})$
B0 (Pure petrodiesel)			
303.15	822.2	2.4498	1320.0
308.15	818.2	2.1896	1310.5
313.15	816.6	2.0865	1300.2
318.15	815.0	2.0187	1281.0
B5 (Petro diesel 95% + Biodiesel 5%)			
303.15	827.8	3.0312	1324.0
308.15	825.1	2.7328	1315.7
313.15	823.3	2.4041	1305.0
318.15	821.6	2.2181	1297.5
B10 (Petro diesel 90% + Biodiesel 10%)			
303.15	829.9	3.0925	1334.5
308.15	828.1	2.8138	1324.0
313.15	825.4	2.4421	1318.5
318.15	823.7	2.1818	1310.0
B15 (Petro diesel 85% + Biodiesel 15%)			
303.15	832.8	3.1280	1340.8
308.15	830.4	2.8757	1332.5
313.15	828.4	2.4617	1324.0
318.15	826.7	2.3612	1316.0
B20 (Petro diesel 80% + Biodiesel 20%)			
303.15	836.8	3..2778	1352.0
308.15	834.0	3.0531	1340.0
313.15	832.2	2.5200	1330.0
318.15	830.4	2.4480	1322.5
B100 (Petro diesel 0 % + Biodiesel 100%)			
303.15	871.2	4.1104	1378.0
308.15	868.7	4.0860	164.5
313.15	865.4	3.9865	1358.3
318.15	861.0	3.9458	1354.0

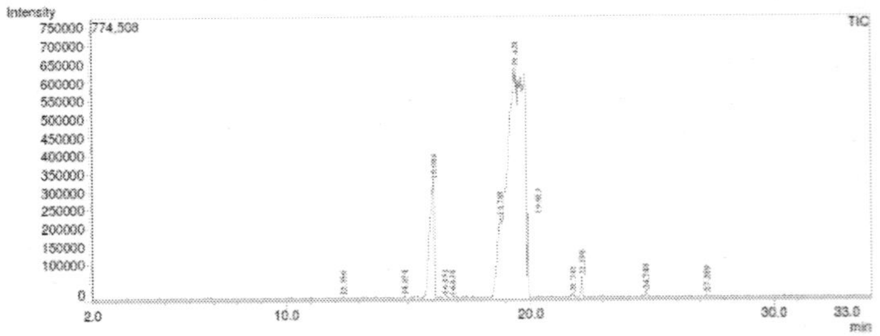

Peak Report TIC

Peak#	R.Time	Area	Area%	Name	Base m/z
1	12.356	14212	0.04	Heptadecanoic acid, methyl ester	74.00
2	14.874	8438	0.03	1,2-Benzenedicarboxylic acid, bis(2-methylpropyl) ester	149.05
3	16.088	3955096	11.89	Hexadecanoic acid, methyl ester	74.00
4	16.551	13312	0.04	1,2-Benzenedicarboxylic acid, dibutyl ester	149.05
5	16.838	10310	0.03	1,2-Benzenedicarboxylic acid, butyl octyl ester	149.05
6	18.765	2719636	8.18	9,12-Octadecadienoic acid, methyl ester	67.05
7	19.424	25611740	77.01	Methyl linoleate	81.05
8	19.913	653138	1.96	Octadecanoic acid, methyl ester	74.00
9	21.742	30696	0.09	11-Eicosenoic acid, methyl ester	69.05
10	22.139	168300	0.51	Eicosanoic acid, methyl ester	74.00
11	24.748	47947	0.14	Docosanoic acid, methyl ester	74.00
12	27.249	25012	0.08	Tetracosanoic acid, methyl ester	74.00

Fig. 1: GC Chromatogram of the Salicornia Biodiesel

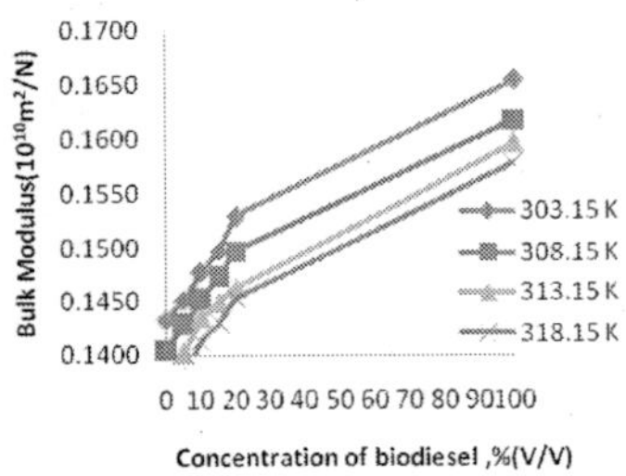

Fig. 2: The Relationship between Bulk Modulus with Concentration of Salicornia Biodiesel

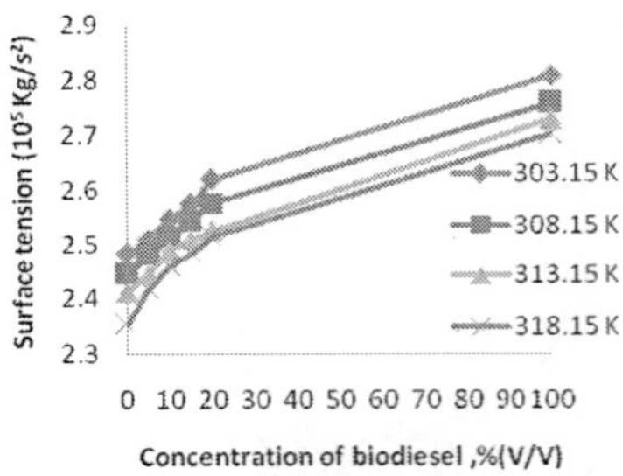

Fig. 3: The Relationship between Surface Tension with Concentration of Salicornia Biodiesel

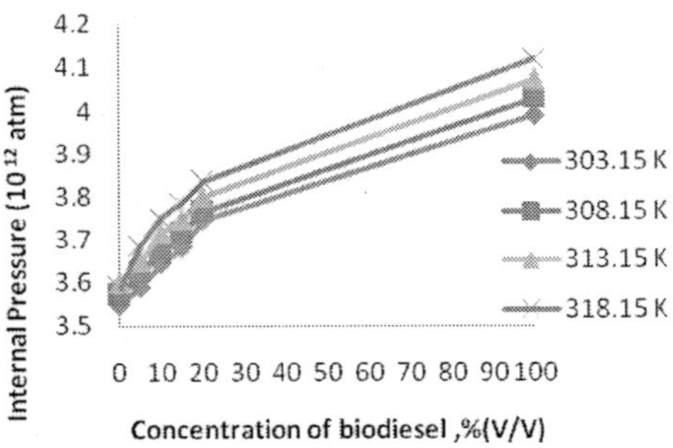

Fig. 4: The Relationship between Internal Pressure with Concentration of Salicornia Biodiesel

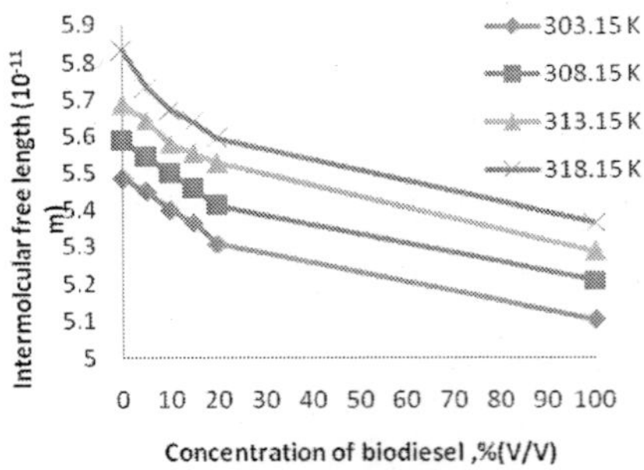

Fig. 5: The Relationship between Intermolecular Free Length with Concentration of Salicornia Biodiesel

It is very well known that bulk modulus of the biodiesel, is generally higher than petro diesel, along with viscosity and sound velocity, play a prominent role in advancing the starting of injection process [9]. It was reported by Szybistetal [10] that bulk modulus of the fuel is related to the free space between the molecules of the medium (viz. intermolecular free length). Furthermore surface tension is a parameter that decides the fuel spray characteristics with viscosity such as droplet size distribution, droplet moment of inertia etc.so surface tension of the liquid is inherently related to the internal pressure of the system .The values of bulk modulus, surface tension and internal pressure increase with increase in biodiesel concentration. In the case of intermolecular free length decrease with increase in concentration of biodiesel.

However the values of bulk modulus and surface tension decrease with increase in temperature, whereas internal pressure and intermolecular free length increases with increase in temperature.

CONCLUSION

The thermodynamic values such as bulkmodulus, surface tension, internal pressure and intermolecular free length are calculated from the accurate measurement of density, viscosity and ultrasonic speed. An attempt is underway to correlate the engine performance, efficiency and emission to physical properties of fuel.

ACKNOWLEDGMENTS

Two of us (K.R. and J.S.) wish to express our sincere gratitude to AICTE authorities for the award of a RPS project.

REFERENCES

[1] Vincete G., Martinez M., Aracil J. Methyl esters of sunflower oil as fuels: Alternative to petroleum-derived diesel fuel. IngQuim 1999; 31:153–9.

[2] Baiju B., Naik M.K., Das L.M., A comparative evaluation of compression ignition engine characteristics using methyl and ethyl esters of karanja oil. Renewable Energy 2009; 34(6): 1616–21.

[3] Ramadhas A.S., Jay Raj S., Muraleedharan C. Biodiesel production from high FFA rubber seed oil. Fuel2005; 84: 335–40.

[4] Malalya Naik, Meher L.C., Naik S.N. and Das L.M. Production of biodiesel from fatty acid karanja (pongamia pinnata) oil. Biomass and Bioenergy 32(2008) 354–57.

[5] Rajagopal K. and Jeyabalakrishnan S.S., J. Chemical thermo dynamics 42 (2010), pp. 984–993.

[6] Rajagopal K. and Jeyabalakrishnan S.S., Chinese J. Chemical Engineering 18 (2010), pp. 659–666.

[7] Rajagopal K. and Jeyabalakrishnan S.S., Chinese J. Chemical Engineering 18 (2010), pp. 425–445.

[8] Chenthianath S., A study on thermodynamic and transport properties of pure liquids and binary mixtures comprising homologous polar aprotic liquid with toluene and 2Methyl-2-Propanol. Thesis Anna University, Chennai Dec (2009).

[9] Szybist J.P., Bochman A.L., Taylor J.D. and Mecormick R.L., Evaulation of 238 formulation strategies to eliminate the biodiesel Nox effect. Fuel Process Technol.2005; 86:1109–1126.

[10] Szybist J., Kirby S. and Boehman A. Nox emissions of alternative diesel fuels: a comparative analysis of biodiesel and FT diesel. Energy and fuels 2005; 19:1484–1492.

Interaction Study of Pyridine and Acetone at 313.15 K

Prashant Dabrase, R.A. Patil[1] and B.M. Suryavanshi[2]

Bhalerao Science College, Saoner

[1]S.S.S.K.R. Innani Mahavidyalaya, Karanja

[2]Department of Physics, Government Institute of Science, Nagpur, M.S.

E-mail: prashantdabrase@gmail.com

ABSTRACT: *The Ultrasonic velocities (υ), density (ρ) and viscosity (η) measurements of pure and binary liquid mixtures of Pyridine and Acetone have been carried out at temperature 313.15 K to evaluate the other thermo dynamical parameters with their excess values such as adiabatic compressibility (β_α), free length (L_f), acoustical impedance (Z), free volume (V_f), internal pressure (π_i) and the excess values, υ^E, β_α^E, L_f^E, Z^E, V_f^E and π_i^E over the entire range of composition. The results thus obtained are discussed for molecular interaction. The non linearity observed in the plots of these ultrasonic parameters and their excess values with the composition range indicates presence of the intermolecular interaction between the components of the unlike molecules of the mixture.*

The nature of excess values of the compressibility β_α^E, free length L_f^E, acoustical impedance Z^E, free volume V_f^E and internal pressure π_i^E conforms about the existence of the molecular association between the components of the mixture.

Keywords: Ultrasonic Velocity, Compressibility, Impedance, Internal Pressure, Free Volume, Excess Values, Molecular Interaction.

INTRODUCTION

Ultrasonic studies have always played an important role in understanding the nature of intermolecular interaction in pure liquid and liquid mixtures. They find application in several industrial and technological processes. Because the Ultrasonic velocity is highly sensitive to molecular structures [1–4] and Ultrasonic investigations of liquid mixtures consisting of polar and non polar components are of highly importance in understanding the physical nature and strength of molecular interaction in the liquid mixtures [5]. For the better understanding of the physical and chemical properties and the inter-molecular interactions between the molecules of the mixture, ultrasonic velocity with density and the viscosity are measured at different concentration. The ultrasonic velocity of liquid is basically related to the bonding forces between the atoms and molecules, it helps to understand the nature of molecular interactions in pure and binary mixtures of the liquids [6–8]. Pyridine is an important liquid used currently in the extraction process for coal to analyze its compounds and in the manufacture of vitamin B_6 and other drugs. These liquid mixtures are of interest to organic chemists to know about the type of bond and number of each molecule in the pyridine complex [9, 10].

In the present study the ultrasonic velocity, density and viscosity measurements have been carried out for pure acetone and pyridine and their binary mixtures for various concentrations at 313.15 K. The variations of different ultrasonic parameters and their excess values with concentration of binary liquid mixtures are studied to understand molecular interaction between unlike molecules of the mixtures.

EXPERIMENTAL

The binary mixtures of various concentrations in mole fraction of the liquids Acetone and Pyridine were prepared at room temperature by obtaining the AR grade liquids with purity of 99.5% commercially and used without further purification and were stored in special air tight glass bottles to avoid air contact. Ultrasonic velocities (υ) for different concentration of binary mixtures and pure liquids were measured using single crystal multi frequency Ultrasonic Interferometer at 1 MHz (Mittal enterprises- model M-81). The measurements of density of pure liquids and their binary mixtures were carried out by using specific gravity bottle on a sensitive mono-pan balance within ±0.1 mg accuracy. The viscosities were measured using suspended level viscometer which was calibrated with doubly distilled water at different temperatures. The experimentally measured density (ρ) in $kgms^{-1}$, ultrasonic velocity (υ) in ms^{-1} and viscosity (η) in NSm^{-2} are used to evaluate various thermodynamic parameters like adiabatic compressibility (β_α), free length (L_f), acoustical impedance (Z), free volume (V_f), internal pressure(π_i) and their excess values as υ^E, η^E, β^E, L_f^E, Z^E, V_f^E and π_i^E by using standard relations as below.

Adiabatic Compressibility

$$\beta_a = \frac{1}{\rho v^2} \qquad \ldots (1)$$

Intermolecular free length (L_f) [8],

$$L_f = K\beta_\alpha^{1/2} \qquad \ldots (2)$$

Acoustic impedance (Z)

$$Z = \square\ \upsilon\ \rho \qquad \ldots (3)$$

free volume (V_f)

$$V_f = \left[\frac{M_{eff}\,v}{\eta\, k}\right]^{3/2} \qquad \ldots (4)$$

The internal pressure (π_i)

$$\pi_i = bRT\left(\frac{k\eta}{v}\right)^{1/2}\left(\frac{\rho^{2/3}}{M_{eff}^{7/6}}\right) \qquad \ldots (5)$$

and their excess values are determined by using the relation,

$$A^E = A_{exp} - A_{id} \qquad \ldots (6)$$

RESULTS AND DISCUSSION

From Table 1 it can be seen that the ultrasonic velocity (υ) and density (ρ) increases with the concentration of Pyridine in Acetone but the adiabatic compressibility (β_α) shows the negative trend than that of velocity at 313.15K temperature. Free length (L_f) also shows similar relation with the concentration as that of compressibility but the viscosity (η) and acoustic impedance (Z) increases with the concentration of pyridine in the mixture. The intermolecular free length depends upon intermolecular attractive and repulsive force [11]. The free volume shows decreasing trend and internal pressure is seen as increasing with increasing concentration of pyridine in the mixture. The plots of these parameters with concentrations also conforms the non linear variation (Figure 1), all the plots are non linear and the most variations can be seen at 0.6 of molar concentration of pyridine in the mixture. The values of adiabatic compressibility (β_α) and free length (L_f) decreases with increase in the mole fraction of Pyridine. The decrease in the values of free length with increasing concentration can be concluded as there is significant interaction between the two liquids [12]. It is also observed that as the concentration of pyridine increases, free volume decreases whereas the internal pressure increases. This suggests the close packing of the molecules, which may be concluded as the increasing magnitude of the interaction [13–15]. The non linear variation in the above parameters with the mole fraction of pyridine is an indication of existence of interaction between the components of the mixture [16, 17]. The viscosity values showing increasing trend is the indication of frictional resistive force that may be due to a change in effective molecular area of the cohesive or adhesive forces or relative random velocity between the components of the mixture. In this case this may be due to the tight packing and close structure of pyridine [10].

The existence of any inter molecular interaction can be supplemented with the help of excess values of the thermo dynamical parameters. The extent of deviation and sign of these functions depends on the strength of interaction between unlike molecules [18, 19]. Here Table 2, presents the data of excess values of the ultrasonic velocity (υ^E), excess viscosity (η^E), excess adiabatic compressibility (β_α^E), excess free length (L_f^E), excess impedance (Z^E), excess free volume (V_f^E) and excess internal pressure (π_i^E) corresponding to the molar concentration of pyridine in the mixture at temperature 313.15 K. The excess velocity here is seen to be positive. The excess compressibility and free length shows negative values and the same trend can be observed in the excess viscosity but the values are more negative it means the strong attractive or repulsive forces are acting between the different components of the mixture. The close perusal of the excess values as shown in the Figure 2, conforms the trend in excess free volume and excess internal pressure is opposite to each other. Excess free volume is positive and excess internal pressure is negative and in both the cases the nature of the cure is nearly flat showing the existence of interaction between the different components of the mixture for all the concentrations of the pyridine in the mixture.

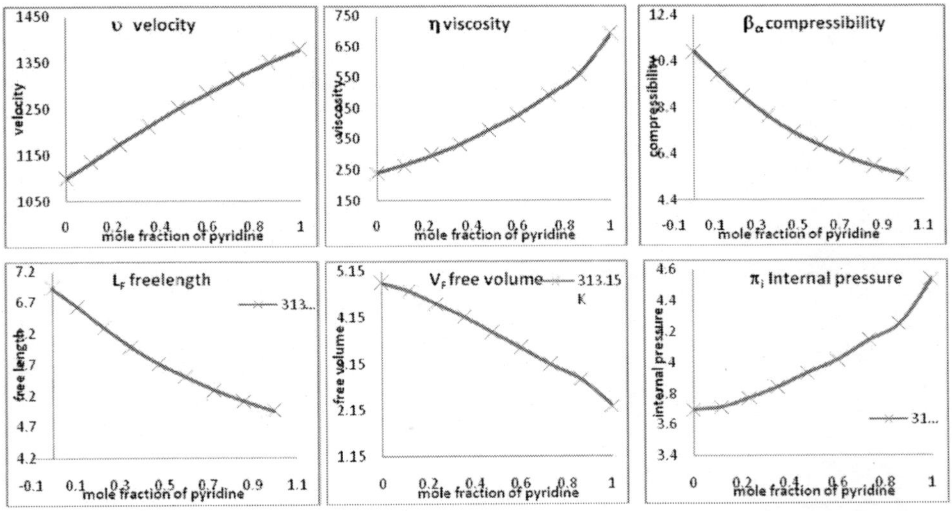

Fig. 1: Plots of Ultrasonic Velocity, Viscosity, (β_α), (L_f), (Z), (V_f) and (π_i) for Acetone + Pyridine Mixture at 313.15 K

Table 1: Values of Density, Ultrasonic Velocity, Viscosity, Adiabatic Compressibility (β_α), Free Length (L_f), Acoustical Impedance (Z), Free Volume (V_f) and Internal Pressure (π_i) for Acetone + Pyridine Mixture at 313.15 K

Mole Fract of Pyridine	*Density* ρ kgm^{-3}	*Velocity* υ ms^{-1}	*Viscosity* η *mPas*	*Compress* $\beta\alpha$ $10^{-10}\,m^2N^{-1}$	*Free length* L_f $X\,10^{-11}\,m$	*Impedance* Z $\times 10^6\,Kgm^{-2}s^{-1}$	*Free volume* V_f $\times 10^{-7}\,m^3$	*Internal pressure* $\pi_i \times 10^8$ *Pa*
0	767.3450	1098.955	240.174	10.7907	6.9394	0.84328	4.89277	3.69326
0.11582	791.9711	1134.233	264.342	9.8149	6.6182	0.89828	4.72475	3.71319
0.2341	819.1132	1173.568	296.5339	8.8642	6.2895	0.96129	4.44627	3.77244
0.35490	847.8668	1211.538	332.8187	8.0352	5.9882	1.02722	4.16186	3.84359
0.47833	874.1163	1249.795	377.9024	7.3244	5.717	1.09247	3.81971	3.93335
0.60446	897.7669	1282.649	427.3347	6.7705	5.4968	1.15152	3.49785	4.01926
0.73338	922.7287	1317.222	490.4197	6.2461	5.2796	1.21544	3.13258	4.14167
0.86520	944.7639	1348.500	559.3288	5.8207	5.0966	1.27401	2.81605	4.25322
1	962.8652	1377.257	690.5037	5.4753	4.9431	1.32611	2.23806	4.53858

Table 2: Excess Values of Ultrasonic Velocity, Excess Viscosity (η^E), Excess Adiabatic Compressibility (β^E), Excess Free Length (L_f^E), Excess Acoustical Impedance (Z^E), Excess Free Volume (V_f^E) and Excess Internal Pressure (π_i^E) for Acetone + Pyridine at 313.15 K Temperature

Mole Fract of Pyridine	υ^E ms^{-1}	$\beta\alpha^E$ $10^{-10}\ m^2N^{-1}$	ηE *mPas*	L_f^E $\times 10^{-11}\ m$	Z^E $\times 10^6 Kgm^{-2}s^{-1}$	V_f^E $\times 10^{-7}\ m^3$	π_i^E $\times 10^8\ Pa$
0	0	0	0	0	0	0	0
0.115808	3.0459	–0.3602	–27.9878	–0.0899	–0.0009185	0.1394	–0.07797
0.234077	9.4653	–0.6822	–49.0578	–0.1826	0.0049804	0.1749	–0.1187
0.354882	13.8139	–0.8691	–67.1769	–0.2427	0.0125878	0.2112	–0.14967
0.478309	17.7206	–0.9239	–77.6765	–0.2674	0.0182353	0.1968	–0.16425
0.604441	15.4718	–0.8072	–85.0451	–0.2359	0.0163883	0.2097	–0.18496
0.733371	14.1645	–0.6464	–80.0193	–0.1957	0.018057	0.1867	–0.17154
0.865191	8.7584	–0.371	–70.4700	–0.1155	0.0129877	0.2201	–0.17141
1	0	0	0	0	0	0	0

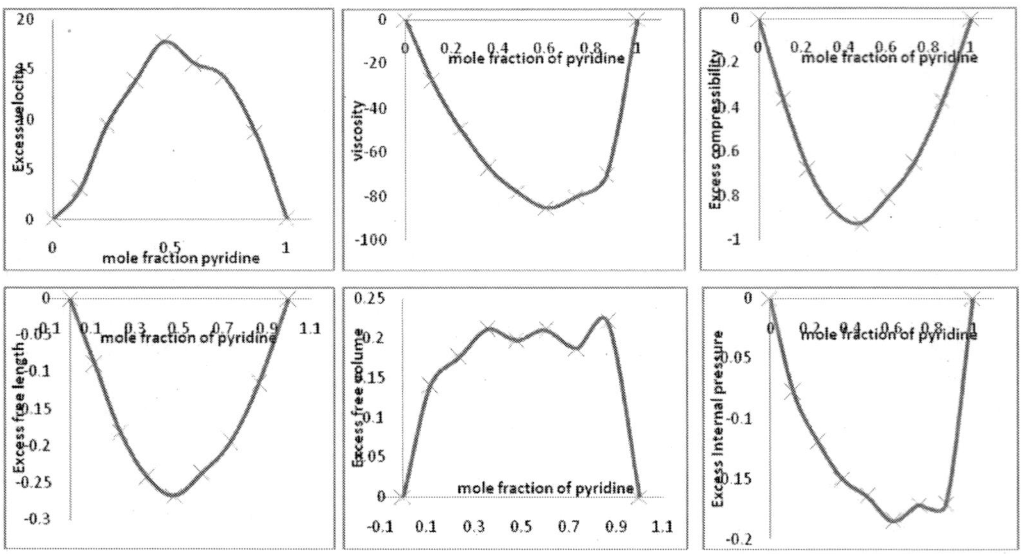

Fig. 2: Excess Values of (υ^E), (η^E), (β_α^E), (L_f^E), (Z^E), (V_f^E) and (π_i^E) for Acetone + Pyridine at 313.15 K

The sign of excess free length plays a vital role in assessing compactness due to molecular interaction through dipole-dipole interaction [20, 21] leading to more compact structure making which enhances excess free length to have negative values. Graphs of excess adiabatic compressibility (β_α^E) and excess free length (L_f^E) are negative which indicates the strong molecular interaction between the components of the mixture. The deviation of excess impedance (Z^E) is found to be positive. The excess ultrasonic velocity (υ^E) for the binary mixture is positive. The positive deviation and non-linear dependence suggests the presence of strong interaction between the components of mixture [22, 23].

CONCLUSION

From the present study it can be concluded that the results are very basic in nature and indicates that the non linear variation of the all the parameters measured at 313.15 K for the mixture of Pyridine in the Acetone indicates existence of interaction between the different molecules of the compounds in the mixture. The non-linearity of the curve is common for all compositions. The negative values of excess compressibility ($\beta_\alpha{}^E$), excess viscosity (η^E), excess free length (L_f^E) and excess internal pressure (π_i^E) and the positive values of excess ultrasonic velocity (υ^E), excess acoustic impedance (Z^E) and excess free volume (V_f^E) for the different combinations of the liquids in the mixture shows the presence of strong dispersive interaction between the components of molecule in the mixture. The excess values of thermodynamic parameters are sensitive to the molecular association present in the liquid mixture. The interaction is strong from 0.2 to 0.8 of concentration of the Pyridine in the mixture.

REFERENCES

[1] H. Eyring and J.F. Kincaid, J. Chem, Phys, (1938), 6, 220–229.
[2] K.S. Pitzer, Thermodynamics, 3rd edition (Mc.Graw-Hill book Co, NY (1995).
[3] A.N. Kannapan, S. Thirumam and R. Palani, J. Phys. Science, (2009), Vol. 20 (2), 97–108.
[4] V.A. Tabhane, S. Ghosh and S. Agrawal, J. Pure and Appl. Ultrasonic, (1900), Vol. 21, 122.
[5] R J Fort and W R Moore, Trans. Farady Society, (1966), 62, 1112.
[6] V. Srinivasulu and P.R. Naidu, J. Pure and Appl. Ultrasonic, (1995), Vol.17, 14–28.
[7] K. Swain, P. Priyadarshani, Indian J.Pure and Appl. Phys.(2010) vol.48, Aug 2010, 539–542.
[8] Jacobson, B. Ultras Velo. of liquids and liquid mixtures, J. Chem. Phys. (1952), 20, 927–928.
[9] P K Gupta and F B Stump, Indian J pure and Appl. Phys., (2010), 48, 326–333.
[10] G Arul and L Palaniappan, Indian J pure and Appl. Phys. (2005), 43, 755–758.
[11] Eyring, H. and Kincaid, J.F, J. Chem,Phys, (1938), 6, 620.
[12] K. Rajgopal and S. Chenthilnath, Indian J.Pure and Appl. Phys., (1981), 19, 358.
[13] T.S. Banipal and A.F. Toor and V.K. Rattan, Indian J Chem. (2000), 39A, 809.
[14] P.S. Naidu and K. Ravindra Prasad, Indian J.Pure and Appl. Phys., (2002), 40, 264.
[15] G. Arul and L. Palaniappan, Indian J pure and Appl. Phys. (2001), 39, 561.
[16] A.N. Kannappan, and V. Rajendran, Indian J.Pure and Appl. Phys., (1999), 73B, 531.
[17] A. Sadasiva Rao, B. Vijaykumar Naidu, J Acoust Soc India, (2000), 28, 303.
[18] O. Redlich, A.T. Kister, Ind. Eng Chem, (1948), 40, 345.
[19] A. Ali and A.K. Nain, Int conf and exhi on Ultrsonics (ICEU-99) (1999), Dec-2, New Delhi.
[20] D. Anbananthan, Jou Acost Soc of India (1979), 7, 123.

[21] Nisha Sharma and B.P. Singh, Acta Ciencia india (2010), Vol. XXXIV P, No. 3, 375.
[22] M.K. Praharaj *et al.*, Journ. Chemical And Pharma. Research, (2012), 4(4), 1910–1920).
[23] A.N. Kannappan, and V. Rajendran, Indian J.Pure and Appl. Phys., (1991), 29, 465–468.
[24] P.B. Dabrase, R.A. Patil, B.M. Suryavanshi, Applied Ultras, (Shree Publi, N Delhi, (2012), 233–241).

Acoustical Behavior of NN Dimethyl-acetamide and Acetone at 313.15 K

R.A. Patil, Prashant Dabrase[1] and B.M. Suryavanshi[2]

S.S.S.K.R. Innani Mahavidyalaya, Karanja (lad)

[1]Bhalerao Science College, Saoner

[2]Department of Physics, Government Institute of Science, Nagpur, M.S.

E-mail: rapatil10@rediffmail.com; prashantdabrase@gmail.com; bm_suryavanshi@yahoo.co.in

ABSTRACT: *The measurements of Ultrasonic velocities (υ), density (ρ) and viscosity (η) of pure and binary liquid mixtures of Acetone and N-N dimethyl acetamide have been carried out over the entire range of composition at temperature 313.15 K to evaluate the other thermo dynamical parameters with their excess values such as adiabatic compressibility (β_α), free length (L_f), free volume (V_f), internal pressure (π_i) and their excess values, υ^E, β_α^E, L_f^E, V_f^E and π_i^E. The results thus obtained are discussed for molecular interaction with varying composition of binary system. The non linearity found in all the plots of these ultrasonic parameters and their excess values with the composition range indicates presence of the intermolecular interaction between the components of the unlike molecules of the mixture.*

The nature of excess values of the β_α, L_f, V_f and π_i conforms about the existence of the molecular association between the components of the mixture.

Keywords: Ultrasonic Velocity, Excess Values, Compressibility, Internal Pressure, Free Volume, Molecular Interaction.

INTRODUCTION

There are many physical methods which play an important role in determining molecular structure and molecular properties of different materials. In recent years the advances in the ultrasonic technique has become a powerful tool in evaluating information about the physical and chemical behavior of molecules of the liquids [1–4]. The ultrasonic studies of the liquids are most preferred in many fields such as pharmaceutical industry, biomedical research, automobile industry, chemical industry, water research and scattering spectroscopy, etc.[5]. The information of density (ρ), ultrasonic velocity (υ) and viscosity (η) of the pure liquids and their mixtures plays very important role in different applications that include surface facilities, pipeline systems and mass transfer operations.

Here in the present study the ultrasonic velocity, density and viscosity measurements have been carried out for determination of ultrasonic parameters such as compressibility (β_α), molecular free length (L_f), free volume (V_f) and internal pressure (π_i) and their excess values (υ^E, β_α^E, L_f^E, V_f^E and π_i^E) for the different composition range of NN-Dimethyl acetamide in acetone at the temperature 313.15 K. The variations of these parameters with concentration of binary liquid mixtures are studied to understand molecular interaction between unlike molecules of the mixtures.

EXPERIMENTAL

The binary mixtures of the chemicals Acetone and NN- dimethyl acetamide for different range of composition were prepared at room temperature and kept in a special airtight glass bottles to avoid air contact. The chemicals used here are of analytical reagent (AR) and used without further purification. The mixtures prepared are used within 24 hours of its preparation. The densities of the binary mixtures and pure liquids were measured using 25 ml specific gravity bottle and a sensitive mono pan balance (K-Roy, K–12 classic) within ±0.1 mg accuracy. The ultrasonic interferometer (Mittal enterprises, New Delhi, India) is a single and direct device to determine ultrasonic velocity in liquids with a high degree of accuracy [6]. Here the measurements of ultrasonic velocities were carried out on single crystal multi frequency ultrasonic interferometer operating at 1MHz (M-81). The constant temperature of the liquid inside the interferometer cell was maintained by circulating water through the outer jacket by electronically controlled thermostat. Accuracy of measurement of ultrasonic velocity was within ±0.01 m/s and the temperature of the test liquids during measurement were maintained within an accuracy of ±0.1°C. The viscosity of the liquids and their mixtures are measured using the suspended level viscometer.

The experimentally measured ultrasonic velocity (υ) measured in ms^{-1}, density (ρ) in kgm^{-3} and viscosity (η) in Nsm^{-2} are used to evaluate various thermo dynamical parameters like–

Adiabatic compressibility (β_α),

$$\beta_\alpha = 1/\upsilon^2\rho \qquad \text{... (1)}$$

Intermolecular free length (L_f)

$$L_f = K\beta_\alpha^{1/2} \qquad \text{... (2)}$$

where, K –is temperature dependant constant 211.25×10^{-8} at 313.15 K and T—absolute temperature [7],

$$V_f = \left[\frac{M_{eff}}{k\eta}\nu\right]^{3/2} \qquad \text{... (3)}$$

where M_{eff} is the effective molecular weight ($M_{eff} = \Sigma m_i X_i$ in which m_i and X_i are the molecular weight and mole fraction of the individual constituents respectively), k is temperature independent constant which is equal to 4.28×10^9 for all liquids.

$$= bRT\left(\frac{k\eta}{\nu}\right)^{1/2}\left(\frac{\rho^{2/3}}{M_{eff}^{7/6}}\right) \quad \ldots (4)$$

where k is a constant, T is the absolute temperature; b is a constant equal to 2 for the liquid and the excess values are determined by using the relation

$$A^E = A_{exp} - A_{id} \quad \ldots (5)$$

A^E – excess parameters of all acoustic parameters, $A_{id} = \sum_{i=1}^{n} A_i X_i$, A_i is any acoustical parameter and X_i – the mole fraction of liquid component.

RESULTS AND DISCUSSION

The measured values of densities (ρ) and ultrasonic velocities (υ) of the pure liquids and their binary mixtures along with viscosity (η) were used to obtain the adiabatic compressibility (β_α), free length (L_f), free volume (V_f) and internal pressure (π_i) at different concentration and temperatures 313.15 K and reported in (Table 1). It is observed that Ultrasonic velocity (υ), density (ρ) and viscosity (η) increases with increase in the mole fraction of NN-dimethyl acetamide, in the binary mixture. The table observes the trends of all given acoustical parameters with variation in concentration at temperature 313.15 K. The Ultrasonic velocity (υ) increases with concentration of mixture and the corresponding density also increases. As observed from the Table 1. The values of viscosity increases with the increasing concentration and the same trend is observed for the internal pressure. The compressibility values shows decreasing trend with the increasing concentration of NN-Dymethyl-acetamide (NN DMA) in the mixture. The same nature as that of compressibility is also observed in the free length and in the free volume values. The increase in ultrasonic velocity may be attributed to the cohesion. The values of adiabatic compressibility (β_α) and free length (L_f) decreases with increase in the mole fraction of N-N dymethyl acetamide and increases with temperature this suggests making and breaking of hydrogen bonding. The greater the attractive force among the molecules, the smaller will be the compressibility thus indicates formation of a complex [8]. The intermolecular free length depends upon intermolecular attractive and repulsive force. Eyring and Kincard have proposed that free length is a predominating factor in determining the variation of ultrasonic velocity of solutions [9, 10]. The decrease in adiabatic compressibility suggests that there is significant interaction between unlike molecules. The internal pressure first decreases and then increases due to various degree of dispersive interactions and coloumbic interactions existing between the component molecules [11].The free length and free volume values shows decreasing trends with increase in the concentration of NN DMA hence it can be concluded that there is significant interaction between the two liquids due to which the structural arrangement is also affected [12].

Table 1: Density, Ultrasonic Velocity, Viscosity, and Other Thermodynamic Parameters Such as (β_α, L_f, V_f and π_i for Acetone + NN-Dimethyl Acetamide Mixtures at Temperature 313.15 K

Mole Fraction NNDMA	*Density* ρ kgm^{-3}	*Velocity* υ ms^{-1}	*Compressibility* β_α $10^{-10}\ m^2N^{-1}$	*Viscosity* η $\mu\ Pas$	*Free Length* L_f $\times 10^{-8}\ m$	*Free Volume* V_f $\times 10^{-16}$ $m^3/mole$	*Internal Pressure* π_i $\times 10^8\ Pa$
0.0000	771.610	1108.372	10.549	241.509	6.861	4.915	3.701
0.1024	791.959	1143.366	9.658	270.059	6.565	4.694	3.699
0.2103	813.891	1184.769	8.753	304.876	6.249	4.449	3.709
0.3240	832.431	1221.158	8.055	344.315	5.995	4.182	3.717
0.4441	853.262	1261.263	7.367	390.075	5.733	3.927	3.732
0.5711	871.849	1302.111	6.764	452.007	5.494	3.563	3.780
0.7056	891.621	1331.429	6.326	514.49	5.313	3.274	3.815
0.8483	908.249	1367.657	5.886	593.028	5.125	2.976	3.853
1.0000	924.980	1400.059	5.515	685.660	4.961	2.680	3.901

In order to highlight the presence of intermolecular interaction between molecules it is essential to study the excess parameters. The deviation of the physical properties of the liquid mixtures from the ideal behavior is a measure of interaction between the molecules which may be due to the either adhesive or cohesive force. Table 2 presents the values of excess velocity (υ^E), excess adiabatic compressibility (β_α^E), excess free length (L_f^E), excess free volume (V_f^E) and excess internal pressure (π_i^E).

The values of excess free length (L_f^E) are supported by the variation of excess compressibility exhibits interaction in the system. The excess compressibility and excess free length are negative which indicates the strong molecular interaction between the components. Negative excess free length indicates the ultrasonic wave's needs to cover larger distance because of dominant nature of interaction between unlike molecules [13]. The excess ultrasonic velocity (υ^E) for the binary mixture is observed to be positive. The positive deviation and non-linear dependence suggests the presence of strong interaction between the components of mixture [14]. According to Fort and Moore, a negative volume is an indication of strong hetero-molecular interaction in the liquid mixtures and is attributed to charge- transfer, dipole-dipole, dipole-induced dipole interactions and hydrogen bonding between the unlike compounds, while a positive sign indicates a weak interaction and is attributed to dispersive force, which are likely to be operative [15,16]. The excess free volume (V_f^E) and excess internal pressure (π_i^E) are also negative which convey that the chances of induced dipole-dipole interactions are overruled and strong dipolar interactions alone are conformed [17].

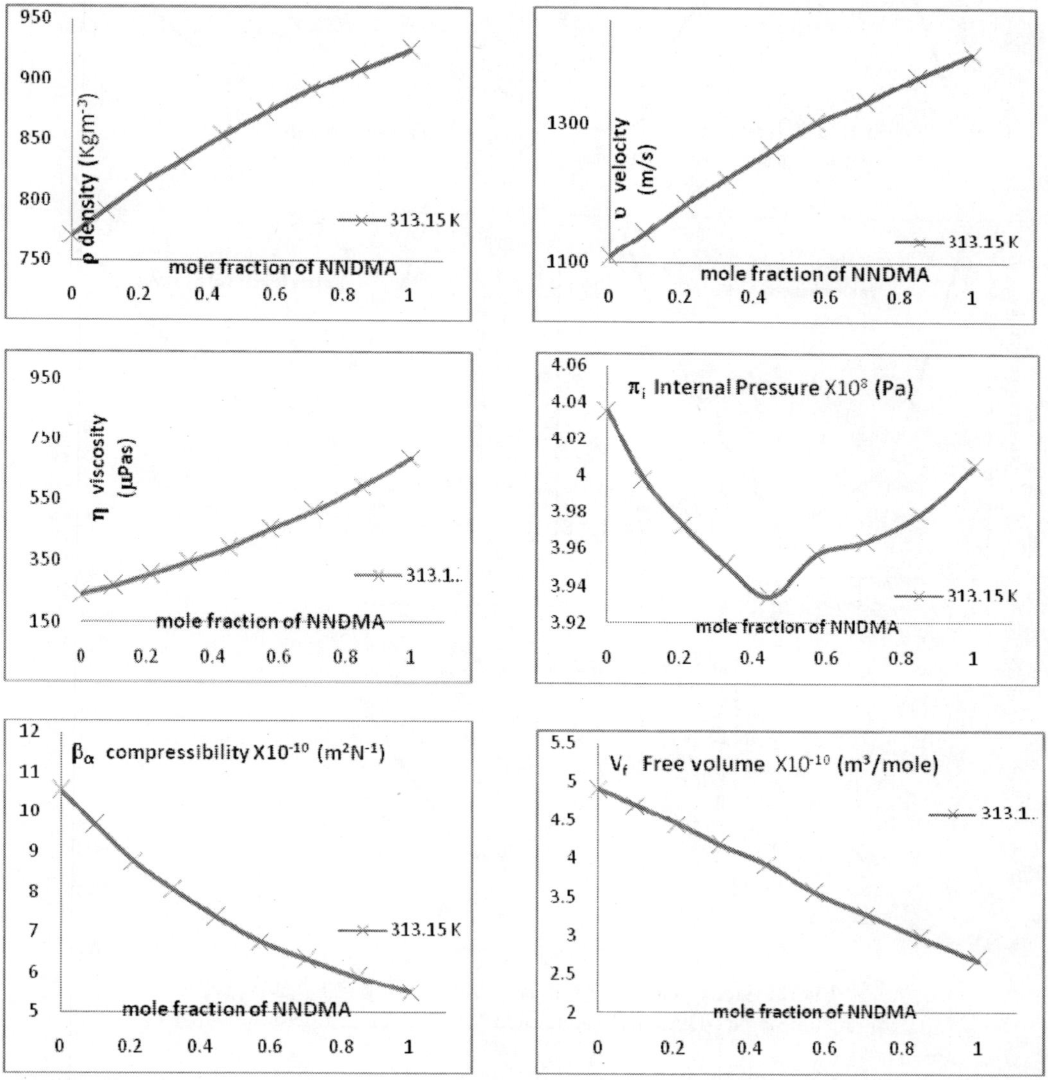

Fig. 1: Density, Ultrasonic Velocity, Viscosity, β_α, L_f, V_f and π_i for Acetone + NN-Dimethyl Acetamide Mixtures at Temperature 313.15 K

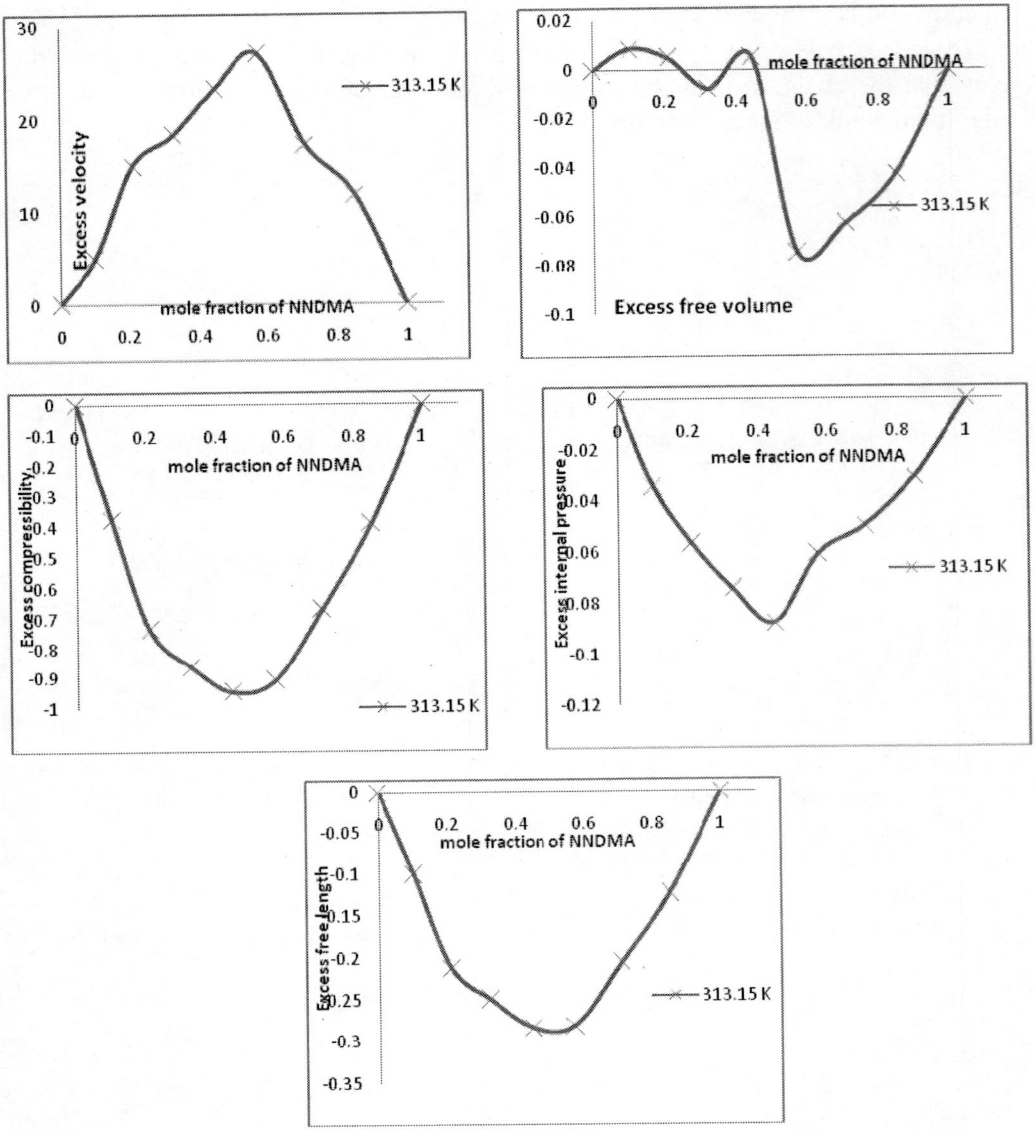

Fig. 2: Excess Values of Ultrasonic Velocity β_α^E, L_f^E, V_f^E & π_i^E for Acetone + NN-Dimethyl Acetamide Mixtures at Temperature 313.15 K.

Thus it is clear from the above evaluated parameters that there is a strong association between the acetone and NN-dimethyl acetamide molecules exists after 0.4 mole fraction of the NNDMA in the mixture.

Table 2: Excess Values of Ultrasonic Velocity and Other Thermodynamic Parameters β_α^E, L_f^E, V_f^E & π_i^E for Acetone + NN-Dimethyl Acetamide Mixtures at Temperature 313.15 K

Mole Fraction NNDMA	*Excess Velocity* υ^E ms^{-1}	*Compressibility* β_α^E $X\ 10^{-10}\ m^2N^{-1}$	*Free Length* L_f^E $\times\ 10^{-8}\ m$	*Free Volume* V_f^E $\times\ 10^{-17}$ $m^3/mole$	*Internal Pressure* π_i^E $\times\ 10^8\ Pa$
0.0000	0	0	0	0	0
0.1024	5.1049	–0.3747	–0.1013	0.0845	–0.0231
0.2103	15.0393	–0.7373	–0.2117	0.0482	–0.0346
0.3240	18.250	–0.8621	–0.2497	–0.0802	–0.0491
0.4441	23.3267	–0.9461	–0.2834	0.0475	–0.0587
0.5711	27.1344	–0.9092	–0.2815	–0.7515	–0.0354
0.7056	17.2221	–0.6703	–0.2069	–0.6307	–0.0276
0.8483	11.8318	–0.3925	–0.1241	–0.4275	–0.0183
1.0000	0	0	0	0	0

CONCLUSION

The ultrasonic velocity, density and viscosity measurements have been carried out for determination of ultrasonic parameters such as compressibility (β_α), molecular free length (L_f), free volume (V_f) and internal pressure (π_i) and their excess values (υ^E, β_α^E, L_f^E, V_f^E and π_i^E) for the different composition range of NN-Dimethyl acetamide in acetone at the temperature 313.15 K.

The present investigations lead us to conclude that the excess values of thermodynamic parameters are sensitive to the molecular association present in the liquid mixture. The negative values of excess compressibility (β_α^E), excess free length (L_f^E) and excess internal pressure (π_i^E) and the positive values of excess ultrasonic velocity (υ^E) in the mixture shows the presence of strong dispersive dipole-dipole interaction between the components of molecule in the mixture. The non-linearity of the curve also supports the interaction exists. The interaction is strong after 0.4 mole fraction of N-N dimethyl acetamide in the mixture of acetone and NNDMA.

REFERENCES

[1] Eyring, H. and Kincaid, J.F, *J. Chem, Phys*, 1938, 6, 220–229.

[2] K.S. Pitzer, Thermodynamics, 3rd edition (Mc.Graw-Hill book co, NY 1995).

[3] A.N. Kannapan, S. Thirumam and R. Palani, J. Phys. Science, 2009, Vol. 20 (2), 97–108.

[4] V.A. Tabhane, S. Ghosh and S. Agrawal, J. Pure and Appl. Ultrasonic, 1999, Vol. 21, 122.

[5] V. Srinivasulu and P.R. Naidu, J. Pure and Appl. Ultrasonic, 1995, 17, 14–28.

[6] K. Swain, P. Priyadarshani, Indian J. Pure and Appl. Phys., vol.48, Aug 2010, 539–542.
[7] Jacobson, B. (1952), Ultras Velo. of liquids and liquid mixtures, J. Chem. Phys., 20,927–.
[8] R. P. Varma and Surendra Kumar (2000), Indian J. Pure and Appl. Phys., 38, 96.
[9] K. Rajgopal and S. Chenthilnath, Indian J.Pure and Appl. Phys.
[10] A.N. Kannappan, and V. Rajendran, Indian J.Pure and Appl. Phys., 1991, 29, 465–468.
[11] M.K. Praharaj, A. Satapathy *et al.* (2012), J. Chem and Pharm. Research, 4(4), 1910–1920.
[12] Eyring, H. and Kincaid, J.F., J. Chem, Phys, 1938, 6, 620–629.
[13] Shrivastava and Dubey (1985), Indian J. Pure and Appl. Phys.
[14] P.B. Dabrase, R.A. Patil, B.M. Suryavanshi, App Ultra, (Shree Publi, N Delhi, 2012, 233–241).
[15] R.T. Fort, W.R. Moore (1966) Trans. Farad. Soc., 62, 1112.
[16] Moh. A. Wahab *et al.* (2002), Bull. Korean Chem. Soc. Vol. 23, No. 7, 953–956.
[17] S. Soundharam (2010), Acta Ciencia India, Vol. XXXII P, No. 2, 285.
[18] Alka Tadlkar, Pravina Pawar and GBichile, J. Chem. Pharm. Research, 2011, 3(3), 165–168.

Electrochemical, Thermodynamical and Vibrational Spectral Analysis of a Peptide in Non-Aqueous Medium

R. Padmavathy, E. Jasmine Vasantharani, N. Radha[1], S. Sujatha[2] and R. Yogarani

Department of Physics, Seethalakshmi Ramaswamy College, Tiruchirappalli, Tamil Nadu
[1]Department of Chemistry, Seethalakshmi Ramaswamy College, Tiruchirappalli Tamil Nadu
[2]Department of Physics, Srimathi Indira Gandhi College, Tiruchirappalli, Tamil Nadu

ABSTRACT: *Ultrasonic velocity measurements are highly sensitive to molecular interactions and can be used to provide qualitative information about the physical nature and strength of the molecular/inter-ionic interactions. Internal pressure is closely related to solubility parameters which determine the way in which the interaction occur in the system and also it can be measured experimentally .Free volume dependent properties have close connection with the molecular structure and it may account interesting features about interactions in the solutions. Suryanarayana and Kuppusamy have formulated quantitative relationship between the internal pressure, free volume and equivalent conductance .Fourier Transform infrared spectroscopy is a powerful tool to identify the presence of various functional groups and assign the corresponding vibrational band of a sample by producing an infrared absorption spectrum. The Glycyl-L-valine peptide acts as an nutritional supplement and pharmaceutical intermediate. The solutions of different molalities were prepared in formamide and experimental studies were made from a very low concentration to a high concentration with different temperatures. In the present work, the results are analyzed on the basis of ultrasonic and the spectroscopic methods to interpret the structural changes taking place in the solution.*

INTRODUCTION

Infrared spectroscopy is one of the vital spectroscopic techniques used by organic and inorganic chemists, for structural elucidation and compound identification. A large number of studies have been made on the molecular interaction in the liquid/solution systems by various physical methods. In recent years ultrasonic technique has become a powerful tool in providing information regarding the molecular behavior of liquids to its ability of characterizing physico-chemical behavior of the medium. To compute the thermodynamic and acoustic parameters of Glycyl-L-Valine in non-aqueous medium the ultrasonic velocity, density and viscosity are measured at varying temperatures of 5°C,

15°C, 25°C, 35°C, 45°C and 55°C Suryanarayana and vekatesan[1] recognized that internal pressure of the medium was closely related to degree of dissociation, ionic mobility, ion pairing, dielectric constant and viscosity.

EXPERIMENTAL TECHNIQUE

The density of the solution is determined using 25 ml specific gravity bottle, using the constant temperature bath. Since the density calculation were made use of the digital balance weighing values with an accuracy ±0.0001 gm/cc. A Cannon Fenske viscometer is used for the viscosity measurements. The accuracy of the viscosity measurements was ±0.5%. The principle of the ultrasonic interferometer was stated by pierce,[2] to detect the variation of the current in the electronic circuit due to acoustic resonance. Variable path interferometer having a frequency of 2 MHz with overall accuracy of 0.1% is used for velocity measurements.

FORMULA USED

Internal pressure (π_i) = $bRT\,[k\eta/u]^{1/2}\rho^{2/3}/M_{eff}^{7/6}$ atms

Free volume (v_f) = $[\,M_{eff}\,U/k\eta]^{3/2}$cc

Equivalent conductance ($\wedge c$) = $k*1000/N$ mho.cm^2

RESULTS AND DISCUSSION

Thermodynamical Parameters

Internal pressure of a solution is a single factor which appears to vary with intermolecular interactions namely, solvation, quantum mechanical forces of dispersion and dielectric forces, solute-solvent interactions which play an important role in transport properties of solution. The interaction depends upon the solvent and solute, structure and presence of temperature. Free volume is one of the most significant parameter in explaining the variations in the physico-chemical properties of liquids. The free space and its dependent properties have close connection with the molecular structure and it may show interesting features about interactions. The increasing trend of π_i with concentration reveals the fact that the solution of Glycyl-L-Valine behaves as a structure maker.[3] The internal pressure decreases with the rise in temperature because when the temperature is increased, there is a tendency for the ions to move away from each other, reducing the possibility for interaction, which may further reduce the cohesive forces and ultimately leads to a decrease in the internal pressure.[4] The variations of free volume with respect to molalities and temperatures Figure (ii) also confirm that there is a strong solute-solvent interaction in Glycyl-L-Valine heteropeptide solutions . When the temperature increases the tendency of molecules to more away from each other increases hence results in the decrease of cohesive forces of system.[5]

Table 1: Internal Pressure (atms)

Molality (m)	*5°C*	*15°C*	*25°C*	*35°C*	*45°C*	*55°C*
0.001	18557	15120	14118	12939	11700	10500
0.005	19666	16403	14315	13596	12236	10880
0.01	19648	16301	14017	13531	12066	10719
0.015	19654	16328	13727	13236	12057	10631
0.02	20028	16172	13716	13111	11954	10611

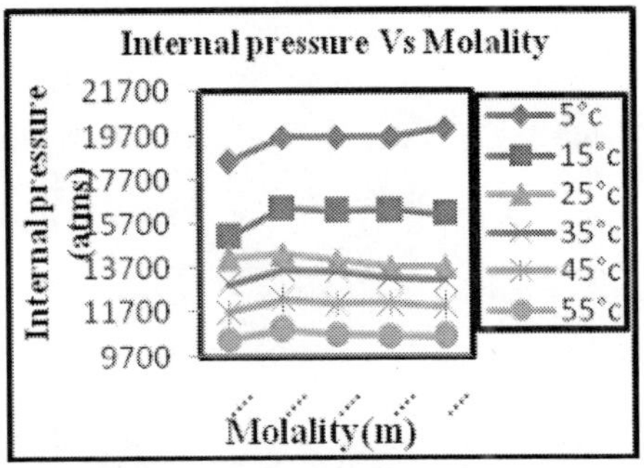

Fig. 1: π_i for Glycyl-L-Valine

Table 2: Free Volume (cc)

Molality (m)	*5°C*	*15°C*	*25°C*	*35°C*	*45°C*	*55°C*
0.001	1.653	2.260	2.988	3.984	4.212	5.478
0.005	0.391	0.544	0.710	0.920	1.007	1.272
0.01	0.230	0.317	0.421	0.552	0.596	0.760
0.015	0.177	0.245	0.325	0.437	0.442	0.594
0.02	0.151	0.220	0.279	0.368	0.383	0.505

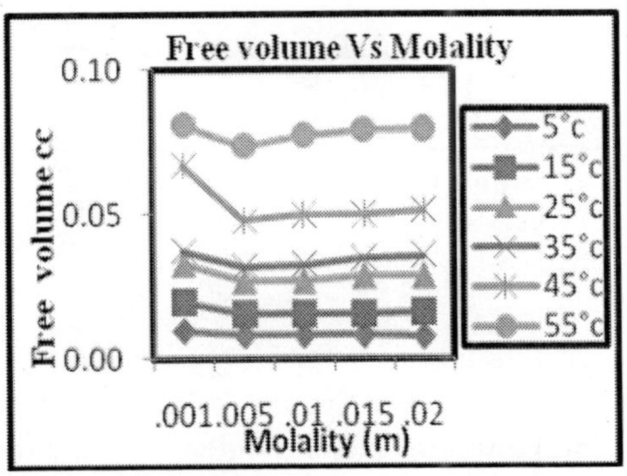

Fig. 2: V_f for Glycyl-L-Valine

Electrochemical Parameters

The computed values of equivalent conductance of the dipeptide solutions are given in Table 3. The variations of Λc with concentration of the solution are as shown in the Figures 3. In general, in all the conducting solutions, the equivalent conductance is found to be high, at low concentration. With increasing the concentration and temperature the equivalent conductance decreases exponentially. The variation of equivalent conductance is found to be very small, at higher concentration. The moving ion in the solution experiences the following retarding effects:

- Movement of the ionic atmosphere in a direction opposite to that of the ion.
- Asymmetry of the ionic atmosphere which imposes certain retardation on the translation motion of the ion.

Table 3: Equivalent Conductance (cm^2mho)

Molality (m)	*5°C*	*15°C*	*25°C*	*35°C*	*45°C*	*55°C*
0.001	0.0096	0.0197	0.0328	0.0372	0.0670	0.0808
0.005	0.0081	0.0154	0.0269	0.0320	0.0478	0.0736
0.01	0.0081	0.0156	0.0269	0.0325	0.0498	0.0771
0.015	0.0081	0.0156	0.0286	0.0346	0.0499	0.0788
0.02	0.0076	0.0160	0.0287	0.0356	0.0511	0.0794

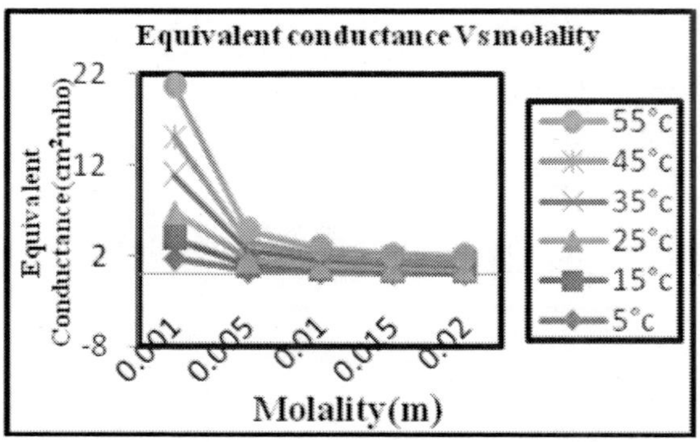

Fig. 3: Λc for Glycyl-L-Valine

IR Spectral Analysis

Formamide is a dipolar solvent and it has a very high dielectric constant and dipole moment. John, Grady, Michael and Mingo[6] established recently the structure of formamide and the hydrogen bond length and energies based on density functional calculation, which are relevant to systems of current interest in molecular chemistry. In the Infra-red spectrum of formamide, OH and NH stretching vibrations are appearing as broad band at 3443 cm^{-1}. The band at 2886 cm^{-1} can be assigned to OH stretching of enolic form of formamide which is due to resonance in formamide dimer. The NH bending absorption also known as amide I band is observed at 1682 cm^{-1}. $\rangle$C=O (carbonyl group), stretching absorption in amide II band occurs at 1391 to 1311 cm^{-1}. The NH bending vibration, i.e. amide II involves coupling between NH bending other fundamental vibrations.

Table 4: Infrared Frequencies of Glycyl-L-Valine

Name of the Sample	γ_{NH} cm^{-1}	γ_{CH} cm^{-1}	$\gamma_{C\text{-}N}$ cm^{-1}	$\gamma_{C=O}$ cm^{-1}	$\gamma_{C=N}$ cm^{-1}	$\gamma_{C\text{-}O}$ cm^{-1}	δ_{CH} cm^{-1}
Formamide	3443	2886	2769, 2699, 2400	1682	1391 1311	1089, 1050	605
Formamide + Glycyl-L-Valine	3407	2887	2769	1682	1391 1311	1051	607

Formamide

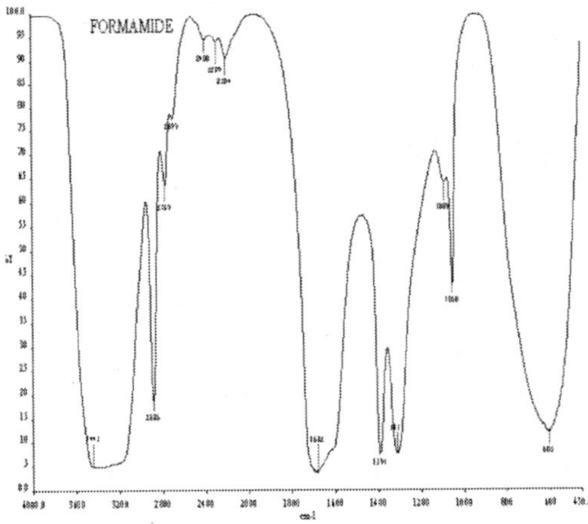

Glycyl-L-Valine + Formamide

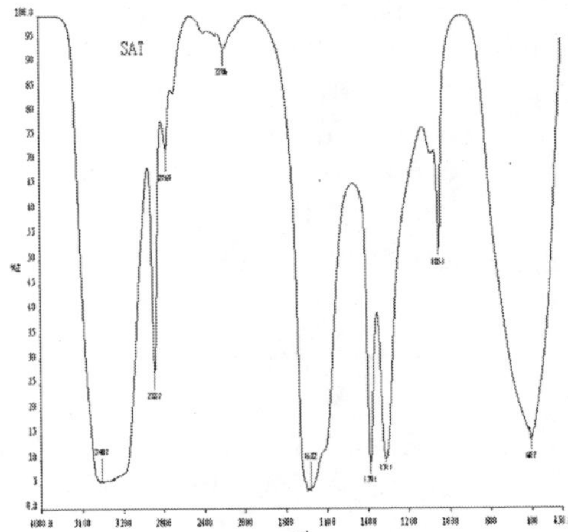

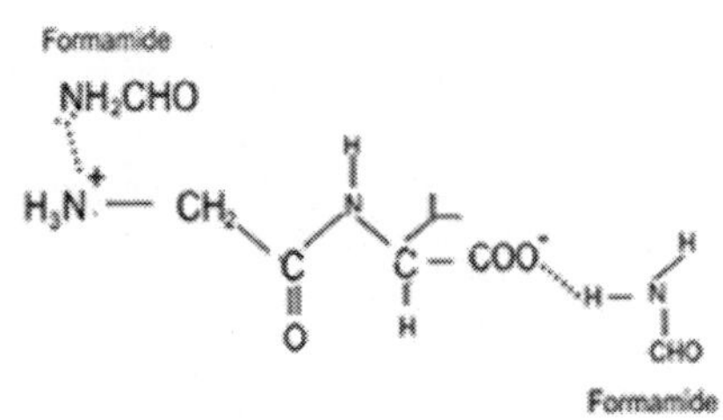

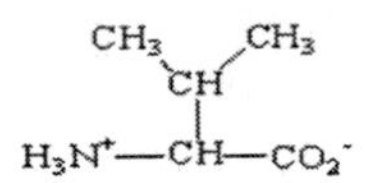

Glycyl-L-Valine Fig (a)

Glycyl-L-Valine in Zwitterionic structure

Fig. 4

In this dipeptide the amide II bands are well separated at 3384 and 3251 cm^{-1}. The amide I band at 1688 cm^{-1} is also sharp. The amide III band at 1339 cm^{-1} is quite sharp. Hence this compound has a linear structure. It is readily solvated by formamide as evinced by the pronounced shift 92 cm^{-1} of the band at 2114 cm^{-1} in the solid state to 2206 cm^{-1} in saturated solution. All the other peaks of the solute have merged with the solvent peak indicating strong solute-solvent interaction. In the polypeptides, the composition of this mode is more complex, since it depends on side chain structure and since NH bending contributes to several modes in the 1400 to 1200 cm^{-1} region. In spite of side chain contributions to the amide III mode, this mode can be used for secondary structure prediction.[7] The solvated structure Figure 4 is aligned for the dipeptide-formamide interaction.

CONCLUSION

Spectroscopic techniques are inevitable, indispensable, indisputable and authentic tools of investigation in any field of research. The IR study of Glycyl-L-Valine in formamide confirms the structure stability of the dipeptide Glycyl-LValine in the polar solvent. Ultrasonic study gives valuable information regarding the molecular interactions in solutions. In the present work Internal pressure, Free volume, equivalent conductance and spectral study have been carried out for the Glycyl-L-Valine in formamide and the behavior of the solute in the solvent reveals as a structure maker. The strong solute-solvent interactions observed in the non-aqueous dipeptide solutions in the present investigation, is based on thermodynamical and thermo-chemical analysis which are in good agreement with spectroscopic study also.

REFERENCES

[1] Suryanarayana C.V. and Venkatesan V.K. Acta Chem. Hung. 19, (1959), 441–451.

[2] Pierce J.E., J. Acou. Soc. Ame. (USA) 33, (1961) 198.

[3] Beulah Mary A., Jesintha Rakini J.H. and Bharathi G., Journal of Acoustical Society of India Col 36, No. 2, April 2009.

[4] Kalyansundaram S., Manuel Stephan A., Gopalan A., Polym J., Mater R. (1995), 323–333.

[5] Jasmine Vasantha Rani E., Padmavathy R., *et al.*, *J. Acoustic. Soc. India*, Vol. XXX NSA (2002), 23–259.

[6] John E. Mc Grady, Michael P. and Mingos D., J.Chem. Soc. Perkin Trans., 2 (1995), 228.

[7] Barth/Biochemica et Biophysica Acta, 1767 (2007), 1073–1101.

Solvation Number, Thermochemical and Infrared Spectroscopic Analysis of a Sulfa Drug Using Ultrasonic Velocity

R. Padmavathy, E. Jasmine Vasantha Rani, N. Radha[1],
S. Santhakumari[2] and Thamarai Selvi

Department of Physics, Seethalakshmi Ramaswamy College, Tiruchirappalli, Tamil Nadu
[1]Department of Chemistry, Seethalakshmi Ramaswamy College, Tiruchirappalli, Tamil Nadu
[2]Department of Physics, Srimathi Indira Gandhi College, Tiruchirappalli, Tamil Nadu

ABSTRACT: *The infrared spectroscopic technique is based upon the simple fact a chemical substance shows marked selective absorption in the infrared region. Measurement of ultrasonic velocity in non-aqueous electrolytic solution gives information about the behavior of solution such as molecular association and disassociation. The attraction and repulsion between the molecules of the components involved show considerable effect upon the physical and chemical properties of a solution such as density, viscosity and ultrasonic velocity. Benzene sulphonamide is one of the sulfa drug used in the treatment of gastrointestinal, duodenayl ulcer, neurological disorder are analyzed for the solution of the sample, in formamide. Adiabatic compressibility, solvation number, appetent molal volume and apparent molal compressibility are computed for Benzene sulphonamide in non-aqueous solution. The results obtained by acoustic and IR study are found to be in good agreement is envisaged.*

INTRODUCTION

IR spectrum is an important record which gives sufficient information about the structure of a compound. The compressibility behavior of solutions provides very useful information related to intermolecular interactions. β, Sn, φ_v, φ_k are computed using the fundamental quantities density, viscosity and ultrasonic velocity.

EXPERIMENTAL SECTION

Ultrasonic velocity was measured with an ultrasonic interferometer at 2 MHz. Accuracy in the velocity measurement was $\pm$ 0.5 ms^{-1}. Density was measured by 25 ml specific gravity bottle. The weighing measurement was carried out by a digital balance with an accuracy $\pm$ 0.0001 gm/cc. Viscosity was measured with Cannon Fenske type viscometer with an accuracy ±0.5%. The FTIR spectrum of this solution is recorded in the region of 4000–400 cm^{-1} using (PERKIN ELMER) model.

COMPUTATION

1. Adiabatic compressibility (β) = $[1/u^2\rho]$ cm^2/dyne
2. Solvation number (Sn) = $n_1/n_2\,[1-\beta/\beta_o]$
3. Apparent molal volume $\varphi_v = [1000/c1\rho_o]\,(\rho_o\text{-}\rho) + (M_l/\rho_0)$ ml/mol
4. Apparent molal compressibility

   $\varphi_k = 1000/m\rho_o\,(\rho_o\beta\text{-}\rho\beta_o) + [\beta_o M/\rho_o]$ (ml/mol.cm^2/dyne)

RESULTS AND DISCUSSION

The velocity of propagation of ultrasonic waves in a solution is sensitive to the thermodynamic fluctuation of pressure, volume and temperature. In the present work, the adiabatic compressibility is found to increase with increasing temperature and concentration for benzene sulphonamide. When the temperature and concentration increase the associated groups of molecules breaker down increasingly and the forces of attraction between the molecules decrease. This leads to an increase in the adiabatic compressibility of the system [1]. This indicates the pre-dominance of association of molecules occurring in the solution. The β increase with increasing concentration happen due to the collection of solvent molecules around molecules around ions supporting weak ion solvent interactions.[2]

The solvation numbers reflect the dynamic situation of the ion as its moves around in the solution.[3] The sign and magnitude of compressibility indicates the Solvation effects. Positive Solvation number of solutions suggests that the compressibility of the solution at high temperature and at low molalities will be less than that of the solvent. The Solvation approach is used to interpret ion-solvent interaction.[4] The ion-solvent interaction may be due to equal to intermolecular interaction energy[5] which results in zero values of solvation number. Negative Solvation number emphasizes that solution are more compressible than solvent. Negative Solvation number for various molalities is reported in literature [6]. The decrease in Solvation number with increasing molality is due to either not enough solvent molecules available for all the ions or preferentially ion- pairing occurred.[7]

The variation of apparent molal volume is noticeable only at lower concentration but it remains constant at higher concentration. It is found that apparent molal volume is negative at lower molalities and it decrease as molality increases, at all temperatures. The decrease in apparent molal volume with an increase in molality suggests that the strong ion-ion interaction[8] taking place in the solution.

The higher values of φ_v suggest that there is strong solute- solvent interaction taking place in the solution. The lower values of φ_v support the strong ion-ion interactions.[9,10] The negative values of φ_k support that the sulfa drug are having structural influence on the solvent. The negative vales of φ_k support weak solute-solvent interactions.[11] The values of φ_k are found to increase with increasing concentration and decrease with increase in temperature for both the systems.

INTERPRETATION OF SPECTRAS

The structural analysis of the solution, by means of radial pair distribution function, showed that FMA can form upto three hydrogen bonds with water.[12] From the IR spectrum of Formamide, a strong N-H stretching absorption occurs at 3143 cm^{-1}. $>C=O$ Stretching absorption known as amide I band is observed at 1690 cm^{-1}. Well defined peaks are observed at 1391, 1314 cm^{-1} these confirms the presence of C-N group.[13]

The NH_2 stretching vibration at 3350 cm^{-1} and 3256 cm^{-1} are of almost equal intensity in the solid state. On dissolution in formamide a red shift at 3420 cm^{-1} is found besides broadening of the band. This large shift and broadening are due to very strong intermolecular hydrogen bonding. In the finger print region 1400–800 cm^{-1} also all the solute vibration are completely replaced by vibration of the solvent formamide there by indicating strong solvent-solute interactions (Figure 5).

Table 1: $\beta \times 10^{-11}$ (cm^2/dynes)

Molality (m)	*5°C*	*15°C*	*25°C*	*35°C*	*45°C*	*55°C*
0.001	3.23	3.29	3.35	3.41	3.46	3.54
0.005	3.26	3.33	3.38	3.44	3.50	3.57
0.01	3.28	3.37	3.43	3.48	3.54	3.60
0.015	3.32	3.40	3.47	3.53	3.57	3.62
0.02	3.36	3.44	3.50	3.56	3.62	3.66

Table 2: Sn

Molality (m)	*5°C*	*15°C*	*25°C*	*35°C*	*45°C*	*55°C*
0.001	–3445	–2688	–2336	–972	–5336	–3570
0.005	–568.7	–128.4	–296.9	226.9	–1007	–539
0.01	–223.1	13.39	–92.52	–73.5	–425.4	–341.6
0.015	–92.00	–167.1	–69.94	109.4	–270.3	–138.9
0.02	80.65	32.49	–32.43	29.92	–147.3	–101.5

Table 3: φ_v (ml/mol)

Molality (m)	*5°C*	*15°C*	*25°C*	*35°C*	*45°C*	*55°C*
0.001	102	353	425	936	1440	1657
0.005	–20	13	46	143	239	286
0.01	–21	–15	–7.2	51	99	128
0.015	–31	–25	–21.8	13	52	77
0.02	–36	–33	–26.7	1.61	27	47

Table 4: φ_k (ml/mol.cm^2/dyne)

Molality (m)	*5°C*	*15°C*	*25°C*	*35°C*	*45°C*	*55°C*
0.001	–28.8	–71.1	–83.91	–176.9	–294.4	–338.5
0.005	1.73	–2.80	–9.36	–25.85	–49.38	–58.32
0.01	3.02	2.82	0.97	–9.82	–20.61	–26.43
0.015	5.09	3.95	3.67	–2.07	–10.96	–15.69
0.02	6.53	5.95	4.67	–0.19	–5.76	–9.75

Table 5: Infrared Frequencies of Benzene Sulphonamide

Name of the Sample	γ_{NH}	γ_{CH}	$\gamma_{C\text{-}N}$	$\gamma_{C=O}$	$\gamma_{C=N}$	$\gamma_{C\text{-}O}$	δ_{CH}
Formamide	3413	2889	2770	1690	1391 1314	1051	604
Formamide+Benzene Sulphonamide	3418	2889	2770	1683	1391 1315	1051	608

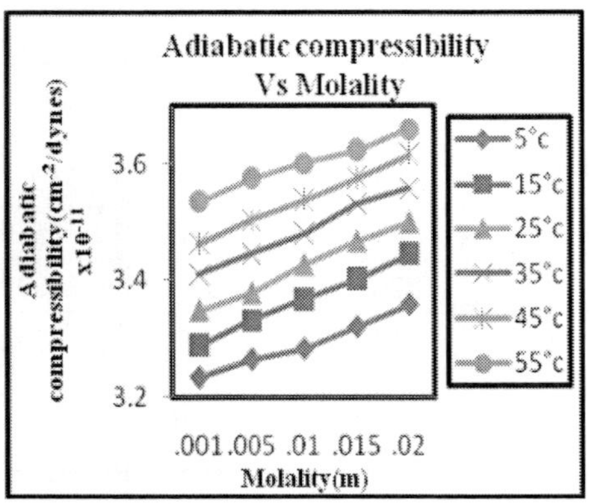

Fig. 1: β for Benzene Sulphonamide

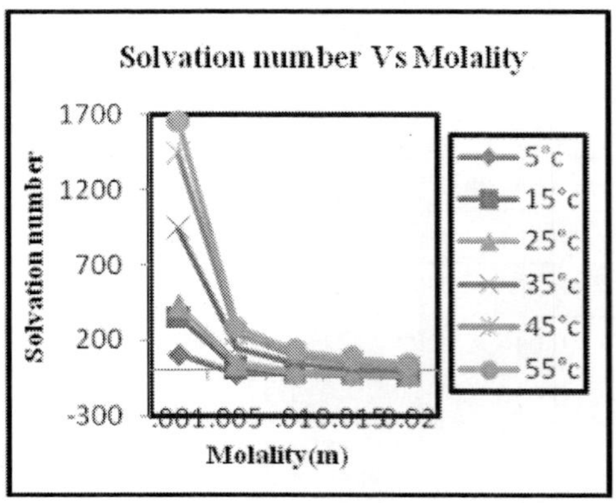

Fig. 2: Sn for Benzene Sulphonamide

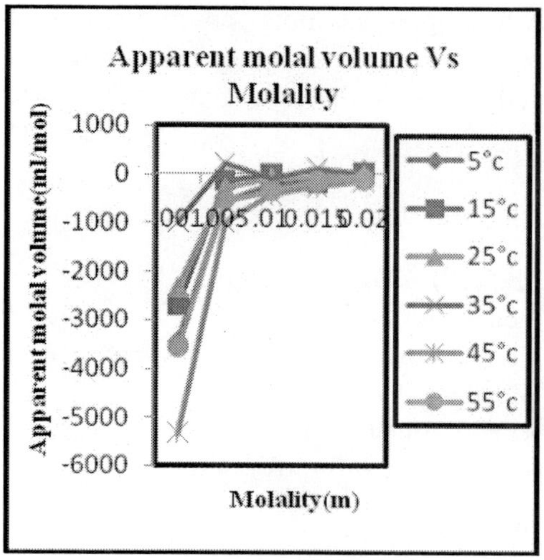

Fig. 3: φ_v for Benzene Sulphonamide

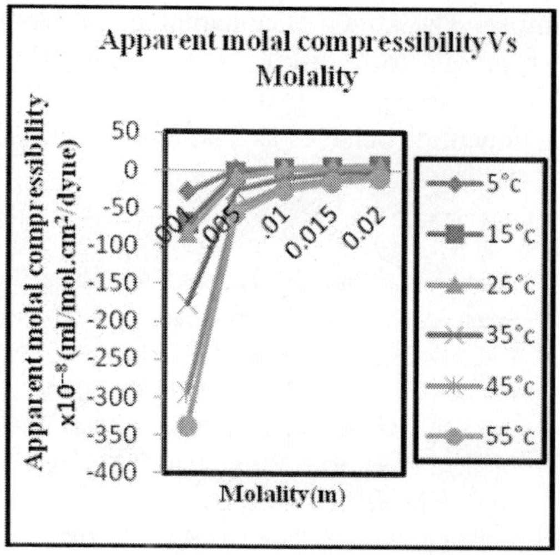

Fig. 4: φ_k for Benzene Sulphonamide

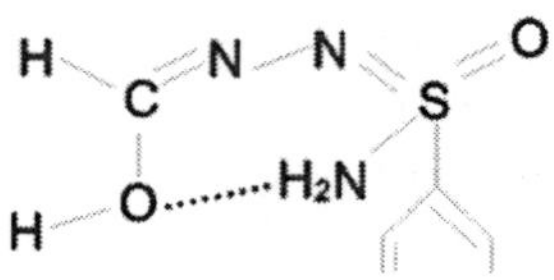

FTIF Spectrum—Formamide

FTIF Spectrum—Formamide + Solute

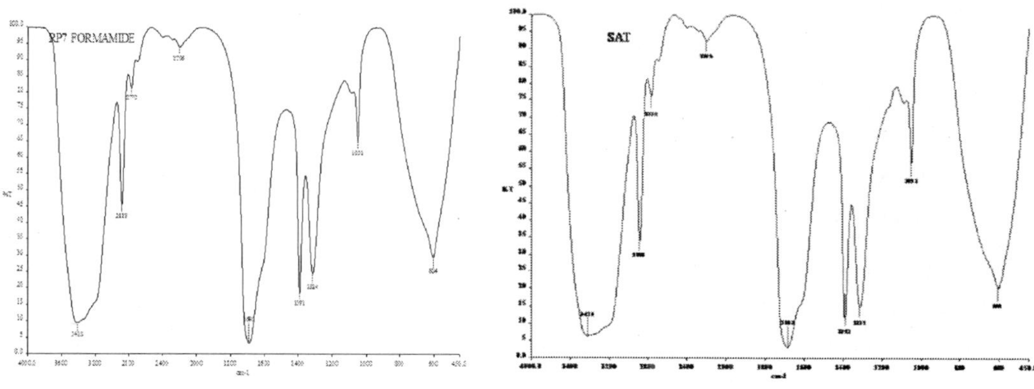

Fig. 5

CONCLUSION

In the present work, a detailed analysis of IR, $\varphi_{v,}$ $\varphi_{k,}$ has been carried out. Intermolecular interaction that is solute-solvent interaction is discussed, based on the values of Solvation number. The study of apparent molal volume and apparent molal compressibility has been proven to be useful in explaining the structural interactions occurring in solution. Both spectroscopic analysis and ultrasonic study reveals that the given salt Benzene sulphonamide behaves as a structure maker.

REFERENCES

[1] Bhullar, K., Bhavneet, K., *et al.* Acoustica, 73, (1991) 291–293.

[2] Sonar, A.N. and Pawar, N.S., *Rasayan J. Chem,* Vol. 3, No. 1, (2010) 38–43.

[3] Padmavathy, R., Jasmine Vasantha Rani, E. and Seethalakshmi, K., *Journal of Acoustic Society of India*, Vol. 38, No. 3, 2011 (140–143).

[4] Padova, J, Isr A. Energy Comm., (1963) 1A-823 (16pp), A-830 (23pp).

[5] Modern Electrochemistry, John O.M., Bockris and Amulya K.N Reedy, A.plenum/rosetta Edition Vol. 1 (1977).

[6] Rajkotia, *et al., J .Pure and Appl. Ultrasonic's*, 21 (1999) 132–135.

[7] Kuppuswami, J., Lakshmanan, A.S., *et al.*, Bull. Chem. Soc. Japan, 38, (1965) 1610.

[8] Sathyavathi, A.V., Acoustica, 28 (1973) 177. . .

[9] Balachandran, Nature, 191 (1961) 164. .

[10] Dhanalakshmi, A., *et al., J. Acous. Soc Ind.,* Vol. 23, 1–4 (2000), 385–388.

[11] Patil, K.J. and Ali, S.T., *Indian Joural of Chem*, Vol. 22A, (1983), 410–412.

[12] Rocha, W.R., Martin, V.M., Coutinho, K., *et al., Theor. Chem. Acc* (2002), 108, 31–37.

[13] Jasmine Vasantha Rani.E, Padmavathy, R., Radha, Kannagi, K. and Lavanya, N., *Journal of Acoustic Society of India*, Vol. 38, No. 2, 2011 (87–91).

Thermodynamic and Spectroscopic Study on Nitro Derivative of Amino Acid Salt Solution Using Ultrasonic Velocity

E. Jasmine Vasantha Rani, R. Padmavathi, N. Radha[1], K. Kannagi[2] and J. Krishnaveni

Department of Physics, SR College, Trichy

[1]Department of Chemistry, SR College, Trichy

[2]Department of Physics, Bharathidasan University Constituent College, Lalgudi

ABSTRACT: *Ultrasonic method is a powerful probe for characterizing the physico chemical properties and the existence of molecular interaction. The measurement of internal pressure is an important study of the thermodynamic properties of liquids. The internal pressure is used in estimation the cohesive forces and also gives an idea of the solubility characteristics. Free volume is one of the significant factors in explaining the variation in the properties of liquids/solution. Infra-red spectrum is an important record which gives sufficient information about the structure of a compound. In the present study, internal pressure and free volume are computed for Nω-Nitro-L Arginine at different concentrations and at different temperatures. This sample has been used in research for studying about coronary constrictor. These parameters have been utilized to study the solute-solvent interaction and to analyze the structure of the sample. The results obtained from acoustic study have been corroborated with the spectroscopic study.*

INTRODUCTION

Ultrasonic investigations of liquid/solution having of polar and nonpolar components are of considerable importance in understanding intermolecular interactions between the component molecules and find applications in several medicine and textile processes [1]. The density, viscosity and speed of sound provide important information about molecular packing, molecular interaction and their strength influenced by the size, shape and the chemical nature of component molecules [2,3,4]. Infrared spectroscopy has established itself as a valuable tool for the determination of organic compounds. In the present study Nω-Nitro-L Arginine is taken as a solute and formamide as solvent. It is an inhibitor of nitrix oxide synthase and hence a vaso-constrictor and coronary constrictor. As such, it finds widespread use as a biochemical tool in the study of nitric oxide and its biological effects [5].

EXPERIMENTAL TECHNIQUES

The density of the solution is measured using 25 ml specific gravity bottle with an accuracy of .0001 gm/cc. Viscosity of a liquid is determined using cannon Fenske viscometer. Ultrasonic velocity is measured with help of variable path interferometer [F-81, mittal enterprizes, New Delhi (2 MHZ)].

The IR spectrum of these solutions was recorded in the region of 4000–400 cm^{-1} using (PERKIN ELMER) model SPECTRUM RXI FTIR spectrometer in SAIF, IIT- CHENNAI.

FORMULA

Internal pressure and free volume are computed using the following formulae:

$$\Pi_i = b\,RT\,(k\,\eta/u)^{1/2}\,\rho^{2/3}\,m^{7/6}\ \text{atms}$$

$$V_f = [M_{eff}u/K\,\eta]^{3/2}\ \text{cc}$$

where, T-absolute temperature, η-viscosity, u-ulrasonic velocity.

RESULTS AND DISCUSSION

Internal pressure is a single factor which varies due to all the internal interaction. The variation of internal pressure and free volume with temperatures and molalities are shown in Figurs 1&2 and the values are tabulated in Tables 1&2.

In Nω-Nitro-L Arginine solution, internal pressure is found to be increasing with respect to increasing molality. Increase in internal pressure suggests that there is increase in cohesive energy due to strong solute-solvent interaction [6]. But, at higher temperature, when the temperature is increased there is a tendency for ions to move away from each other, hence there is a reduction in internal pressure[7]. This behavior indicates solute is acting as structure maker. It is also confirmed by decreasing values of free volume at all concentrations as expected [7]. The present work supports that structure of the solvent is enhanced appreciably in the solution. A systematic quantitative study of the data shows that at a given concentration of any solution, the variation of Π_i with V_f due to a change in temperature can be represented by the equation.

$$\Pi_i(V_f)^{\alpha} = K$$

A plot between the values of log Π_i and $\log(1/V_f)$ results in a family of parallel straight lines as shown in Figure 3 with a slope of nearly 0.270. This shows that in this solution the relation $\Pi_i(V_f)^{0.270} = K$, holds good as in the case of other electrolytes [9]. The above relation has a remarkable signification in its generality of application. In the case of gaseous system, pressure, free volume and temperatures are the important thermodynamic parameters describing the system. Since the equation of $\Pi_i(V_f)^{\alpha}$ = constant, similar to the gas equation, is obeyed here, $\Pi_{i,}V_f$, temperature turn out to be the important thermodynamic parameters of a liquid medium.

Formamide contains a hydrogen atom attach to the CO group, which is acidic in nature. This proton is found down field in NMR spectra at $\sim\delta \approx 8_{ppm}$. The NH_2 group is found at $\delta \approx 6_{ppm}$, indicating that all the protons can be involved in hydrogen bonding [9]. In the FTIR spectrum the NH groups absorbs at 3417 cm^{-1}, a small amount of enolic [10] form of formamide is also evident from that IR spectra of the solvent (Figures 1 and 2). The carbonyl group is found at 1694 cm^{-1} due to the amide 1 stretching vibration. This is the most important signature peak of a carbonyl group indicating hydrogen bonding patterns in formamide.

Normally the carbonyl group stretching vibrations is found at 1720 cm^{-1} and any shift towards lower frequency indicates bond lengthening[11], arising due to hydrogen bonding between the solvent molecule as well as solute and solvent molecules.

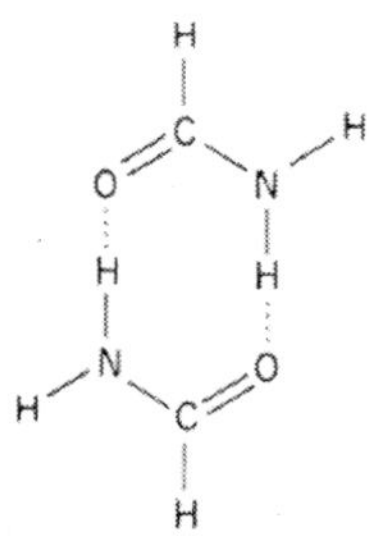

Fig. 1: Dimer Form

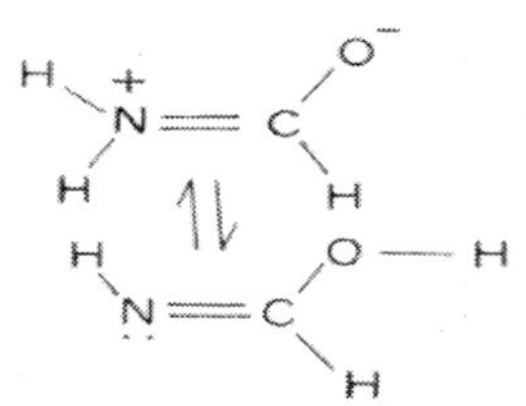

Fig. 2: Enolic Form

In the IR spectrum (Table 3), the primary NH_2 band of arginine is observed at 3416 cm^{-1} in the solution and formamide. A series of absorptions at 2205–2888 cm^{-1} indicate the presence of protonated guanidino group. The NO_2 asymmetric and symmetric stretching vibration are observed at 1314 and 1391 cm^{-1}. There is a blue shift of 3416 cm^{-1} band to 3401 cm^{-1} for NH_2 stretching mode. The NH_2 bending vibration at 1684 cm^{-1} is not affected. The δ_{ch} vibrations affected (8 cm^{-1} to 40 cm^{-1}) on dilution in formamide. This shift supports that there is a strong binding between the solute and solvent (Structures 1 & 2).

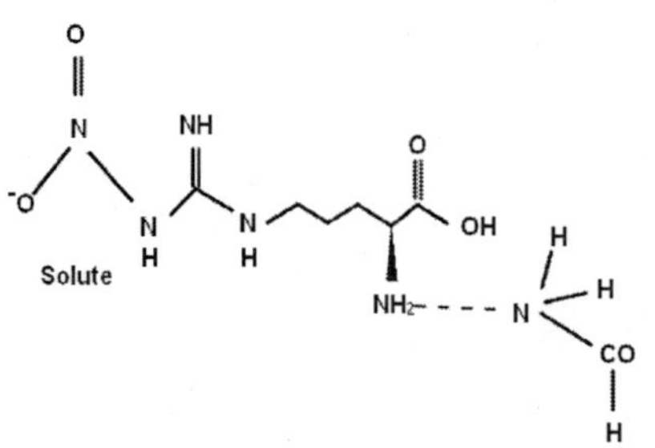

Structure 1

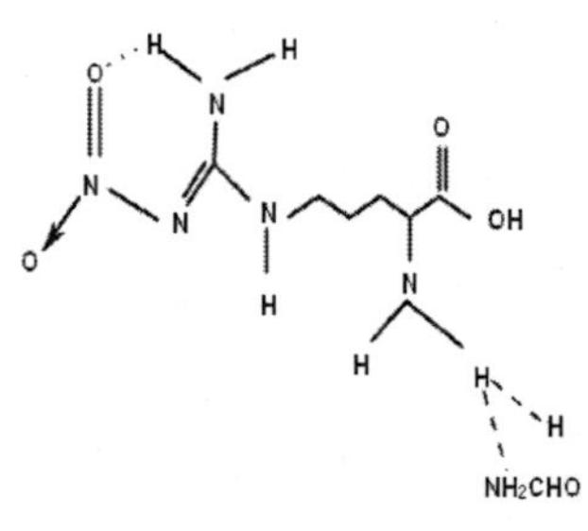

Structure 2

Table 1: Internal Pressure (atms)

Molality (m)	*5°C*	*15°C*	*25°C*	*35°C*	*45°C*	*55°C*
0.001	19641	16658	14255	13015	11767	10757
0.005	19849	16982	14403	13211	11853	10902
0.01	20078	17052	14497	13342	11993	10997
0.015	20163	17338	14633	13401	12083	11056
0.02	20519	17465	14687	13638	12152	11235

Table 2: Free Volume (cc)

Molality (m)	*5°C*	*15°C*	*25°C*	*35°C*	*45°C*	*55°C*
0.001	0.0081	0.0148	0.0256	0.0365	0.0535	0.0759
0.005	0.0079	0.0139	0.0248	0.0348	0.0524	0.0728
0.01	0.0076	0.0137	0.0243	0.0338	0.0505	0.0710
0.015	0.0075	0.0130	0.0236	0.0333	0.0493	0.0696
0.02	0.0071	0.0127	0.0233	0.0316	0.0484	0.0665

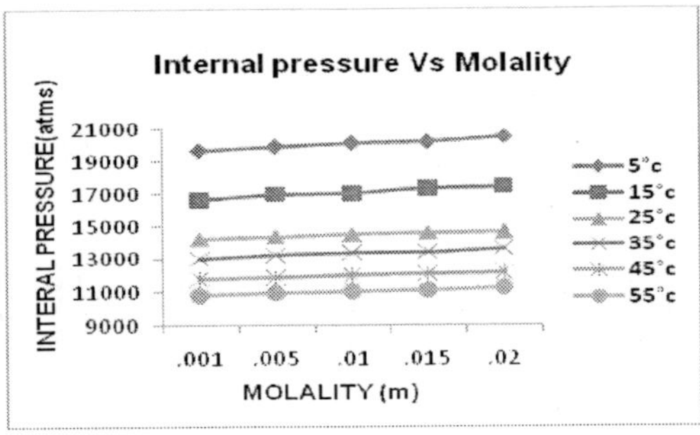

Fig. 3

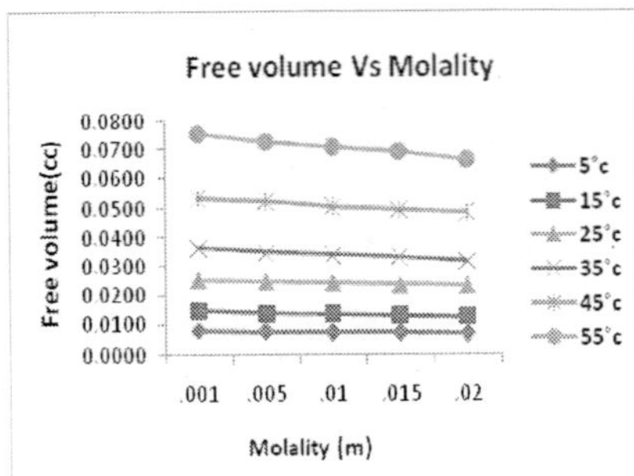

Fig. 4

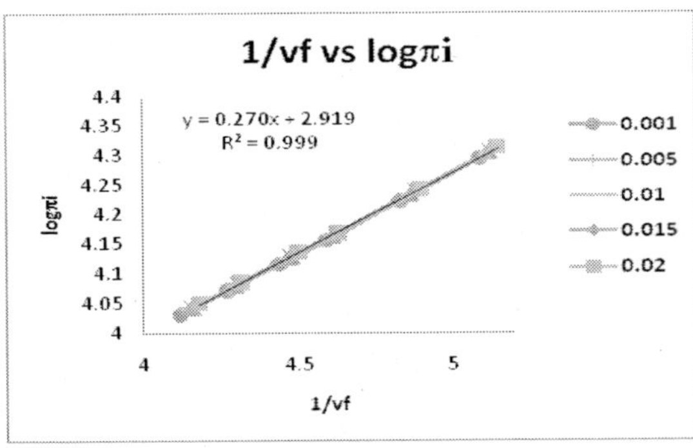

Fig. 5

Table 3: Signature Peaks of Solvent, L-Arginine and Nω-Nitro-l-Arginine Solution

Wavenumber cm^{-1}	γ_{NH}	γ_{CH}	$\gamma_{C\text{-}N}$	$\gamma_{C=O}$	*Amide II Band*	$\gamma_{C\text{-}O}$	δ_{CH}
Formamide	3417	2889	2774	1694	1392 1312	1051	610
sample	3416	2888	2771 2205	1694	1391 1314	1051	617
0.005 molality	3409	2890	2773	1694	1392 1311	1051	618
0.01 molality	3401	2885	2770 2201	1694	1392 1311	1051	651

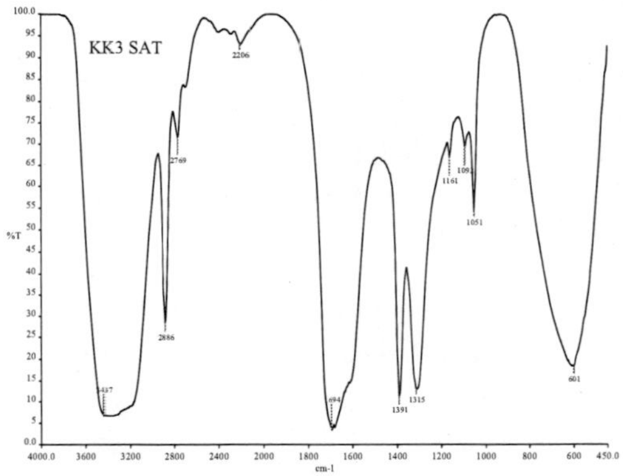

Fig. 6

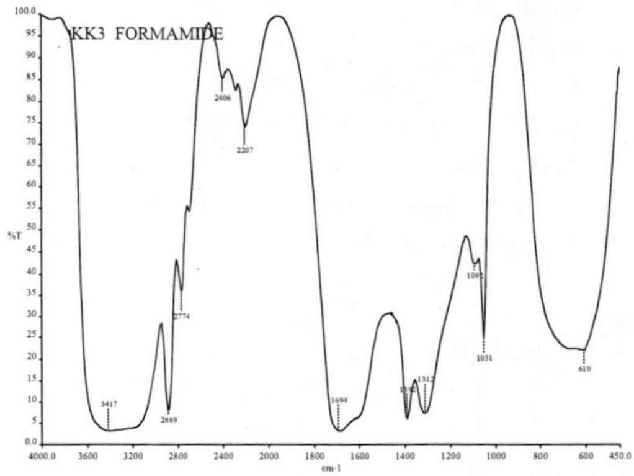

Fig. 7

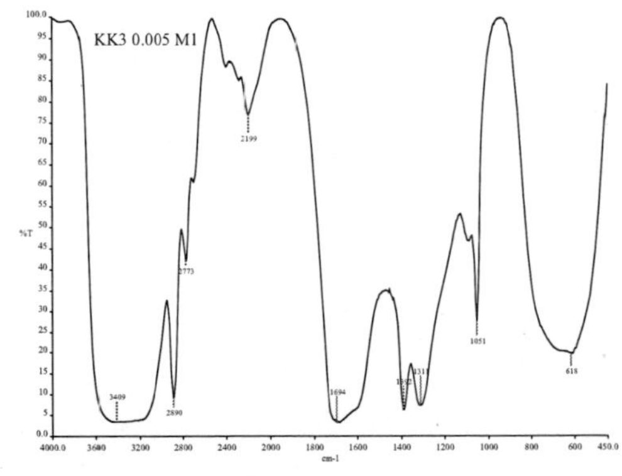

Fig. 8

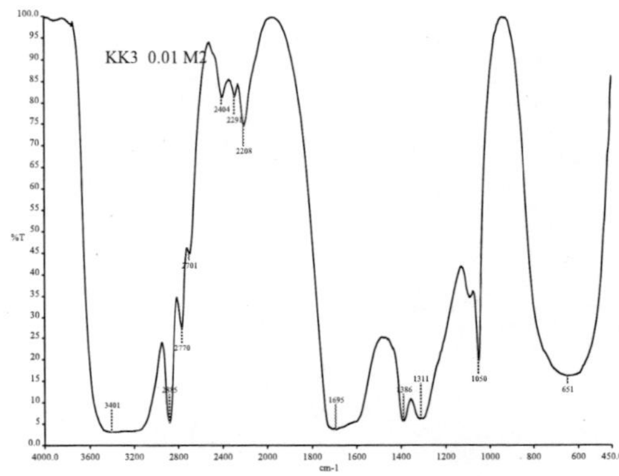

Fig. 9

CONCLUSION

Acoustic study on Nω-Nitro-L-Arginine solution predicts the strong solute-solvent interaction predominating in the solution. This behavior is also confirmed from the IR spectral study.

REFERENCES

[1] Amalandu pal and Gurcharan, 1990, syal *et al.* 1996, thiyagarajan and palniappan (2008).

[2] Reid, C.R., Prausnitz J M and Poling B E, MG Graw Hill, New York, 4th Edn. (1987).

[3] Roy, M.N. and Hazra, D.K., North Bengal University Rev. (1997), 8, 54.

[4] Aminabhavi, T.M., Patil, V.B., Avalaguppi, M.I., Oretego J.D. and Hansen, K.C., *J Chem Engg. Data* (1996), 41, 526.

[5] Bansinath M; Arbabha B; Turndorf H; Garg U C. *Neuro chemical research* 18 (10): 1063–6 (1993).

[6] G. Ravichandran, A. Srinivasa Rao and T.K. Nambinarayanan, Indian journal of pure and applied physics 32 (1994).

[7] Jasmine vasantha rani, E. Suhashini Ernest, R. Padmavathy and N. Radha (2005). An investigation of solvation number, Electro chemical and Thermodynamical properties of calcium lactate solution using ultrosonic velocity.

[8] Schafer M., Pdfer Nc, O Omens J., Blunk D., R.A. In proceedings of the 54th conference on mass spectrometry and Allied topics seattle WA (2006).

[9] Pillai, S.O. and Natarajan, S., *Ind J. Chem.,* Vol. 27–A(1988), 232–236.

[10] Infrared Spectroscopy of Arginine Cation Complexes: Direct Observation of Gas-Phase Zwitterions, J. Physics. Chemistry. A 2007, 111, 11759–11770. Journal of Chemical and Pharmaceutical Research *J. Chem. Pharm. Res.,* (2011), 3(3): 587–596.

[11] Spectroscopic and thermodynamic analysis of some sodium salts in non-adequous medium. R. Padmavathy, E. Jasmine, K. Geetha, V. Solachi, N. Radha proceedings International Conference on Spectro Physics (INCONS 2005).

Solvation Effect and FTIR Spectral Analysis of L-NAME in Non-Aqueous Medium Using Ultrasonic Velocity

E. Jasmine Vasantha Rani, R. Padmavathi, N. Radha[1], K. Kannagi[2] and S. Vijayashree

Department of Physics, SR College, Trichy

[2]Department of Chemistry, SR College, Trichy

[1]Department of Physics, Bharathidasan University Constituent College, Lalgudi

ABSTRACT: *Infrared spectroscopy is one of the most powerful analytical techniques which offer the possibility of chemical identifications. Spectroscopy and acoustics are found to be the two important fields to study the molecular interactions in the liquid system. Solvation number (s_n) of solutions depends on the number of moles of the solutes and solvent molecules present in the solution. Compressibility measurements are highly accurate and yield interesting information regarding various molecular interactions occurring in the solution. Density (ρ), viscosity (η) and ultrasonic velocity (U) have been measured for Nω-Nitro-L-Arginine Methyl Ester Hydrochloride (L-NAME) from 5°C to 55°C at different concentrations. The sample is used to reduce the relaxation induced in carotid glands. The measured parameters are used to calculate adiabatic compressibility (β_{ad}) and solvation number (s_n) in the present study. The results obtain from solvation effect of the non aqueous solution are confirmed from the spectroscopic results.*

INTRODUCTION

There are many approaches and spectroscopic techniques used to determine the structure-function relationship of biomolecules [1]. The spectroscopic techniques and ultrasonic measurements have been found to be most powerful tool in the investigation of structure and molecular interaction occurring in the solutions. The sample (L-NAME) is widely used in medical field. It blocks the increase in renal blood flow [2]. L-NAME treatment suggesting a direct action on venous smooth muscle [3].

L-NAME was purchased from sigma-Aldrich chemicals [4] (AR Grade)and used without further purification. In the present investigation ultrasonic velocity, density and viscosity of L-NAME in formamide have been carried out. Adiabatic compressibility (β_{ad}) solvation number (s_n) and thermo chemical properties have been evaluated and discussed in terms of solute-solvent interactions [5, 6]. These results are compared with infrared spectroscopic study.

EXPERIMENTAL DETAILS

Non-aqueous solutions of L-NAME were prepared using formamide as a solvent. Variable path of fixed frequency 2 MH_Z was used to measure the ultrasonic velocity of the solution. A 25 ml specific gravity bottle was used to measure the density of the solutions with an accuracy of .0001 gm/cc. The measurements were made from 5°C to 55°C at different concentrations. Infrared absorption spectra were measured using FT-IR spectrometer equipped with a KBr beam splitter. The FTIR spectrum of this solution was recorded in the region of 4000–400 cm^{-1} using (PERKIN ELMER) model SPECTRUM RXI FTIR spectrometer in SAIF, IIT- CHENNAI.

FORMULA

In the present study adiabatic compressibility (β_{ad}), apparent molal volume (φ_v), apparent molar compressibility φ_k and solvation number (s_n) have been calculated from 5°C to 55°C using the formulae, Adiabatic compressibility (βad) = $\frac{1}{U^2\rho_0}$

The apparent molar volume (φ_V) = $\frac{1000}{(c_1\ \rho_0\ )(\rho_0 - \rho)} + \frac{M}{\rho_0}$

Apparent molar compressibility φ_k = $\frac{1000}{(c_1\ \rho_0\ )(\rho_0\beta - \beta_0\rho)} + \frac{M}{\rho_0}$

Solvation number (Sn) = $\frac{n_1}{n_2\left(1 - \frac{\beta}{\beta 0}\right)}$

where U-ultrasonic velocity, ρ & ρ_0-densities of the solute and solvent, β_0 and β-adiabatic compressibility of the solvent and solution, n_1 and n_2—the number of moles of solvent and solute.

RESULTS AND DISCUSSION

Adiabatic compressibility, apparent molal volume, apparent molal compressibility and solvation number are tabulated in Tables 1–4 and the variations are shown in Figures 1–4. The decrease in adiabatic compressibility values with rise in temperature and also increase in β with respect to concentrations may be explained in terms of the changes occurred in solution structure around zwitterions and ions (structure 1). This may be due to strong solute interaction [7]. At 25°C L-NAME exhibits negative solvation number at room temperatures. At low temperature the solvation number is positive. The solvation number decreases with increasing concentration at all temperature ranges.

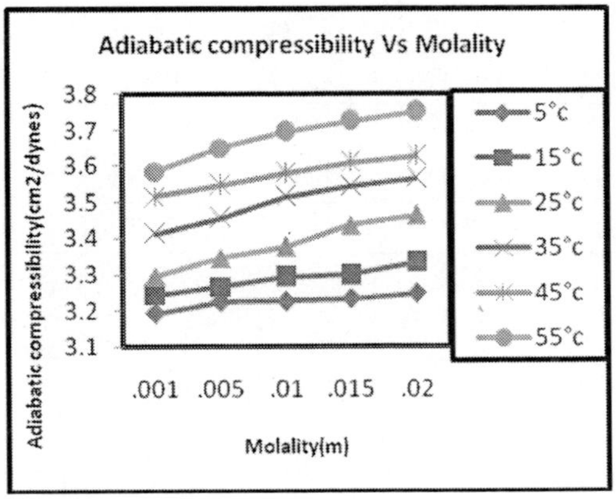

Fig. 1

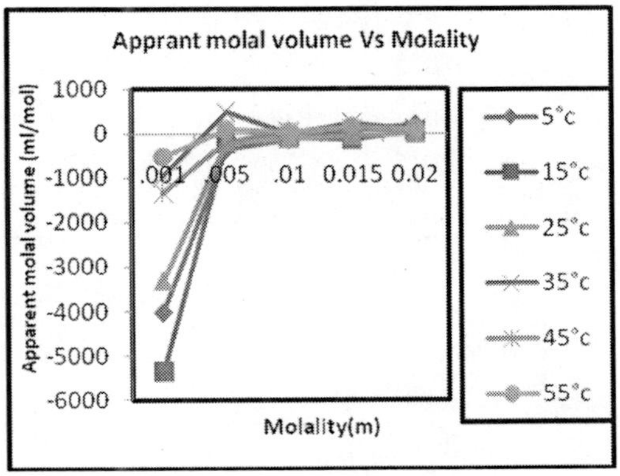

Fig. 2

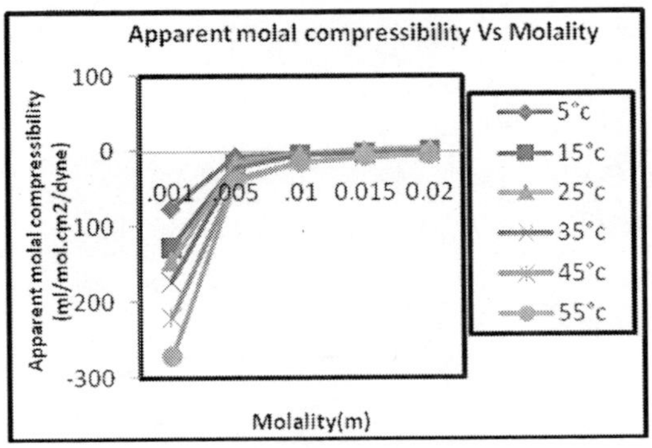

Fig. 3

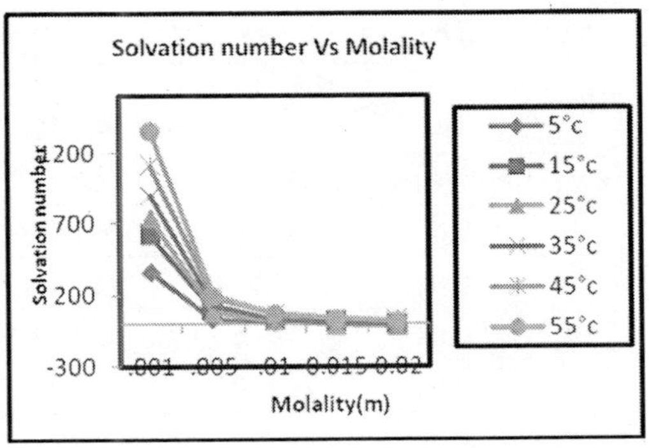

Fig. 4

Table 1: Adiabatic Compressibility × 10^{-11} (cm^2/dyne)

Molality (m)	*5°C*	*15°C*	*25°C*	*35°C*	*45°C*	*55°C*
0.001	3.19	3.24	3.29	3.41	3.52	3.59
0.005	3.22	3.27	3.35	3.46	3.55	3.65
0.01	3.23	3.29	3.38	3.51	3.58	3.69
0.015	3.23	3.3	3.44	3.54	3.61	3.72
0.02	3.25	3.33	3.46	3.57	3.63	3.75

Table 2: Apparent Molal Volume (ml/mol)

Molality (m)	*5°C*	*15°C*	*25°C*	*35°C*	*45°C*	*55°C*
0.001	–4023.84	–5357.8	–3323.03	–974.59	–1346.8	–529.97
0.005	–376.19	–217.38	–184.24	486.42	–213.01	93.27
0.01	–158.11	–60.05	–101.55	2.62	68.45	30.1
0.015	62.45	–128.19	–36.92	216.93	74.42	157.95
0.02	194.81	71.93	13.08	97.92	87.7	98.59

Table 3: Apparent Molal Compressibility × 10^{-8} (ml/mol.cm^2/dyne)

Molality (m)	*5°C*	*15°C*	*25°C*	*35°C*	*45°C*	*55°C*
0.001	–76.37	–128.63	–145.89	–172.08	–220.14	–272.28
0.005	–6.8	–17.86	–15.9	–22.55	–37	–39.55
0.01	–3.07	–5.59	–4.51	–5.54	–14.12	–14.85
0.015	–1.03	–3.69	1.5	–0.72	–7.29	–7.16
0.02	0.5	–0.21	2.86	0.43	–4.27	–4.07

Table 4: Solvation Number

Molality (m)	*5°C*	*15°C*	*25°C*	*35°C*	*45°C*	*55°C*
0.001	363.67	623.11	744.54	898.53	1110.08	1357.82
0.005	32	96	84.33	129.01	186.74	200.29
0.01	14.68	30.17	22.99	29.53	74.02	75.11
0.015	7.03	18.26	–8.99	7.91	38.9	38.86
0.02	0.76	2.49	–15.54	–0.44	23.62	22.25

Table 5: Signature Peaks of Solvent, L–NAME and it's Solution

Wavenumber cm^{-1}	*γNH*	*γCH*	*γC-N*	*γC=O*	*Amide II Band*	*γC-O*	*δCH*
Formamide	3422	2889	2771	1685	1390, 1313	1051	605
sample	3409	2888	2771 2208	1684	1391 1314	1051	610
0.005	3433	2891	2773 2208	1683	1391 1312	1052	608
0.01	3405	2886	2770 2207	1694	1391 1311	1050	609
0.015	3409	2888	2772	1684	1314	1092	618

At low molalities φ_v exhibits negative values. However at .02 m φ_V is positive at all temperature values. Above room temperature it shows positive values. This behavior suggests that at low concentration there is strong solute-solvent interaction occurring in the solution. Negative values of apparent molal compressibility also revels the fact that the apparent molal compressibility of the solution is higher than that of solvent there is a possibility of Hydrogen bond formation in the solution [8]. It is also confirmed from the IR spectral study.

In the FTIR spectra of the solid sample (structure 2), the primary NH_2 vibration is observed at 3409 cm^{-1}. A series of vibration in the region 2208–2888 cm^{-1} assigned to the NH_3^+ in the molecules. The NO_2 asymmetric and symmetric vibrations are observed at 1684 and 1314 cm^{-1} and NH_2 bending vibration is found at 1391 cm^{-1}. The δ_{OH} bending vibration occurs at 610 cm^{-1} as a sharp peak. On dilution the γNH_2 band is shifted from 3409 cm^{-1} in the solid to 3433 cm^{-1} (0.005 m) in formamide solution. This red shift indicates the H-bonding interaction between NH_2 of the solute and C=O of formamide. It is evident by a 10 cm^{-1} shift of $\gamma c = o$ at 0.01 m molality. The NO_2 asymmetric and symmetric stretching vibrations at 1314 and 1319 cm^{-1} are not shifted much on diution indicating presence of a intramolecular (structure 3&4) H-bonded structure. The spectral analysis and the dilution effect indicates the presence of H-bonding between >C=O of formamide and primary NH_2 of the aminoacid.

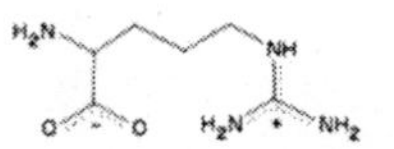

structure 1

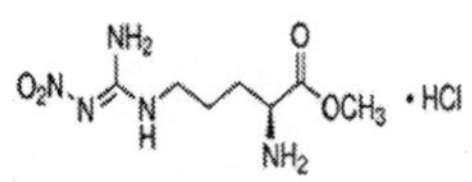

structure 2

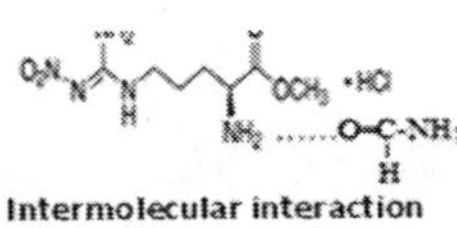

structure 3

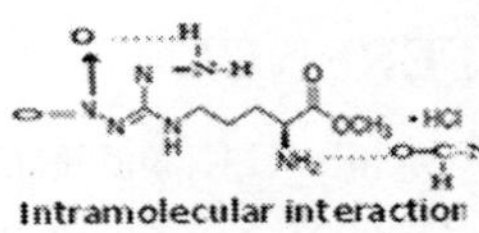

structure 4

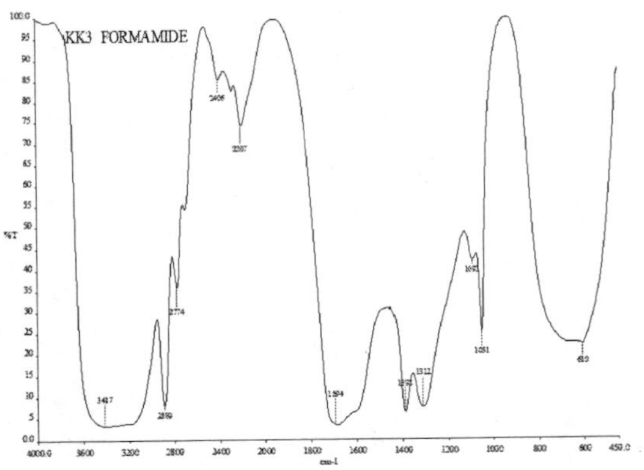

Fig. 5

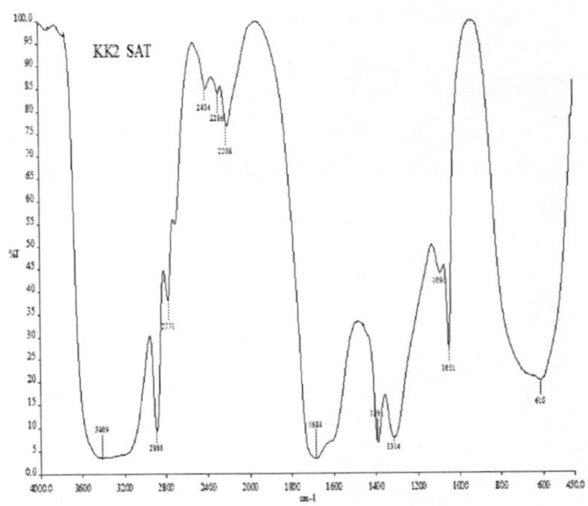

Fig. 6

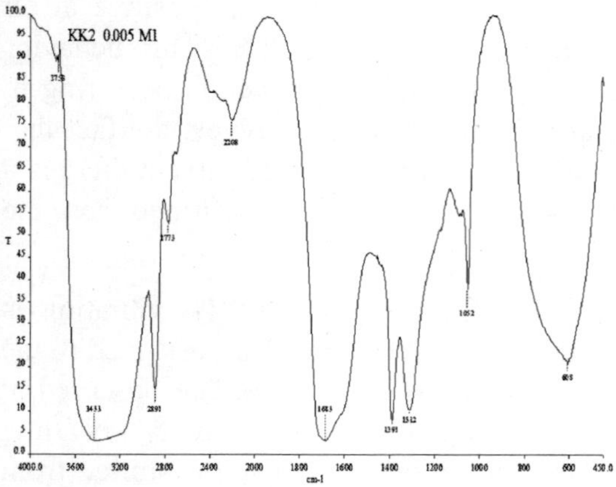

Fig. 7

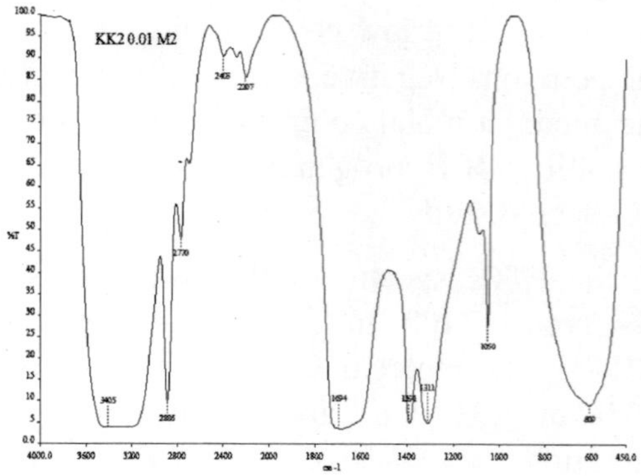

Fig. 8

CONCLUSION

From the present investigation, L-NAME in formamide behaves as structure promoter and also there is a strong hydrogen bond formation between zwitterions and solvent molecule. It is also confirmed by IR spectral study.

REFERENCES

[1] Bichile, G.K. and Kanhekar, S.R.; JCPRC 5, (2012) 4(1), pp. (78–86).

[2] David H. Sigmon; England Gonzalez-Feldman; Maria. A; *JAM. Soc Nephrol*, 15: 1413–1420, (2004).

[3] Ruslin M. Moore; Steven A. and Sedrish, *Vet Res.* (2005), 69(2), 116–122.

[4] Konstantin Tsogi; Tae Y.U. Lee; Young Soo Lee; Han Geuk Seo; *Mol Pharmacol*, 76: 173–182, (2009).

[5] Pooja, P., Adroja; Gami, S.P. and Patel, J.P., *J. Indian Chem. Soc.* (2010), 87, 655–660.

[6] Ernest, S., Glinski, J., Jezowska, *Acta Phy. Pol. Article*, 55 (1979), 501.

[7] Sharma B.K, *Journal of Pure and Applied Phy.*, 23 (1983), 509–513.

[8] Rasayan, *J. Chem.*, Vol. 4, No. 2, 431–436 (2011).

Evaluation of Ultrasonic and Volumetric Studies of Four Homologous α-Aminoacids in Aqueous Xylose Solution at T = 298.15 K

K. Rajagopal and J. Johnson[1]

Government College of Engineering, Tirunelveli, Tamil Nadu

[1]Joe Suresh Engineering College, Tirunelvel, Tamil Nadu

E-mail: Jacobjohn_2710 @ Yahoo.co

ABSTRACT: *Values of density (ρ) and ultrasonic speed (u) have been determined for four amino acids (glycine,L-alanine,L-valine and L-leucine) in aqueous xylose solutions (0.05,0.10,0.15,0.20)M at T = 298.15 K. Apparent molar volume* (V_φ)*, partial molar volumes* (V_φ^0)*, transfer volumes* (ΔV_φ^0) *and hydration number* (n_H) *are evaluated using density data. Apparent molar compressibility* (K_φ)*, partial molar compressibility* (K_φ^0)*, transfer compressibility* (ΔK_φ^0) *and hydration number* (n_H) *have been calculated using ultrasonic speed data. The molecular interactions are explained on the basis of acoustical and volumetric properties in the binary mixtures. The analysis of ultrasonic and densimetric parameters shows the presence of strong interaction between the zwitterionic end groups of the amino acids with OH groups of xylose.*

Keywords: Amino Acids, Xylose, Apparent Molar Volume, Apparent Molar Compressibility.

INTRODUCTION

The properties of carbohydrates solutions are of considerable interest in various aspects of basic researches and applications. It is well known fact that saccharides are typical non electrolytes with several hydroxyl groups and these polyhydroxy compounds help in stabilizing the native confirmations of globular proteins. Hydration behavior of saccharides has been found to be related the number or configuration of hydroxyl groups [1–3]. In living organisms, interactions of saccharides with model molecules of proteins play a key role in understanding the thermodynamic behavior of biochemical process in the body system. Therefore studies on carbohydrates-proteins are very important in immunology, biosynthesis, pharmacology and medicine. Literature survey [4, 5] shows that experimental and theoretical works have been reported on thermodynamics of amino acids in aqueous carbohydrate solutions.However, thermodynamic studies (especially on speed of sound) of the interaction between carbohydrate and protein in solutions are rare.

In continuation of our thermo dynamical studies of amino acids in aqueous/non-aqueous/ drug/electrolyte solution [6–9], we in this paper, report data on density and ultrasonic speed of homologous amino acid in aqueous xylose solution. The main objective of these studies is to investigate the solute-co solute interactions in such solution.

EXPERIMENTAL

The data on density and ultrasonic speed are experimentally measured using single stem pyconometer and multifrequency ultrasonic interferometer (M-84, Mittal Enterprises, India) at a frequency of 2MHz [6]. The values of V_φ^0, K_φ^0 and n_H for water fairly agrees well with literature values are shown in Tables 3, 4 and 5 [10–15].

RESULTS AND DISCUSSION

The apparent molar volume (V_φ) and apparent molar compressibility (K_φ) are calculated using the following equations.

$$V_\varphi = (M/\rho) - 1000\,(\rho - \rho_0)/m\,\rho\,\rho_0 \quad \text{... (1)}$$

$$K_\varphi = \beta_s M/\rho + 1000\,(\beta_s \rho_0 - \beta_0\,\rho)/m\,\rho\,\rho_0 \quad \text{... (2)}$$

Table 1 shows that V_φ increases with increase in concentration of xylose as well as of all amino acids studied at T = 298.15 K. It is seen further from Table 2 that the K_φ values of the amino acid in aqueous xylose solutions are negative indicating the water molecules around ionic charged groups of amino acids that are less compressible than the water molecules in the bulk solution [16]. This supports further the conclusion that the interaction between OH groups of xylose and the zwitterions of amino acids which are localized at the head groups. This increases the electrostriction of water caused by the charged centers of amino acids resulting in a decrease in volume, thereby, increasing the compressibility of the solution.

The partial molar quantities (Y_φ^0) are obtained using the equation:

$$Y_\varphi = Y_\varphi^0 + S_y m \quad \text{... (3)}$$

Where Y_φ^0 (Y_φ^0 denotes V_φ^0 or K_φ^0) is the infinite dilution value that is equal to the partial molar property at infinite dilution, S_y (S_y denotes $S_v$or S_k) is the experimental slope [17]. Equation (3) is fitted to Y_φ (Y_φ denotes V_φ or K_φ) values by the method of the least squares to evaluate V_φ^0 and K_φ^0 values. The increase in V_φ^0 in Table 3 for all the amino acid concentration indicates that the solute-cosolute interactions increase with increase in xylose concentration [15]. From Table 4, it is found that K_φ^0 values are negative and increase linearly with the increase in concentration of xylose solutions. This indicates that the solute–solvent interactions increase on increasing concentration of xylose solutions.

The transfer parameters ΔV_φ^0 and ΔK_φ^0 are calculated using equation (4),

$$\Delta Y_\varphi^0 = Y_\varphi^0 \text{ (in aqueous xylose)} - Y_\varphi^0 \text{ (in pure water)} \quad \text{... (4)}$$

Table 1: Density, ρ, and Apparent Molal Volume, V_φ, of Amino Acids in Aqueous Xylose Solution at T = 298.15 K

m_A ($mol\cdot kg^{-1}$)	$m_S = 0\ mol\cdot kg^{-1}$		$m_S = 0.05\ mol\cdot kg^{-1}$		$m_S = 0.10\ mol\cdot kg^{-1}$		$m_S = 0.15\ mol\cdot kg^{-1}$		$m_S = 0.20\ mol\cdot kg^{-1}$	
	$10^{-3}\rho/(kg\cdot m^{-3})$	$10^6 V_\varphi/(m^3\cdot mol^{-1})$	$10^3 \rho/(kg\cdot m^{-3})$	$10^6 V_\varphi/(m^3\cdot mol^{-1})$	$10^{-3}\rho/(kg\cdot m^{-3})$	$10^6 V_\varphi/(m^3\cdot mol^{-1})$	$10^{-3}\rho/(kg\cdot m^{-3})$	$10^6 V_\varphi/(m^3\cdot mol^{-1})$	$10^{-3}\rho/(kg\cdot m^{-3})$	$10^6 V_\varphi/(m^3\cdot mol^{-1})$
Glycine Density										
0	0.99704		0.99979		1.00250		1.00517		1.00780	
0.02	0.99768	43.02	1.00043	43.04	1.00314	43.06	1.00581	43.08	1.00844	43.10
0.04	0.99832	43.11	1.00107	43.14	1.00377	43.19	1.00644	43.21	1.00907	43.23
0.06	0.99895	43.20	1.00169	43.26	1.00440	43.29	1.00707	43.31	1.00969	43.36
0.08	0.99958	43.28	1.00232	43.35	1.00502	43.41	1.00769	43.45	1.01031	43.46
0.10	1.00020	43.37	1.00294	43.44	1.00564	43.51	1.00830	43.58	1.01092	43.60
L-Alanine										
0	0.99704		0.99979		1.00250		1.00517		1.00780	
0.02	0.99763	59.90	1.00037	59.91	1.00308	59.98	1.00575	59.99	1.00838	60.01
0.04	0.99820	60.09	1.00095	60.13	1.00365	60.17	1.00632	60.18	1.00894	60.20
0.06	0.99877	60.26	1.00151	60.33	1.00421	60.39	1.00688	60.40	1.00950	60.42
0.08	0.99933	60.43	1.00206	60.53	1.00477	60.56	1.00742	60.61	1.01004	60.65
0.10	0.99988	60.61	1.00261	60.70	1.00531	60.75	1.00796	60.81	1.01058	60.86
L-Valine										
0	0.99704		0.99979		1.00250		1.00517		1.00780	
0.02	0.99758	90.24	1.00033	90.26	1.00303	90.29	1.00570	90.32	1.00833	90.35
0.04	0.99811	90.44	1.00085	90.47	1.00356	90.49	1.00622	90.55	1.00884	90.60
0.06	0.99863	90.63	1.00137	90.69	1.00407	90.75	1.00672	90.77	1.00934	90.80
0.08	0.99915	90.82	1.00188	90.90	1.00457	90.96	1.00722	91.00	1.00983	91.04
0.10	0.99965	91.00	1.00237	91.13	1.00505	91.22	1.00771	91.22	1.01032	91.24
L-Leucine										
0	0.99704		0.99979		1.00250		1.00517		1.00780	
0.02	0.99753	107.12	1.00027	107.14	1.00297	107.22	1.00564	107.25	1.00826	107.27
0.04	0.99800	107.37	1.00074	107.45	1.00344	107.49	1.00610	107.52	1.00872	107.54
0.05	0.99823	107.50	1.00096	107.58	1.00366	107.62	1.00632	107.64	1.00894	107.69
0.06	0.99846	107.63	1.00119	107.70	1.00389	107.74	1.00654	107.79	1.00915	107.80
0.08	0.99891	107.88	1.00162	108.01	1.00432	108.07	1.00697	108.10	1.00957	108.14

m_A, molality of amino acids. m_S, molality of xylose.

Table 2: Ultrasonic Speed, *u*, and Apparent Molar Compressibility, K_φ, of Amino Acids in Aqueous Xylose Solution at T = 298.15 K

m_A $(mol{\cdot}kg^{-1})$	$u/(m{\cdot}s^{-1})$	$10^{15} K_\varphi/(m^3{\cdot}mol^{-1}{\cdot}Pa^{-1})$	$u/(m{\cdot}s^{-1})$	$10^{15} K_\varphi/(m^3{\cdot}mol^{-1}{\cdot}Pa^{-1})$	$u/(m{\cdot}s^{-1})$	$10^{15} K_\varphi/(m^3{\cdot}mol^{-1}{\cdot}Pa^{-1})$	$u/(m{\cdot}s^{-1})$	$10^{15} K_\varphi/(m^3{\cdot}mol^{-1}{\cdot}Pa^{-1})$	$u/(m{\cdot}s^{-1})$	$10^{15} K_\varphi/(m^3{\cdot}mol^{-1}{\cdot}Pa^{-1})$
	Glycine									
	$m_S = 0\ mol{\cdot}kg^{-1}$		$m_S = 0.05\ mol{\cdot}kg^{-1}$		$m_S = 0.10\ mol{\cdot}kg^{-1}$		$m_S = 0.15\ mol{\cdot}kg^{-1}$		$m_S = 0.20\ mol{\cdot}kg^{-1}$	
0	1496.40		1497.30		1500.20		1503.50		1506.20	
0.02	1497.50	–28.17	1498.38	–27.24	1501.29	–27.10	1504.60	–26.94	1507.31	–26.82
0.04	1498.50	–26.56	1499.40	–26.23	1502.32	–26.08	1505.64	–25.93	1508.35	–25.67
0.06	1499.50	–25.95	1500.35	–25.10	1503.28	–24.97	1506.60	–24.74	1509.33	–25.62
0.08	1500.40	–24.84	1501.23	–23.98	1504.19	–23.99	1507.51	–23.70	1510.23	–23.47
0.10	1501.30	–24.14	1502.03	–22.79	1505.01	–22.82	1508.34	–22.56	1511.04	–22.20
L-Alanine										
0	1496.40		1497.30		1500.20		1503.50		1506.20	
0.02	1497.70	–25.06	1498.58	–24.41	1501.49	–24.21	1504.80	–24.03	1507.51	–23.88
0.04	1498.90	–23.66	1499.82	–23.59	1502.74	–23.42	1506.06	–23.25	1508.78	–23.11
0.06	1500.10	–22.97	1501.03	–22.89	1503.97	–22.80	1507.30	–22.64	1510.02	–22.41
0.08	1501.30	–22.54	1502.21	–22.21	1505.16	–22.13	1508.50	–21.95	1511.20	–21.52
0.10	1502.40	–21.61	1503.35	–21.52	1506.32	–21.47	1509.64	–21.11	1512.36	–20.79
L-Valine										
0	1496.40		1497.30		1500.20		1503.50		1506.20	
0.02	1498.18	–25.19	1499.07	–24.45	1501.98	–24.18	1505.29	–23.89	1508.00	–23.64
0.04	1499.90	–24.08	1500.81	–23.79	1503.72	–23.38	1507.04	–23.08	1509.73	–22.39
0.06	1501.60	–23.39	1502.51	–23.03	1505.42	–22.55	1508.73	–22.08	1511.41	–21.38
0.08	1503.20	–22.19	1504.16	–22.17	1507.07	–21.67	1510.36	–21.04	1513.00	–20.10
0.10	1504.90	–22.00	1505.80	–21.50	1508.71	–20.96	1511.94	–20.04	1514.48	–18.64
L-Leucine										
0	1496.40		1497.30		1500.20		1503.50		1506.20	
0.02	1498.47	–25.08	1499.36	–24.30	1502.27	–23.93	1505.58	–23.58	1508.28	–23.00
0.04	1500.50	–24.23	1501.39	–23.53	1504.31	–23.22	1507.60	–22.45	1510.29	–21.73
0.05	1501.50	–23.80	1502.40	–23.25	1505.31	–22.77	1508.59	–21.92	1511.25	–20.88
0.06	1502.50	–23.48	1503.39	–22.82	1506.30	–22.34	1509.57	–21.41	1512.20	–20.20
0.08	1504.50	–22.98	1505.36	–22.06	1508.26	–21.49	1511.46	–20.15	1513.98	–18.31

m_A, molality of amino acids. m_S, molality of xylose.

Table 3: Partial Molar Volume, V_φ^0, and Transfer Partial Molar Volume, ΔV_φ^0,/of Some Amino Acids in Aqueous Xylose Solution at T = 298.15 K

Amino Acids	$m_S = 0\ mol{\cdot}kg^{-1}$	$m_S = 0.05\ mol{\cdot}kg^{-1}$		$m_S = 0.10\ mol{\cdot}kg^{-1}$		$m_S = 0.15\ mol{\cdot}kg^{-1}$		$m_S = 0.20\ mol{\cdot}kg^{-1}$	
	$10^6\ V_\varphi^0/$ $(m^3{\cdot}mol^{-1})$	$10^6\ V_\varphi^0/$ $(m^3{\cdot}mol^{-1})$	10^6 $\Delta V_\varphi^0/(m^3{\cdot}mol^{-1})$	$10^6\ V_\varphi^0/$ $(m^3{\cdot}mol^{-1})$	$10^6\ \Delta V_\varphi^0/$ $(m^3{\cdot}mol^{-1})$	$10^6\ V_\varphi^0/$ $(m^3{\cdot}mol^{-1})$	$10^6\ \Delta V_\varphi/$ $(m^3{\cdot}mol^{-1})$	$10^6\ V_\varphi^0/$ $(m^3{\cdot}mol^{-1})$	$10^6\ \Delta V_\varphi^0/$ $(m^3{\cdot}mol^{-1})$
Glycine	43.20, 43.14[a]	43.25	0.050	43.29	0.096	43.33	0.129	43.35	0.154
L-alanine	60.26, 60.23[a]	60.32	0.061	60.37	0.110	60.40	0.142	60.43	0.169
L-valine	90.63, 90.65[a]	90.69	0.062	90.74	0.116	90.77	0.144	90.81	0.180
L-leucine	107.50, 107.48[a]	107.58	0.075	107.63	0.129	107.66	0.160	107.69	0.191

m_S, molality of xylose. [a]Ref. [10].

Table 4: Partial Molal Compressibility, K_φ^0, Experimental Slope, S_k, and Transfer Partial Molar Compressibility, ΔK_φ^0, of Amino Acids in Aqueous Xylose Solution at T = 298.15 K

Amino Acid	$10^{15}\ K_\varphi^0/$ $(m^3{\cdot}mol^{-1}{\cdot}Pa^{-1})$	$10^{18}\ S_k/$ $(kg{\cdot}m^3{\cdot}mol^{-2}{\cdot}Pa^{-1})$	$10^{15}\ K_\varphi^0/$ $(m^3{\cdot}mol^{-1}{\cdot}Pa^{-1})$	$10^{18} S_k/(kg{\cdot}m^3{\cdot}mol^{-2}{\cdot}Pa^{-1})$	$10^{15}\ \Delta K_\varphi^0/$ $(m^3{\cdot}mol^{-1}{\cdot}Pa^{-1})$	$10^{15}\ K_\varphi^0/$ $(m^3{\cdot}mol^{-1}{\cdot}Pa^{-1})$	$10^{18} S_k/$ $(kg{\cdot}m^3{\cdot}mol^{-2}{\cdot}Pa^{-1})$	$10^{15}\ \Delta K_\varphi^0/$ $(m^3{\cdot}mol^{-1}{\cdot}Pa^{-1})$
$m_S = 0\ mol{\cdot}kg^{-1}$			$m_S = 0.05\ mol{\cdot}kg^{-1}$			$m_S = 0.10\ mol{\cdot}kg^{-1}$		
Glycine	–28.86, –26.95[b], –27.00[c]	0.49	–28.41	0.56	0.45	–28.19	0.53	0.67
L-alanine	–25.57, –25.03[c], –25.69[d]	0.40	–25.07	0.36	0.50	–24.83	0.34	0.74
L-valine	–25.85, –30.00[d]	0.41	–25.25	0.38	0.60	–24.99	0.41	0.86
L-leucine	–25.68, –31.30[d]	0.35	–25.05	0.37	0.63	–24.79	0.41	0.89
$m_S = 0.15\ mol{\cdot}kg^{-1}$				$m_S = 0.20\ mol{\cdot}kg^{-1}$				
Glycine	–28.08	0.55	0.78	–27.99	0.57	0.87		
L-alanine	–24.73	0.36	0.84	–24.68	0.39	0.89		
L-valine	–24.94	0.49	0.91	–24.92	0.61	0.93		
L-leucine	–24.73	0.57	0.95	–24.72	0.78	0.96		

m_S, molality of xylose; [b]Ref.[11].; [c]Ref.[12]; [d]Ref.[13].

Table 5: Values of Hydration Number, n_H, of Some Amino Acids in Aqueous Xylose Solution at T = 298.15 K

Amino Acid	n_H									
	From Volume Using Eqn. 5					*From Compressibility Using Eqn. 6*				
	$m_S = 0$ $mol \cdot kg^{-1}$	$m_S = 0.05$ $mol \cdot kg^{-1}$	$m_S = 0.10$ $mol \cdot kg^{-1}$	$m_S = 0.15$ $mol \cdot kg^{-1}$	$m_S = 0.20$ $mol \cdot kg^{-1}$	$m_S = 0$ $mol \cdot kg^{-1}$	$m_S = 0.05$ $mol \cdot kg^{-1}$	$m_S = 0.10$ $mol \cdot kg^{-1}$	$m_S = 0.15$ $mol \cdot kg^{-1}$	$m_S = 0.20$ $mol \cdot kg^{-1}$
Glycine	2.10, 2.63[e]	2.08	2.07	2.06	2.06	3.56, 3.26[f]	3.51	3.48	3.47	3.46
L-Alanine	2.47, 2.89[e]	2.44	2.38	2.35	2.30	3.16, 2.89[f]	3.10	3.07	3.05	3.05
L-Valine	3.63, 3.43[e]	3.62	3.60	3.60	3.59	3.19	3.12	3.09	3.08	3.08
L-Leucine	3.53, 4.96[e]	3.51	3.49	3.49	3.48	3.17	3.09	3.06	3.05	3.05

m_S, molality of xylose; [e]Ref.[14]; [f]Ref.[15].

where $\Delta Y_\varphi{}^0$ denotes $\Delta V_\varphi{}^0$ and $\Delta K_\varphi{}^0$ and the results are given in Tables 3 and 4 respectively. The apparent molar compressibilities of transfer ($\Delta K_\varphi{}^0$) in Table 4 are found to be positive that indicates the dehydration of solute and cosolute is high. [18]

The hydration number n_H may be evaluated from the volumetric data and compressibility data calculated using the following s equations [19, 12]

$$n_H = V_\varphi{}^0(\text{elect})/(V^0{}_E - V^0{}_B) \quad \dots (5)$$

$$n_H = -K_\varphi{}^0\,(\text{elect})/V^0{}_B\,K^0{}_B \quad \dots (6)$$

Furthermore, the decrease in n_H values with the increase in the concentration of xylose shows that xylose has dehydration effect on amino acids are given in Table 5. Thus the compressibility results are fairly complimenting the volumetric results.

CONCLUSION

The reported values of partial molar volumes and partial molar compressibility indicate the presence of strong solute-cosolute interactions in the solution. The compressibility parameters complement the results obtained from volumetric studies. The hydration number variations prove that xylose has dehydration effect on amino acids. The zwitterionic end groups of the amino acids with OH group of aqueous xylose are much stronger than the hydrophobic groups.

REFERENCES

[1] J.F. Back, D. Oakenfull and M.B. Smitj, *Biochemistry*, 18 (1979), 5191.
[2] H. Uedaira, H. Uedaira, *Bull. Chem. Soc. Japan*, 53 (1980), 2451.
[3] P.K. Banipal, T.S. Banipal, B.S. Lark and J.C. Ahulwalia, *J. Chem. Soc. Faraday Trans.*, 1,93 (1993), 81.
[4] Y. Chen, H. Liu, R. Lin and H. Zhang, *J. Sol. Chem*, 36 (2007), 1321.
[5] H. Liu, R. Liu and H. Zhang, *J. Solution Chem.*, 32 (2003), 977.
[6] K. Rajagopal and S. Edwin Gladson, *J. Chem. Thermodyn.*, 43 (2011), 852–867.
[7] K. Rajagopal and S. Edwin Gladson, *J. Solution Chemistry* (in Press, DOI: 10.1007/s10953-012–9820–9)
[8] K. Rajagopal, S.S. Jayabalakrishnan, *J. Chem. Thermodyn*, 42 (2010), 984–993.
[9] K. Rajagopal, S.S. Jayabalakrishnan, *Chinese J. Chem. Engg.,* 18 (4), (2010), 659–666.
[10] R. Bhat and J.C. Ahluwalia, *J. Phys. Chem.*, 89 (1985), 1099–1105.
[11] Ana Soto, Alberto Arce, Mohammad, K. Khoshkbarchi, *Biophysical Chem.* 74(1998), 165–173.
[12] F.J. Millero, A.L. Surdo and C. Shin, *J. Phys. Chem.*, 82 (1978), 784–792.
[13] T.S. Banipal and G. Singh, *Thermochim. Acta*, 412 (2004), 63–84.
[14] H. Zhao, *Biophys. Chem.*, 122 (2006), 157–183.
[15] R.K. Wadi and P. Ramaswami, *J. Chem. Soc. Faraday Trans.* (1997), 93(2), 243–247.

[16] A. Soto, A. Arce and M.K. Khoshkbarchi, *J. Solution Chem.*, 33 (2003), 11–21.
[17] A. Pal, S. Kumar, *J. Mol. liq.,* 121 (2005), 148–155.
[18] T.S. Banipal, G. Sehgal, Thermochimica Acta, 262(1995), 175–183.
[19] J. Wang, Z. Yan, K. Khno and J. Lu, *Biophys. Chem.*, 80 (1999), 179–188.

Theoretical Evaluation of Ultrasonic Velocity in Binary Liquid Mixtures

B. Nagarjun, A.V. Sarma[1], G.V. Rama Rao[2] and C. Rambabu[3]

Department of Physics, G.V.P. College of Engineering (A), Visakhapatnam

[1]Department of Physics, Andhra University, Visakhapatnam

[2]Department of Physics, D.A.R.College, Nuzvid

[3]Department of Chemistry, Acharya Nagarjuna University, Guntur

ABSTRACT: *Theoretical values of ultrasonic velocities in binary liquid mixtures of ethyl benzoate + NN-dimethyl formamaide(NNDMF), ethyl benzoate + NN-dimethyl acetamide (NNDMAc) and ethyl benzoate + NN-dimethylaniline (NNDMA), have been evaluated using various theoretical relations of Nomoto, Van Dael and Vangeel, Rao's specific velocity, Junjie and impedance relation. In the chosen binary systems there is good agreement between experimental and theoretical values calculated by Junjie's theory, impedance relation and Nomoto relation at temperatures 303.15 K, 308.15 K, 313.15 K and 318.15 K. The deviation in the variation of* U^2_{exp}/U^2_{imx} *from unity has also been evaluated for explaining the non-ideality in the mixtures.*

INTRODUCTION

Ultrasonic velocity of sound waves in a medium is fundamentally related to the binding forces between the molecules. Ultrasonic sound velocities of the liquid mixtures consisting of polar and non-polar components are of considerable importance in understanding intermolecular interaction between component molecules and they find applications in several industrial and technological processes. Comparison of theoretical values of ultrasonic velocities with those obtained experimentally in the present binary liquid mixtures is expected to reveal the nature of interaction between component molecules in the mixture.[1–5] Such theoretical study is useful in bringing a comprehensive theoretical model for the liquid mixtures. The aim of the present investigation is to compare the ultrasonic sound velocity in binary liquid mixtures from various theoretical relations of Nomoto[6], Van Dael and Vangeel[7], impedance relation[8], Rao's specific velocity[9], and Juniie[10]. An attempt has been made to compare the merits of the relations for the binary liquid mixtures investigated at different temperatures. The deviation in the variation of U^2_{exp}/U^2_{imx} from unity has also been evaluated for explaining the non-ideality in the mixtures.

THEORETICAL

Nomoto Equation

On assuming the additivity of molar sound velocity (R) and no volume change on mixing, Nomoto established the following relation for a liquid mixture,

$$R = M/\rho\ U^{1/3} \qquad \dots (1)$$

Where U and ρ are determined experimentally and M is the mean molecular weight in a binary liquid mixture,

$$M = (X_1M_1 + X_2M_2) \qquad \dots (2)$$

where M_1 and M_2 are molecular weights of constituent components.

Simple manipulation yields the following relation

$$U_{Nomoto} = [(X_1R_1+X_2R_2)/(X_1V_1 + X_2V_2)]^3 \qquad \dots (3)$$

The Van Dael and Vangeel Equation

The ideal mixing theory advanced by Van Dael and Vangeel[5] in the light of assumptions made by Blandamer and Waddington[11], yield the following relation for adiabatic compressibility $(\beta_{ad})_{imx}$,

$$(\beta_{ad})_{imx} = \Pi_1\gamma_1/\gamma_{imx}\ (\beta_{ad})_1 + \Pi_2\ \gamma_2/\gamma_{imx}\ (\beta_{ad})_2 \qquad \dots (4)$$

where Π_1, Π_2 are the volume fraction of species 1 and 2, γ_1 and γ_2 are ratios of specific heats of the respective species. This relation holds well, if the mixture is ideal and if $\gamma_1 = \gamma_2 = \gamma_{imx}$. Using the additional assumption that $V_1 = V_2$ the above equation can be transformed in to a linear combination of mole fraction X_1 and X_2,

$$(\beta_{ad})_{imx} = X_1(\beta_{ad})_1 + X_2(\beta_{ad})_2 \qquad \dots (5)$$

On the basis of this equation, Van Dael obtained the relation for ultrasonic velocity in liquid mixtures as,

$$1/(X_1M_1+X_2M_2)\times 1/U^2_{imx} = X_1/M_1U_1^2 + X_2/M_2U_2^2 \qquad \dots (6)$$

where U_{imx} is the ideal mixing ultrasonic velocity in liquid mixture. U_1 and U_2 are ultrasonic velocity in species.

The Impedance Relation

Impedance relation

$$U = \sum X_i Z_i/\sum X_i\rho_i \qquad \dots (7)$$

where X_i mole fraction, ρ_i is the density of the mixture and Z_i is the acoustic impedance.

The Rao's Specific Velocity Method Relation

Rao's specific velocity method,

$$U = (\sum X_i r_i \rho)^3 \quad \dots (8)$$

where X_i mole fraction, U_i is the ultrasonic velocity, ρ_i is the density of the mixture, r_i is the Rao's specific sound velocity = $Ui^{1/3}/\rho_i$ and Z_i is the acoustic impedance.

The Jungie Equation

Jungie equation

$$U_J = (X_1M_1/\rho_1 + X_2M_2/\rho_2)/[\{X_1M_1 + X_2M_2\}^{1/2}\{X_1M_1/\rho_1U_1^2 + X_2M_2/\rho_2U_2^2)\}^{1/2}] \quad \dots (9)$$

where M_1, M_2 are molecular weights of constituent components, ρ_1 and ρ_2 are the densities of constituent components.

RESULTS AND DISCUSSION

The experimental values of ultrasonic velocity along with the values calculated theoretically using the relations of Nomoto, Van Dael ideal mixing, impedance, Rao's specific velocity and Junjie for Ethyl benzoate +NNDMF, Ethyl benzoate +NNDMAc, Ethyl benzoate +NNDMA binary systems, at the temperature 303.15 K are given in Table 1. The validity of different theoretical formulae is checked by percentage deviation for all the mixtures and is given in Table 2.

In all the three systems Ethyl benzoate+NNDMF, Ethyl benzoate+NNDMAc and Ethyl benzoate+NNDMA there is good agreement between experimental and theoretical values calculated by Nomoto relation and impedance relation. Both the theories provide the best results than other theories at all the temperatures. However, higher deviations are observed in Rao's specific velocity method, VanDael ideal mixing relation and Jungie equation.

The agreement between experimental and theoretical velocities of Nomoto relation in most of the binary systems, suggests that R is additive property in all those systems. Higher deviations in some intermediate concentration range suggest the existence of strong tendency for the association between component molecules, in this concentration range as a result of dipole-dipole interactions.[12,13] These values vary non-linearly with compositions suggesting that there is association in some compositions. This is in good agreement with the conclusions drawn by Pandey *et al.*[14] and Kaulgud[15]. The percentage deviations, i.e. $\Delta U_{Nomoto,}$ $\Delta U_{idealmixing}$ and ΔU_{IR} show maxima at an intermediate concentration at all the temperatures in all the three binary liquid mixtures.

Figures 1, 2 and 3 represent the variation of U^2_{exp}/U^2_{imx} with mole fraction of ethyl benzoate. It is observed to be similar in all the systems with maximum at approximately

0.5 in the case of Ethyl benzoate + NNDMF and Ethyl benzoate + NNDMAc system and at approximately 0.6 in the case of Ethyl benzoate + NNDMA system at all the temperatures. Thus, the linearity of molar sound velocity and additivity of molar volumes, as suggested by Nomoto, impedance relation and Junjie in deriving the empirical relations (1) and (7) have been truly observed in the aforementioned binary liquid systems.

Table 1: Experimental and Theoretical Values of Velocity at 303.15 K

X_1 Molefraction	U_{exp} ms^{-1}	U_{Nomoto} ms^{-1}	U_{idmx} ms^{-1}	U_{ir} ms^{-1}	U_j ms^{-1}
Ethyl Benzoate (X_1) + NN-dimethyl Formamide (X_2)					
0.0000	1438.53	1438.52	1438.53	1438.53	1484.36
0.0568	1433.50	1429.19	1415.29	1432.78	1465.76
0.1193	1427.94	1419.90	1393.25	1426.54	1447.63
0.1885	1421.80	1410.64	1372.68	1419.73	1429.93
0.2654	1414.94	1401.43	1354.00	1412.28	1412.67
0.3514	1407.25	1392.26	1337.76	1404.09	1395.82
0.4484	1398.50	1383.13	1324.81	1395.05	1379.38
0.5584	1388.25	1374.04	1316.40	1385.02	1363.31
0.6843	1376.50	1364.99	1314.55	1373.82	1347.62
0.8298	1362.89	1355.97	1322.67	1361.23	1332.29
1.0000	1347.00	1347.00	1347.00	1347.00	1317.30
Ethyl Benzoate (X_1) + NN-Dimethyl Acetamide (X_2)					
0.0000	1438.00	1438.01	1438.00	1438.00	1469.64
0.0674	1432.95	1428.73	1420.46	1431.17	1452.06
0.1400	1427.32	1419.49	1403.97	1423.95	1434.90
0.2181	1421.01	1410.29	1388.68	1416.30	1418.14
0.3026	1414.06	1401.13	1374.82	1408.20	1401.76
0.3943	1406.27	1392.01	1362.66	1399.59	1385.75
0.4940	1397.57	1382.93	1352.58	1390.42	1370.09
0.6030	1387.19	1373.89	1345.08	1380.65	1354.78
0.7225	1375.40	1364.89	1340.89	1370.20	1339.81
0.8542	1362.17	1355.92	1341.03	1359.01	1325.16
1.0000	1347.00	1347.00	1347.00	1347.00	1310.82

X_1 Molefraction	U_{exp} ms^{-1}	U_{Nomoto} ms^{-1}	U_{idmx} ms^{-1}	U_{ir} ms^{-1}	U_j ms^{-1}
Ethyl Benzoate (X_1) + NN-Dimethyl Aniline (X_2)					
0.0000	1470.50	1470.50	1470.50	1470.50	1450.13
0.0900	1463.99	1457.73	1457.44	1458.36	1433.71
0.1820	1455.87	1445.13	1444.54	1446.17	1417.66
0.2761	1446.91	1432.60	1431.81	1433.94	1401.95
0.3724	1436.85	1420.14	1419.24	1421.66	1386.58
0.4709	1425.42	1407.75	1406.82	1409.33	1371.53
0.5717	1413.22	1395.44	1394.56	1396.96	1356.80
0.6750	1398.88	1383.19	1382.45	1384.54	1342.37
0.7807	1383.14	1371.02	1370.48	1372.07	1328.24
0.8890	1366.09	1358.92	1358.67	1359.56	1314.40
1.0000	1347.00	1347.00	1347.00	1347.00	1300.83

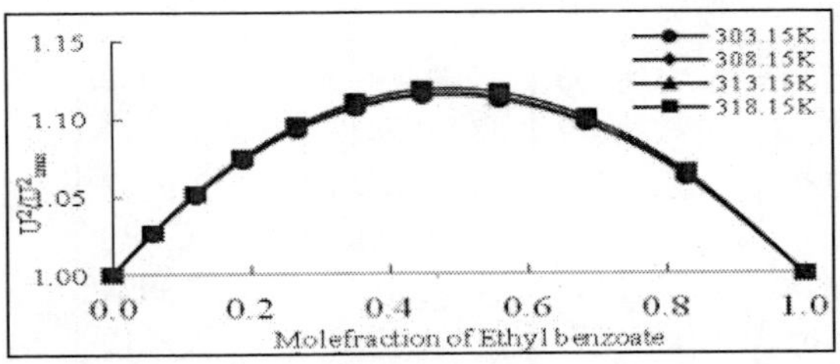

Fig. 1: Variation of U^2/U^2_{imx} with Molefraction of Ethyl Benzoate for Ethyl Benzoate + NNDMF System

Table 2: Percentage Deviation between Experimental and Theoretical Values of Velocity at 303.15 K

X_1 Molefraction	$\%U_N$	$\%U_{imx}$	$\%U_{IR}$	$\%U_{Rao}$	$\%U_J$	U^2/U^2_{imx}
Ethyl Benzoate (X_1) + NN-Dimethyl Formamide (X_2)						
0.0000	–0.0000	0.0000	0.0000	0.0000	3.1860	1.0000
0.0568	–0.3009	–1.2704	–0.0502	2.6579	2.2504	1.0259
0.1193	–0.5634	–2.4298	–0.0984	5.0396	1.3784	1.0504
0.1885	–0.7846	–3.4545	–0.1457	7.0164	0.5721	1.0728
0.2654	–0.9544	–4.3066	–0.1878	8.6702	–0.1600	1.0920
0.3514	–1.0650	–4.9375	–0.2243	9.7397	–0.8117	1.1066
0.4484	–1.0990	–5.2692	–0.2465	10.0603	–1.3673	1.1143
0.5584	–1.0240	–5.1757	–0.2331	9.1138	–1.7965	1.1121
0.6843	–0.8363	–4.5001	–0.1947	7.1520	–2.0979	1.0965
0.8298	–0.5072	–2.9506	–0.1212	4.1135	–2.2452	1.0617
1.0000	0.0000	0.0000	–0.0000	0.000	–2.2049	1.0000

X_1 Molefraction	$\%U_N$	$\%U_{imx}$	$\%U_{IR}$	$\%U_{Rao}$	$\%U_J$	U^2/U^2_{imx}
Ethyl Benzoate (X_1) + NN-Dimethyl Acetamide (X_2)						
0.0000	0.0000	0.0000	–0.0000	0.0000	3.5894	1.0000
0.0674	–0.2949	–0.8716	–0.1246	2.9210	2.6117	1.0177
0.1400	–0.5491	–1.6362	–0.2366	5.4589	1.7029	1.0335
0.2181	–0.7547	–2.2749	–0.3312	7.5077	0.8688	1.0471
0.3026	–0.9146	–2.7750	–0.4146	9.1240	0.1045	1.0579
0.3943	–1.0142	–3.1012	–0.4752	10.0711	–0.5769	1.0650
0.4940	–1.0478	–3.2195	–0.5115	10.1046	–1.1714	1.0676
0.6030	–0.9588	–3.0352	–0.4714	9.1861	–1.6247	1.0636
0.7225	–0.7646	–2.5089	–0.3780	7.3157	–1.9565	1.0521
0.8542	–0.4584	–1.5520	–0.2316	4.2481	–2.1624	1.0318
1.0000	0.0000	0.0000	–0.0000	0.000	–2.2049	1.0000
Ethyl Benzoate (X_1) + NN-Dimethyl Aniline (X_2)						
0.0000	0.0000	0.0000	0.0000	0.0000	2.6952	1.0000
0.0900	–0.4276	–0.4477	–0.3848	1.0777	1.6558	1.0090
0.1820	–0.7380	–0.7782	–0.6664	1.9942	0.7634	1.0157
0.2761	–0.9893	–1.0437	–0.8966	2.7440	–0.0378	1.0212
0.3724	–1.1628	–1.2255	–1.0570	3.3924	–0.7314	1.0250
0.4709	–1.2398	–1.3051	–1.1289	3.8817	–1.3011	1.0266
0.5717	–1.2585	–1.3207	–1.1508	4.0566	–1.7874	1.0269
0.6750	–1.1211	–1.1746	–1.0250	3.7606	–2.0952	1.0239
0.7807	–0.8761	–0.9151	–0.8003	3.0464	–2.2751	1.0186
0.8890	–0.5245	–0.5431	–0.4781	1.8266	–2.3301	1.0110
1.0000	0.0000	0.0000	–0.0000	0.000	–2.2049	1.0000

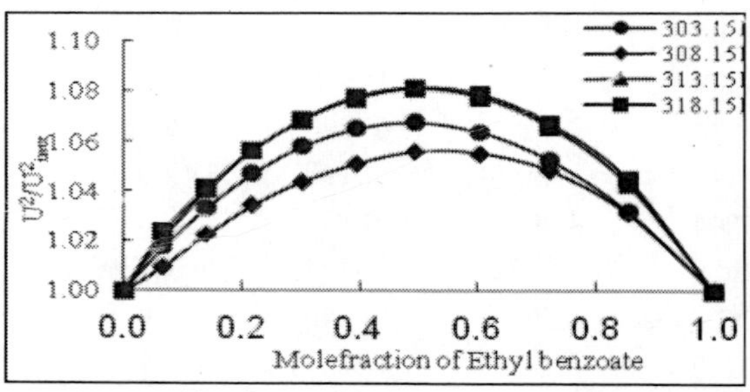

Fig. 2: Variation of U^2/U^2_{imx} with Molefraction of Ethyl Benzoate for Ethyl Benzoate for Ethyl benzoate + NNDMAc System

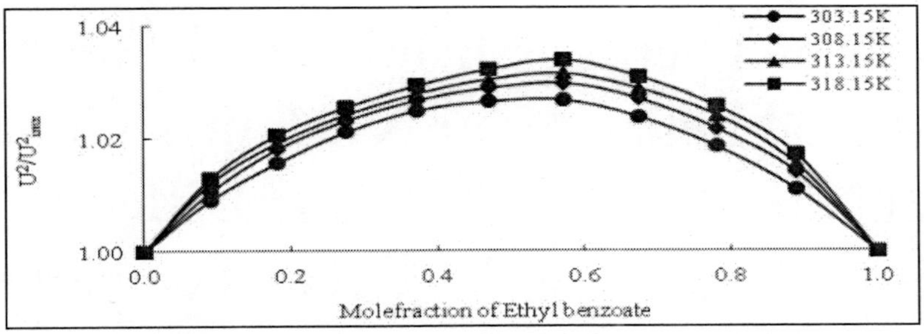

Fig. 3: Variation of U^2/U^2_{imx} with Molefraction of Ethyl Benzoate for Ethyl Benzoate + NN Dimethyl Aniline System

REFERENCES

[1] Dominguez M., Martin S., Santafe J., Artigas H. and Royo F.M., *Thermo Chim Acta*, 381 (2002), 181.

[2] Gnana Kumari P., Venkatesu P., Prabhakar Rao M.V., Ming-Jer Lee, and Ho-mu *Lin, J Chem Thermodyn*, 41 (2009), 586.

[3] Ritzoulls G., Messopolinou D., Dolulami S. and Panayiotou C., *J Chem Eng Data*, 45 (2000), 636.

[4] Ali A., Tewari K., Nain A.K. and Chakravorthy V., Phys Chem Liq, 38 (2000), 459.

[5] Meyyappan M. and Sethupathy, *J Pure Appl Ultrason*, 25 (2003), 1.

[6] Nomoto O., *J Phys Soc.,* Japan, 4 (1949) 280 and 13 (1958) 1528 and *J Chem Phys*, 21 (1953), 950.

[7] Van Dael, W. and Vangeel, E., *Pro Int Conf on Calorimetry and Thermodynamics,* Warsa (1955), 555.

[8] Jacobson B., *Acta Chem Scandin*, 6 (1952), 1485 and *J Chem Phys*, 20 (1952), 927.

[9] Schaaffs W., Molekularakustik, Springer-Verlag, Berlin, 1963.

[10] Junjie Z., *J China Univ Sci Techn*, 14 (1984), 298.

[11] Blandamer M. and Waddington D., *J Phys Chem*, 74 (1970), 2569.

[12] Shipra Baluja and Falguni Karia, *J Pure Appl Ultrason,* 22 (2000), 82.

[13] Schaffs W., Z Phy, 114 (1939) 110 and ibid, 115 (1940), 69.

[14] Pandey J.D., Dey R. and Dwivedi D.K., Pramana, 52 (1999), 187.

[15] Kaulgud M.V., Acustica, 10 (1960), 316.

Study of Thermo Chemical and Thermo Dynamical Behavior for N-Methyl Thiourea in Non-Aqueous Media at Different Concentration and at Different Temperatures

E. Jasmine Vasantha Rani, R. Padmavathy, M. Ragamathunnisa[1], M. Mathi Mozhi and M. Malini

Department of Physics, Seethalakshmi Ramaswami College, Trichirappalli, Tamil Nadu

[1]Department of Physics, Government Arts College for Women, Pudukkottai, Tamil Nadu

E-mail: jasstephen@yahoo.com; mars.phy@gmail.com

ABSTRACT: *Thiourea and its derivatives are versatile intermediates for the synthesis of modified thermosetting resins, plant protection agents, corrosion inhibitors and thiazole drugs. In the present work, for N-methyl thiourea fundamental parameters such as ultrasonic velocity, viscosity and density have been measured. Using these parameters thermo chemical and thermo acoustical behavior have been studied by calculating apparent molal volume, apparent molal compressibility, internal pressure and free volume. The study has been carried out at different temperature and at different concentrations. Result obtained from ultrasonic study is compared with spectroscopic study.*

INTRODUCTION

The ultrasonic velocity measurements find wide applications in characterizing the physico-chemical behavior of liquid mixtures [1–3] and in the study of molecular interactions . Ultrasonic velocity of a liquid is related to binding forces between the atoms or the molecules. Ultrasonic velocities have been adequately employed in understanding the nature of molecular interaction in pure liquids[4]. The method of studying the molecular interactions from the knowledge of variations of thermodynamic parameters and their excess values with composition gives an insight into the molecular process.[5–7] Thioureas are a group of compounds possessing a wide spectrum of biological activities. Various N-substituted isoquinolinyl–N-substituted phenyl thioureas have been found to be useful for the treatment and prophylaxis of anxiety, mania, depression, migraine and defects associated with AIDS. Recent bioactivity of thiourea exhibits antithyroidal, antibacterial and hypoglycemic activities. Some thiourea derivatives possess valuable biological pharmacological activities such as, anti-HIV,[8–11] and anticancer properties.[11–13] N-methyl thiourea (NMTU) is an amide group which is solvated in formamide is taken for the present investigation. Formamide is used as cryopreservation of tissues and other manufacture of drugs.

EXPERIMENTAL DETAILS

The chemicals used in present work are analytical (AR) reagent grade and spectroscopic (SR) reagent grade of minimum assay of 99.9% obtained from E-Merck. An electrically maintained constant temperature bath has been used to circulate water through the double walled measuring cell made up of steel containing the experimental solution at the desired temperatures. The accuracy in the temperature measurement is ±0.1 K. Since density is one of the prime parameters characterizing many physical properties of a liquid medium. The density of the solution is measured using specific gravity bottle. The viscosity of the solution is measured by the Ostwald's viscometer. Density, viscosity and ultrasonic velocity are determined for N-Methyl thiourea in formamide from 278 K to 328 K at different molalities using compressibility method. From these obtained data thermo chemical and acoustical parameters are calculated using the following formulae.

Internal Pressure [π_i]

$$\pi_i = b\,RT\,(k\eta/u)^{1/2} * (\rho^{2/3}/M_{eff}^{7/6})\ \text{atms}$$

Free Volume [V_f]

$$V_f = (M_{eff} * u/k\eta)^{3/2}\ \text{cc}$$

Apparent Molal Volume

$$\varphi_v = \frac{1000}{c_1\rho_0}(\rho_0 - \rho) + \frac{M_1}{\rho_0}\ \text{ml mol}^{-1}$$

Apparent Molal Compressibility

$$\varphi_k = \frac{1000}{c_1\rho_0}(\rho_0\beta - \rho\beta_0) + \beta_0\frac{M_1}{\beta_0}\ \text{ml mol}^{-1}\ \text{cm}^2\ \text{dyne}^{-1}$$

ρ_0 and ρ are the density of the solvent and solution in gm/cc.

β_0 and β are the compressibility of the solvent and solution in cm^2/dynes.

c_1 – the molal concentration

k – Constant equal to 4.28×10^9

u – ultrasonic velocity in cm/sec

b – cubic constant (2)

T – temperature

M_1 – the molecular weight of the solute in gm.

η – Viscosity of the solution in poise

ρ – density of the solution in gm/cc

R – gas constant (8.314×10^7)

K – constant depends on temperature.

M_{eff} – Effective molecular weight of the solution in gm.

The results of various parameters are given in the Table 1 and Table 2. The variation of the parameters with respect to molalities is represented in Figures 1–4.

Table 1

Parameters	*Molality (m)*	*Temperature in °C*					
		5°C	*15°C*	*25°C*	*35°C*	*45°C*	*55°C*
Internal pressure (π_i) atms	.001	20326	17058	14177	13459	12008	10874
	.005	20884	17408	14609	13800	12064	10903
	.01	20975	17196	14624	13731	12212	10042
	.015	21006	17039	14677	13454	12354	10666
	.02	20505	17092	14710	13550	12379	10524
Free Volume (V_f) cc	.001	0.0073	0.0137	0.0260	0.0330	0.0503	0.0736
	.005	0.0067	0.0127	0.0236	0.0304	0.0495	0.0725
	.01	0.0066	0.0132	0.0235	0.0308	0.0474	0.0928
	.015	0.0065	0.0136	0.0231	0.0325	0.0457	0.0769
	.02	0.0070	0.0133	0.0228	0.0318	0.0452	0.0798

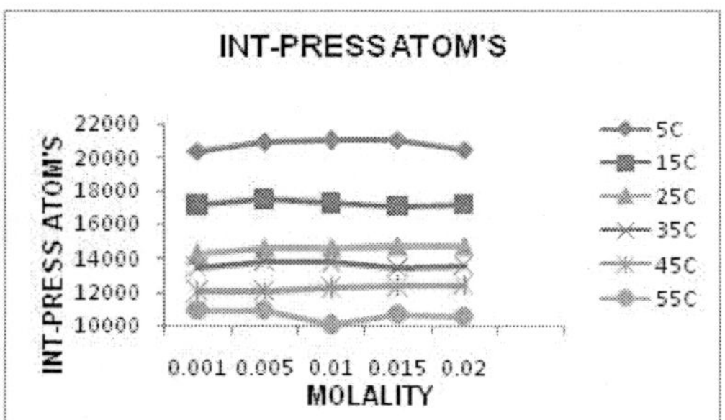

Fig. 1

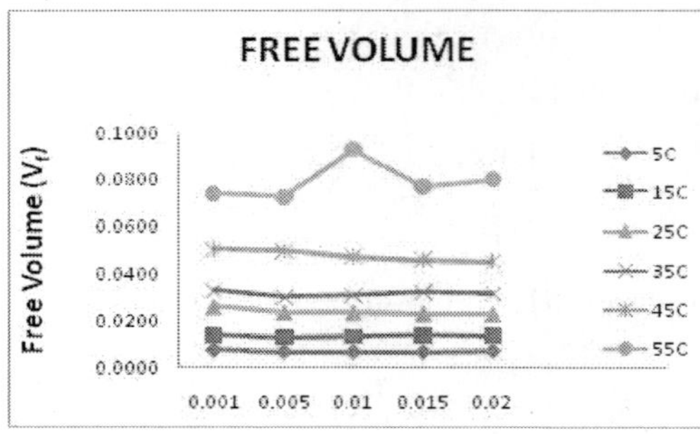

Fig. 2

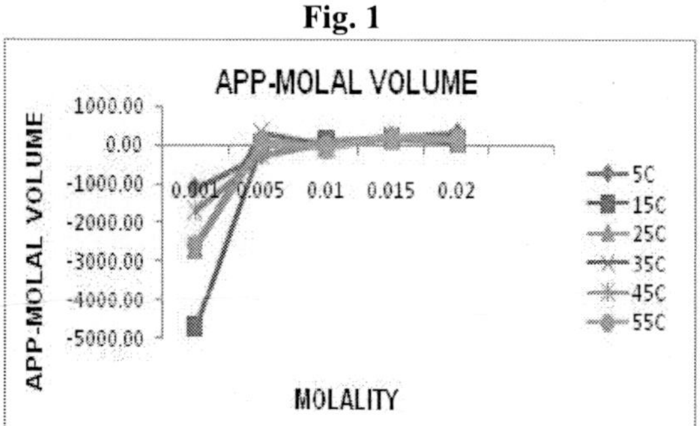

Fig. 3

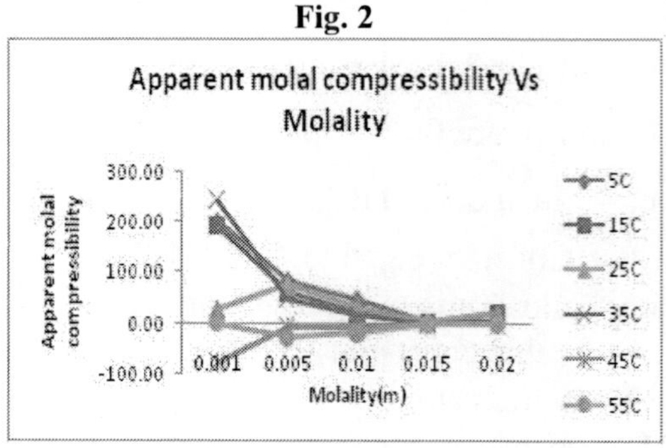

Fig. 4

Table 2

Parameters	*Molality (m)*	*Temperature in °C*					
		5°C	*15°C*	*25°C*	*35°C*	*45°C*	*55°C*
Apparent Molal Volume ml mol^{-1}	.001	–1110.33	–4712.83	–2672.81	–2634.23	–1686.82	–2554.30
	.005	–261.15	7.80	0.32	295.22	–283.54	78.94
	.01	37.30	108.88	–49.50	–33.88	80.64	–78.05
	.015	188.67	162.95	172.05	190.62	184.88	187.30
	.02	303.94	92.33	96.40	92.50	126.09	117.67
Apparent molal Compressibility ml mol^{-1} cm^2 dyne^{-1}	.001	0.0073	0.0137	0.0260	0.0330	0.0503	0.0736
	.005	0.0067	0.0127	0.0236	0.0304	0.0495	0.0725
	.01	0.0066	0.0132	0.0235	0.0308	0.0474	0.0928
	.015	0.0065	0.0136	0.0231	0.0325	0.0457	0.0769
	.02	0.0070	0.0133	0.0228	0.0318	0.0452	0.0798

Spectroscopic Analysis

FTIR spectroscopy is especially used for detection of functional group in organic compounds. FTIR spectrum was recorded between 4000 cm^{-1} to 400 cm^{-1} using PERKIN ELEMER SPECTRUM RX1 spectrometer. The observed spectrum of formamide, thiourea and thiourea in formamide are shown in Figure (5–7).

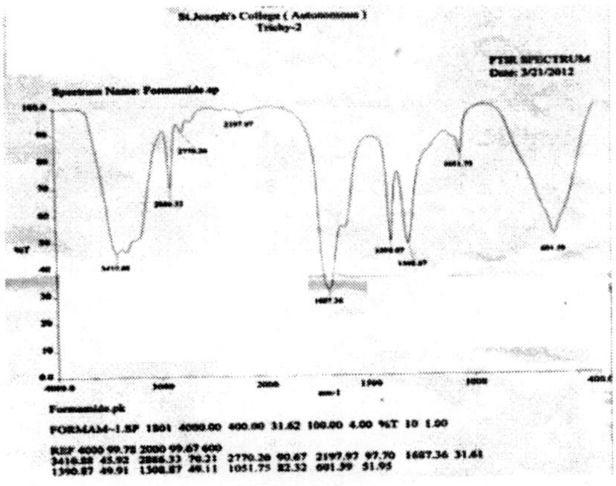

Fig. 5

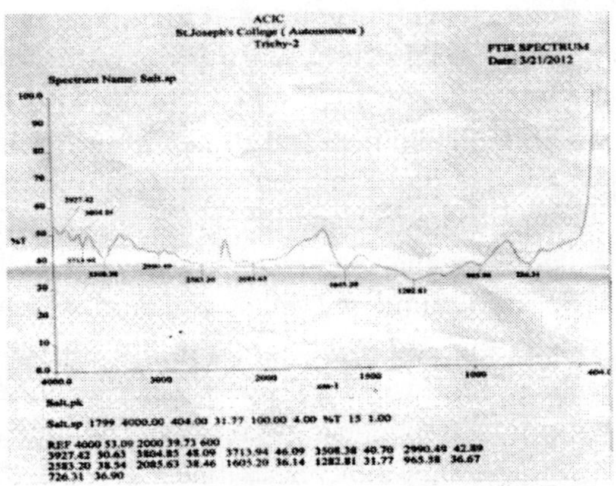

Fig. 6

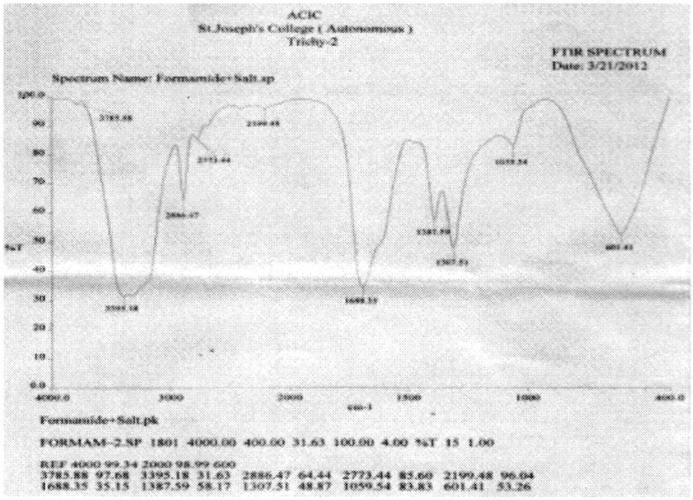

Fig. 7

RESULTS AND DISCUSSIONS

Ultrasonic Analysis

Internal pressure, free volume is reliable and easily measurable parameters, which are fundamentally responsible for the structure of liquid or a liquid solution. Internal pressure is closely related to the solubility parameters, which determine the way in which occur in the system. The reduction in internal pressure may be due to the breaking up of the structure of the solvent medium due to the addition of the solute in the solvent [13] this type of character is observed in all the concentrations as the temperature increases. Internal pressure is known to be an inverse function of free volume. The value of free volume is found to be positive and increase with increasing of temperature. The values of free volume are the resultant of contributions from several opposing effects [14]. In the present work, the positive values of free volume in all the systems show the existence of dispersive forces between the component molecules of the mixtures [15]. Apparent molal volume is the parameter through which molecular interaction in solution can be discussed. In the present study, the apparent molal volume Φ_v of NMTU exhibits negative values for .001 concentration, but for other lower concentrations negative and positive values are present and for higher concentrations it is found to be positive. High values of Φ_v denote the tendency of breaking the structure of the solvent.As the concentration increases the value of Φ_k increases linearly for lower temperature, but for other temperature this linearity is not found. As the concentration of the solute increases and a large portion of the solvent molecules are elecrostricted, the amount of bulk solvent decreases causing the Φ_k to decrease [16].

Spectroscopic Analysis

In Formamide, the following peaks are observed. The stretching vibration of-NH2 group in formamide is observed at 3411 cm^{-1}. The –CH stretching vibration occurs at 2886 cm^{-1}. The CO stretching vibration is found at 1687 cm^{-1}. The bands at 1391 cm^{-1} and 1309 cm^{-1} are assigned to amide II asymmetric and symmetric modes. Bending vibrations (δ_{CH}, δ_{CO}) are observed at 602 cm^{-1}. In N-metyl thiourea the broad band at 3508 cm^{-1} assigned to the amino groups. At 2991 cm^{-1} –CH3 group stretching vibration occurs. The symmetrical distributed vibration at 2583 cm^{-1} and 2086 cm^{-1} are due to the presence of different type of C=NH and C=NCH_3 groups. At 1605 cm^{-1} the =C=S stretching vibration is located. The amide II vibration, the asymmetric and symmetric modes occur as a broad band at 1283 cm^{-1}, 966 cm^{-1} and 726 cm^{-1} may be due to δ_{CS}, δ_{CH}, δ_{NH} modes. When N-methylthiourea is dissolved in formamide the following spectral changes are observsd. The solution spectra is predominated by that of formamide, The NH stretching vibration has blue shifted from 3441 cm^{-1} to 3395 cm^{-1}. This shift may be due to lengthening of N-H bond due to H-bonding interactive with the solute C=S bond. The C=S vibration at 1605 cm^{-1} found in the solute spectrum has merged with the solvent peak and has shifted by 1.5 cm^{-1}. The asymmetric amide stretching frequency has blue shifted by 3 cm^{-1} indicating solvation at NH group. The >C–O vibration at 1059 cm^{-1} has red shifted by 8 cm^{-1} while the >C=N vibration by 3.24 cm^{-1}. These results indicate that solvation has occurred by H-bonding at –NH_2 of formamide. This study also confirms that the solute-solvent interaction occurs in the solution is through inter- molecular H-bonding.

CONCLUSION

Ultrasonic investigation is done for the solvation of N-methyl thiourea with formamide. By computing the thermochemical and thermo acoustical parameters, the varation of these parameters with various concentrarions at different temperatures are analysied. From thermochemical and thermoacoustical behavior, the breaking structure of the solvent is notified. This spectroscopic study also confirms that the solute–solvent interaction occurs in the solution is through inter-molecular H-bonding.

REFERENCES

[1] Kinocid, *J.Am. Chem., Soc.,* (1929) SI, 2950

[2] Mehra and Sajnami, Ind. J. Pure and Appl.Phy., (2000) 38, 760.

[3] Fort R.J. and Moore W.R., Trans Faraday Soeiety, (1965) 61, 2102.

[4] Kasare S.B. and Patdai B.A., Indian J. Pure and Appl. Phys., (1987) 25, 180.

[5] Jain D.V.S. and Dhar N.S., *Ind., J. Tech.*, (1992) 310, 620.

[6] Clier M. and Kesanovil D. Hydrogen bonding edited by Ha, dn, Zi, Paragamon press, (1957) London.

[7] Largeman R.J. and Dundbar W.S., J. Phys., Chem., (1945) 49, 428.

[8] Bell F.W., Cantrell A.S., Högberg M., Jaskunas S.R., Johansson N.G., Jordan C.L., Kinnick M.D., Lind P., Morin J.M., Noreen R., Öberg B., Palkowitz J.A., Parrish C.A., Pranc P., Sahlberg C., Ternansky R.J., Vasileff R.T., Vrong L., West S.J., Zhang H., Zhou X.X.. Phenethylthiazolethiourea (PETT) compounds, a new class of HIV-1 reverse transcriptase inhibitors. 1. Synthesis and basic structure-activity relationship studies of PETT analogs. J Med Chem 1995; 38(25): 4929–4936.

[9] Uçkun F.M., Pendergrass S., Maher D., Zhu D., Tuel-Ahlgren L., Mao C., Venkatachalam T.K. N'-[2-(2-thiophene) ethyl]-N'-[2-(5-bromopyridyl)] thio-urea as a potent inhibitor of NNI-resistant and multidrug-resistant HIV-1.Bioorg Med Chem Lett 1999; 9(24): 3411–3416.

[10] Küçükgüzel İ., Tatar E., Küçükgüzel Ş.G., Rollas S., De Clercq E. Synthesis of some novel thiourea derivatives obtai ned from 5-[(4-aminophenoxy) methyl]-4-alkyl aryl-2,4-dihydro-3H-1,2,4-triazole-3-thiones and evaluation as antiviral/anti-HIV and anti-tuberculosis agents. Eur J Med Chem 2008; 43(2): 381–392.

[11] Todoulou O.G., Papadaki-Valiraki AE, Filippatos EC, Ikeda, S, De Clercq E. Synthesis and anti-myxovirus activity of some novel N,N′-disubstituted thioureas. Eur J Med Chem 1994; 29(2): 127–131.

[12] Kaulgud M.V., Indian journal of chemistry (1990) Vol. 29A, 439–444.

[13] Treszc Zanowicz A.J., Kiyohara O. and Benzon G.C., J. Chem. Thermodyn (1981), 13,253.

[14] Palani R., Saravana S and Kumar R., Rasayan J. Chem., (2009), Vol. 2, No. 3, 622–629.

[15] Millero F.J., Curry R.W., Drost-Hansen W, *J. chem. Data* (1969), 14, 422.

Section–VI

Vibration

Vibration Analysis of Internal Combustion Engine—Modeling and Optimization

T. Ramachandran and K.P. Padmanaban[1]

PSNA College of Engineering and Technology, Dindigul

[1]SBM College of Engineering and Technology, Dindigul

E-mail: ramji_kkp@yahoo.com; padmarubhan@yahoo.co.in

ABSTRACT: *Internal combustion (IC) engine produces intolerable vibrations during the operation, when the vibration forces transferred to the supporting structures the reliability and the driving comfortness are much reduced. It is necessary to reduce the vibrations from the engine to the engine mounts to satisfy customer requirements. The vibration from the engine is due to the inertial and unbalanced forces of the reciprocating and rotating components in the engine. In this work a mathematical model will be developed to determine the engine inertial and unbalanced forces and the model will be solved using a computer program to determine the displacements and velocity at the engine supports.*

Keywords: Vibration, IC Engine, Runge-Kutta Method, Balancing Masses.

INTRODUCTION

Engine vibrations are caused due to the reciprocating and rotating masses of the engine. The variations of inertial forces are due to the combustion and the compression differences of the piston cylinder arrangement during their operation The engine inertial forces leads to the unbalanced forces of the engine and they are quiet varying with respect to speed, fuel supply and combustion characteristics of the fuel. Though the supply and combustion of fuel is also the reasons of the vibration, they are not in the controllable sources for the vibration. Because, they are the parameters decide the performance of an engine. To predict the vibration output of an engine and to minimize the possible durability and consumer perceived quality problems associated with engine vibration, a robust and accurate design and simulation model is needed. Such model should execute quickly and requires relatively simple inputs.

The engine mounts should have characteristics of high stiffness and high damping in the low-frequency range and of low stiffness and low damping in the high-frequency range. Hydraulic mounts do not perfectly satisfy such requirements. Although hydraulic mounts greatly increase damping at low frequencies, they also degrade isolation performance at higher frequencies. Also hydraulic mounts are not cost effective, they had complexity in design and low reliability. Though various types of hydraulic mounts have been developed for the vehicle mount systems, it is still reported that the rubber mounts show significant importance in ride comfort and reduced noise levels. Rubber mounts can be designed for the necessary elastic stiffness rate characteristics in all directions for proper vibration isolation and they are compact, cost-effective, and maintenance free. Also the rubber mounts offer a trade-off between static deflection and vibration isolation. Rubber mounts have been successfully used for vehicle engine mounts for many years.

In this work, a 4-cylinder engine has been considered. The mass of balancing discs and their lead angles can control and reduce the engine vibrations and they are optimized using genetic algorithm. To demonstrate the genetic algorithm, the mass of balancing discs and their lead angles are considered as design variables. The objective function is the function of reaction forces at the engine mounts, inertial forces of the reciprocating parts, unbalanced forces from the crank shaft. The developed model is simulated at a particular speed for specific time and the design parameters of the engine are optimized such that objective function is minimized.

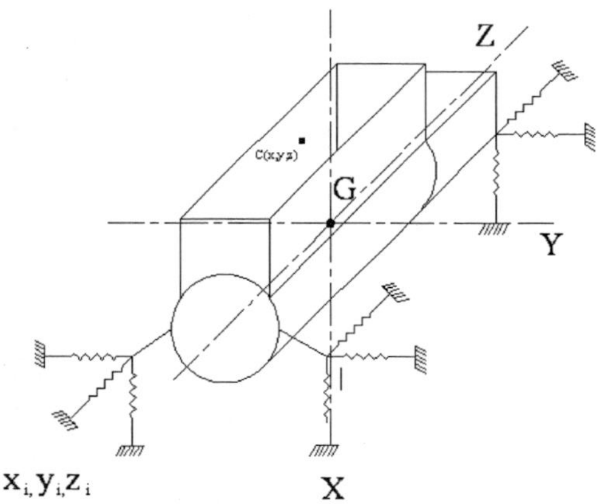

Fig. 1: Rigid Body Model of the Engine

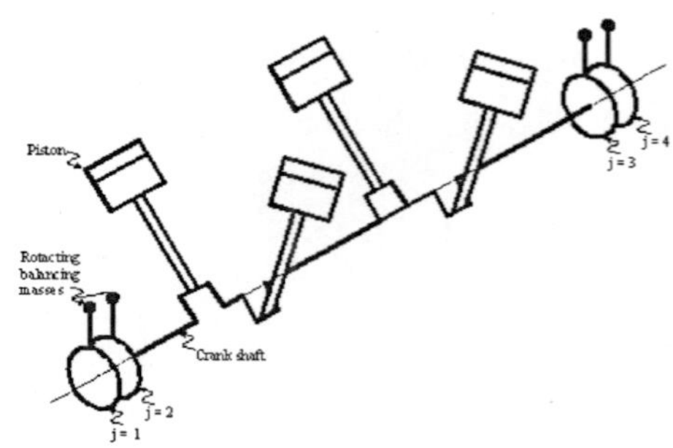

Fig. 2: Crank Shaft with Balancing Masses

KINEMATIC MODELING OF THE ENGINE

Considering an internal combustion engine (shown in Figure 3) mounted on its chassis through three rubber mounts, i.e., two at the front of the engine and the other at the rear end of the engine. The engine is idealized as a rigid body of mass 'M' attached to rigid chassis by means of elastic supports. The origin 'G' of the fixed global coordinate system 'G_{xyz}' is located at the centre of the mass 'C' of the engine when in static equilibrium. The z-axis is parallel to the crankshaft and the x-axis is in the vertical direction. Under the idling conditions, the centre of mass will of course not necessarily coincide with the origin. In certain applications, it is of paramount importance to go to extremes in order to eliminate acoustical radiation. In such cases, the counter weights or balancing masses are geared to rotate about the crankshaft at the same angular velocity as the crank may be justified.

Kinematics and Dynamics of Piston and Connecting Rod

The engine vibration can be divided into two categories: vibration of engine parts relative to each other, called internal vibration, and the movement of the engine as a whole, called external vibration. Unbalanced forces and moments cause the external vibration. As for the vibration isolation, an engine is normally regarded as a rigid-body having six degrees of freedom (DOF) about orthogonal axis through its center of gravity: linear vibrations along each axis and rotations about each axis . In the classical determination of diesel excitations, only three modes are usually of consideration:

- Vertical oscillations on the X axis due to unbalanced vertical forces,
- Rotation about the Z axis due to unbalanced vertical forces in different trans verse planes,
- Rotation about the Y axis due to cyclic variations in torque.

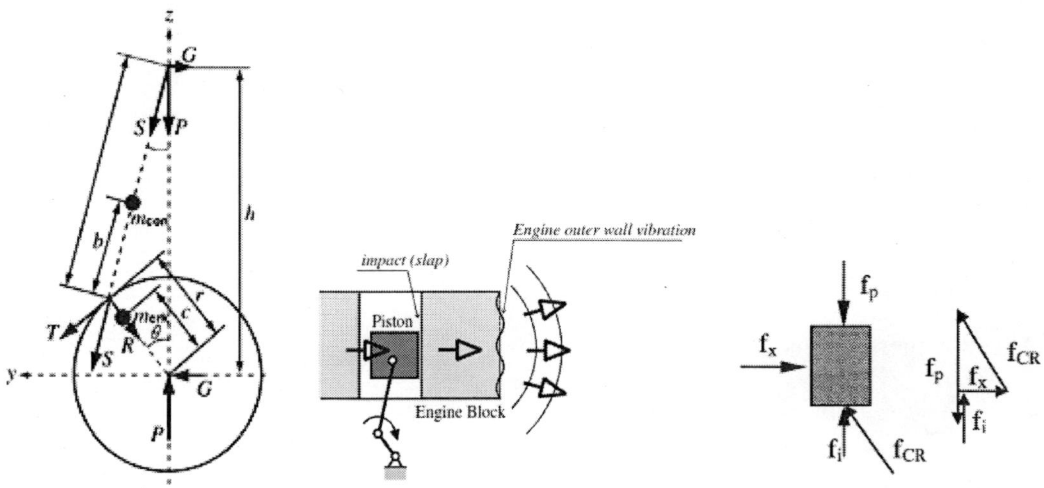

(a) (b) (c)

Fig. 3: (a) Kinematics of Piston and Connecting Rod, (b) Typical Slap Mechanism, (c) Resolved Forces in Cartesian Coordinates

The variation of torque, which balances the torque from the shaft and must be absorbed by elements such as torsional elastic couplings outside the engine frame, will not be considered here. For in-line engines force couples about Z-axis due to unbalanced vertical forces in different transverse plane are fully balanced. In general the rotating masses are carefully balanced but periodic forces due to the reciprocating masses cannot be avoided. A crankshaft, connecting rod and piston assembly, see Figure 1, is subjected to a periodic force in the line of action of the piston given approximately by

$$f_i = m_p R\omega^2\left(\cos\theta + \frac{R}{L}\cos 2\theta\right) \qquad f_x = (f_p - f_i)\tan\beta \qquad \tan\beta = \frac{\sin\beta}{\sqrt{1-\sin^2\beta}} \cong \frac{R}{L}\sin\theta\left[1 + \frac{1}{2}\left(\frac{R}{L}\right)^2 \sin^2\theta\right]$$

for constant crankshaft rotational speed(ω) and where m_p is the effective mass of the piston and includes that fraction of the mass of connecting rod. R and L are respectively the crank radius and connecting rod length. This inertia force associated with piston reciprocating motion is of most concern in classical analysis of engine vibration isolation. From the force diagram as shown in Figure 2, the lateral piston- slap force can be written as:

$$f_x \cong \frac{R}{L}\sin\theta\left[\frac{\pi}{4}D^2 p - m_p R\omega^2\left(\cos\theta + \frac{R}{L}\cos 2\theta\right)\right]$$

Another one is called inertia-controlled impact which occurs in those parts of the cycle for which the pressure forces are negligible. Hence, the lateral piston-slap force Figure 1c for one cylinder is given by:

$$f_x \cong f_{xp} + f_{xi} \qquad f_{xp} \cong \frac{\pi}{4}D^2 p \frac{R}{L}\sin\theta \qquad f_{xi} \cong -m_p\omega^2 R\left(\cos\theta + \frac{R}{L}\cos 2\theta\right)\frac{R}{L}\sin\theta$$

The force exerted by the entire engine—the sum of all force of all cylinders and is given by:

$$\sum f_i = m_p\omega^2 R\left(\sum \cos\theta_i + \frac{R}{L}\cos 2\theta_i\right) \qquad \sum f_x = \left\{\sum\left[\frac{\pi}{4}D^2 p_i - m_p\omega^2 R\left(\cos\theta_i + \frac{R}{L}\cos 2\theta_i\right)\right]\frac{R}{L}\sin\theta_i\right\}$$

Then the inertia based piston slap is given by:

$$\sum f_{xi} = -m_p\omega^2 R\sum\left(\cos\theta_i + \frac{R}{L}\cos 2\theta_i\right)\frac{R}{L}\sin\theta_i = 0$$

The unbalanced forces at the balancing discs is given by:

$$f_{xbj} = m_j r_j \omega^2 \cos(\Delta_j - \theta) \quad f_{ybj} = m_j r_j \omega^2 \sin(\Delta_j - \theta)$$

Using the above procedure, the reciprocating mass exerted vertical inertia force and piston-slap imposed force on the mounts located at the chassis of a 4-stroke, 4-cylinder in-line diesel engine are calculated for a mass range of 0.2–04kg and for the lead angle range of 0–360°.

RESULTS AND DISCUSSION

In this model and optimization problem the engine balancing mass discs and their lead angles are considered as the parameter for the model and the displacements at the engine block. The model is programmed in the Matlab with optimization code. The model is simulated to a specified time of 0.04s and the force and displacement values are obtained for the different orientation of the piston connecting rod. In the optimization at the GA the initial population was generated as a set of binary strings of specified string length and the string is converted in to the real values to be used as input for the mathematical model. Based on the objective function obtained for the generated populations the reproduction was performed through the calculation of the fitness function and by the selection (roulette wheel) operators. In the binary strings obtained in the reproduction are reformed to form the new chromosomes in the crossover process.

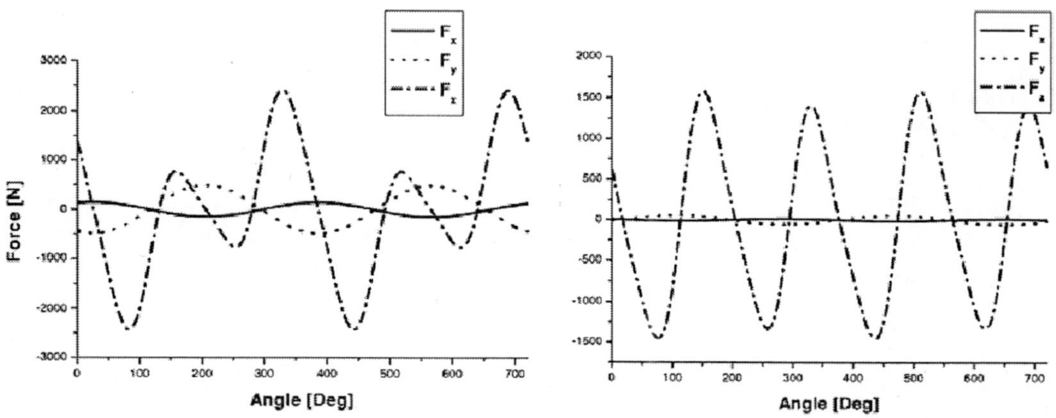

(a) Before Optimization (b) After Optimization

Fig. 4: Engine Forces

After the single point cross over at the selected chromosomes the mutation at the selected points of the chromosomes were performed. The final sets of chromosomes are converted in to the real variables and fed in to the mathematical model to obtain the displacements for different sets of the new chromosomes. The genetic process was performed for the different combinations of cross over and mutation probabilities and minimum displacements at the block were obtained from different runs of the GA. The variables corresponding to the minimum displacement was used in the mathematical model and the inertial and slap forces at the engine cylinder wall at every angle are obtained (Table 1). The slap forces at the cylinder wall were comparatively less and hence displacement Figure 4 at block after the optimization was performed.

Table 1: Optimized GA Results of Displacement for Different Crossovers and Mutations

GA Run	*Least engine Mount Resultant Displacement (mm)*	*Mass of Balancing Discs (kg)*				*Lead Angle (degrees)*			
		m_1	m_2	m_3	m_4	Δ_1	Δ_2	Δ_3	Δ_4
1.	0.0022	0.30	0.36	0.24	0.37	164	218	220	162
2.	0.0022	0.36	0.39	0.30	0.39	198	194	130	142
3.	0.0027	0.26	0.39	0.40	0.40	88	178	188	296
4.	0.0020	0.30	0.40	0.38	0.37	166	194	162	206
5.	0.0021	0.35	0.36	0.33	0.33	144	186	210	162

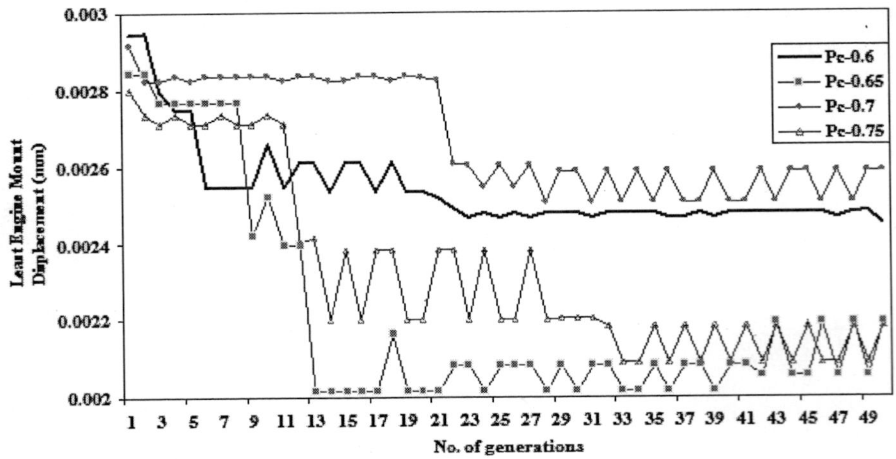

Fig. 3: GA Convergence for Different Runs

CONCLUSION

The dynamics analysis of diesel engine through investigation of the piston-slap force by considering the friction and connecting rod effects was conducted. Analytical calculations for kinematics and dynamics analysis were applied. Kinematics analysis was based on the vector analysis and dynamics analysis was based on the Cartesian coordinate principles. The design variables are optimized using GA to minimize the piston slap forces from the moving parts to the engine block. Although this analysis has led to predictions of the vibrations due to piston slap, the present work makes no attempt to assess the importance of piston-slap relative to that of other potential vibration sources.

One may reasonably expect piston slap to dominate the "mechanical noise" of reciprocating compressors with quiet accessories, where there is no "combustion noise". However, for reciprocating machines in general one requires experimental data in order to determine whether and under what conditions piston-slap noise predominates. A possible means for studying this dominance would consist of comparing experimentally observed dependences of noise and vibrations on various engine parameters with dependences implied by the development presented here.

However, it must be pointed out that the present analysis is not entirely complete, and that it may fail to consider some parameters conceivably of considerable importance. The most important likely discrepancy between the present analysis and reality is associated with the tacit assumptions that were made that all forces acting on the piston act through its center of gravity, and that the wrist-pin centerline passes through this center of gravity. Thus, rotation or cocking of the piston during its travel across the clearance space has been neglected, although such rotations may have possibly marked effects on the character as well as on the severity of the piston-slap impacts. Other shortcomings may be associated with the neglect of piston-ring friction, of oil film cushioning, of angular accelerations of the crankshaft, of clearances at the crankshaft bearings, and of clearances and friction torques at the wrist-pin and crank-pin. However, one may entertain the reasonable hope that these omissions are of minor importance.

REFERENCES

[1] Zheng-Dong Ma, Noel C. Perkins, "An efficient multibody dynamics model for internal combustion engine systems", *Multibody System Dynamics,* Vol. 10, 2003, 363–391.

[2] Rajendran S., Narasimhan M.V., "Effect of inertia variation due to reciprocating parts and connecting rod on coupled free vibration of crank shaft". *ASME Journal of Engineering for Gas Turbine and power,* Vol. 19, 1997, 257–263.

[3] Ohadi, A.R., Maghsoodi G., "Simulation of engine vibration on non-linear hydraulic engine mounts", *ASME Journal of Vibrations and Acoustics,* Vol. 129, 2007, 417–42.

[4] Chung-Ha Suh, Clifford G.Smith,"Dynamic simulation of engine mount system". *SAE Paper#971940.*

[5] Tetsuro Butsuen, Masaaki Ookuma, Akio Nagamatsu,"Application of direct system identification method for engine rigid body mount system", *SAE Paper#860551.*

[6] Sudhir Kaul, Anoop K. Dhingra, Timothy G. Hunter,"Frame flexibility effects on engine mount optimization for vibration isolation in motor cycle", *ASME Journal of Vibrations and Acoustics,* Vol. 129, 2007, 590–600.

[7] Conti P, Bretl J, " Mount stiffness and inertia properties from modal test data", *ASME Journal of Vibrations and Acoustics-Stress and Reliability in Design,* Vol. 111, 1989, 134–141.

[8] Snyman J A, "An improved version of the original leap frog dynamic method for unconstrained minimization: LFOPI ", *Appl. Math. Modelling,* Vol. 7, 1983, 216–218.

[9] Niccolo Baldanzini, Davide Carprioli, Marco Pierini, "Designing the dynamic behaviour of an engine suspesion system through genetic algorithm", *ASME Journal of Vibrations and Acoustics,* Vol. 123, 2001, 480–486.

[10] James E. Bernard, John M. Starkey, "Engine mount optimization", *SAE#830257.*

[11] Tsuneo Tanaka, Mitsuo Iwahara, Tetsuya Sakai, "The optimization of engine vibration reduction by simulation analysis", *SAE#962203.*

[12] Ashrafiuon H, Natraj C, "Dynamic Analysis of Engine-Mount Systems", *ASME J. Vibr. Acoust,* Vol. 114, 1992, 79–83.

[13] Ashrafiuon H,"Design Optimization of Aircraft Engine-Mount Systems", *ASME J. Vibr. Acoust.*, Vol. 115, 1993, 463–467.

[14] Kim G, Singh R, "A Study of Passive and Adaptive Hydraulic Engine Mount Systems With Emphasis on Non Linear Characteristics", *J. Sound Vib.*, Vol. 179(3), 1995, 427–453.

[15] Jun-Hwa Lee, Kwang-Joon Kim,"An Efficient Technique for Design of Hydraulic Engine Mount via Design Variable-Embedded Damping Modeling", *ASME Journal of Vibrations and Acoustics,* Vol. 127, 2005, 93–99.

Wave Theory Approach for Developing the Mathematical Model of the Right Angled Joint

V.H. Patil and D.N. Manik

Department of Mechanical Engineering, Indian Institute of Technology, Bombay, Mumbai

ABSTRACT: *In this study, vibrational energy distribution within the audible frequency range is considered between two plates. The two plates are connected to form a welded joint (line connection), riveted and bolted joint (point connection). The mathematical model for the joints is developed using a spring and dashpot model. The wave approach is used to develop the mathematical model of the coupled plate. The bending wave transmission and reflection efficiencies are determined. The limiting value of high stiffness joint is determined, which represents the weld joint like a rigid joint. The bolt and rivet joints are modelled, with spring dashpot in only one direction and in another direction the joint is considered as rigid. After studying the stiffness effect, the mathematical model is updated to include damping at the joint. The damping has smoothening effect on these efficiencies.*

INTRODUCTION

Joints are an important part of any complex structure. The waves propagating from one component to other pass through these joints. The joint has its own stiffness and damping, which particularly attenuates the incident wave and then transmits the rest to the other connected component. Leung and Pinnington[1] applied wave theory to a right angled joint with compliance. They have found the transmission and reflection coefficients for the joints. The stiffness and damping at the joint are the major factors which influence energy transfer between the coupled systems. Cremer[2] and others have studied the attenuation of an incident wave through a joint. Ungar[3] has given detailed classification of the damping present in a system, which is responsible for the energy loss. Yoshimura and Okushima[4] developed a method to identify the quantitative values of the rigidity and damping coefficients at the joint surface by interactive use of theoretical simulation and experiment. Mead[5] worked on aircraft structures having riveted and bolted joints.

JOINT MODEL

Figure 1 shows the right angle joint model with spring dashpot attached. For the cases of a joint with springs in all three degrees of freedom, wave approach is used to analyse the joint. In the Figure 1 k_1, k_2 are linear stiffness and k_3 is rotational stiffness. Similarly, c_1, c_2 and c_3 are the viscous dampers. The plate 1 is in $x-z$ plane and plate 2 is in $y-z$ plane. Both the plates are symmetric about the z axis, in z plane .The coordinate system is taken at the mid of the both plates in z plane and at the joint location. The coordinate system represents $x=0$, $y=0$, $z=0$ at the joint.

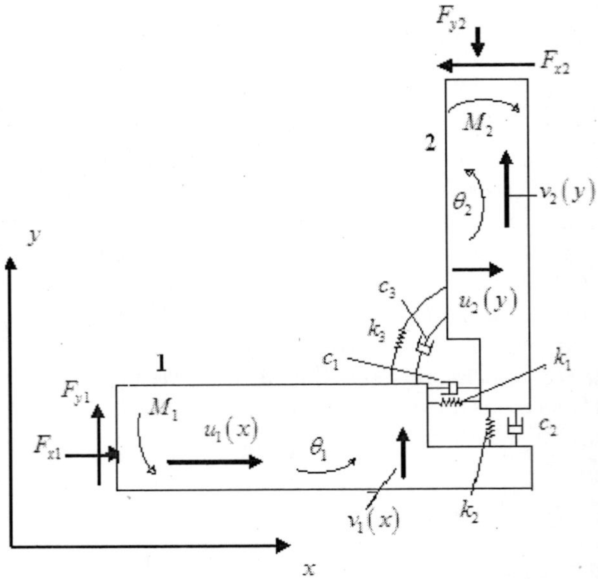

Fig. 1: Linear Spring-Dashpot Model of Joint

Consider the incident wave of amplitude A_1 travelling towards the joint in plate 1. Therefore, the wave field in the component 1 is composed of three parts; incident wave, reflected wave and near field wave at the joint. The transverse displacement $v_1(x)$ is given as,

$$v_1(x) = A_1\left(e^{-jk_{b1}x} + r_B e^{jk_{b1}x} + r_{B,n} e^{k_{b1}x}\right) \qquad \dots (1)$$

In plate 2, the transmitted bending wave and near field wave at the joint gives the transverse displacement $u_2(y)$ as

$$u_2(y) = A_1\left(t_B e^{-jk_{b2}y} + t_{B,n} e^{-k_{b2}y}\right) \qquad \dots (2)$$

where r_B , $r_{B,n}$, t_B and $t_{B,n}$ are the bending wave reflection coefficient, bending wave near field reflection coefficient, bending wave transmission coefficient and bending wave near field transmission coefficient respectively. Here, k_{b1}, k_{b2} are the bending wave numbers for plate 1 and 2 respectively. As the bending wave goes through a corner, longitudinal wave will be induced in both the components as,

$$u_1(x) = A_1 r_L e^{jk_{L1}x}\,;\; v_2(y) = A_1 t_L e^{jk_{L2}y} \qquad \text{... (3)}$$

where, r_L, t_L are the longitudinal wave reflection and transmission coefficients respectively. Also k_{L1} and k_{L2} are the wave numbers, generating longitudinal wave in component 1 and 2 respectively.

By equating forces and moments at the joint, boundary conditions developed for the determination of coefficients are,

$$\begin{aligned} F_{x1} &= F_{x2} = Z_x(u_1 - u_2); \\ F_{y1} &= F_{y2} = Z_y(v_1 - v_2); \\ M_1 &= M_2 = Z_\theta(\theta_1 - \theta_2). \end{aligned} \qquad \text{... (4)}$$

The transmission coefficients t_B, t_L and reflection coefficients r_B, r_L are determined from the equations (1 to 4).

$$r_B = \frac{[(1-j)(Z_2+\beta_2)-2]}{[(1+j)(Z_2+\beta_2)+2]} - \psi t_B \frac{[(1+j)(Z_1+\beta_1)+2][1+j(Z_2+\beta_2)]}{[1+j(Z_1+\beta_1)][2+(1+j)(Z_2+\beta_2)]}; \qquad \text{... (5)}$$

$$t_B = \frac{2[Z_1+\beta_1+Z_2+\beta_2]-2j[1-(Z_1+\beta_1)(Z_2+\beta_2)]}{D}. \qquad \text{... (6)}$$

where

$$\begin{aligned} D = {} & X\{-1+j-2(Z_1+\beta_1)-(Z_2+\beta_2)-(1+j)(Z_1+\beta_1)(Z_2+\beta_2)\} \\ & +\psi\{-1+j-(Z_1+\beta_1)-2(Z_2+\beta_2)-(1+j)(Z_1+\beta_1)(Z_2+\beta_2)\} \\ & +\psi Z_3\{-(Z_1+\beta_1)(Z_2+\beta_2)+2j-(1-j)(Z_1+\beta_1)-(1-j)(Z_2+\beta_2)\}. \end{aligned} \qquad \text{... (7)}$$

where,

$$\beta_1 = (m_2 C_{B2})/(m_1 C_{L1});\; \beta_2 = (m_1 C_{B1})/(m_2 C_{L2});\; Z_x = (k_1/j\omega)+c_1\,;$$

$$Z_y = (k_2/j\omega)+c_2\,;\; Z_\theta = (k_3/j\omega)+c_3\,;\; Z_1 = m_2 C_{B2}/Z_x\,;\; Z_2 = m_1 C_{B1}/Z_y\,;$$

$$Z_3 = B_1/(C_{B1} Z_\theta);\; \psi = \frac{k_{b2}^2 B_2}{k_{b1}^2 B_1};\; X = k_{B2}/k_{B1}.$$

The efficiencies for the transmitted and reflected bending waves and longitudinal waves are

$$\tau_B = X\psi |t_B|^2\,;\; \rho_B = |r_B|^2 \qquad \text{... (8)}$$

Weld Joint (Rigid Joint)

The weld joint mathematical model is obtained by setting k_1, k_2, k_3 to infinity in Eqs. (5) to (8). Therefore, dimensionless terms Z_1, Z_2, Z_3 become zero.

Bolt/Rivet Joint (Joint with compliance in only one direction)

Similarly, for the bolt/rivet joint mathematical model is obtained by setting $X = \psi = 1, \beta_1 = \beta_2 = \beta$ and $Z_1 = Z_3 = 0, u_1 = u_2, \theta_1 = \theta_2$. in Eqs. (5) to (8).

RESULTS

Here the two mild steel plates, each of size 300 X 300 X4 mm were considered to form a weld joint, bolt/rivet joint. The critical frequency f_{cr}, for 4 mm thickness plate is 3210 Hz. The transmission and reflection efficiencies are plotted against frequency in Figures 2 & 3, without considering the damping at the joint.

From Figure 2, the maximum value of τ_B for low stiffness value to very high stiffness is approximately 0.48. This indicates that a maximum up to 48% of the supplied energy is transmitted by a flexural wave. As the joint stiffness increases, the frequency for attenuation of τ_B increases. Reflection efficiency ρ_B plotted against frequency in Figure 3, almost 50% to 90% of supplied energy is reflected back by the softer joint. A very stiff joint 10 GN/m to ∞ N/m, reflect less amount of energy compared to softer joint. The peaks in the reflection efficiency are observed for stiffness value of 1 MN/m and 100 MN/m

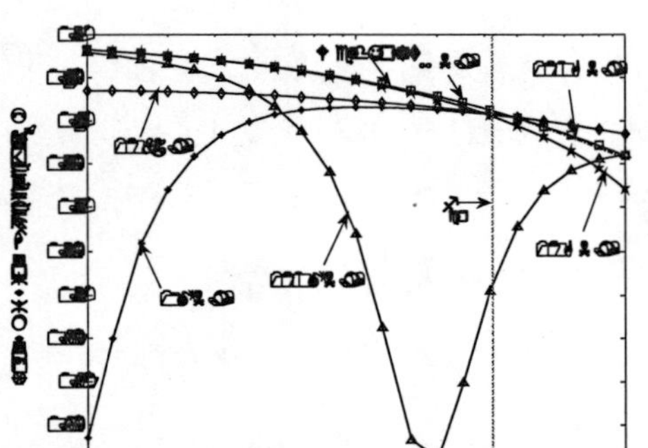

Fig. 2: Bending Wave Transmission Efficiency (τ_B) for Various Stiffness Values

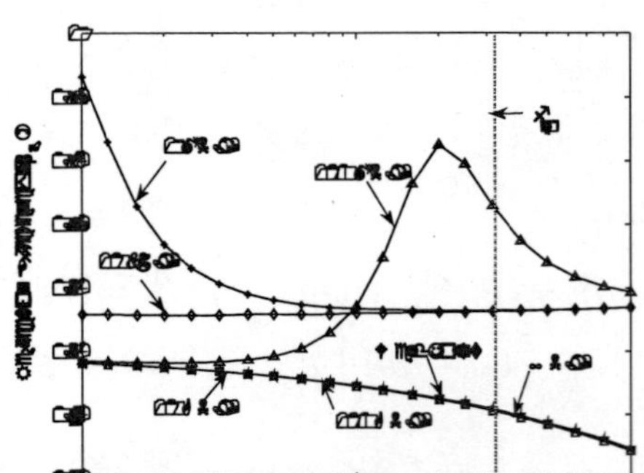

Fig. 3: Bending Wave Reflection Efficiency (ρ_B) for Various Stiffness Values

in the frequency range at the same location where dips are observed in Figure 2. From Figures 2 and 3, the efficiency plots for $k_y = \infty$ N/m resemble the efficiency plot in for a rigid joint. Therefore 100 GN/m stiffness is taken as joint stiffness value for the weld joint.

EFFECT OF DAMPING

The effect of damping on the τ_B and ρ_B efficiencies of the coupled plates is shown in Figures 4 and 5. Here, the assumption of viscous damping is used to represent the joint damping. The curve becomes smooth as damping increases. At very high damping, τ_B never becomes zero as compared to lower damping value of zero and 100 Ns/m. Similarly, the peak in the ρ_B decreases and becomes smooth with the increase in the damping.

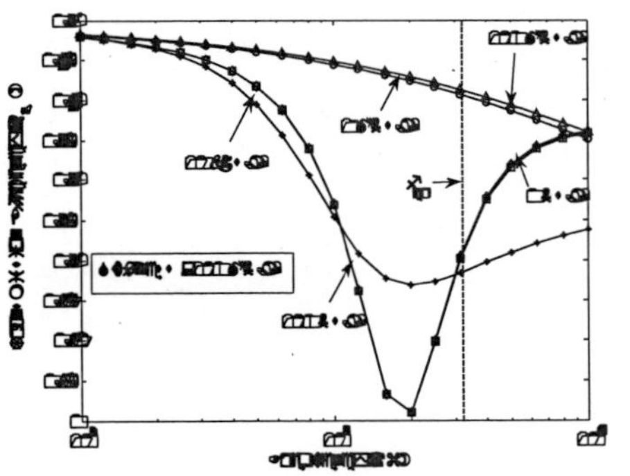

Fig. 4: Effect of Variation of Damping on Bending Wave Transmission Efficiency (τ_B)

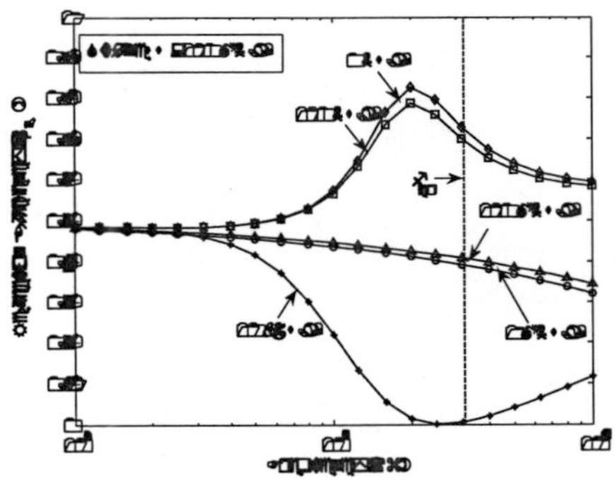

Fig. 5: Effect of Variation of Damping on Bending Wave Reflection Efficiency (ρ_B)

CONCLUSION

In conclusion, the limiting value of stiffness (100 GN/m) for weld joint is obtained. For riveted and bolted joints, stiffness has more effect on the energy transfer between the coupled systems and the effect of damping is more at low stiffness of the joint. The damping can be controlled by application of lubricating layer between the two faces. The bending wave transmission efficiencies determined can be used to determine the coupling loss factor of the various joint connections.

REFERANCES

[1] Leung, R. and Pinnington, R., Wave propagations through right angled joints with compliance—flexural incident wave, *Journal of Sound and Vibration*, 142(1), 31–46 (1990).

[2] Cremer, L., Heckl, M. and Ungar, E.E., *Structure-Borne Sound*, Berlin: Springer-Verlag (1987).

[3] Ungar, E.E., The status of engineering knowledge concerning the damping of built- up structures, *Journal of Sound and Vibration,* 26(1), 141–154 (1973).

[4] Yoshimura, M. and Okushima, K., Measurement of dynamic rigidity and damping property for simplified joint models and simulation by computer, *Annals of CIRP* 25, 193–198, (1977).

[5] Mead, D.J., The damping, stiffness a fatigue properties of joints and configuration representative of aircraft structures, *WADC-University of Minnesota Conference Acoustical Fatigue,* 235–261 (1961).

Minimization of Downtime in Industries by Acoustical Vibration Control

D. Joslin Vijaya and S. Charles[1]

Department of Mechanical Engineering,
Dhanalakshmi Srinivasan Engineering College, Perambalur
[1]Dhanalakshmi Srinivasan College of Engineering, Navakkarai Post, Coimbatore
E-mail: joslinvijayamech@rediff.com

ABSTRACT: *Several methods can be used to monitor the condition monitoring of a machine. They are Aural, Visual, Operational variables, Temperature, Wear debris and Vibration techniques. In aural method a skilled technician, having an intimate knowledge of machines, can identify a failure simply by listening to the sound produced by a damaged machine. Sometimes a microphone or a stroboscope is used to hear the machine noise. The measured vibration levels can be extrapolated in order to predict whenthe vibration levels reach unacceptable values and when the machine must be serviced fully. Hence the periodic measurement of vibration characteristics of machinery and structures becomes very essential to ensure adequate safety margins. The occurrence of frequent resonant conditions during the operation of machinery is also avoided. By means of this the operating speeds of machinery and the productivity rate are tremendously increased which leads to the agility in manufacturing automation. The status of the product quality and of the machine is displayed in the simplest possible way, for examplea green/red light showing the status is good/bad and short messages in case of a bad process or machine situation. This requires fully automatic diagnostic algorithms and also a database to store the records and perform a statistical long-term analysis. The time signal, FFT-spectra, cumulative FFT-spectra, etc. allows the performance of standard signal processing and analysis.*

OBJECTIVES OF THE PAPER

The days when engineers, operators and maintenance staff interact only with the process automation system are over. Plant staff will be required to interact with the production management systems and the process automation systems, virtually on a simultaneous basis. Production management automation allows optimization of production activities, ranging from the scheduling of job orders to the traceability of finished products. Computerized production management is also referred to as a "Manufacturing Execution System" (MES). One of the objectives of MES is zero breakdown. In recent years, the digital analyzers have become quit popular for real-time signal analysis. Real-time analyzers are especially useful for machinery health monitoring since a change in the noise can be observed at the same time the change in the machine occurs. Thus the calculation process must not take more time than the time taken to collect the signal data.

Most of the prime movers have vibration problems due to the inherent unbalance. The unbalance may be due to faulty design or poor manufacture. When the machine component is subjected to vibration it may fail because of material fatigue resulting from the cyclic variation of the induced stress. The vibration causes more rapid wear of machine parts such as bearings and gears and also creates excessive noise. In machines, vibration causes fasteners such as nuts to become loose. In metal cutting processes, vibration can cause chatter, which leads to a poor surface finish. Whenever the natural frequency of a machine or a structure coincides with the external excitation, there occurs resonance, which leads to excessive deflections and failure.

Because of the devastating effects that vibrations can have on all machines and structures, vibration testing has become a standard procedure in the design and development of most engineering systems. Many basic problems of vibrations are nonlinear. In nonlinear systems, phenomena may occur that are theoretically impossible to solve in linear systems. Then the mathematical theory of non linear vibrations were developed. The advent of high speed digital computers made it possible to treat even complex systems.

METHODOLOGY ADOPTED

Sound which can be either low, soft or loud is created by vibration. When an object moves rapidly back and forth, the air molecules around it are disturbed. The disturbance moves away from the object in waves, which is what is interpreted as sound. Often sensors like nono-power tilt, shock and vibration sensors, calibrated OEM board sensor, internal sensor systems for harsh environment. These sound waves are also measured by means of acoustical instruments such as sound level meters, noise dosimeter, sound calibrators, measurement microphones, octave band, third-octave band analyzers and signal analyzers. Spectrum or frequency analyzers can be used for signal analysis. This is a device that analyses a signal in the frequency domain by separating the energy of the signal into various frequency bands. The separation of signal energy into frequency bands is accomplished through a set of filters. The analyzers are usually classified according to the type of filter employed. For example, if an octave band filter band is used, the spectrum analyzer is called an octave band analyzer. The basic component of a spectrum analyzer is the band pass filter. A band pass filter is a circuit built by using resistors, inductors and capacitors that permits the passage of frequency components of a signal over a frequency and rejects all other frequency components of the signal. For a good band pass filter, the ripples within the band will be minimum and the slopes of the filter skirts will be steep to maintain the actual band width. There are two types of band pass filters: the constant percent bandwidth filters and constant bandwidth filters. Since the output impedance of transducers is not suitable for direct input into the signal analysis equipment, signal conditioners, in the form of charge or voltage amplifiers, are used to match and amplify the signals before signal analysis. The response signal, after conditioning, is sent to an analyzer for signal processing. A commonly used analyzer is FFT analyzer. This receives analog voltage signals from an amplifier, filter and digitizer for computations. It computes the discrete frequency spectra and these signals are used to find the natural frequency, damping ratios and mode shapes either in numerical or graphical form.

Many instruments are available now a days to custom sound and vibration measurement and analysis systems. The instruments are used for audio testing, acoustic measurements, environmental noise testing, vibration analysis, noise vibration and harshness measurements, noise and vibration control, machine condition monitoring, rotating machine evaluation, etc.

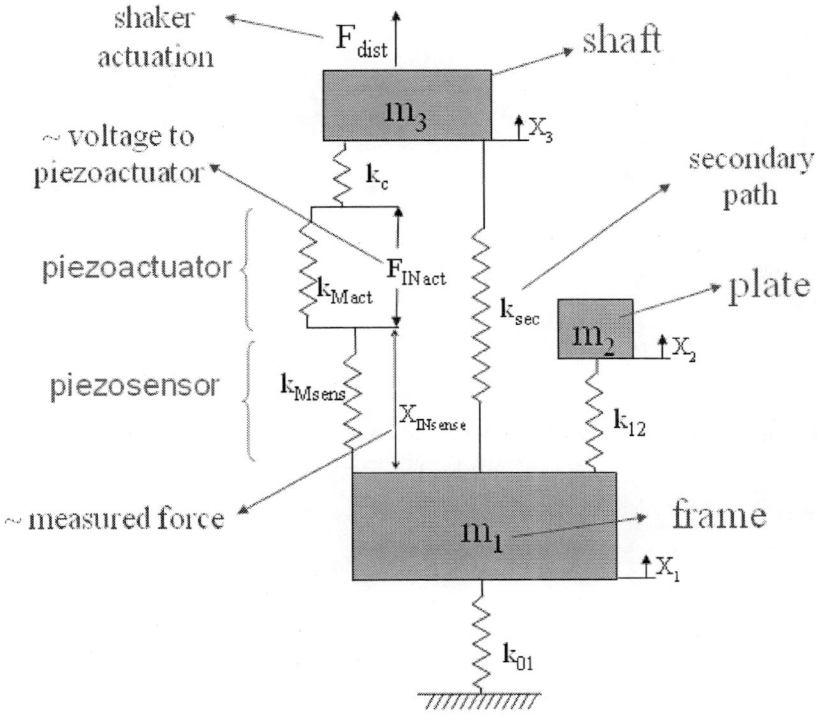

Simplified Lumped Parameter Model of the Experimental Test Bed

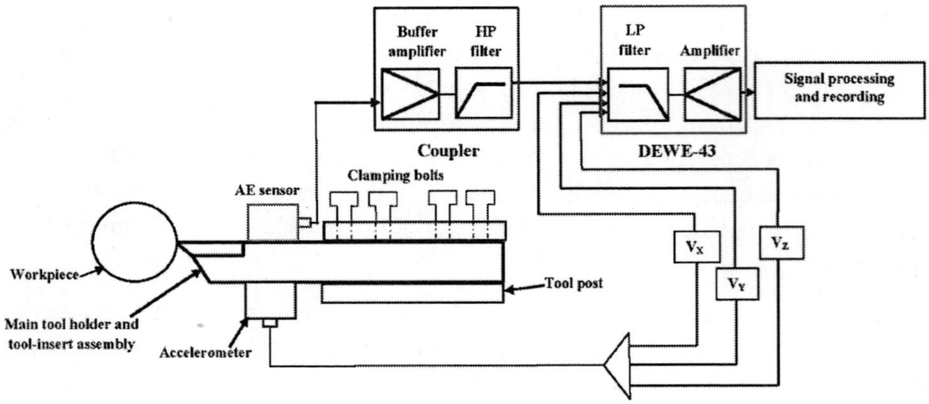

Experimental Setup to Capture the Acoustic Emission and Vibration Signal from the Cutting Tool in Turning

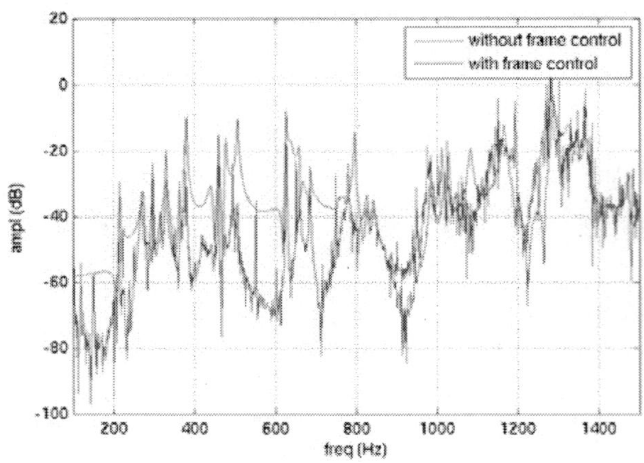

Theoretical Reduction of the Plate Acceleration when the Frame Acceleration would Be Ideally Controlled to Zero

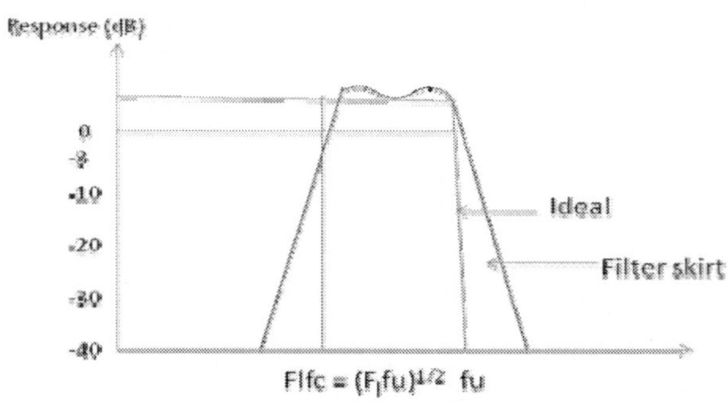

Response of a filter

The MATLAB commands to detect threshold crossings are based upon the use *ofthejindand interpl* functions as shown below.

```
% create a new vector, time shifiedby one data point
Shifidata = [data(l);data([l:length(data)-l])J;
Yofindthe time points which cross the threshold with positive going slope.
Z = find(data>threshold&shifidata<threshold);
% interpolate on each zero crossingtofindan accurate
estimate of actual
% crossing time
n = length(z);
Jork = l:n
% construct the short 3ptx andy vectors for interpolation
at each revolution
y = [data(z(k)-l,data(z(k)),data(z(k)+l)];
x = [(z(k)-l)*dt,z(k)*dt.(k)+1)*dt];
s(k) = interp 1 ty.x.threshold.spline'):
end.
```

CONCLUSION

The partticular feat5ure of acoustic emission (AE) is its ability to directly detect the processes associated with wear and degradation (including friction, impacts, crushing, cracking, turbulance etc.). Another convenient feature of AE signals is that they generally have a high signal to noise ratio (SNR) which means that the signals related to machine condition can be clearly seen and are not buried in other consequential signals. This is a direct consequence of the high frequency and resonant nature of an AE sensor response. AE sensors, signals and positive consequences in practical terms. In the frequency domain a modified peak ratio was proposed and was proven to a better indicator than the original peak ratio references. Likewise many more properties could be enumerated.

REFERENCES

[1] Journal of Sound and Vibration.
[2] ASME Journal of Vibration and Acoustics.
[3] Journal of the acousticval Society of America.
[4] Noise and Vibration worldwide.
[5] Mechanical vibrations by Singarasu S. Rao.
[6] Thoery of machines by Shigley.
[7] Thoery of machines by Thomas Bevan.

Vibration Isolation System for Hydrophones

V. Vinothkrishnan, S. Usha, D. Thomas, and D.D. Ebenezer

Naval Physical and Oceanographic Laboratory, Thrikkakara, Kochi

ABSTRACT: *The acceleration sensitivity of a bush-baffle-hydrophone system is measured in air. The system consists of a hydrophone mounted on a baffle that is mounted on a base structure using bushes. The frequency dependent output from the hydrophone when the structure is forced to vibrate is measured. The vibration transfer function of each component of the system is also measured. The acceleration sensitivity of the system in water is calculated. The measurements and calculations are used to identify the components of the vibration isolation system that affect sonar system performance.*

Keywords: Baffles, Bushes, Hydrophone, Transmissibility, Acceleration Sensitivity.

INTRODUCTION

Sonar self-noise is reduced by mounting the hydrophones on vibration isolators that reduce the vibration transmitted to the hydrophone from the structure. In passive surveillance systems on submarines, the vibration isolator often consists of a baffle on which hydrophones are mounted and bushes used to mount the baffle on the structure. In addition, the hydrophone itself has compliant mounts.

The frequency-dependent acceleration sensitivity of the bush-baffle-hydrophone is of interest because the signal to noise ratio (SNR) of the sonar is dependent on this sensitivity and the acceleration spectrum of the platform at the array location. The operating band of the sonar is determined by the SNR and the vibration isolators play a vital role in the system.

The acceleration sensitivity of radially polarized piezoelectric ceramics used in hydrophones has been theoretically determined and experimentally verified for excitations that are along the axis [1, 2, and 3]. The sensitivity of compliantly mounted hydrophones to acceleration along three axes has been measured.

The bush, the baffle, and the hydrophone have resonances. At frequencies much below its resonance, each component behaves like a mass and transfers the vibration without attenuation. Near its resonance, the vibration is amplified and transmitted by each component. At frequencies that are much greater than the first resonance frequency but not close to other resonance frequencies, the component provides good isolation.

In this paper, the acceleration sensitivity of a bush-baffle-hydrophone system and a hydrophone in air and acceleration transfer functions for the bush and the baffle are measured. The acceleration sensitivity in water is then calculated using the measured data. The measurements and calculations are used determine the resonances in the components of the vibration isolation system that have a significant effect on the sonar performance.

VIBRATION RESPONSE

Structural components such as baffles and bushes are described in terms of vibration transfer functions. The acceleration, A_{exc}, of the excited face and the acceleration, A_{free}, of a free face are measured. The transfer function $H(\omega) = A_{free}/A_{exc}$ is equal to the displacement, velocity, and acceleration transfer functions for linear systems where $\omega = 2\pi f$ is the angular frequency.

The response of a hydrophone to vibration is described in terms of its acceleration sensitivity. The acceleration of the base on which a hydrophone is mounted, A_{exc}, and the open-circuit voltage, V, generated by the hydrophone are measured. The frequency-dependent acceleration sensitivity is defined as $\mathbb{A}_{hyd} = V/A_{exc}$.

3. Air Measurements

The multilayer sonar baffle is comprised of several alternate layers of rubber and thin titanium sheets sandwiched between thick top and bottom titanium plates. Hydrophones are mounted on the top plate. The bottom plate is mounted, using a fixture, on a base structure using studs that pass through bushes – as shown in Fig. 1a. The acoustic response of a hydrophone mounted on this baffle has been theoretically and experimentally studied [4].

The system is excited using a vibration exciter (Saraswathy Dynamics 1000 kgf). Two uni-axial charge-type accelerometers (B&K) are fixed using instant adhesive. The outputs

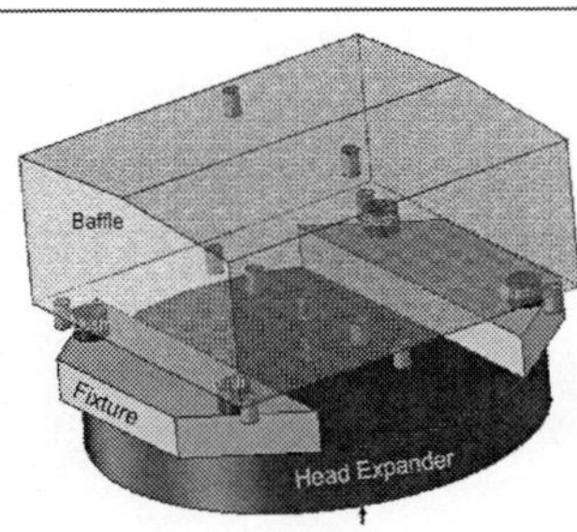

Fig.1. (a) Schematic showing baffle, bushes, and fixture on vibration exciter.

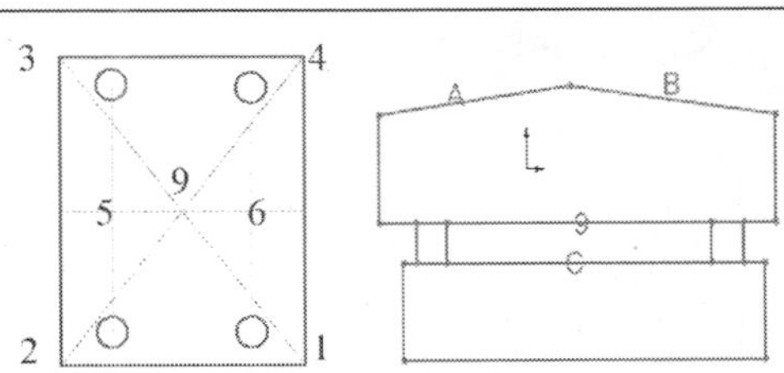

Fig.1. (b) View of bottom plate including mounting holes, elevation of baffle, and the points at which accelerometers are mounted.

from the control accelerometer (on the head expander of the exciter or on the bottom plate of the baffle) and another accelerometer (at various locations) are fed to an Econ 4 channel controller. The output from a hydrophone is fed to a conditioning amplifier (Nexus 2693) and then to the Econ system.

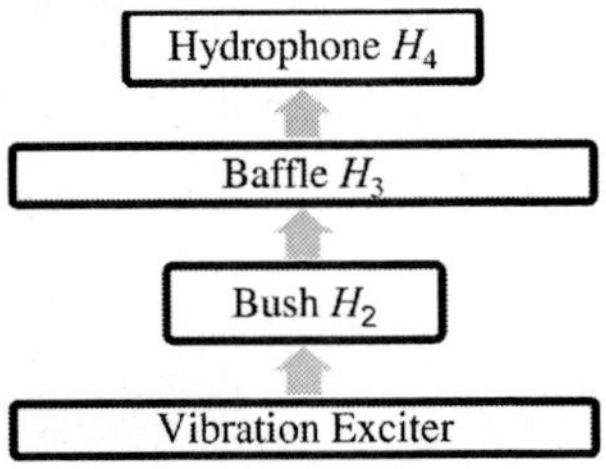

Fig. 2. Block diagram of the bush-baffle-hydrophone system. H_n n=1 to 5 are transfer functions where $H_1 = H_2H_3$ and $H_5 = H_2H_3H_4$.

The points at which the accelerometers are mounted are shown in Fig. 1b. The control accelerometer is mounted at point C on the head or at point 9 on the bottom-plate of the baffle. The measurement accelerometer is mounted at one of the points: 1-4 (corners of the bottom-plate), 5-9 (centre-line of the bottom-plate), A or B (top-plate of the baffle).

A block diagram of the experimental set-up is shown in Fig. 2. The acceleration sensitivity of the system in air is defined as $H_5 = \mathbb{A}_s = V/A_C$ where A_C is the acceleration at point C on the vibration exciter. All measurements are carried out by exciting the system with a harmonic sine source sweeping from normalized frequency of 0.01 to 1.0 with 0.5 g_{RMS} acceleration.

First, the acceleration sensitivity, $H_4 = \mathbb{A}_{hyd}$, of the hydrophone alone is measured and shown in Fig. 3. $\mathbb{A}_{hyd}$, at very low frequency, is about -40 dB ref V/g; has a peak of about -15 dB at 0.2; and then decreases. At frequency > 0.4, $\mathbb{A}_{hyd}$ is less than -30 dB ref V/g.

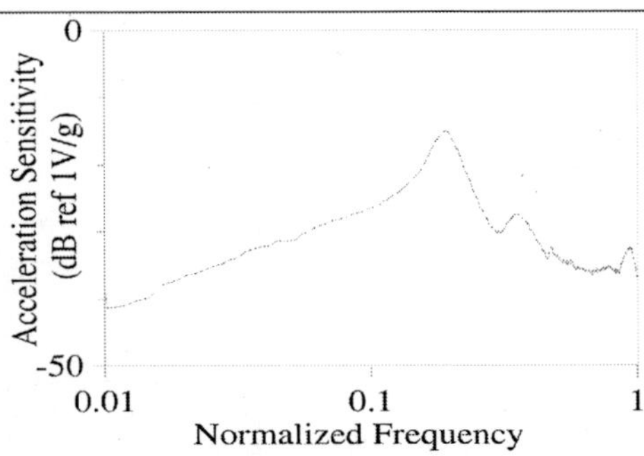

Fig. 3. Acceleration sensitivity, $H_4 = \mathbb{A}_{hyd}$, of hydrophone.

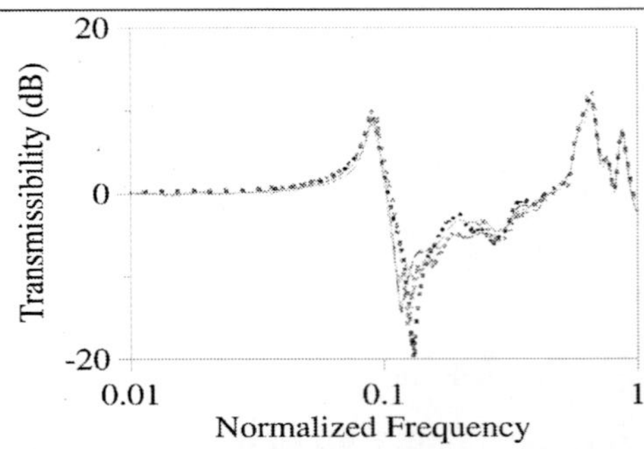

Fig. 4. Transfer functions of bushes, H_2, with accelerometers at corners 1, 2, 3, and 4.

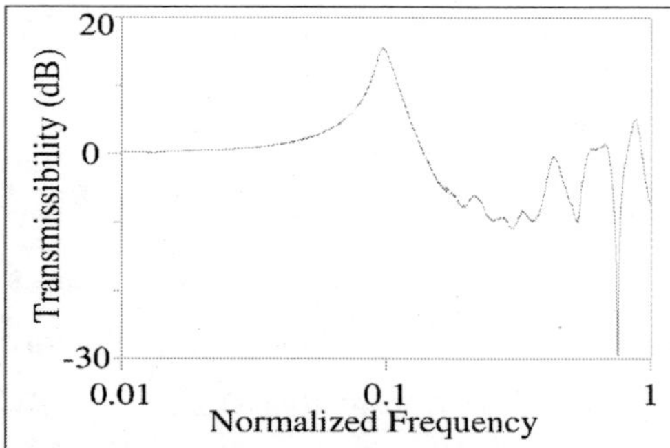

Fig. 5. Transfer functions of bushes, H_2, with accelerometer at center point 9.

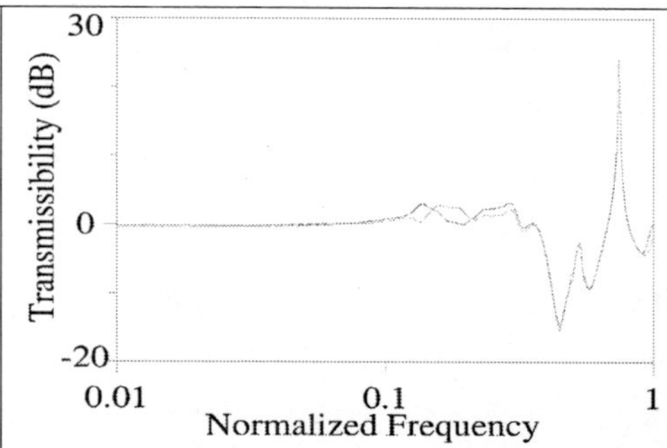

Fig. 6. Transfer functions of the baffle, H_3, with accelerometers at points A and B.

Next the system is mounted on the vibration exciter using a fixture as shown in Fig. 1a. The measured transfer function of the fixture is unity. The measured transfer functions of the bushes, H_2, are shown in Figs. 4 and 5 with accelerometers at various points. Vibration at the corner points 1 to 4 is nearly the same but differs from that at the center point 9 because of flexure of the bottom plate. The results in Fig. 5 are used in subsequent calculations. There are peaks at about 0.09, 0.65, and 0.9 and amplification occurs in 3 frequency bands.

Next, the transfer function of the baffle H_3 is measured. The acceleration at Point 9 on the bottom-plate of the baffle is controlled and the accelerations at points A and B on the top-plate are measured simultaneously. The results are shown in Fig. 6. The points A and B are symmetrically placed and have very nearly the same vibration levels. The first peak at about 0.3 is 3 dB. The dominant peak at 0.75 is 25 dB.

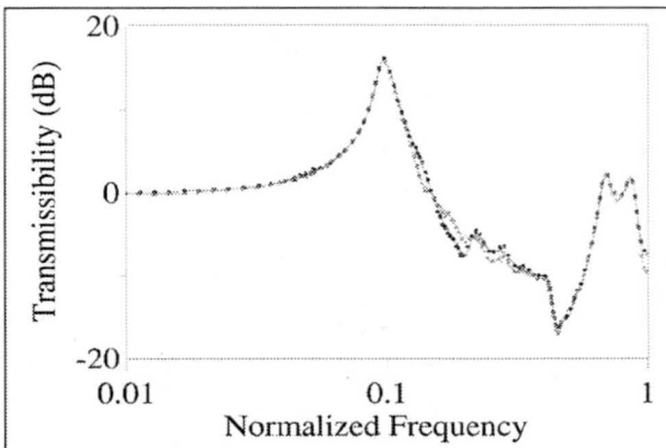

Fig. 7. Transfer functions of bush-baffle system. Solid line: measured H_1. Dashed line: calculated $H_1' = H_2 H_3$.

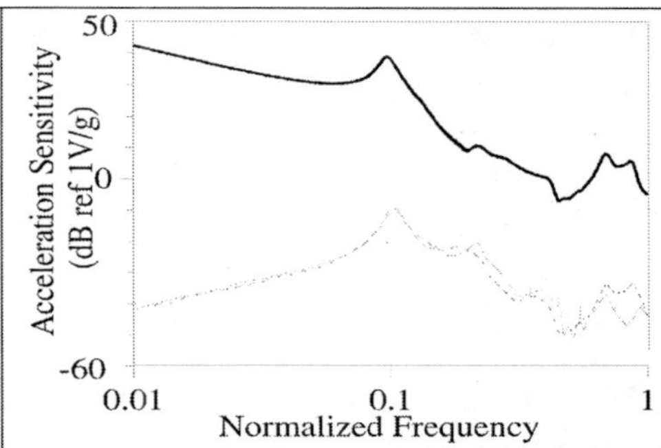

Fig. 8. Acceleration sensitivity in air and water. Thin solid line: measured: $H_5 = \mathbb{A}_s$. Dashed line: calculated $H_5' = H_2 H_3 H_4$. Thick solid line: $\mathbb{A}_w$.

Next, the vibration transfer function of the bush-baffle system, H_1, is measured. The acceleration at point C on the head expander of the exciter is controlled and the accelerations at points A and B on the top-plate are measured. The results are shown in Fig. 7 using solid lines. The measured transfer functions, H_1, are very nearly the same for both points confirming that the bushes, baffle, and mounting are symmetric. The calculated transfer functions, H_1', of the bush-baffle obtained using the measured transfer functions of the bush alone (Fig. 3) and the baffle alone (Fig. 6) are shown in Fig. 7 using dashed lines. There is very good agreement between the measured and calculated results and it is not possible to distinguish the four functions. It is seen from Fig. 7 that the first resonances in H_2 (at 0.1) and H_3 (at 0.2) result in a broad resonance in H_1 with a peak at 0.1. The peaks at 0.9 in H_2 and at 0.75 in H_3 are also seen in H_1.

Next, the output from the hydrophone is measured with the control accelerometer at point C. This yields the acceleration sensitivity of the bush-baffle-hydrophone system due to structure-borne vibrations: $H_5 = \mathbb{A}_s = V/A_C$, shown in Fig. 8. $\mathbb{A}_s$ has peaks at about 0.1 and 0.75 that are caused by the bush resonances seen in Fig. 5. The peak at 0.2 is due to the hydrophone and the peak at 0.9 is due to the baffle. The acceleration sensitivity of the bush-baffle-hydrophone, H_5', calculated using the measured values of H_2, H_3, and H_4 is in good agreement with the measured value H_5. This shows that the measurements are internally consistent. The bush-baffle provides isolation by about 7dB in the 0.2 to 1 band (except near 0.75). This shows that the bush-baffle provide vibration isolation in this frequency band. After the peak at about 0.9, there is a sharp decrease in $\mathbb{A}_s$ and it drops below -60 dB ref V/g when the normalized frequency is greater than one.

4. Underwater Analysis

The face of the baffle vibrates with acceleration $A_A = H_1 A_C$ in the direction normal to its surface and generates water-borne pressure waves. The pressure is expressed as $P(\omega) = \rho c A_A(\omega)/\omega$ where ρ is the density of water and c is the speed of sound in it. The voltage generated by hydrophone due to the pressure field is $V_{hyd} = MP(\omega)$ where M is the free-field acoustic sensitivity of the hydrophone. The acceleration sensitivity of the bush-baffle-hydrophone due to only the water-borne waves is $\mathbb{A}_w = V_{hyd}/A_C = M\rho c H_1/\omega$ and is inversely proportional to frequency at low frequency. It is shown in Fig. 8 and it is much greater than $\mathbb{A}_s$.

5. Discussion and Conclusions

The measured acceleration sensitivity, $\mathbb{A}_s$, of a bush-baffle-hydrophone system in air and the calculated sensitivity in water, $\mathbb{A}_w$, shown in Fig. 8, are used to draw the following conclusions. The voltage generated by the hydrophone in response to structure borne vibrations (acceleration sensitivity in air) has very little effect on the underwater system performance but the vibration transfer function of the bush-baffle H_1 has a strong effect. The resonances at 0.1 and 0.9 in the bush and 0.75 in the baffle result in peaks in $\mathbb{A}_w$. Reduction in H_1 in general and at the resonances in particular will reduce the noise in the sonar system. Reduction at resonances can be achieved by increasing internal losses in the bush and baffle but this can increase the transmissibility at other frequencies. The spectrum of the base excitation on-board the platform should also be measured and used to design the vibration isolation system.

ACKNOWLEDGEMENT

The authors thank Dr. R. Ramesh and Mr. Rijo Mathews Abraham for useful discussions, Mr. Kurian Isac for guidance and support, and others who made it possible to conduct the experiments.

REFERENCES

[1] D.D. Ebenezer, "Acceleration sensitivity of radially polarized piezoceramic cylinders," Proc. 3rd Intl. Cong. Air and Structure Borne Sound and Vibration pp. 883–890 (Montreal, Canada, 1984).

[2] D.D. Ebenezer, "Three-port parameters and equivalent circuit of radially polarized piezoelectric ceramic cylinders of finite length," J. Acoustical Soc. America, 99, pp. 2908–2912 (1996).

[3] G.D. Hugus III, "An analysis of vibration sensitivity in hydrophone design," NRL, Washington DC, ADA 092809; TR4295 (1980).

[4] D.D. Ebenezer and Pushpa Abraham, "Effect of multilayer baffles and domes on hydrophone response," J. Acoustical Soc. America, 99 pp. 1883–1893 (1996).

Genetic Algorithm based Neural Network for Classification of Rolling Element Bearing Condition

G.S. Vijay, Srinivasa P. Pai[1] and N.S. Sriram[2]

Department of Mechanical and Manufacturing Engineering, MIT, Manipal, Karnataka

[1]Department of Mechanical Engineering, NMAMIT, Nitte, Karnataka

[2]Department of Mechanical Engineering, VVIET, Mysore, Karnataka

E-mail: vgs.mpl@gmail.com

ABSTRACT: *In this paper, the performance of Multi Layer Perceptron Neural Network (MLPNN) based on features selected using Fisher's Criterion (FC) and Genetic Algorithm (GA) has been compared for rolling element bearing condition classification. Vibration signals have been acquired from four conditions (normal bearing, bearing with defect on inner race, bearing with defect on outer race and bearing with defect on ball element) of a deep groove ball bearing (6205) from a customized bearing test rig under different conditions of load and speed. Several statistical features in time domain and frequency domain were extracted from denoised vibration signal. GA was also used to optimize the learning rate, momentum factor and number of neurons in the hidden layer. MLPNN based on GA has been found to perform better than that based on features selected using FC, thereby helping in effective classification of rolling element bearing condition.*

INTRODUCTION

Rolling element bearings (REBs) are widely used in industrial and domestic applications. In industrial applications, these bearings are considered as critical mechanical components and a defect in such a bearing, if not detected in time can cause malfunction and may lead to catastrophic failure of the machinery [1]. Different methods are available for detection and diagnosis of bearing defects. They can be broadly classified as vibration and acoustic measurements, temperature and wear debris analysis. One of the commonly used techniques for bearing fault detection is vibration measurement [1, 2]. The vibration signals measured can be analysed in time domain, frequency domain or time-frequency domain. Vibration signals collected from rolling element bearings have lot of noise because of varying compliance and presence of defects [1]. The REB vibration signals due to defects alone are weak, compared to the harmonics of the resonant frequency of the system (noise). There is a need to remove or reduce this noise and extract dominant features in order to perform fault diagnostics. Conventional noise reduction or signal pre-processing techniques like Adaptive Noise Cancellation (ANC) and High Frequency Resonance Technique (HFRT) or envelope technique need information about resonant frequencies of the rolling element bearing system [2]. Also the bearing vibration signals are not really cyclostationary (they are rather pseudo cyclostationary) [3, 4]. Wavelet based denoising is increasingly being used in condition monitoring applications, in order to increase the data's reliability and improve accuracy of signal analysis and to increase the signal-to-noise ratio (SNR) [4]. Wavelet transform due to its sparsity, locality and multi-resolution nature has emerged as an effective tool for denoising. The Artificial Neural Network (ANN) has been widely used in fault diagnostics [2]. Different architectures have been studied in bearing fault diagnostics. Multi Layer Perceptron Neural Network (MLPNN) with back propagation is one of the widely used ANN type [6–9]. While designing an ANN architecture, the focus is always on developing a compact network structure, with minimum number of nodes in the input layer and hidden layer and achieve an overall good generalization ability (in terms of high accuracy on test data). Thus there is a need to reduce the dimensionality of the features extracted from the bearing vibration signals, in order to select the most sensitive features, so that there is dimensionality reduction of the ANN inputs and lead to a compact network. There are several dimensionality reduction techniques (DRT) like Principal component analysis (PCA), Linear Discriminant Analysis (LDA), Fisher's Criterion (FC), Singular Value Decomposition, Genetic Algorithm (GA), etc. Efforts are also needed to optimize the design of MLPNN architecture in terms of learning rate, momentum factor, number of neurons in the hidden layer etc. GA can be effectively used for optimizing these parameters and also select the most sensitive features [11].

In this context, this paper presents results of comparison of MLPNN trained and tested using features selected from denoised vibration signals from REBs from four conditions namely normal, defect on inner race, outer race and ball. The features selected using FC and GA has been compared. GA has also been used to optimize MLPNN architecture.

EXPERIMENTAL FRAMEWORK

Experiments were conducted on a customized test rig as shown in Figure 1. The test bearing has been mounted on the left side of the horizontal shaft. A 6205 deep groove ball bearing was tested under four conditions

Namely-normal (N), defect on inner race (IR), defect on the outer race (OR) and defect on the ball (B). Experiments were conducted for a radial load of 0 and 1.7 kN and a shaft speed of 622 rpm. Vibration signals (acceleration) were acquired through the Y accelerometer at a sampling rate of 48000 samples/s for a time period of 5.08 s, and were processed by the data acquisition system (DAQ) hardware and stored in the computer for further analysis. For each experiment, a data vector of size 250000 was generated, which was divided into bins of 5000 samples, such that it would encompass the probable occurrence of defect characteristics in one cycle.

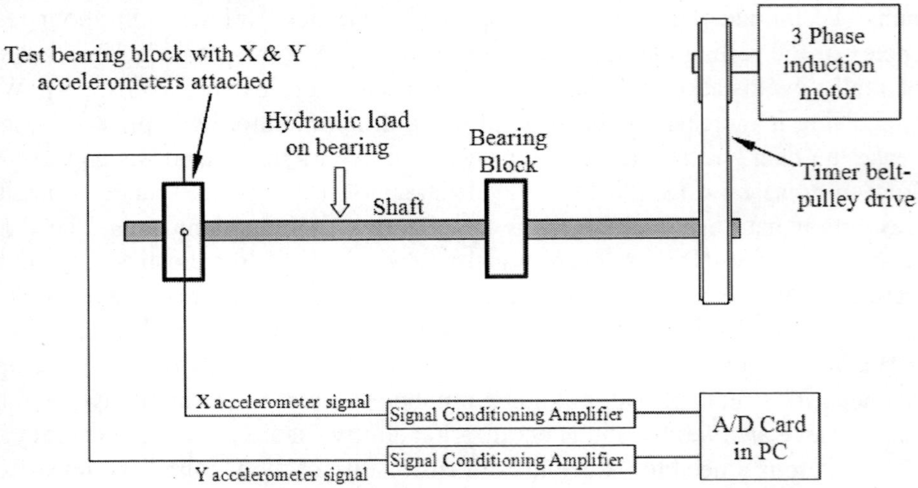

Fig. 1: Bearing Test Rig Used in the Work

BEARING CONDITION MONITORING SCHEME

Denoising Scheme

In this work, 'K-hybrid thresholding rule' (K-HTR), proposed by Hong and Liang [5] has been used. The raw bearing vibration signals obtained from the test rig has been decomposed to four levels using Discrete Wavelet Transform (DWT). Daubechies 8 is used as the mother wavelet. The K-HTR scheme is able to retain the impulsive nature of the REB vibration signals in the denoised signal, thereby increasing its kurtosis value. The details of K-HTR scheme and its associated equations are available in [5].

Feature Extraction

The denoised vibration signals with a vector size of 250000 × 1 were divided into 50 non-overlapping bins each with 5000 data. From each bin, 30 features have been extracted, 17 time domain features and 13 frequency domain features, details of which are available in [10]. This formed a single pattern. For four conditions of bearing, two loads and one speed condition, a total of 400 patterns were extracted. The feature set matrix consisted of 30 features × 400 patterns. Each feature was normalized, so as to get values between 0 and 1. The patterns were thoroughly mixed, out of which 75% were used in the training data set and the remaining 25% in the test data set.

Dimensinality Reduction

Fisher's Criterion (FC) has been used for dimensionality reduction. The Fisher discriminant distance $J_k^{P,Q}$ between the two bearing conditions P and Q for the kth feature set can be computed using Eq. (1) [12].

$$J_k^{P,Q} = \frac{\left| Mean(t_k^P)\text{-}Mean(t_k^Q) \right|^2}{\left[Std(t_k^P) \right]^2 + \left[Std(t_k^Q) \right]^2} \quad \dots (1)$$

where, t_k^P and t_k^Q are the kth feature set for the bearing conditions P and Q respectively. For four bearing conditions, the Fisher Discriminant Power (FDP) is calculated as the sum of the pair wise distances [12] as given in Eq. (2).

$$F_k = J_k^{N,IR} + J_k^{N,OR} + J_k^{N,B} + J_k^{IR,OR} + J_k^{IR,B} + J_k^{OR,B} \quad \dots (2)$$

A distance based threshold method (threshold θ method) given in Eq. (3) is used for selecting the sensitive features.

$$\theta \approx \sum_{k=1}^{s} F_k^* \Big/ \sum_{k=1}^{f} F_k \quad \dots (3)$$

where, 'f' is the total number of features extracted (30 in this work) and 's' is the number of selected features such that, the sum of 's' largest FDPs divided by the sum of all the FDPs is approximately equal to θ.

GA is a popular global optimization tool that can optimize the solution to a given problem by using the concepts of evolutionary computing. GA can been used to simultaneously select sensitive features [11] and optimize ANN simulation parameters, viz., number of neurons in hidden layer, learning rate and momentum rate.

Multilayer Perceptron

Multilayer Perceptron (MLP) is a widely used ANN architecture which can be applied to a wide variety of engineering problems. It consists of an input layer, where the number of neurons is equal to the number of features selected (5 in this work), a hidden layer, whose number of neurons are fixed by trial and error and using GA for optimizing the same and an output layer with four neurons, representing four conditions of the bearing, which has been represented using a binary scheme. The activation functions of the hidden and output layer were sigmoid. The different schemes discussed above, along with MLPNN were implemented using customized MATLAB programs. The MLPNN was trained using back-propagation algorithm. The learning rate and momentum rate were set by trial and error and also optimized using GA.

RESULT AND DISCUSSION

FC selected 5 features out of a set of 30 based on the threshold θ. Table 1 shows the performance of the MLPNN to which the features selected by FC were given as input. It is clear that for 25 neurons, with learning rate of 0.03 and momentum rate of 0.0019 the performance of the MLPNN was the best with a training error of 0.0035, an accuracy of 96.33% on the training data and an accuracy of 92% on the test data.

Table 1: MLP Performance with Features Selected Using FC as Inputs

Neurons	*Training Error*	*Training Accuracy (%)*	*Test Accuracy (%)*	*Learning Rate*	*Momentum Rate*
15	0.0049	93.33	89	0.030	0.0019
20	0.0051	92.67	89	0.030	0.0019
25	**0.0035**	**96.33**	**92**	**0.030**	**0.0019**
30	0.0062	90.67	88	0.030	0.0019
35	0.0078	87.33	85	0.030	0.0019

GA in this paper is not only used to select the most sensitive features, but also to optimize MLPNN training parameters viz., (i) number of neurons in the hidden layer (ii) Learning rate, and (iii) Momentum rate. The GA program creates a population consisting of eight real valued genomes (chromosomes). Length of each genome is 8. Fitness values of 8 genomes are computed. Parents for crossover are selected based on Normalised Geometrical Rank selection method. Crossover offsprings are produced with a probability of 0.6 using multi-point crossover. Some of the crossover offsprings are mutated with a probability of 0.2. Fitness of each offspring produced by crossover and mutation is computed. The average of the classification accuracies for training and test itself is the fitness value of a genome. Genomes are arranged in the descending order of their fitness values (Good genomes first and worst genomes last). The GA is set to run for a maximum of 150 generations or until the maximum fitness value attained is 99%. The MLPNN is set to run for a maximum of 2000 epochs or a minimum training error (MSE) of 1×10^{-4}.

Table 2 compares the performance of the MLPNN with the GA and FC selected features as input. It is clear that the performance of the MLPNN in both the cases is similar in terms of the accuracies on the training data and the test data. However, the GA-based MLPNN performs slightly better than the FC-based MLPNN in terms of the dimension of the hidden layer, viz., 17 for GA-based MLPNN and 25 in case of FC-based MLPNN. Further, the number of neurons in the hidden layer, learning rate and momentum rates used by the GA-based MLPNN are the values optimized by the GA, whereas in the FC based MLPNN, these values have been fixed by trial and error, i.e., the values which gave best performance of the MLPNN have been selected.

Table 2: Comparison of MLP Performance for Features Selected by the GA and the FC

Parameters	*MLP Performance for Features Selected by*	
	GA	*FC*
Features selected	14, 15, 17, 24 and 28	1, 5, 17, 20 and 21
Best Fitness Value (%)	95.83	94.17
Training accuracy (%)	96.67	96.33
Test accuracy (%)	95.00	92.00
MSE reached	0.0046	0.0035
Number of neurons	17	25
Learning Rate	0.0175	0.0300
Momentum Rate	0.0030	0.0019
Number of generations	150	–

CONCLUSION

In this paper, the MLPNN has been used for classification of REB based on features selected using FC and using GA. GA has also been used to optimize the MLPNN architecture in terms of number of neurons in the hidden layer and simulation parameters viz., learning and momentum rate. GA-based selection of the input features takes much longer time as it depends on the performance of the MLPNN for computing the fitness value, whereas the FC-based selection of input features is faster as it is independent of the performance of the MLPNN. MLPNN based on GA gave a comparable performance with that using features selected using FC. Hence MLPNN using GA can be effectively used to classify REB condition.

REFERENCES

[1] Tandon N., Choudhury A., A review of vibration and acoustic measurement methods for the detection of defects in rolling element bearings, Tribology International 32 (1999), pp. 469–480.

[2] Patil M. S., Jose Mathew, Kumar Rajendra P. K., Bearing signature analysis as a medium for fault detection: A review, Journal of Tribology 130 (2008), pp. 014001.1–014001.7.

[3] Robert B. Randall, Jerome Antoni, Rolling element bearing diagnostics—A tutorial, Mechanical Systems and Signal Processing 25 (2011), pp. 485–520.

[4] Z. K. Peng, F. L. Chu, Application of the wavelet transform in machine condition monitoring and fault diagnostics: a review with bibliography, Mechanical Systems and Signal Processing 18 (2004), pp. 199–221

[5] Hoonbin Hong, Ming Liang, K-Hybrid: A Kurtosis-Based Hybrid Thresholding Method for Mechanical Signal Denoising, Journal of Vibration and Acoustics, Transactions of the ASME Vol. 129 (2007), pp. 458–470.

[6] C. Castejon, O. Lara, J. C. Garcıa-Prada, Automated diagnosis of rolling bearings using MRA and Neural Networks, Mechanical Systems and Signal Processing 24 (2010), pp. 289–299.

[7] Chun-Chieh Wang, Yuan Kang, Ping-Chen Shen, Yeon-Pun Chang, Yu-Liang Chung, Applications of fault diagnosis in rotating machinery by using time series analysis with Neural Network, Expert Systems with Applications 37 (2010), pp. 1696–1702.

[8] Kankar, P.K., Satish Sharma, C., Harsha, S.P., Fault diagnosis of ball bearings using machine learning methods, Expert Systems with Applications 38 (2011), pp. 1876–1886.

[9] Jafar Zarei, Induction motors bearing fault detection using pattern recognition techniques, Expert Systems with Applications 39 (2012), pp. 68–73.

[10] Vijay G.S., Srinivasa Pai P., N.S. Sriram, "Bearing diagnostics—A Radial Basis Function Neural Network approach", Proceedings of the 24th International Congress on Condition Monitoring and Diagnostics Engineering Management, Advances in Industrial Asset Integrity Management (COMADEM 2011), Stavanger, Norway, 30th May – 1st June, 2011, ISBN 0-9541307–2–3, pp. 843–854.

[11] B. Samanta, K.R. Al-Balushi, S. A. Al-Araimi, Artificial Neural Networks and Genetic Algorithm for bearing fault detection, Soft Computing 10 (2006), pp. 264–271.

[12] Gary G.Y., Kuo-Chung Lin, Wavelet packet feature extraction for vibration monitoring, IEEE Transactions on industrial electronics 47 3 (2000), pp. 650–667.

ID FAN Motor Bearing Damage Detection Using Co-Ordinated Condition Monitoring at DTPS

Hemant Bari, Atul Deshpande and Suhas Patil

Department of Maintenance Planning, Condition Monitoring Cell, Reliance Infrastructure, Dahanu Thermal Power Station, 2x250MW, Dahanu, Mumbai.

E-mail: hemant.bari@relianceada.com;
atul.deshpande@relianceada.com; suhas.patil@relianceada.com

CASE STUDY: MOTOR BEARING FAILURE OF ID FAN - 1B

Pre-Maintenance History

1. On 25.11.2010, during Unit overhaul, it was decided to replace ID Fan 1B motor with the serviced spare motor as per schedule.
2. On 09.12.2010, when the fan was taken in service, vibrations of Motor, Hydraulic Coupling and Fan were within limit. Also, all the operating parameters were normal.
3. On 22.12.2010, abnormal sound noticed from NDE side of Motor. Vibration readings were taken and found to be well within limits. Noise level measured for reference which was 94 dB.

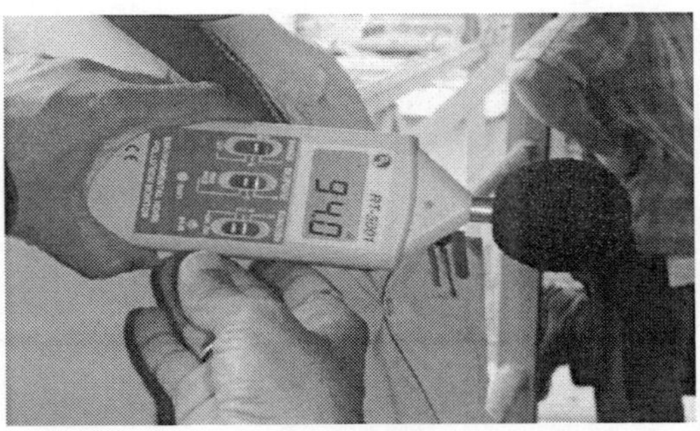

4. The abnormal sound level was constant for next 10 days. No variations in vibrations were noticed. On 05.01.2011, sound level increased to 99.1 dB though no deviation observed in Motor vibration.
5. On 06.01.2011, Motor NDE & DE horizontal vibration readings deviated from 0.5 to 0.8 and 0.6 to 0.8 mm/sec (RMS) respectively.

6. On 10.01.2011, Motor NDE & DE horizontal vibrations increased to 1.3 &1.2 mm/sec respectively and vertical vibrations increased to 1.1 & 1.3 mm/sec respectively.

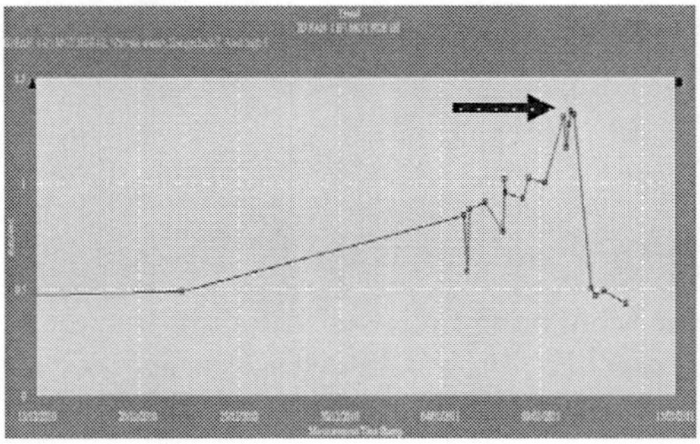

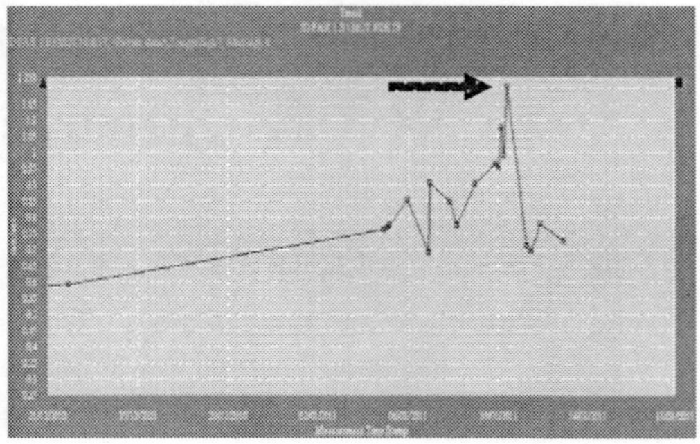

7. On 10.01.2011 Noise level also increased from 99.1 to 101.9 dB

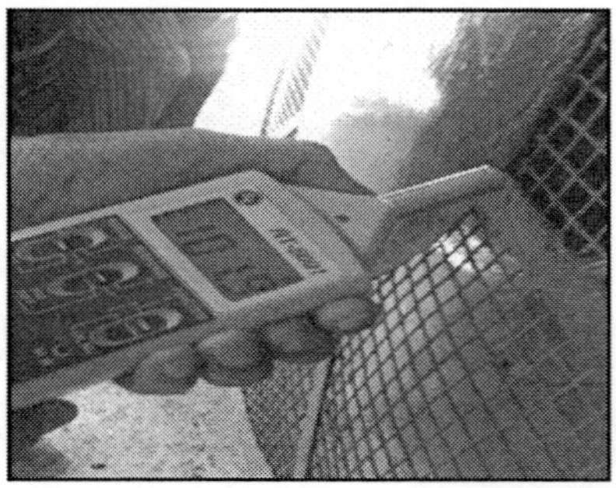

8. In view of Steep increasing trend of Noise from 94 to101.9 dB, it was decided to stop ID FAN for Motor replacement.

1.2. Vibration Diagnosis of ID Fan 1B Motor

On 10.01.2011, Motor NDE & DE horizontal vibration increased from 0.5 & 0.6 to 1.3 &1.2 mm/sec & vertical vibration increased from 0.7 & 0.6 to 1.1 & 1.3 mm/sec (RMS) respectively.

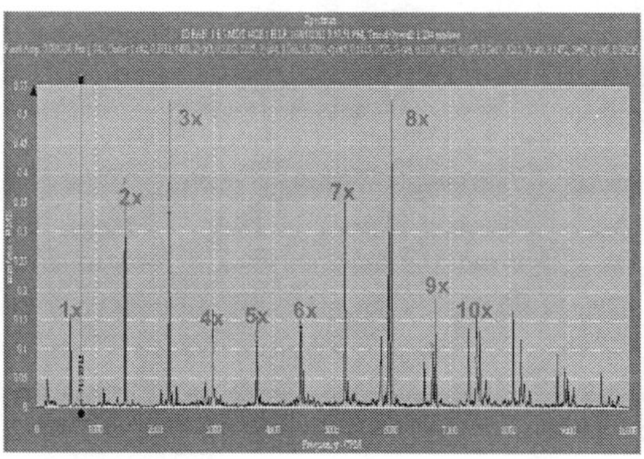

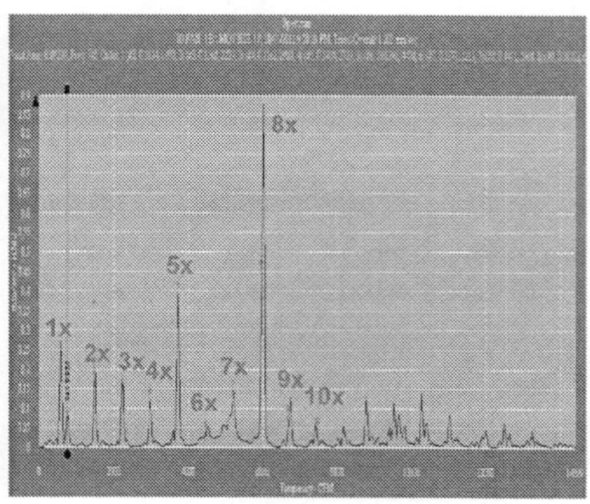

. Spectrum of Motor NDE (Hori. & Vert.) showing multiples of running speed frequency

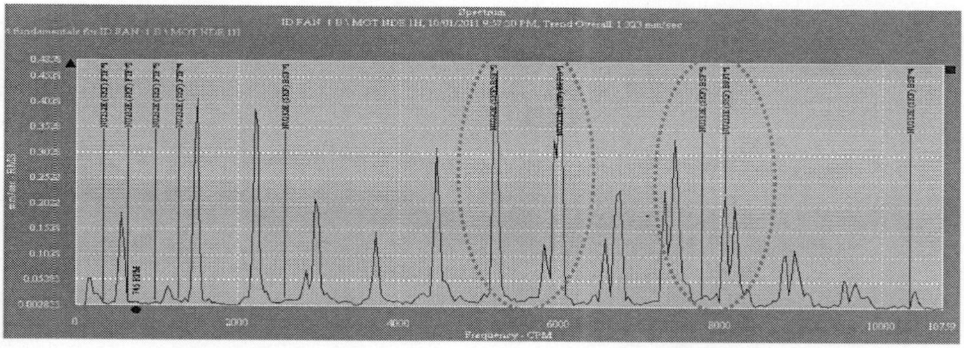

. Spectrum of Motor NDE (H) showing bearing defect frequencies

1.3. Frequency Analysis

1. Initially the Motor NDE side noise level increased from 94 .0 to 101.9 db.
2. Increase in the amplitude of Motor NDE & DE bearing vibrations was also noticed from 0.5 & 0.6 to 1.3 &1.2 mm/sec respectively.
3. Spectrum analysis revealed that there was increase in **multiples of running speed harmonics from 1x, 2x, 3x ….. up to 10x frequency.**
4. It was suspected that Motor bearings were loose on shaft.
5. Also, the spectrums showed **Bearing Defect Frequencies** at BSF, BPFO & BPFI which clearly pinpointed bearing defects.

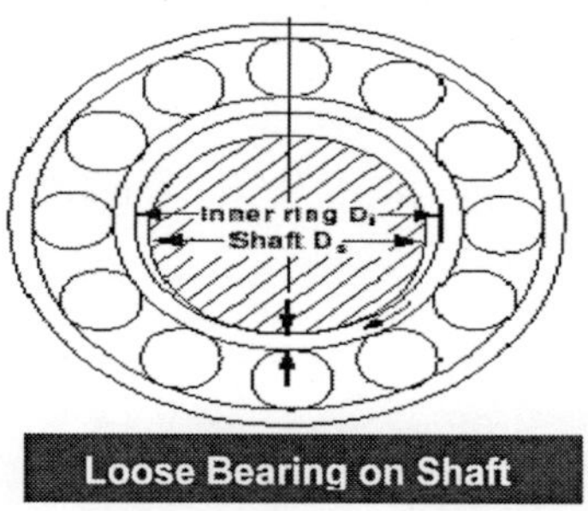

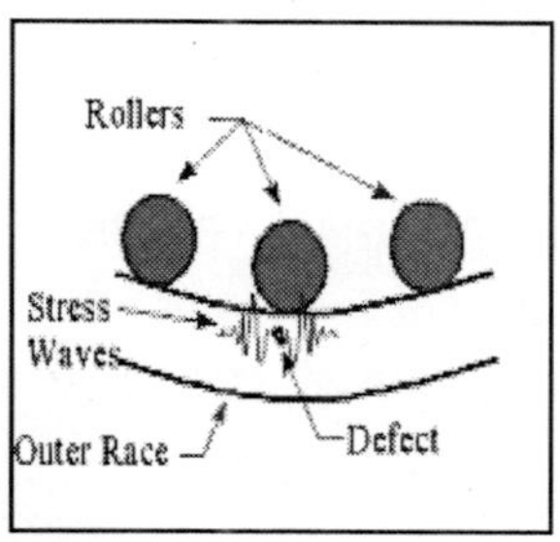

6. When these frequency stress waves' events occur at repeatable time intervals corresponding to a bearing defect freq. then it is likely that a bearing defect is present.
7. Hence it was concluded that vibration of both Motor bearings increased due looseness of bearings on shaft & the deterioration of bearings was in progress.

1.4. Observations during Maintenance work

The bearing clearances were found on higher side than the original readings. There was a clearance of more than 0.5mm at NDE bearing inner race area. Also scoring observed on NDE side bearing housing cap.

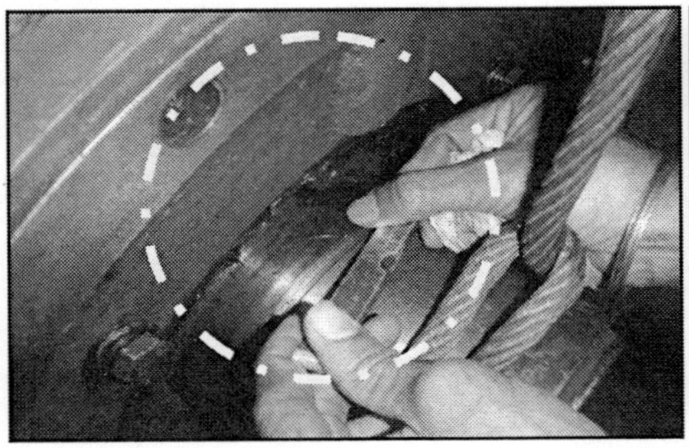

1. Inner race at NDE side was cracked and got loosened over the shaft.

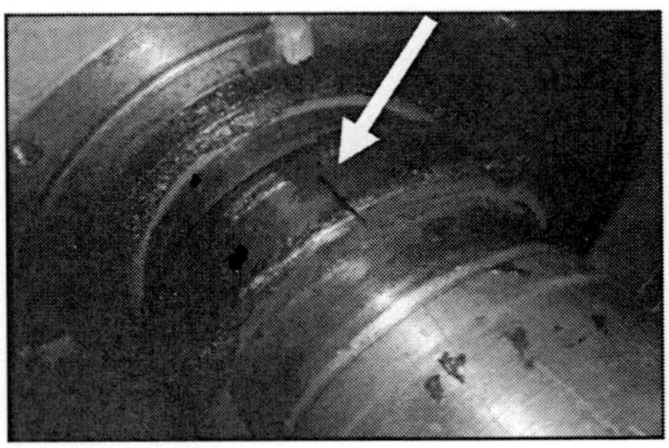

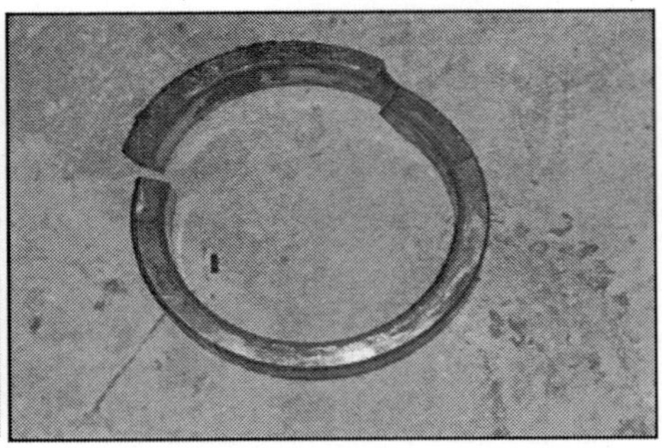

2. Scoring was observed over the shaft at NDE side due to relative motion between inner race of bearing and shaft. Due to this, the diameter of shaft was found 159.34 mm at bearing area instead of 160.03 mm.

3. Scoring was observed on both DE & NDE bearings rollers.

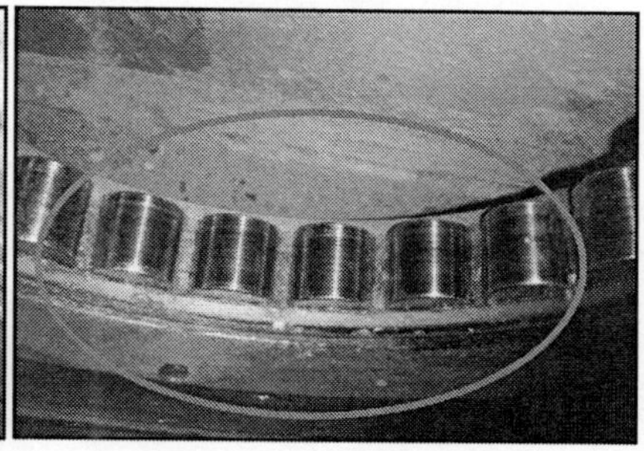

4. Thorough inspection of Motor Winding & Stator Rotor was done & found OK.

   No Scoring marks observed.

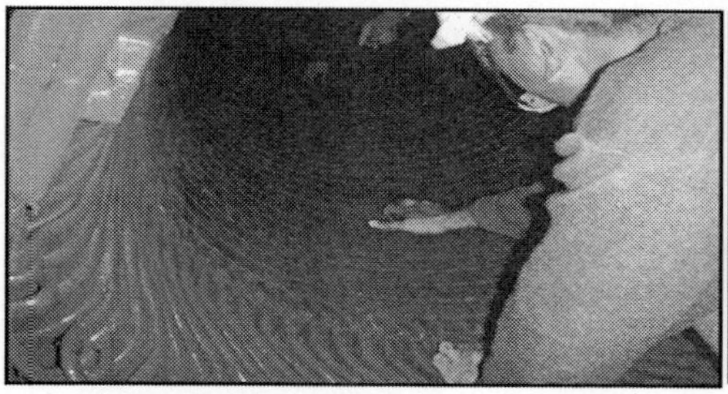

1.5. Maintenance work carried out

1. Gear coupling from Motor DE side was removed.
2. Motor was replaced with the spare serviced Motor & Gear coupling was mounted on motor shaft. ID fan alignment done with rest of the drive.
3. ID fan was boxed-up and made available for operation & was then taken in service
4. At full load, all vibrations & Noise level found in well within limit

1.6. Post Maintenance Vibration Readings

<table>
<tr><td colspan="3">EQUIPMENT : ID FAN 1B</td><td colspan="2">DATE : 10 to 11 January 2011</td></tr>
<tr><td colspan="5">LOCATION:- UNIT – 1 - VIBRATION REPORT OF ID FAN 1B</td></tr>
<tr><td colspan="2">MEASUREMENTS</td><td colspan="3">OVERALL VIBRATION LEVELS</td></tr>
<tr><td rowspan="3">Point</td><td rowspan="3">Position</td><td colspan="3">Velocity
(mm/sec) RMS</td></tr>
<tr><td>Initial Readings</td><td colspan="2">After Replacing Motor</td></tr>
<tr><td></td><td>11/01/11</td><td>11/01/11</td></tr>
</table>

			Motor @ full Load		Decoupled Motor Trial	Motor @ full Load
Motor	(NDE)	H	1.3 (1.0 g)		0.3 (0.7g)	0.5 (0.6g)
		V	1.2		0.5	0.6
		A	1.0		0.3	0.4
Motor	(DE)	H	1.2 (1.3 g)		0.3 (1.2g)	0.4 (1.0g)
		V	1.3		0.4	0.5
		A	1.0		0.5	0.4
Noise			101.9 dB			89.7 dB

1.7. Results and Discussion

Vibration analysis & Noise Monitoring helped in diagnosing the exact root cause of abnormity at very early stage. The early detection of Motor bearing looseness on shaft of HT auxiliary ID FAN – 1B avoided costly catastrophic failure. The application saved the plant from huge losses in terms of generation, unplanned breakdown time and associated maintenance & spares costs

Cost Saved by preventing catastrophic failure of 'ID Fan 1B':

1. Generation loss Calculation

0.09 MUs/hr (1 MU = 1000000 Units)

Total saving for 48 hrs =0.09x48x1000000

= 4320000 Units

Cost per unit = 3.5 Rs

Total Saving = 3.5 x 4320000

=1,50,00,000 Rs

2. Motor cost = 65,00,000/- Rs

3. Total Cost saving = Generation loss + Motor cost

= 1,50,00,000 + 65,00,000

= **2,15,00,000/- Rs**

3. Conclusion

- Co-ordinate Condition Monitoring helped in diagnosing the exact root cause of abnormity at very early stage of HT auxiliary ID Fan1B.
- It has been shown that early diagnosis has resulted in timely action and savings in cost and avoidance of unplanned shutdown.
- This has saved the plant from huge losses in terms of generation, unplanned breakdown equipment and associated maintenance & spares costs.

Reference

[1] Rao J S, 2000, *Vibratory Condition Monitoring of Machines*, Narosa Publishing House.

Damping Studies on Polymer Nano Composite Materials

S. Sathiyakumar, P.V.S. Ganesh Kumar, K. Venkata Subbaiah[1] and B. Parvatheeswara Rao[1]

Naval Science and Technological Laboratory, Visakhapatnam, Andhra Pradesh

[1]Andhra University, Visakhapatnam, Andhra Pradesh

E-mail: sci.sathiya@gmail.com

ABSTRACT: *Polymer matrix composites reinforced by nano fillers are rapidly gaining popularity in vibration damping applications. All materials exhibit a certain amount of damping, but materials such as metals, cements, polymers and their composites are good for vibration damping. In marine applications where high strength and less weight are the desired parameters, the polymers serve as suitable damping materials. Suitable nano fillers in appropriate concentrations further enhance both the mechanical strength as well as damping ability of these materials due to their better inter phase properties. The properties of rubber are improved using fillers in the size of nano level as reinforcement. In this study, Neoprene rubber doped with nano silica and montmorillonite clay is used to prepare the polymer nanocomposite samples by Triple Roll Method. These nano materials are prepared with various concentrations from 1–5% of both nano clay and silica. The samples were prepared as per the ASTM standard for damping studies by Oberst beam method and Damping ratio is calculated using the half power bandwidth method. Results were analyzed experimentally. The highest damping ratio is obtained for the 5% filler concentration in both the series of samples. Further, suggestions were made at the end on percentage of filling to obtain better damping properties in these materials and also to carry out future studies for improved damping characteristics in the light of the present understanding.*

Keywords: Damping, Triple Roll, ASTM, Oberst, Half Power Bandwidth.

INTRODUCTION

Nanocomposites are composites in which at least one of the phases shows dimensions in the nanometer range (1 nm = 10^{-9} m) [1]. These materials are reported as materials for 21st century because of their design uniqueness and combinations of properties which are not found in conventional composites.

The number of published papers containing words such as nanoscience, nanotechnology, nanomaterials, etc., doubled in 1.6 years [2] in the late 1990s. Also, a literature survey made by the authors reveals that about 13.420 papers (of which 4028 contain the keywords nanocomposite and polymer in Web of Science-ISI: updated on 10 February 2009) have been published on nanocomposites in the last two decade (1988–2008). Similarly, patents with complete document on nanocomposites account for about 4663 during the same period as per Scirus (www.scirus.com).

The improvements of physicochemical and thermomechanical properties of polymer composite materials has been a major research interest in the last few decades. Polymer nanocomposites (Mohammad, 2006; Xiangling, 2003) represent a new class of materials alternative to conventional filled polymers [3]. The main challenge in fabrication of these polymer nanocomposites for structural applications is uniform dispersion of nanoparticles in the polymer matrix. However, good dispersion for nanoparticles in polymer composite materials is extremely difficult to achieve, since nanoparticles tend to aggregate together during fabrication. The degree with which the nanoparticles can be homogeneously dispersed in the polymer matrix would significantly influence the thermal, mechanical and optoelectronic properties of the material. Researchers have used several techniques for dispersing nanoparticles may include: (1) mechanical agitation, such as ball milling or magnetic stirring, (2) ultrasonic vibration, (3) shear mixing (4) non-contact mixing (5) using the dispersing agent. The process of curing of these polymers is an important factor in improving the mechanical, thermal, electrical and optical properties.

DAMPING

Damping is an important characteristic of structural mechanical vibrating systems. A reasonably accurate input damping data is needed to design the system under different vibratory loadings. Inaccurate estimation of damping may lead to failure of systems during service [4]. It is thus appropriate to select proper method and loading to accurately determine the damping in a material or structure. In general, damping can be classified into two categories i.e., material damping and system damping. Material damping is defined as “the inherent property of any material to dissipate energy in a volume of macro continuous media”. System damping relates to “energy dissipation in total structure”. It includes energy dissipation effects of joints, fasteners and interfaces in addition to energy dissipation due to material. Mechanical vibration damping can be defined as the ratio of energy dissipated (D) to maximum strain energy (W) per cycle of vibration. [4]

Damping is normally measured as damping ratio that is ratio of viscous damping coefficient to that of critical damping coefficient. Damping coefficient is basically defined as ratio of damping force and velocity of vibratory motion.

EXPERIMENTS

Samples of desired size i.e. length, breadth and thickness of the sample dimensions were fixed approximately at 300 mm × 25 mm × 3 mm as per ASTM standards. The GFRP (10 in no) samples were prepared and they are shown Figure 1.

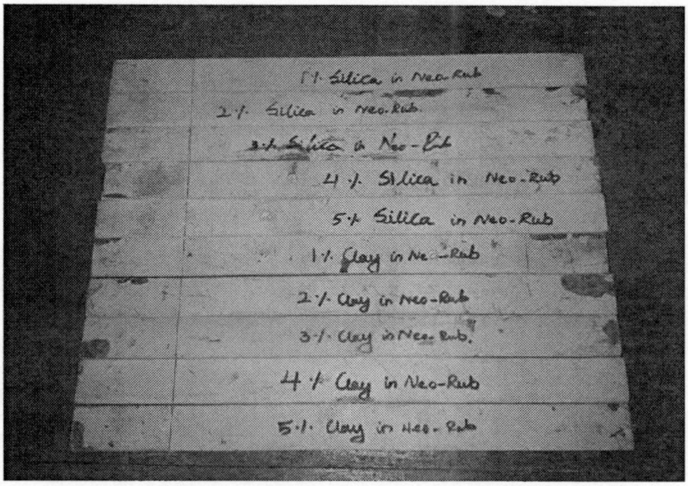

Fig. 1: Samples of Nano Filled GFRP Materials

Samples are made of polymer nano composites filled with nano Clay and Silica particles of 1% to 5% concentrations. The vibration damping measurements on the samples were performed by Impact Hammer Method [IHM] and Logarithmic Decrement Method. The specimens were excited using the impulse excitation with a modally tuned hammer in the frequency range of interest. The accelerometer measures the vibration and produces an electrical signal that is amplified by the charge amplifier which then was acquired by signal analyzer. Block diagram of the experiment and its working principal is shown in Figure 2a. In this figure, a = 50 mm, b = 100 mm, c = 250 mm and clamped length is 50 mm. Figure 2b depicts typical impact hammer test on a beam sample.

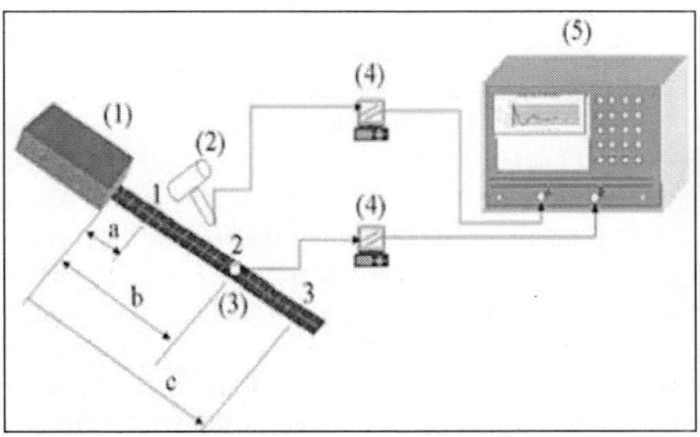

1. Cantilever beam, 2. Impact hammer with piezoelectric cells,
3. Accelerometer, 4. Power amplifier, 5. Spectrum analyzer

Fig. 2a: Block Diagram of Typical Test Setup

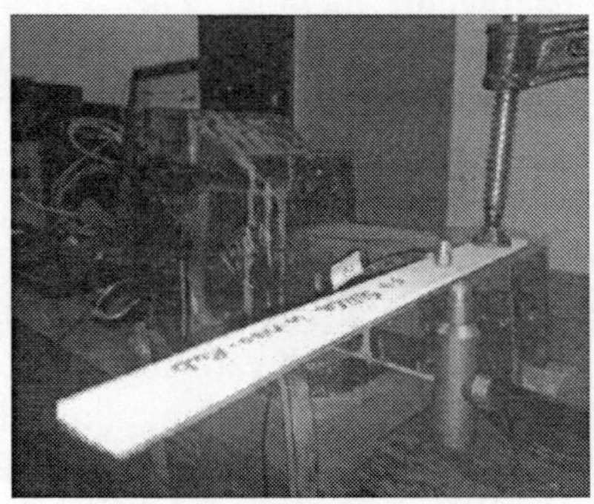

Fig. 2b

Logarithmic Decrement Method

One of simplest methods to determine the damping in a system is to measure the rate of decay of free oscillations (as shown in Figure 3a). The logarithmic decrement is defined as the natural logarithm of the ratio of any two Successive amplitudes in the same direction. Theoretically damping factor is determined by the equation (1),

$$Xn//Xn+1 = \exp\left(2\pi\xi/(\sqrt{1}-\xi 2)\right) \quad \ldots (1)$$

Where ξ = damping factor

X_n = displacement at the end of first oscillation

X_{n+1} = displacement at the end of second oscillation

Figure 3b depicts typical time decay curve obtained in experiment.

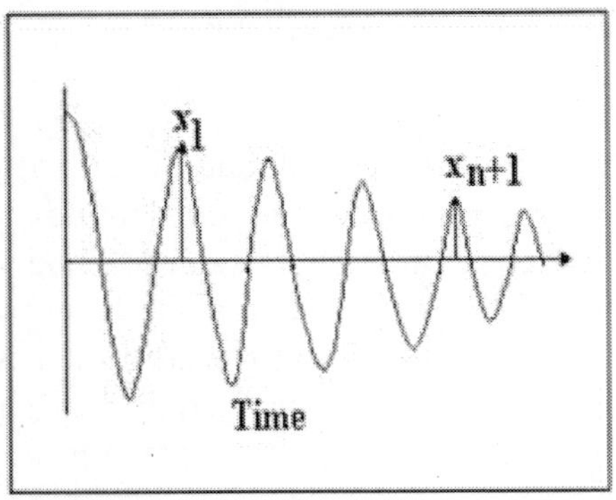

Fig. 3a: Decay of Free Oscillations

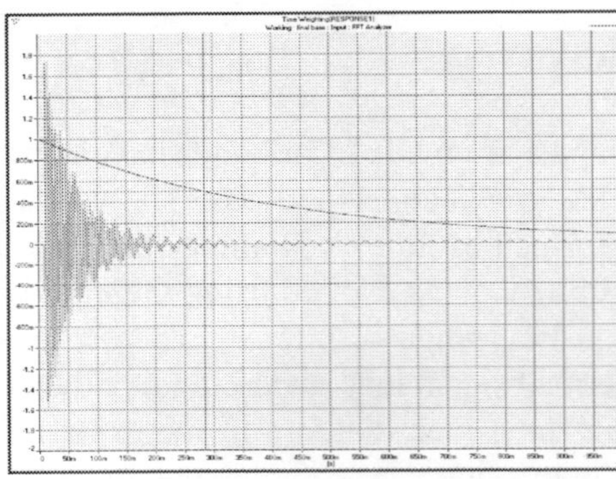

Fig. 3b: Typical Experimental Time Decay

As the beam is excited over a frequency range, resonance condition is encountered at certain frequency for different modes. That frequency is treated as natural frequency of the cantilever beam for that mode [5]. Damping ratio is calculated using the half power bandwidth method. The response spectra were experimentally obtained for all the samples. The observed frequency response functions (FRFs) spectra for neoprene rubber nanocomposites filled with nano clay as well as nano silica are shown in Figures 4a&b. The corresponding variations of natural frequencies and damping ratios as a function of nano filler concentration are also shown in Figures 5a&b, respectively. It can be clearly seen from the spectra that among all the filler concentrations, the 5% clay and 5% silica filled neoprene rubber beams have shown highest amplitudes. However, it appears that there is not much variation in the natural frequency with respect to the filler concentration in clay samples. At the same time, the percent damping ratio is seen increasing with increasing concentration of the filler, particularly in mode 2 for the clay nanoparticles. Moreover, the clay filled samples exhibit higher percentage of damping ratio compared to silica filled samples in both the modes. Interestingly, the highest damping ratio is obtained for the 5%

filler concentration in both the series of samples, thus stresses the need for continue the investigations beyond 5% filler concentrations further.

Results of the natural frequencies and the corresponding observed and estimated damping ratios for mode 2 and mode 3 are also presented in Table 1 for comparison.

Table 1: Data Showing Natural Frequencies, Experimental Damping Ratios in Modes 2 and 3 for Silica and Clay Filled Polymer Nanocomposites at Different Concentrations

S. No	Material	Modes	Natural frequencies Hz	Damping ratio
1.	1% silica	2 3	360 708	1.26 1.19
2.	2% silica	2 3	356 700	1.25 1.20
3.	3% silica	2 3	352 692	1.30 1.23
4.	4% silica	2 3	352 700	1.47 1.20
5.	5% silica	2 3	344 676	1.50 1.36
1.	1% clay	2 3	356 700	1.38 1.34
2.	2% clay	2 3	364 716	1.51 1.24
3.	3% clay	2 3	352 700	1.57 1.23
4.	4% clay	2 3	360 712	1.45 1.24
5.	5% clay	2 3	352 700	1.62 1.35

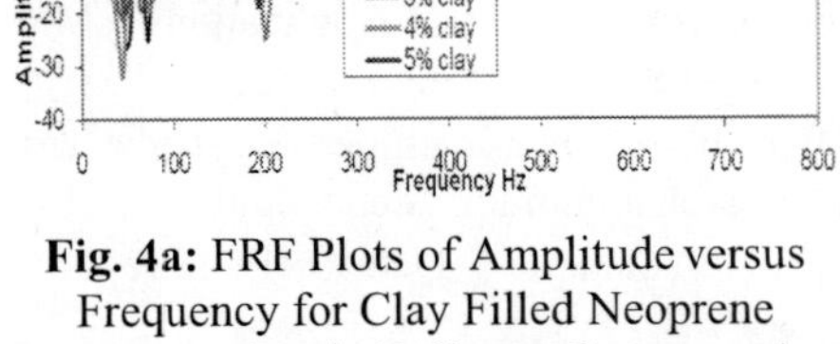

Fig. 4a: FRF Plots of Amplitude versus Frequency for Clay Filled Neoprene Nanocomposites in Different Concentrations

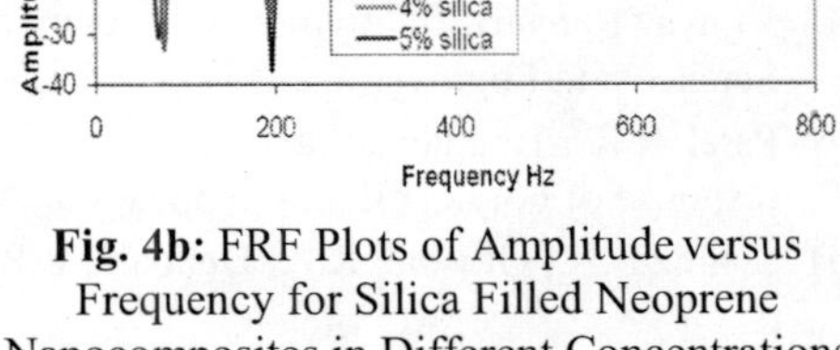

Fig. 4b: FRF Plots of Amplitude versus Frequency for Silica Filled Neoprene Nanocomposites in Different Concentrations

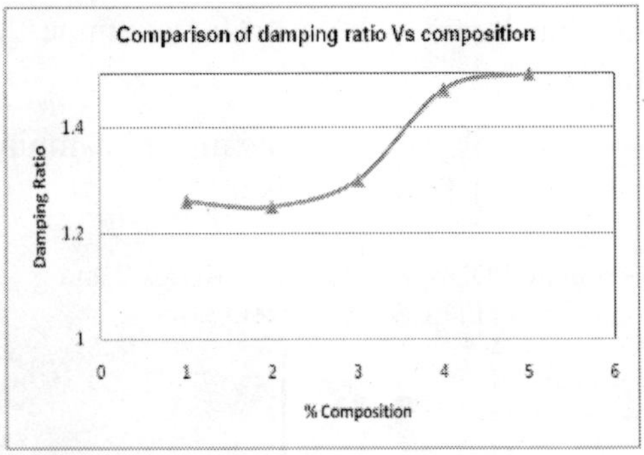

Fig. 5a: Variations in Damping Ratio as a Function of Silica Concentration in Mode 2 for Silica Filled Neoprene Nano Composites

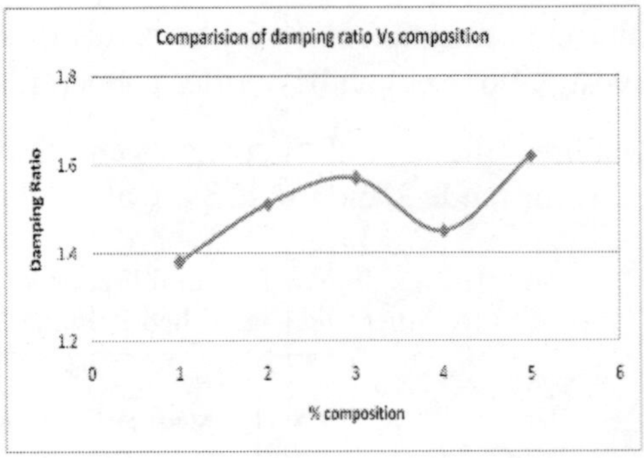

Fig. 5b: Variations in Damping Ratio as a Function of Silica Concentration in Mode 2 for Clay Filled Neoprene Nano Composites

The results of the natural frequencies and the corresponding observed and estimated damping ratios for mode 2 and mode 3 are presented in Table 1 for comparison.

CONCLUSION

1. Damping characteristics of E-glass/epoxy nanocomposite beams suitably filled with 1%–5% concentrations of nano silica and nano clay particles were studied. Both experimental and calculated results matched. It was observed that 5% nano silica and 5% nano clay filled polymer showed higher damping compared to other concentrations. However, these compositions are to be studied for other mechanical properties before using them in real life applications.
2. Overall, it is observed that 5% clay showed high damping in both 2nd and 3rd modes indicating that this concentration showed optimal damping among the studied nano fillers.

REFERENCES

[1] Pedro Henrique Cury Camargo, Kestur Gundappa Satyanarayana and Fernando Wypych "Nanocomposites: Synthesis, Structure, Properties and New Application Opportunities" *Materials Research*, Vol. 12, No. 1, 1–39, 2009.

[2] Braun T., Schubert A. and Sindelys Z., "Nanoscience and nanotechnology on the balance" *Scientometrics*, 1997, 38(2): 321–325.

[3] Vijaya Rangari, "Polymer Nanocomposite Materials for Structural Applications" Materials Science and Engineering, Tuskegee University, AL United States of America.

[4] Patel R.K., Bhattacharya B. and Sumit Basu, "Effect of interphase properties on the damping response of polymer nano-composites", Mechanics research communications, 2008, 35, 115–125.

[5] Suarez S.A., Gibsonb R.F., Deobald, L.R., Experimental Techniques. (Oct. 1984), 19–24.

SECTION–VII

Noise

Aerodynamic and Aeroacoustic Analysis of Centrifugal Blower for Noise Reduction

S. Rama Krishna, Vommi Krishna[1], A. Rama Krishna[2] and K. Ramji[2]

Department of Mechanical Engineering, G.V.P. College of Engineering (A), Visakhapatnam, Andhra Pradesh

[1]Department of Mechanical Engineering, ANITS, Visakhapatnam, A.P.

[2]Department of Mechanical Engineering, AUCE (A), Visakhapatnam. A. P.

E-mail:shinagamsai@gmail.com

ABSTRACT: *Centrifugal blowers, which are designed for high rotating frequencies, generate high levels of noise. Thus noise reduction is a key parameter in the design of centrifugal blowers. The objective the present study is to estimate the aerodynamic and aeroacoustic parameters like pressure, velocity and noise of centrifugal blower with three different types of impellers namely forward, backward and radial by both experimental and numerical analysis. Reverse engineering method is used to get the dimensions of centrifugal blower. The modelling of the blower is performed by using Solid modelling software 2010 and blower is meshed with a three dimensional tetra mesh by using HYPERMESH 9.0. Numerical analysis is performed using CFD-Fluent software package for the three types of impellers. These numerical results are compared with experimental results for validation. From the study it is observed that backward impeller generates less noise and more discharge than forward and radial impellers.*

Keywords: Centrifugal Blower, Forward, Backward, Radial Impellers, Hyper Mesh, CFD, Fluent, Aerodynamic and Aeroacoustic Analysis.

INTRODUCTION

A centrifugal blower is a roto-dynamic blower that uses a rotating impeller to increase the pressure of a fluid. With the growing importance of blowers in industries, several researches have focused their work on subject. Different problems associated with the blower like noise and vibration reduction have been received due to concern. Son *et al.* [1] presented the effects of bell mouth geometries on the flow rate of centrifugal blowers were numerically simulated using a commercial Computational Fluid Dynamics (CFD) code, Fluent. Ramesh Kumar *et al.* [2] observed that mechanical malfunctions such as, rotor unbalance and shaft misalignment are the most common causes of vibration in rotating machineries. Chen-kang and Hsieh [3] explained the performance analysis and optimized design of backward-curved airfoil centrifugal blower. CFD package FLUENT is used to simulate four backward curved air foil centrifugal blowers. Kolla *et al.* [4] proposed to carry out a study to evaluate the effectiveness of composites in reducing noise levels of the casing. Jianfeng *et al.* [5] conducted experimental study to reduce the noise of the centrifugal fan, whose impeller has equidistant forward-swept blades. Two new impellers with different blade spacing were designed for noise reduction. Moreland *et al.* [6] explained the housing effect of centrifugal blower. The sound power spectrum for a centrifugal blower operating at free delivery is characterized by enhancement at various frequencies owing to acoustical resonances in the blower housing.

Noise, generated by a centrifugal blower, can be divided according to its origin, into aerodynamically induced noise and vibration-induced noise. The contribution of the individual noise source to the total emitted noise is hard to determine, but it is crucial for the design of noise reduction measures. In order to reduce the noise of the centrifugal blower in a broad range of operating conditions, an identification of noise sources needs to be performed. The present blower is of 5 HP capacity with 12 blades is rotating at constant speed of 2660 rpm. The present work is aimed to determine the aerodynamic and aero acoustic analysis of centrifugal blower experimentally and comparing with numerical analysis by using FW-H equation coupled with LES computer code based on cell-centered finite volume method (FVM) on unstructured meshes for viscous flow field around blower.

METHODOLOGY

Experimental Setup

The experimental set up consists of centrifugal blower driven by motor and it is mounted on an adjustable bed. An orifice meter is fixed in the outlet pipe line to measure the actual discharge. A set of pitot tube and thermometer is provided at the outlet to measure the velocity and temperature. The pitot tube can also be used to measure the velocity profile at the blower inlet. U-tube Manometers are provided to measure the pressure difference across the orifice meter in the pitot tube and the delivery pressure. Three types of interchangeable impellers with radial vanes, backward curved vanes and forward curved vanes are provided with test rig to study the different effects of vanes types. Figure 1 shows the experimental set up for aerodynamic measurements.

The noise of centrifugal blower is measured by Bluer Kjaer Sound Level Meter (SLM)-2260 for backward, forward and radial impeller. The B&K 2260 investigator is a precision hand-held SLM with real-time displays of 1/3-octave and octave band spectra and statistical distributions. Also included is a PC compatible filing system and large memory. The 2260 conforms to the BS EN 61672 Class 1 standard. Figure 2 shows the experimental set up for aeroacoustic measurements.

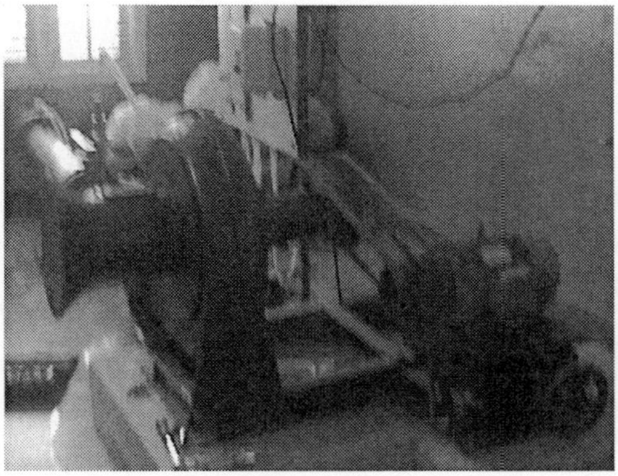

Fig. 1: Experimental Set up for Aerodynamic Measurements

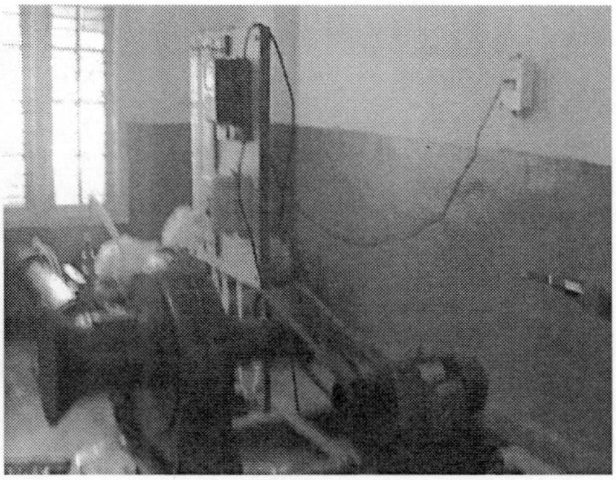

Fig. 2: Experimental Setup for Aeroacoustic Measurements

Modelling and Meshing of Centrifugal Blower

Reverse engineering method is used to obtain the dimensions of the centrifugal blower with forward, radial, and backward curved impeller blades and modelled using Solid works 2010. In order to model the centrifugal blower it is necessary to model the parts of the blower which are spiral casing, impellers, pulley, shaft, hub, suction arrangement, supporting structure then it was assembled to forward, radial, and backward curved impeller blades positioned in housing between two housing faces that are spaced apart along the impeller axis. A bell mouth shaped inlet is attached to the one of the facing and the housing is substantially closed with exception of the inlet and outlet. The impeller is attached to the hub which is keyed to the shaft. The solid model is imported as surface set

Fig. 3: Solid Model of Blower

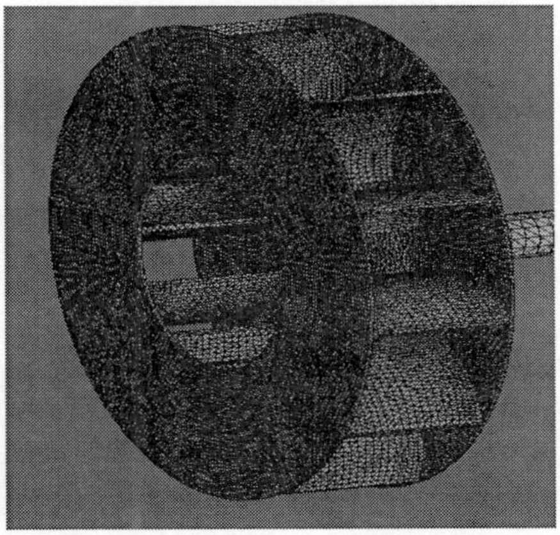

Fig. 4: Meshed Model of Backward Impeller

to HYPERMESH 9.0 in ANSYS environment. The solid model is repaired by toggle section option the gaps are stitched. After repairing, the closed surface set was converted into solid. Each component was loaded in the separate component collector. The model was tetra meshed in volume tetra mesh by considering various values like element size, featured angles of the elements and by selecting gradual mesh. After convergence total 439506, 441314 and 435689 elements are for entire blower with backward radial and forward impellers respectively. Figures 3 and 4 shows the solid model and meshed model of backward impeller.

Numerical Simulations

The numerical simulations have been carried out with a finite volume code method using Fluent. The turbulent nature of the flow is incorporated through the LES. LES is chosen as viscous model as it is used because, it needs time dependent solution for aeroacoustic solution and it is not highly dependent to geometrical conditions. For unsteady-state simulations, the Multiple Reference Frames (MRF) model is used. Surfaces that rotate relatively are defined as "moving wall". Moreover, as they are dependent on the fluid around them and as they rotate, they are defined as "relative to adjacent cell zone" and "rotational motion". Cylinder walls are defined as "stationary wall" and the inlet and outlet are defined as "pressure inlet" and "pressure outlet". Fluid zone in the inner volume is defined as "MRF" and rotates with 2660rpm in x-direction.

RESULTS AND DISCUSSION

The aerodynamic and aeroacoustic properties of blower with radial, forward and backward impellers are predicted with numerical simulations (CFD-Fluent software) and same is compared with the measurements. Table 1 shows the aerodynamic results comparison. From this table it is observed that blower with backward impeller develop maximum discharge (0.399 m^3/sec) and pressure (2079.5 Pa) compared with other forward and back and impellers. For all noise measurements carried out at 1m from the inlet of blower. The measurement position is aligned to the impeller axis. Figure 5 shows

Table 1: Aerodynamic Results Comparison

Vane Type	*Experimental Results*			*Numerical Results*			*Error in Discharge (%)*
	Pressure (N/m^2)	*Velocity (m/sec)*	*Discharge (m^3/sec)*	*Pressure (N/m^2)*	*Velocity (m/sec)*	*Discharge (m^3/sec)*	
Backward	2079.5	29.6	0.399	103.8	99.4	0.453	13.53
Forward	1471.38	28.95	0.346	92.2	90.1	0.386	11.56
Radial	1481.19	25.66	0.378	101.5	98.6	0.414	9.53

the noise prediction graph at inlet of blower with backward impeller from measurements. It is clearly shows that maximum peak is observed at fundamental blade passing frequency (532 Hz) and some of other peaks are observed due to interaction between casing and impeller noise. Table 2 shows the aeroacoustic results comparison. The results shows that blower with backward impeller was generating less noise (92.2dB) compared with other forward and radial impellers. Figure 6 shows the noise prediction graph at inlet of blower from numerical simulations with Fluent software.

Table 2: Aeroacoustic Results Comparison

Vane Type	*Experimental Results (dB)*		*Numerical Results (dB)*		*Error (%)*	
	Inlet	*Outlet*	*Inlet*	*Outlet*	*Inlet*	*Outlet*
Backward	83.3	81.2	92.2	90.1	10.68	10.96
Forward	89.2	86.8	101.8	99.4	14.13	14.51
Radial	85.4	83.6	98.1	96.1	14.87	14.95

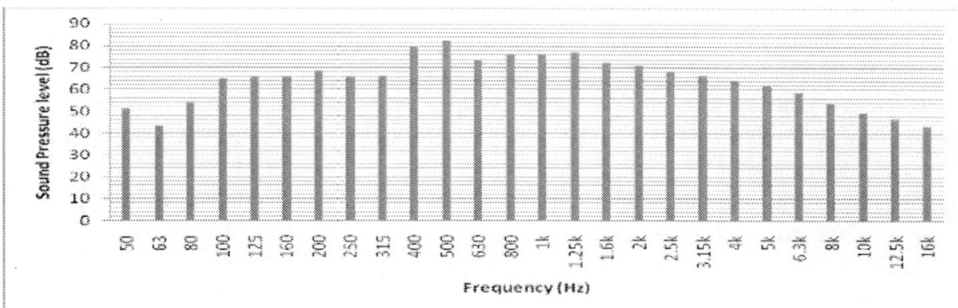

Fig. 5: Noise Prediction Graph at Inlet of Blower with Backward Impeller from Measurements

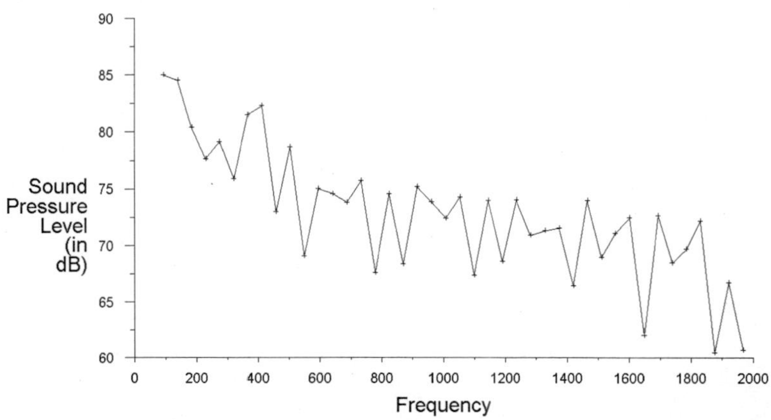

Fig. 6: Noise Prediction Graph at Inlet of Blower from Numerical Simulations with Fluent Software

CONCLUSIONS

The aerodynamic and aeroacoustic parameters of centrifugal blower with forward, backward and radial impellers with are measured and these results are compared with CFD-Fluent software. Errors between measured and predicted results are less than 15%. So CFD analysis results are close to the experimental results. From the analysis it is concluded that backward impeller generates less noise and high pressure when compared to other forward and radial impellers.

REFERENCES

[1] Pham Ngoc Son, Jaewon Kim and E.Y. Ahn, 2011, Effect of bell mouth geometries on the flow rate of centrifugal blower, J. of Mech. Sci. and Tech., 25 (9), pp. 2267–2276.

[2] G.R. Rameshkumar, B.V.A. Rao, and K.P. Ramachandran, 2010, Condition monitoring of forward curved centrifugal blower using coast down time analysis, Hindawi Publishing Corporation, International Journal of Rotating Machinery, Article ID 962804, 12 pages, doi:10.1155/2010/962804.

[3] Huang Chen-kang and Hsieh Mu-En, 2009, Performance analysis and optimized design of backward curved airfoil centrifugal blowers, American society of heating, refrigeration and air condition engineers.

[4] S. Kolla, Y. Anil Kumar and S. Rajesh, 2009, Noise reduction of air blower casing using composites, Vibration Problems ICOVP-2007, Springer Proceedings in Physics, Vol. 126, pp. 215–222, DOI: 10.1007/978-1-4020-9100-1_22.

[5] Jianfeng M.A., Datong Q.I., Yijun M.A.O., 2008. Noise reduction for centrifugal fan with non-isometric forward swept blade impeller, Front. Energy power engg. China, 2(4), pp. 433–437 doi 10.1007/s11708-008-0082-6.

[6] J.B. Moreland, 1974, Housing effect on centrifugal blower noise, J. Sound and Vibration, 36(2), pp. 191–205.

Performance Analysis of Speckle Noise Filters for Medical Ultrasound Images

P. Thirumoorthy and K.M. Prabusankarlal[1]

Department of Electronics and Communication,
Government Arts and Science College, Paramakkudi

[1]Department of Electronics and Communication,
K.S.R. College of Arts and Science, Tiruchengode

E-mail: tmoo@rediffmail.com; kmsankar@gmail.com

ABSTRACT: *Ultrasound is a relatively inexpensive, noninvasive, and portable technique for imaging the internal anatomy for medical diagnosis. The Speckle noise, an inherent nature of ultrasound often contaminates the image, significantly degrades the quality by discriminating fine details which may contain vital information and complicates the diagnostic decisions. The Speckle noise removal is the general pre-processing step for feature extraction, analysis, and recognition of ultrasound imaging. This paper analyzes the performance of different techniques proposed for the suppression of this multiplicative noise which is very difficult to be removed. Statistical quantity measures such as Mean Square Error (MSE), Root Mean Square Error (RMSE) and Peak Signal-to-Noise Ratio (PSNR) are used to measure and compare the quality of the enhanced images.*

Keywords: Ultrasound, Prostate, Medical Image, Speckle Noise, Speckle Filter, SNR, PSNR, MSE, RMSE.

INTRODUCTION

Ultrasonic plays an important role in medical diagnosis by revealing the internal anatomy of organs to medical practitioners for accurate prediction of abnomabilities which may be treated if detected at the earliest [1]. The main problem faced by them is the noise introduced due to the consequence of the coherent nature of the wave transmitted during the ultrasonic process. These noises tend to corrupt the quality of the image and an inferior quality image may lead to incorrect diagnosis [2]. Speckle noise affects the ultrasound images which are a complex phenomenon, which degrades image quality with a backscattered wave appearance which originates from many microscopic diffused reflections that passing through internal organs and makes it more difficult for the observer to discriminate fine detail of the images in diagnostic examinations [3]. Suppressing this speckle noise from an ultrasonic image by retaining the thin and small image features is the primary step in processing medical ultrasonic images [4]. For the early detection of cancers, such as breast cancer, prostate cancer, lung Cancer and cervical cancer, Ultrasound imaging method is suitable to diagnose and progenies. The accurate detection of region of interest in ultrasound image is crucial.

In this paper, different methods proposed for suppressing and removing speckle noise are analyzed. Using statistical measures such as Signal-to-Noise Ratio (SNR), Peak Signal-to-Noise Ratio (PSNR), and Root Mean Square Error (RMSE) the performance of the speckle filters are compared and evaluated with different medical images. This paper is organized as follows: Section 2 describes problems associated with noise generated during ultrasound process. Section 3 describes various spatial filtering techniques for denoising the speckle noise in the ultrasound medical image. Section 4 discusses about our experimental analysis and discussion. Section 5 concludes this paper and provides direction for further research.

PROBLEM DESCRIPTION

A digital image is defined as a two dimensional function $f(x, y)$, where x and y are spatial coordinates and the amplitude of f at any pair of coordinates (x, y) is called the intensity or grey level of the image [5]. The noise contaminates the data sets collected by image sensor and the region of interest in the image is degraded, which makes the original image not suitable for applying image processing techniques and analysis for medical diagnosis. As a preprocessing step, an image enhancement technique is mandatory before image is processed and analyzed. The noise removal in the image is still a challenging problem and so an efficient filtering method is necessary for removing noise in the images which introduces artifacts and causes blurring of the image. The speckle noise [6] is observed in ultrasound images where images are corrupted by a random granular pattern that delays the interpretation of the image content and reduces the chances of detection of features of interest. Noise is generally grouped into two categories viz., image data independent noise [7] and image data dependent [8] noise.

Speckle Noise

Speckle noise is a kind of data dependent noice found in the ultrasonic image is caused by errors in data transmission [10]. The corrupted pixels by speckle noise are either set to the maximum or minimum value which appears as something like a snow in image or have single bits flipped over. Speckle noise has the characteristic of multiplicative noise [11] and it follows a gamma distribution and is given as $F(g)=\left[\frac{g^{\propto-1}}{(\propto-1)!a^{\propto}}e^{\frac{g}{a}}\right]$. Where, variance is $a^2\propto$ and g is the gray level.

SPECKLE FILTERING TECHNIQUES

The primary goal of an image enhancement technique is to adjust the digital image so that the resultant image is more suitable than the original image for further processing and analysis [12]. The image processing function in a spatial domain can be expressed as $g(x, y) = T(f(x, y))$. Where, T is the transformation function, $f(x, y)$ is the pixel value of input image, and $g(x, y)$ is the pixel value of the processed image. Different filtering techniques such as Median Filter [13], FIR Filter [14], DCT Filter [15] and FFT filter are used for denoising the different medical images and four other filters are also proposed.

Entropy Filter

The proposed entropy filter filters image by replacing every value by the information entropy of the values in its range r neighborhood [17]. The entropy *H(L)* of the belief over L is defined as,

$$H(L) = -\sum_{l} Bel(L=l)\log Bel(L=l)$$

The s denote the measurement, the change of the entropy of Bel(L) given s is defined as

$$\Delta H(L \mid s) := H(L \mid s) - H(L)$$

The term $H(L \mid s)$ is the entropy of the belief Bel(L) *after* incorporating the measurement s.

Range Filter

This filter uses the local-intensity subrange of pixel intensity values within a window [18]. It uses the morphological functions to determine the maximum and minimum values in the specified neighborhood and uses the padding behavior of these morphological functions,

$$h(x) = k_r^{-1}(x) \int_{-\infty}^{\infty} \int_{-\infty}^{\infty} f(\xi) s(f(\xi), f(x)) d\xi$$

The $s(f(\xi), f(x))$ measures the photometric similarity between the pixel at the neighborhood center x and that of a nearby point.

Standard Deviation Filter

The standard deviation filter [12] calculates the standard eviation for each group of pixels in the subimage, and assigns this value to the center pixel in the output image. The objective function of standard deviation filter is given by,

$$S_w = \sqrt{\frac{1}{n^2 \sum_{r=1}^{n} \sum_{c=1}^{n} (x_{rc} - \bar{x})^2}}$$

where, $n \times n$ is the total number of pixels in a subimage, w denotes the indices of the subimage, x_{rc} is the value of the pixel at row r and column c in the subimage and x is the mean of pixel values in the window.

Tophat Filter

The Tophat filter is proposed for speckle noise removal which performs morphological top-hat filtering on grayscale or binary input image [19]. This algorithm computes the morphological opening of the image and then subtracts the result from the original image. In Fourier space, a top hat filter selects a band of signal of desired frequency by the specification of lower and upper bounding frequencies. Let $f: E \rightarrow R$ be a grayscale image, mapping points from a Euclidean space or discrete grid E (such as R^2 or Z^2) into the real line. Let $b(x)$ be a grayscale structuring element. The white top-hat transform of f is given by: $T_w(f) = f - f \circ b$, Where, the o denotes the opening operation.

EXPERIMENTAL ANALYSIS AND DISCUSSION

The proposed algorithms have been implemented using MATLAB 7.8. The performance of various spatial enhancement approaches are analyzed and discussed. The images are categorized into three groups; Breast, Prostate and cervix. More than 240 images are taken as samples and these images are fed into appropriate functions of all the eight filters. The parameters related with the image are commonly set in the filter functions for all images. Over 240 ultrasonic medical images in three categories are fed to all the eight filter algorithms and to evaluate the performance of filters three standard measures are taken; the Mean Square Error (MSE), Root Mean Square Error (RMSE), and Peak Signal-to-Noise Ratio (PSNR). The PSNR computes the peak signal-to-noise ratio, in decibels, between the input ultrasound image and output enhanced image which illustrates quality measurement between the original and enhanced image. The higher the PSNR, the better the quality of the enhanced image. To compute the PSNR, the MSE is calculated first using the following equation,

$$MSE = \frac{\sum_{M,N}(I_1(m,n) - I_2(m,n)^2)}{M * N}$$

Where, M and N are the number of rows and columns in the images. The RMSE is given by $RMSE = \sqrt{MSE}$. The following equation computes the PSNR.

$$PSNR = 10log_{10}\left[\frac{R^2}{MSE}\right]$$

where, R is the maximum fluctuation in the input image data type. Since we defined them as an 8-bit unsigned integer data type, R is 255. The Table 1 shows the calculated values using core i3 based PC with MATLAB 7.8. The kernel size of images is kept at 3×3.

The original ultrasound image and filtered images obtained by various filtering techniques are shown in Figure 1. Medical ultrasonic images of the breast, prostate (not shown) and cervix (not shown) are filtered and their performance parameters are analyzed. If the value of MSE is low and the value PSNR are larger and the enhancement approach is better. The Table 1 shows the performance analysis of the different filtering techniques we have used. It is observed from the tables in terms of high PSNR and visual observation of the enhanced image, the proposed tophat filter removes the speckle noise better than other existing and proposed approaches.

Ultrasound Breast Images

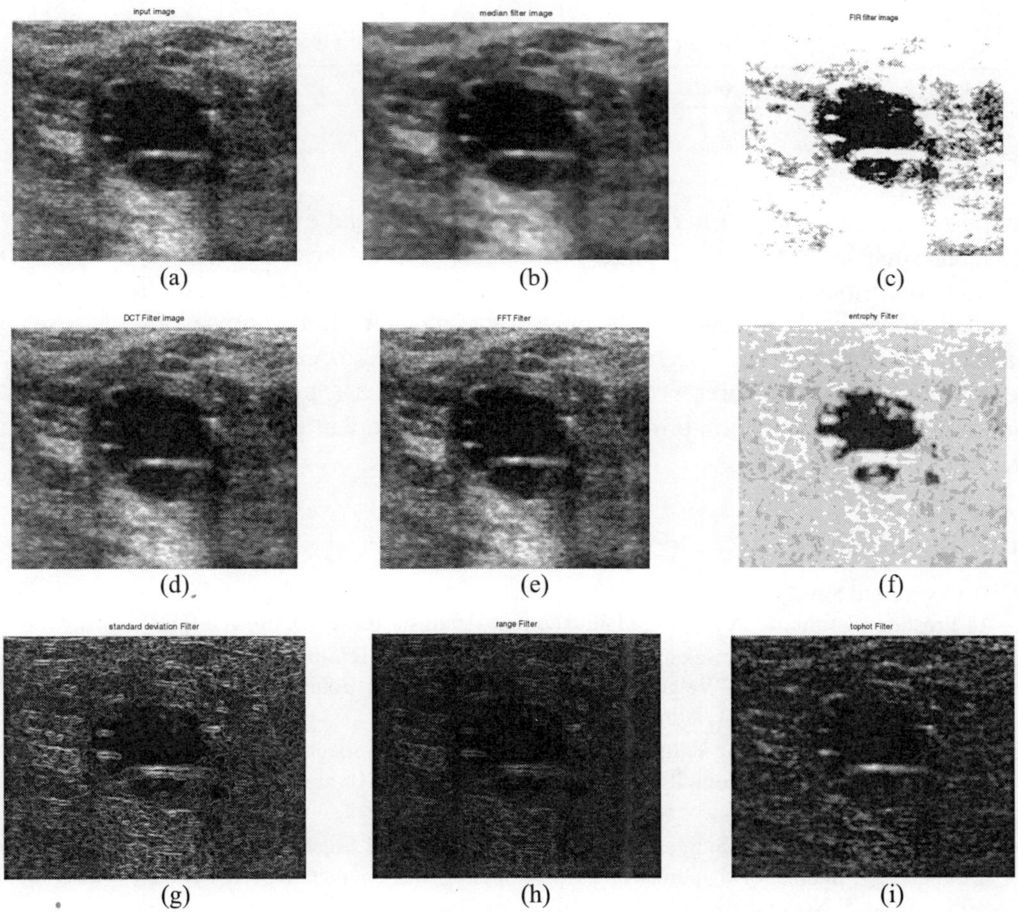

Fig. 1: Ultrasound Breast image (a)Input image (b) Median Filter (c) FIR Filter(d) DCT Filter(e) FFT Filter(f) Entropy Filter(g) Range Filter(h) Standard Deviation Filter (i) Tophat Filter

Table 1: Filter Performance for Ultrasound Breast, Prostate and Cervix images

Filter	*Breast Image*			*Prostate Image*			*Cervix Image*		
	MSE	*RMSE*	*PSNR*	*MSE*	*RMSE*	*PSNR*	*MSE*	*RMSE*	*PSNR*
Median Filter	8196	90.53	8.99	7536	86.81	9.36	5892	76.75	10.42
FIR Filter	14512	120.46	6.51	12635	111.19	7.11	23765	154.16	4.37
DCT Filter	9453	97.22	8.37	6755	82.18	9.83	5763	75.9	10.52
FFT Filter	5432	73.7021	10.78	4537	67.35	11.56	5547	74.47	10.69
Entropy Filter	8237	90.75	8.97	9459	97.25	8.37	7845	88.57	9.18
Range Filter	5841	76.42	10.46	5661	75.23	10.60	6687	81.77	9.87
SD Filter	7829	88.48	9.19	8228	90.7	8.97	8438	91.85	8.86
Tophat Filter	4627	68.02	11.47	3959	62.92	12.15	5287	72.71	10.89

CONCLUSION

The performance of noise filters are compared analyzed and evaluated using quantitative measures such as MSE, RMSE and PSNR and also in term of visual quality of the images. Some methods removed speckle noise better but the quality of the image too degraded. Since the finest details of the images are crucial for further processing such as edge detection and segmentation process, they have to be preserved. Very few algorithms removed speckle noise as well as preserved the details of the breast, prostate and cervix medical images during the evaluation process, among them the tophat filter performed better than others.

REFERENCES

[1] Yanong Zhu, Stuart Williams, Reyer Zwiggelaar, Computer Technology in Detection and Staging of Prostate Carcinoma: A review, Medical image Analysis 10(2006), pp. 178–199.

[2] Sudha.S, Suresh.G.R and Sukanesh.R Speckle Noise Reduction in Ultrasound Images by Wavelet Thresholding based on Weighted Variance, International Journal of Computer Theory and Engineering, Vol. 1, No. 1, April 2009, 1793–8201.

[3] Hossein Rabbani, Mansur Vafadust, Purang Abolmaesumi, Speckle Noise Reduction of Medical Ultrasound Images in Complex Wavelet Domain Using Mixture Priors, IEEE transactions on biomedical engineering, vol. 55, No. 9, september 2008.

[4] Arpit Singhal, Mandeep Singh, Speckle Noise Removal and Edge Detection Using Mathematical Morphology, International Journal of Soft Computing and Engineering (IJSCE) ISSN: 2231–2307, Vol.1, Issue 5, November 2011.

[5] Gonzalez R.C. and Woods R. 'Digital Image Processing', Addison-Wesley Publishing company, 1992.

[6] Image Processing Fundamentals-Statistics, "Signal to Noise Ratio", 2001.

[7] Yunyi Yan and Baolong Guo. Application of Wavelet Neural Network (WNN) and Gradient Descent Method (GDM) in Natural Image Denoising. Journal of Computational Information Systems, 2(2):625–631, 2006.

[8] Robert D Nowak, "Wavelet based Rician noise Removal", IEEE Transaction on image processing, Vol. 8, no. 10, pp. 1480, October 1999.

[9] Jain.A.K, fundamental of digital image processing. Englewood cliffs, NJ Prentice-Hall, 1989.

[10] Matias Barber, Francisco Grings, Pablo Perna, Marcela Piscitelli, Martin Maas,Cintia Bruscantini, Julio Jacobo-Berlles, and Haydee Karszenbaum. Speckle Noise and Soil Heterogeneities as Error Sources in a Bayesian Soil Moisture Retrieval Scheme for SAR Data. IEEE journal of selected topics in applied earth observations and remote sensing, vol. 5, no. 3, June 2012.

[11] Bala Prakash.K, Venu Babu.R, VenuGopal.B, Image Independent Filter for Removal of Speckle Noise. IJCSI International Journal of Computer Science Issues, Vol. 8, Issue 5, No 3, September 2011.

[12] Thangavel, K, Manavalan.R, Laurence Aroquiaraj.I, Removal of Speckle Noise from Ultrasound Medical Image based on Special Filters: Comparative Study, ICGST-GVIP Journal, ISSN 1687–398X, Volume (9), Issue (III), June 2009.

[13] Lim, Jae S., Two-Dimensional Signal and Image Processing, Englewood Cliffs, NJ, Prentice Hall,1990.

[14] Fliege, N. J. Multirate Digital Signal Processing: Multirate Systems, Filter Banks, Wavelets. West Sussex, England: John Wiley and Sons, 1994.

[15] Haitham Hassanieh, Piotr Indyk, Dina Katabi, and Eric Price, "Simple and Practical Algorithm for Sparse Fourier Transform," Acm-Siam Symposium on Discrete Algorithms (SODA), Kyoto, January 2012.

[16] Sivaranjani, R., Performance Analysis of Transform Based Image Compression Algorithms. CIIT International Journal of Digital Image Processing, December 2011.

[17] Gonzalez, R.C., R.E. Woods, S.L. Eddins, Digital Image Processing Using MATLAB, New Jersey, Prentice Hall, 2003, Chapter 11.

[18] Tomasi C., Manduchi.R, Bilateral Filtering for Gray and Color Images. Proceedings of the 1998 IEEE International Conference on Computer Vision, Bombay, India.

[19] Ming Zeng, Jianxun Li, Zhang Peng, The design of Top-Hat morphological filter and application to infrared target detection, Infrared Physics and Technology, Vol. 48, Issue 1, April 2006, pp. 67–76.

Vibration Monitoring and Piezoelectric Energy Harvesting

Himanshu Porwal and J. Srinivas

Department of Mechanical Engineering, NIT-Rourkela
E-mail: himanshu011090@gmail.com; srin07@yahoo.co.in

ABSTRACT: *This work reports analytical modeling and analysis results of a piezoelectric bimorph symmetric laminated cantilever beam for energy harvesting application. Initially, a single degree of freedom lumped-parameter model of the system is presented. It incorporates electromechanical coupling. Steady-state response of the base excited energy harvester model and time-varying open-circuit voltage generated by the system are arrived by solving the coupled differential equations. Studies are carried-out to illustrate the effect parameters such as load resistance on the output power. Furthermore, in order to harvest a constant amount of power in frequency mismatching conditions, a two-degree of freedom frequency tunable damped system is proposed and outputs of the resulting system are described.*

Keywords: Bimorph Cantilever, Lumped-Parameter Model, Base Excitation, Frequency Bandwidth, Parasitic Secondary System.

1. Introduction

Emerging wireless sensor technology has grown the applications in health monitoring of mechanical, aerospace and civil structures. The limited lifespan of batteries has motivated research on generation of alternative energy to wireless sensors from ambient sources. In particular, vibration-based energy harvesting techniques were developed based on various different transduction mechanisms like electromagnetic induction, piezoelectricity, electrostatic generation, dielectric elastomers and so on. These devices are essentially micro-scale resonant structures having beams with a proof (tip) mass. Piezoelectric (PZT) cantilever beams are widely used structures, because of their simplicity in construction, high flexibility, low natural frequency and more deflection. In practice, the cantilever PZT bimorph beams are often mounted on a vibrating host structures to generate AC voltage output for powering an electrical load. There are many experiments as well as mathematical models [eg.,1–3] on the piezoelectric energy harvesters in order to investigate the energy harvesting performance. One major limitation of piezoelectric energy harvester is that the maximum power output occurs only when it operates at resonance, which is very difficult to guarantee due to manufacturing tolerances and excitation frequency changes. To overcome this, several innovative approaches were considered, such as tuning the resonance frequency of harvester so that it matches with frequency of ambient vibration at all times. Alternatively, enhancing bandwidth using multimodal energy harvesting methods [4–8] and nonlinear oscillation techniques [9–12] would also be helpful. Other approaches like optimizing parameters of the energy harvester [13–14] and impedance matching technology have also been proposed to improve the power output of the energy harvester.

In present article, a two-degree of freedom (DOF) piezoelectric energy harvesting model is presented to overcome the bandwidth issue. Initially single DOF piezoelectric cantilever beam with proof mass is studied using lumped-parameter model. By considering a 2-DOF model and parametric studies, it is shown that one can design a harvester to have two close peaks with effective magnitude contributing in the better harvesting performance.

2. Analytical Modeling

The piezoelectric cantilever beam energy harvester considered is assumed to be in the bimorph configuration, as shown in Fig. 1(a), in which two identical piezoelectric layers are attached to the substrate beam. The piezoelectric layers can be connected in series or in parallel, but only the series connection is considered in this paper. The piezoelectric material generates electrical charge through the thickness (3-direction) of the piezoelectric layer as the beam vibrates in the direction-3 with strain in the direction-1. The piezoelectric effect that couples the mechanical and electrical properties is described using the linear constitutive equations for piezoelectric materials

$$\begin{bmatrix} \sigma_1 \\ D_3 \end{bmatrix} = \begin{bmatrix} c_{11}^E & -e_{31} \\ e_{31} & \varepsilon_{33}^S \end{bmatrix} \begin{bmatrix} S_1 \\ E_3 \end{bmatrix} \tag{1}$$

In these equations, D_3 is the electric displacement, while σ_1 and S_1 are the axial stress and strain of the piezoelectric material, respectively. The electric field in each piezoelectric layer E_3 is assumed to be uniform throughout the entire layer and is defined as ratio of voltage potential difference of each layer v(t)/2 and the thickness t_p: $E_3=v(t)/2t_p$. The elastic modulus for the piezoelectric material c_{11}^E is measured at constant electric field, while the permittivity of the piezoelectric material ε_{33}^S is measured at constant stress. The piezoelectric coupling coefficient e_{31} is a product of the piezoelectric constant d_{31} and the elastic modulus: $e_{31}=d_{31}c_{11}^E$. Since a single degree of freedom model provides the fundamental insights into the electromechanical coupled characteristics of the harvester, it is first reviewed. Schematic of the harvester model under base excitation is shown in Fig.1(b).

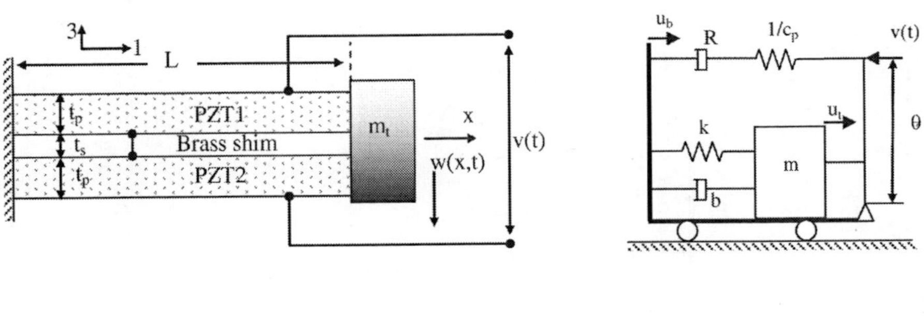

(a) Piezoelectric cantilever (b) Equivalent SDOF system

Fig.1.

The system has equivalent mass m, damping co-efficient b and stiffness k subjected to small base excitation. Here, R is load resistance, $1/c_p$ is capacitive nature of piezoelectric model in which equivalent capacitance is $c_p = \frac{\varepsilon_{33}^S wL}{2t_P}$ and θ is an electromechanical coupling coefficient of the system, v(t) is voltage across R. Writing equation of motion in terms of relative displacement: $u = u_t - u_b$, we obtain:

$$m\ddot{u} + b\dot{u} + ku = -m\ddot{u}_b - \theta v(t) \quad (2)$$

The equivalent mass for a uniform cantilever beam loaded at its free end with tip mass m_t is given by: $m = m_t + \frac{33}{140} m_b$, with $m_b = \rho AL$ is the mass of the beam. The equivalent stiffness k can be written as: $k = \frac{3EI}{L^3}$. Here, the effective unit mass and flexural rigidity of the beam in terms of width w and thickness are given by [15]:

$$(\rho A) = w\,(\rho_S t_S + 2\rho_P t_P) \quad (3a)$$

$$EI = c_s w \frac{t_s^3}{12} + 2c_{11}^E w\left(\frac{t_P^3}{12} + t_p\left(\frac{t_P + t_s}{2}\right)^2\right) \quad (3b)$$

Equation relating the electrical parameters obtained from Kirchhoff's Law as:

$$-\theta\dot{u} + c_p\dot{v}(t) + \frac{v(t)}{R} = 0 \quad (4)$$

Equations (2) and (4) are coupled by the parameter 'θ'. Assuming harmonic base motion and taking Laplace transforms of eqs.(2) and (4),we get

$$ms^2\hat{u} + bs\hat{u} + k\hat{u} + \theta\hat{v} = -s^2 m\hat{u}_b \quad (5)$$

$$-\theta s\hat{u} + c_p s\hat{v} + \frac{\hat{v}}{R} = 0 \quad (6)$$

where s is Laplace variable, and the cap denotes a variable of function in the Laplace domain. Solving equations (5) and (6) and setting $s = j\Omega$, we get the displacement and voltage magnitudes as:

$$U(\Omega) = \frac{m\Omega^2 U_b\left(\frac{1}{R} + jc_p\Omega\right)}{\left((k - m\Omega^2) + jb\Omega\right)\left(\frac{1}{R} + jc_p\Omega\right) + j\Omega\theta^2} \quad (7)$$

$$V(\Omega) = \frac{jU_b\theta\Omega^3}{\left((k - m\Omega^2) + jb\Omega\right)\left(\frac{1}{R} + jc_p\Omega\right) + j\Omega\theta^2} \quad (8)$$

By exploiting two degrees of freedom on adding a parasitic subsystem as shown in Fig.2, the two peaks in wide frequency bandwidth may be observed.

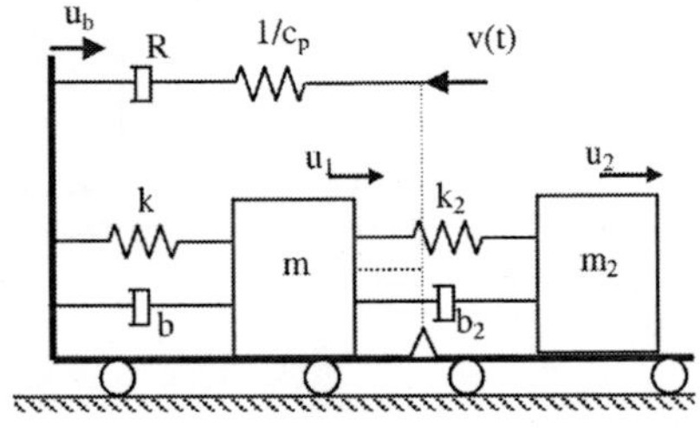

Fig.2. Two degree of freedom piezoelectric energy harvesting model

The equations of motion of the system can be written first in terms of parameters u_1, u_2 and u_b and then by setting $y=u_1-u_2$ and $u= u_1-u_b$, the governing equations of the model can be written as:

$$m_2\ddot{y} + b_2\dot{y} + k_2y = -m_2\ddot{u} - m_2\ddot{u_b} \tag{9}$$

$$m\ddot{u} + b\dot{u} + ku - b_2\dot{y} - k_2y + \theta v + m\ddot{u_b} = 0 \tag{10}$$

$$-\theta\dot{u} + c_p\dot{v}(t) + \frac{v(t)}{R} = 0 \tag{11}$$

Applying Laplace transform for these equations, we obtain:

$$m_2s^2\hat{u} + (m_2s^2 + b_2s + k_2)\hat{y} + m_2s^2\,\hat{u}_b = 0 \tag{12}$$

$$(ms^2 + bs + k)\hat{u} - (b_2s + k_2)\hat{y} + \theta\hat{v} + ms^2\hat{u}_b = 0 \tag{13}$$

$$-\theta s\hat{u} + \left(c^s s + \frac{1}{R_l}\right)\hat{v} = 0 \tag{14}$$

After solving these equations,

$$\hat{u} = -\frac{-(b_2s+k_2)m_2s^2 + ms^2(m_2s^2+b_2s+k_2)}{\{(m_2s^2+b_2s+k_2)(ms^2+bs+k)+(b_2s+k_2)m_2s^2\}\left\{c_ps+\frac{1}{R}\right\}+\theta^2 s(m_2s^2+b_2s+k_2)} \tag{15}$$

$$\hat{v} = \frac{\theta s}{\left(c_ps+\frac{1}{R}\right)}\hat{u} \tag{16}$$

Substituting $s=j\Omega$ and simplifying, we can obtain the frequency responses for displacement u and voltage v just like earlier case. Here, we define the nondimensional parameters: $\mu=m_2/m_1$ and $\alpha=\omega_2/\omega_1$, where $\omega_1 = \sqrt{\frac{k_1}{m_1}}$ and $\omega_2 = \sqrt{\frac{k_2}{m_2}}$ to select the values of secondary system parameters m_2 and k_2.

3. Results and Discussion

In this section, a numerical example is considered to compare the performance of a single DOF normal piezoelectric energy harvester with that of a combined two DOF tuned system. In order to illustrate the outputs of the model, the dimensions and properties of biomorph-harvester considered here are depicted in Table1.

Table 1. Dimensions and properties of cantilever [16]

Parameter	Substrate (s)	Piezoelectric (p)
Beam length(L),(mm)	20	20
Beam width(w),(mm)	6.4	6.4
Thickness t,(mm)	0.14	0.26
Elastic Modulus c,(GPa)	100	66
Densities,(kg/m^3)	8700	7800
Piezoelectric Constant,d_{31}(m/V)	---	190e-12
Absolute permittivity ε_{33}^S	---	15.93nF/m
Electromechanical coupling term, θ	---	3.1623e-5

The other constant parameters considered are: tip-mass m_t=0.03375 mg, system damping ratio $\xi=b/2\sqrt{km}$ =0.01 and the base acceleration U_b=0.1 mm/s^2. All the steps are implemented in MATLAB. Fig.3 shows the relative displacements of mass m of piezoelectric beam for different values of the parasitic secondary system: ξ_2 and μ. It is seen that the natural frequencies have not changed much with slight increase of mass ratio μ. It was also observed that as mass ratio μ decreases, left peak also reduces and right peak increases. The results show that the additional spring-mass system can magnify the tip-displacement of the piezoelectric beam when a proper mass ratio is selected. Fig.6 shows the variation of output power in frequency domain. It is seen that just like single DOF harvester model, power drastically increases with load resistance.

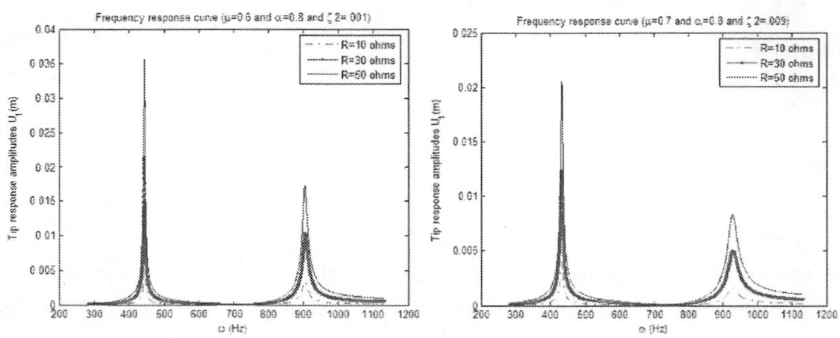

(a) ξ_2=0.001, α=0.8, μ=0.6 (b) ξ_2=0.009, α=0.8, μ=0.7

Fig.3. Tip response amplitudes of the 2-DOF harvester model

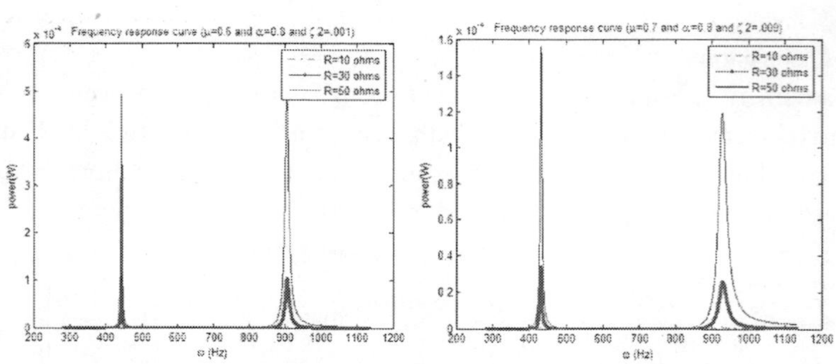

(a) ξ_2=0.001, α=0.8, μ=0.6 (b) ξ_2=0.009, α=0.8, μ=0.7

Fig.4. Output power variation at one of the set conditions.

4. Conclusion

A two-degree of freedom wide band energy harvester design has been proposed in this paper to enhance the workability of conventional piezoelectric energy harvesters. For smaller peak distances, the mass ratio μ should be as low as possible, which facilitates also in decrease of total weight. The magnitudes of the first peak increases with mass ratio, while that of the second peak decreases. Even magnitudes are smaller than that of single DOF piezoelectric harvester, both the peaks have significant contribution to energy harvesting at lower mass ratios.

REFERENCES

[1] M. Renaud, P. Fiorini and C.V. Hoof, 'Optimization of a piezoelectric unimorph for shock and impact energy harvesting', Smart Materials and Structures, vol. 16, pp. 1125–1135, 2007.

[2] A. Erturk and D.J. Inman, 'On mechanical modeling of cantilevered piezoelectric vibration energy harvesters', J. Intelligent Material Systems and Structures, vol.19, pp.1311–1320, 2008.

[3] A. Erturk and D.J. Inman, 'An experimentally validated bimorph cantilevered model for piezoelectric energy harvesting from base excitations', Smart Materials and Structures, Vol. 18, pp. 1–18, 2009.

[4] Y. Tadesse, S. Zhang and S. Priya, 'Multimodal energy harvesting system: piezoelectric and electromagnetic', J. Intelligent Material Systems and Structures, Vol. 20, pp. 625–632, 2009.

[5] V.R. Challa, M.G. Prasad, Y. Shi and F.T. Fisher, 'A vibration energy harvesting device with bidirectional resonance frequency tunability', Smart Materials and Structures, vol. 17, p. 015035, 2009.

[6] I.H. Kim, H.J. Jung, B.M.Lee and S.J. Jang, 'Broadband energy-harvesting using a two-degree of freedom vibrating body', Applied Physics Letters, Vol. 98, p. 214102–1, 2011.

[7] L. Tang and Y. Yang, 'A multiple-degree-of-freedom piezoelectric energy harvesting model', J. Intelligent Material Systems and Structures, DOI: 10.1177/1045389X12449920, pp. 1–17, 2012.

[8] W.A. Ashtari, M. Hunstig, T. Hemsel and W. Sextro, 'Frequency tuning of piezoelectric energy harvesters by magnetic force', Smart Materials and Structures, Vol. 21, p. 035019, 2012.

[9] R. Viguie, M. Peerers, G. Kerschen and J.C. Golinval, 'Energy transfer and dissipation in a Duffing oscillator coupled to a nonlinear attachment', J. Computational and Nonlinear dynamics, Trans. ASME, Vol. 4, p. 041012–1, 2009.

[10] D.A.W. Barton, S.G. Burow and L.R. Clare, 'Energy harvesting from vibrations with a nonlinear oscillator', J. Vibration and Acoustics, Trans. ASME, Vol. 132, pp. 021009–1–7, 2010.

[11] R. Ramian, M.J. Brennan, B.R. Mace and I. Kovacle, 'Potential benefits of a nonlinear stiffness in an energy harvesting device', Nonlinear Dynamics, Vol. 59, pp. 545–558, 2010.

[12] T.W. Ma, 'Opportunities for using nonlinear oscillators to enhance energy harvesting from impulsively loaded structures', Proc. IMechE, Part:I, J. Systems and Control Engineering, vol.225, p.467, 2011.

[13] Y.B. Liao and H.A. Sodano, 'Model of a single mode energy harvester and properties for optimal power generation', Smart Materials and Structures, Vol. 17, p. 065026, 2008.

[14] A.Hu and M. Long, 'Finite element analysis and structural parameters optimization of piezoelectric vibration energy harvester', Applied Mechanics and Materials, Vols. 44–47, pp. 1465–1469, 2011.

[15] A.M. Wickenheiser, "Design optimization of linear and nonlinear cantilevered energy harvesters for broadband vibrations", J. Intelligent Material systems and Structures, Vol. 22, pp. 1213–1225, 2011.

[16] K.H. Mak, S.M. William, A.A. Popov and C.H.J. Fox, "Performance of a cantilever piezoelectric energy harvester impacting a bump-stop", J. Sound and Vibration, Vol. 330, pp. 6184–6202, 2011.

Noise Control of a Hand Drill Using Hybrid Muffler

R. Praveena, Gyanadutta Swain and B. Venkatesham

Indian Institute of Technology Hyderabad, Yeddumailaram, Andhra Pradesh

E-mail: venkatesham@iith.ac.in

ABSTRACT: *Extensive use of flexible hand tools in workshops and domestic applications are common. These tools are designed for multipurpose and low cost. This requirement makes designer to choose a compact and integrated system. So, it involves lot of challenges to design a quieter product. This paper emphasises in reducing the noise levels of a hand drill. The performance and noise characteristics of instrument are tested experimentally, followed by some analytical calculations on various subassemblies of the product to find out the dominant noise source. Experimental results show that fan is the major noise source and this characteristic is being observed at high frequency region. After conducting parametric studies a suitable hybrid muffler of short length is mounted on the instrument to reduce the noise levels to a considerable value. The concept of noise reduction is verified with the prototype of a muffler by measuring its insertion loss.*

Keywords: Hand Drill, Fan Noise Reduction and Hybrid Muffler.

INTRODUCTION

Acoustic noise problems exist in wide variety of applications. Sound pressure levels depend on the part of the total mechanical or electrical energy transformed into acoustical energy. All noise control regulation acts emphasize in developing a quieter product. In the noise management data sheets for electric drills, by department of commerce, government of western Australia it is mentioned that the quietest form of hand drill is the cordless variety that produces about 78 dB (A) and a normal hand drill produces about 90 dB (A) to 94 dB (A) and hammer hand drills produce in excess of 100 dB (A) [1].The Internal components of hand drill such as fan, gear arrangement, motor and bearing act as various sources for noise which produce high sound pressure levels that can cause noise induced hearing loss if a person is exposed to it for a longer time duration.

DESCRIPTION OF HAND DRILL

A hand drill consists of an electric motor that rotates a replaceable drill bit to make holes in wood, plastic, or metal. Various components such as motor, bearing and fan are mounted on a single rotating shaft. A gear arrangement connects the output shaft with the input shaft for speed reduction. Chuck is mounted on the output shaft where drill bit is mounted. Figure 1 shows different components of a typical hand drill.

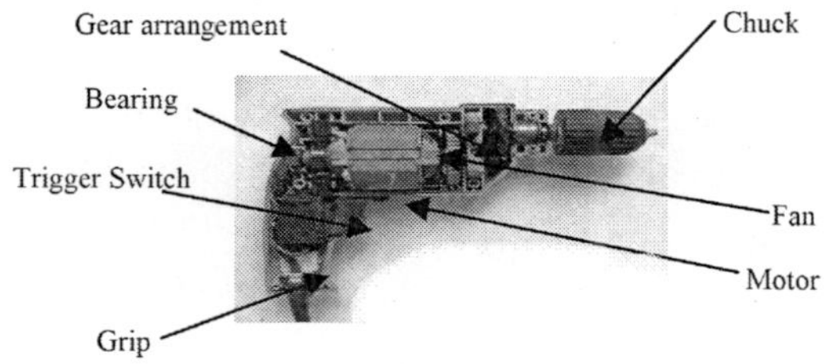

Fig. 1: Various Components of a Hand Drill

METHODOLOGY FOR QUIETER DESIGN

Methodology that must be followed to determine the best solution for quieter product involves the following steps:

(1) Identification of potential noise sources, (2) Analysis of noise sources and noise generation mechanism, (3) Source Ranking, (4) Noise propagation and (5) Develop noise control measures for the major sources based on noise propagation mechanism.

Potential Noise Sources

The first step is to list all the potential noise generating sources and then classify into two broad categories as: (a) Air borne noise: fan, windage (b) Structure borne noise: gears, bearings, electro–magnetic forces and commutator.

Analysis of Various Noise Sources

Fan: Fan is made of plastic material and is integrated with the input shaft on which the motor is mounted. The number of fan blades are 34. Rotating speed of the fan measured using the tachometer is 24500 rpm.

Blade passing frequency of fan is calculated as:

$$f_b = \frac{\text{Number of Blades x Rotatingspeed}}{60} = \frac{34 \text{ x } 24500}{60} = 13883.3 \text{ Hz}$$

Gear: A helical gear arrangement is provided in the instrument. Gear meshing frequency is calculated from the measured speed and number of teeth of the gear. Number of teeth on driver and driven are 6 and 47, respectively.Speed of driven gear is 2950$\pm$ 5% rpm. The driver speed ranges from 21952–24263 rpm. Gear meshing frequency

$$f_g = \frac{\text{Number of teeth X RPM}}{60} = 2195.2\text{Hz} - 2426.3\text{Hz}$$

Motor: The drive unit is a 2 pole DC motor with a rated voltage of 220 V and working speed of 24500 rpm.

Bearing: It is a ball bearing designated as 626ZZ.It is a double chrome steel shielded ball bearing having seven number of balls with a runout speed of 34000 rpm. Bearing can withstand a dynamic load upto 2.34 kN.

Experimental Analysis

Noise generated due to drilling operations is not considered in this study. The focus is only on the noise generated from the instrument. The generated noise is measured using class A sound level meter. Measurements are taken at 1 meter distance at various angles. Both A weighting and C weighting at faster and slower time scale has been measured and the difference of measurements had been taken to check for any low frequency sources. Figure 2 shows the A weighted sound pressure measurement values at various angles with respect to hand drill orientation. A-weighted Sound pressure levels of each individual source components are also estimated for source ranking.

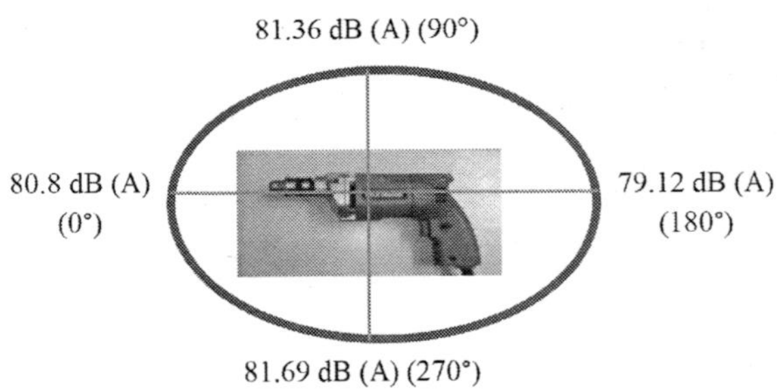

Fig. 2: Directivity of a Hand Drill

Figure 3 shows the frequency spectrum and Table 1 gives the various sound pressure parameters at 0°. It is observed from the spectrum that the dominant sound pressure levels are at high frequencies and also it is broad band. Sound source is steady based on measured parameter value.So, this infers that fan acts as the primary noise source.

Source Ranking

To achieve effective noise control solution source ranking [2] is done to find out the dominant noise source. Figure 4 shows the source ranking. It is observed that the dominant noise source is air-born and also a steady source. Contribution of structure–

bornnoise is not significant in the current model. Multiple fan noise control solutions are available in the literature [2]. Noise reduction at source is ideal. However providing a solution at the source is not possible due to logistic constraints. So it has been decided to use a path control solution. One of the best solutions to control flow noise is by means of a muffler [3].An extensive literature is provided for designing a muffler [4].

Table 1: Various Sound Pressure Parameters at 0°

LAF_{max}	LAS_{max}	LAF_{min}	LAS_{min}	LA_{eq}
81.64	81.13	78.23	79.3	80.78

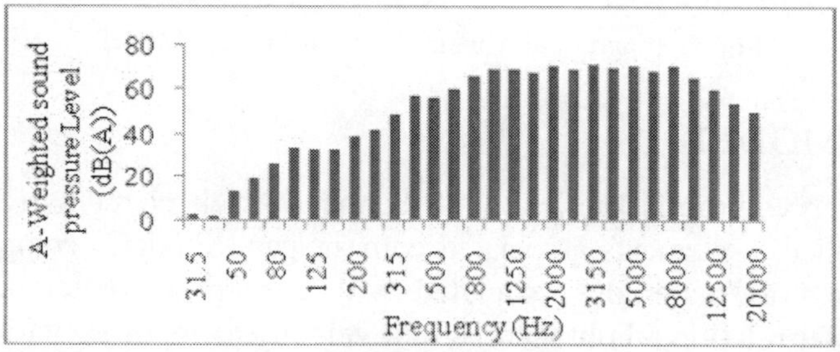

Fig. 3: One-Third Octave Sound Pressure Spectrum of a Hand Drill

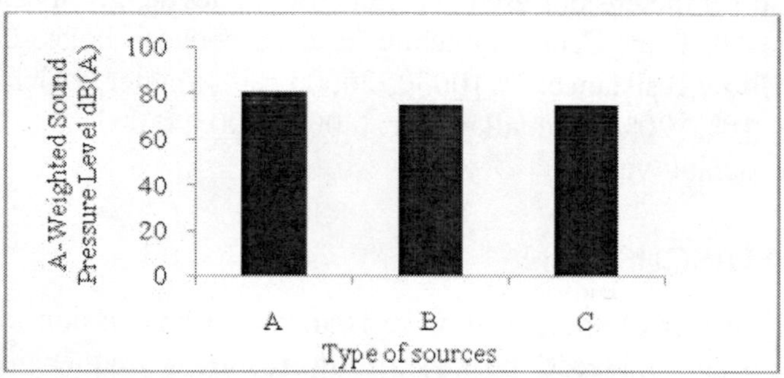

Fig. 4: Source Ranking, A-Fan, B-Gear train, C- Motor and Bearing

A hybrid muffler has been proposed as a solution. It has both reactive and dissipative solutions. Constraints like weight of the muffler, space limitations, back pressure and cost have been considered while designing a muffler. Figure 5a and 5b shows the schematic diagram and also isometric view of a muffler model with all dimensions in mm. Proposed hybrid muffler consists of a side inlet and end outlet with absorptive lining. Extensive parametric studies are conducted before developing the prototype. Computational model of proposed model is developed using commercial software LMS virtual lab [5]. Air gap in this model is not symmetric with respect to centre.

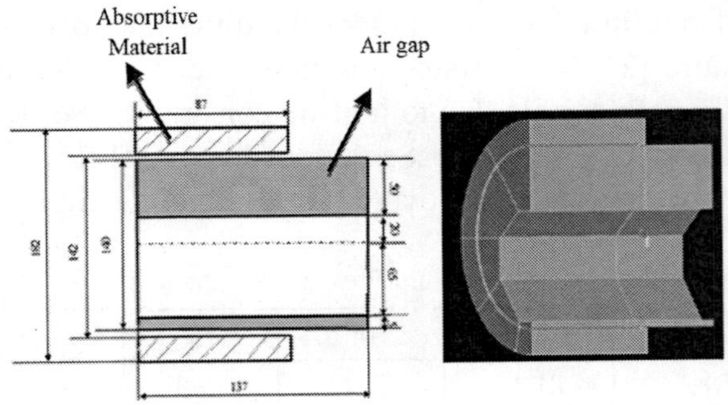

(a) Schematic (b) Isometric

Fig. 5: Diagram of Muffler View of Muffler Model

NUMERICAL MODEL

Completed muffler volume is meshed with 8 noded brick elements (SOLID 185).Air properties are applied to elements associated with air gap. Similarly complex velocity and density are applied to materials associated with absorptive material. Perforation is modelled as transfer relation admittance. A unit velocity as input excitation and anechoic termination ($z = \rho c$) as outlet condition has been modelled. Dimensions of muffler are given in Figure 5a; other dimensions used in parametric studies are given below. Thickness of the perforated sheet: 0.001m, Centre to centre distance between holes: 0.0075m are kept constant. Specific Flow resistance: 0, <u>10000</u>, 20000 rayls /meter, Mach number: 0, <u>0.1</u>, 0.2., Porosity: 5%, <u>10%</u>, 20%, Hole diameter: 0.002, <u>0.003</u>, 0.004 meter. The underlined values are taken as default values.

RESULTS AND DISCUSSION

Figure 6 shows the effect of specific flow resistance on transmission loss. It is inferred that as the flow resistance increases peaks shift towards lower frequencies. The resistance to the sound propagation in the absorptive material increases as the flow resistivity increases, so the peaks shift towards low frequencies. Similar behaviour is observed in reference [6].When SFR=0, muffler behaviour is not exactly same as concentric tube resonator (CTR) which is due to asymmetry in air gap. It is observed from computational results that as the Mach number increases peaks shift towards lower frequencies. Significant change in high frequencies is observed. The maximum achievable transmission loss is controlled by Mach number. As flow velocity increases flow noise contributes as self-generated noise. Figure 7 shows the variation of porosity. As the porosity increases the peaks are shifted towards higher frequency [6].It can be observed that high frequency region can be improved by selecting a higher porosity value. Figure8 shows a comparison of concentric tube resonator model with side inlet and end outlet model. The presence of uneven air gap in the model and short length leads to amismatch in the peak frequency values calculated by analytical method. So, existing simple analytical models have their own limitation for this application.

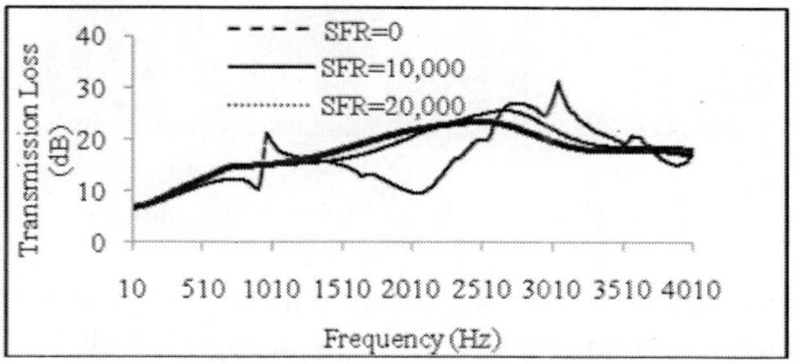

Fig. 6: Effect of Specific Flow Resistance on Transmission Loss

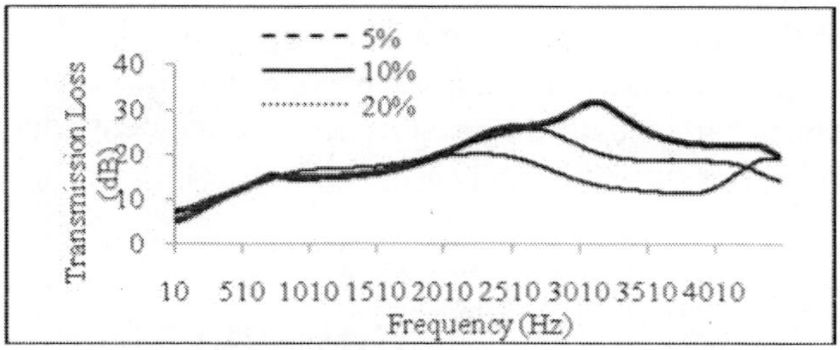

Fig. 7: Effect of Porosity on Transmission Loss

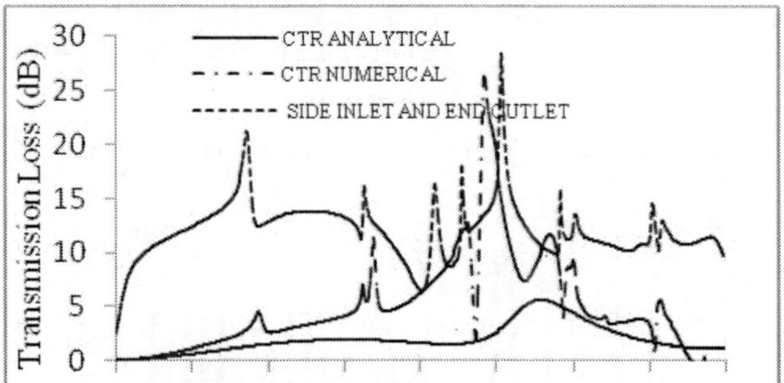

Fig. 8: Comparison of Side Inlet and End Outlet with CTR Model

Based on these parametric studies a prototype is made. Parametric studies suggest developing a muffler with high SFR absorptive material, higher porosity in perforated sheet are best for higher frequency noise control and also effective over a broader frequency range. The insertion loss of the muffler is measured. Figure 9 shows the muffler prototype that is mounted on hand drill. Specifications are outer diameter 0.18 m, inner diameter: 0.14 m porosity: 0.35, hole diameter: 0.005 m, centre to centre distance between holes: 0.0075.

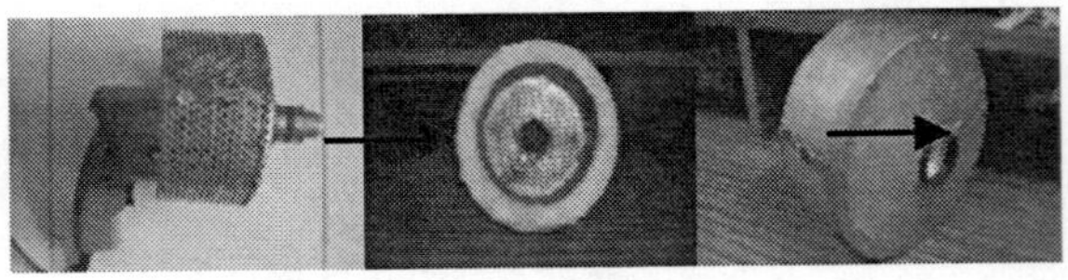

(a) (b) (c)

Fig. 9: Hybrid Muffler in Different Orientation

Figure 10 shows a comparison of sound pressure with and without muffler. Table2shows a comparison of sound pressure levels at various angles with respect to hand drill orientation as shown in Figure 1. Figure 11 shows the insertion loss spectrum. Its hows that the muffler is effective at higher frequencies. There are certain limitations in the current study. The length of the muffler is much smaller compared to the diameter due to which the effective length for sound attenuation is reduced. Since the cross section dimensions of the muffler are much larger than the wavelength, the incident sound field contains a very large number of higher order modes [7].

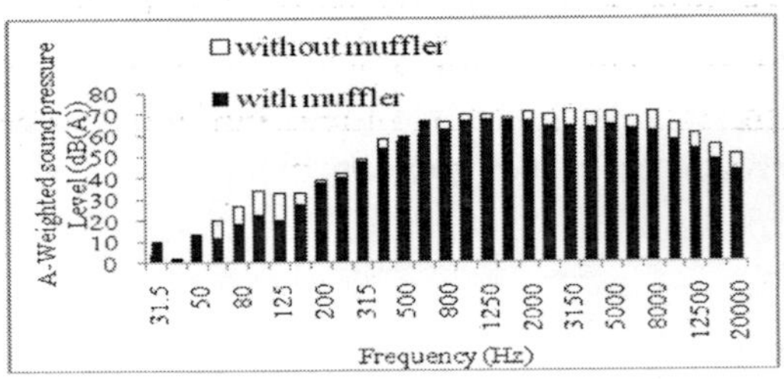

Fig. 10: A-weighted SPL with and Without Muffler

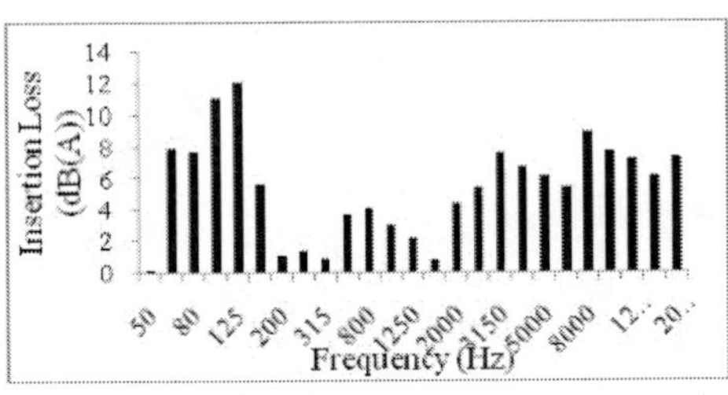

Fig. 11: Insertion Loss Spectrum of Prototype Muffler

This might result in an entrance loss [7] which needs to be characterized. The muffler has been modelled with an uneven air gap, so the amount of air flow is different along the circumference of the hand drill. Due to this there are challenges in validating with the analytical model.

Table 2: A-weighted Sound Pressure Level with and Without Muffler in Different Orientations

	Without Muffler LA_{eq} (dB(A))	*With Muffler LA_{eq} (dB(A))*	*Noise Reduction SPL dB(A)*
0°	80.8	76.29	4.51
90°	81.69	76.65	5.04
180°	79.12	77.17	1.95
270°	81.36	76.5	4.86

SUMMARY

Noise sources in the hand drill machine have been identified based on acoustical and vibration measurement and fan has been identified as the major noise source. This is done based on the concept of source ranking. A passive noise control technique has been proposed in the form of a hybrid muffler to control fan noise. A prototype has been prepared based on extensive parametric study using numerical models. An attempt has been made to reduce the sound from the hand drill by using muffler concept. This concept helped to get insertion loss of 4–5 dB (A).This paper described the complete noise control approach by integrating analytical, computational and experimental methods. Similar approach can be extended to other industrial products. This model can be further improved with an even air gap muffler by optimising the muffler dimensions with considerable back pressure and thereby achieving more reduction in noise levels.

REFERENCES

[1] Noise management data sheets http://www.commerce.wa.gov.au/worksafe/content/safety_topics/noise/Further_information/Electrical_drills.html

[2] David A. Bies and Colin H. Hansen, Engineering Noise Control Theory and practice 2009,Taylor & Francis.

[3] KIM, H.-S., Hong, J.-S., Sohn, D.-G., Oh, J.-E., Development of an Active Muffler System for Reducing Exhaust Noise and Flow Restriction in a Heavy Vehicle, Department of Mechanical Engineering, Institute of Noise Control Engineering, Auburn University.

[4] M.L. Munjal Acoustics of Ducts and Mufflers 1987, John Wiley.

[5] LMS Virtual lab Rev. -10, User's manual.

[6] M.L. Munjal and B. Venkatesham, Analysis and design of an annular air gap lined duct for hot exhaust systems, Proceedings of the IUTAM Symposium on Designing for Quietness, Dec 2000, Kluwer Publications, Holland.

[7] IstvanL. Ver, Leo L. Beranek. Noise and vibration control engineering principles and applications 2006, John Wiley & Sons.

Noise Pollution in Wards of a Hospital in Tiruchirapalli City

V. Bhanumathy, R. Geetha and M. Sangeetha

Department of Physics, Seethalakshmi Ramaswami College, Tiruchirapalli

ABSTRACT: *Environmental noise present in hospitals is a common stressor and is recognized as a serious health hazard for patients, doctors and medical care workers. The aim of the work is to assess the existing noise levels in a 112 bed hospital situated in the heart of Tiruchirapalli city. The plan of the study was to measure the variations of noise in various wards of the hospital and appraise the main sources of noise. Noise levels in 12 places, 6 inside the wards and 6 outside the wards were measured at two and a half hour intervals over a period of 12 hours. The equivalent sound level and peak sound pressure levels were monitored by a digital sound level meter Quest suite professional 2900, during a week in May 2012. The highest mean* $L_{eq}(A)$ *was 67.7 dB(A) and maximum noise level was 80.6 dB(A) during the measurement time in female ward and the lowest mean* $L_{eq}(A)$ *62.1 dB(A) and minimum 55.8dB(A) in the general ward. Maximum noise levels were recordrd in the reception outside the wards of the hospital.The observed noise levels exceeded the recommended CPCB & WHO guidelines for hospitals. It is found that ward structures, location and various activities like conversation of patients, nurses, visitors and traffic resulted in different levels of noise in different wards. The study concludes that noise is a problem in the city hospital where the study was undertaken.*

Keywords: Noise Levels, dB(A), L_{eq}(A), CPCB and Peak Sound Pressure.

INTRODUCTION

Hospitals are extremely noisy, and noise levels in most hospitals far exceed recommended guidelines. Many studies have revealed that extended exposure to noise pollution may cause auditory and non-auditory disorders, such as temporary or permanent hearing loss, sleep disruption (Freedman *et al.*), cardiovascular dysfunction, speech interference and mental health distortion (Ising and Kruppa). Conversation among patients, staff and visitors, sounds of doors, carts, phones, beepers, movement of hospital equipment etc., make up the "Sound environment" which is a risk to healthcare. Whether by accident or incident, the accumulation of noise and unwanted distraction add up to a negative healthcare experience. The United States Environmental Protection Agency (USEPA) recommended a guideline value for continuous background noise as 45 dB(A) during the day and 35 dB(A) at night in patients' room. The International Noise Control and the World Health

Organizations (WHO) have recommended that noise levels in hospital areas should be 35–40 dB in the day time and 30–40 dB(A) in the evening.. In India, noise around hospitals is around 85 dB(A) in major cities(The Hindu Dec 10, 2010). The present study was undertaken to assess noise levels, patterns and sources on wards in a 112 bed room hospital at the heart of Thiruchirapalli city.

MATERIALS AND METHODS

The study was conducted after obtaining institutional ethical clearance at the 112 bedded hospital with a bed occupancy rate of 85% and a daily attendance of about 350 outpatients. The hospital is located in a commercial area at the heart of Tiruchirapalli city housing many shopping complexes with hectic activity from morning 7a.m to evening 9p.m.Many two wheelers and automobiles ply near the hospital and traffic hold up is a frequent event. Impatient drivers honk their horns unmindful of the presence of a big hospital. Sound levels in 12 places inside the hospital were recorded using a Digital Sound Level Meter (Quest suite Professional 2900) in dB (A) units, representing the sound level measured with the A-Weighting network. The instrument incorporates a type-1 microphone and records sound ranging from 50–110 dB(A). Fast mode was used to record the readings. The Sound Level Meter was placed at body level for measurements. The readings were taken throughout the day at five minute interval between successive recordings. Observations were done in one week during the morning 7.30–10.00 a.m., forenoon 10.30–1.00 p.m., afternoon 1.30–4.00 p.m., and evening 4.30–7.00 p.m. The parameter used in the discussion of noise pollution is the equivalent continuous noise level (L_{eq}) expressed in dB(A) which is the mean rate at which energy is received by the human ear during the period monitored. The other variables obtained were the maximum (L_{max}) and minimum (L_{min}) at each place. The data were tabulated using Microsoft excel 2010 and descriptive statistics carried out using the statistical package SPSS for windows, version 14.0.

RESULTS AND DISCUSSION

The values of equivalent sound levels (L_{eq}), Standard Deviation (SD), maximum (L_{max}) and minimum (L_{min}) sound levels inside and outside the wards during the measurement periods are shown in Tables 1&2 and L_{eq} pattern over the day are displayed in Figures 1 &2. On all monitored wards the average equivalent sound levels (L_{eq}) were 65.73±2.75 dB(A), 67.57± 1.71 dB(A), 68.04±1.80 dB(A) and 66.75±3.37 dB(A) during morning, forenoon, afternoon and evening respectively.

Inside the wards, ICU and general ward were the quietest (63.74 & 64.24 dB(A)), as the general ward had partitions for individual patients and visitors were restricted inside the ICU. Triage being the casuality hall, had more patients and attendantsand tended to be the noisiest (69.82 dB(A)).

The present study reveals that the average equivalent sound levels (L_{eq}) in all places in the chosen hospital exceeded the WHO guidelines. The heavy inflow of about 350 patients and more than 100 students everyday with no restriction on the number of visitors are the contributing factors to such high noise levels. It is astonishing to note that even the lowest recorded noise level was 20 dB(A) more than the recommended WHO level of 40 dB(A). Such high levels could slow down the recovery of patients, sick children and might contribute to undesirable physiological and behavioral effects on the children admitted.

Table 1: Equivalent Continuous Noise Level (L_{eq}), Standard Deviation (SD), Maximum (L_{MAX}) and Minimum (L_{MIN}) Readings dB(A) Inside the Wards in the Hospital

Time	*Ward*	*Noise Levels dB(A)*			
		L_{eq}	*SD*	L_{MAX}	L_{MIN}
7.30–10.00	Male ward	65.46	1.41	80.37	56.80
	Female ward	66.14	1.02	79.91	60.21
	General ward	63.03	1.00	77.24	57.21
	ICU	61.41	1.40	78.10	50.84
	Single room	67.79	2.30	82.29	62.47
	Triage	67.34	1.62	81.23	63.00
10.30–13.00	Male ward	67.30	1.43	81.89	58.49
	Female ward	67.63	1.15	82.14	60.46
	General ward	63.56	2.76	79.67	56.97
	ICU	64.80	2.03	81.70	52.16
	Single room	66.34	0.83	78.99	60.80
	Triage	67.89	2.11	83.41	59.19
13.30–16.00	Male ward	67.07	1.02	82.53	58.87
	Female ward	67.70	1.24	80.56	59.84
	General ward	64.40	3.29	78.26	57.34
	ICU	64.07	1.02	80.60	51.74
	Single room	67.97	1.86	81.36	62.76
	Triage	68.96	1.41	84.03	61.91
16.30–19.00	Male ward	64.03	6.56	79.54	55.46
	Female ward	65.16	6.71	78.69	57.83
	General ward	62.09	3.49	76.41	55.81
	ICU	63.36	3.09	82.44	48.03
	Single room	67.97	2.18	83.36	60.43
	Triage	67.80	1.57	82.29	62.14

The highest L_{eq}(72.51dB(A)) and maximum (91.20 dB(A)) sound levels were recorded in the pharmacy followed by reception and corridor,outside the patients' wards. Pharmacy was the nosiest place in the entire hospital during the whole day except in the morning 65 dB(A) because of heavy influx of patients, visitors, movement of ambulance and vehicles as this is situated very close to the city road. Noise contribution on the ward corridors came from movement of hospital equipments and conversation of nurses as nursing stations were close to all the corridors of the hospital.

Table 2: Equivalent Continuous Noise Level (L_{eq}), Standard Deviation (SD), Maximum (L_{MAX}) and Minimum (L_{MIN}) Readings dB(A) Outside the Wards in the Hospital

Time	*Sites*	*Noise levels dB(A)*			
		L_{eq}	*SD*	L_{MAX}	L_{MIN}
7.30–10.00	Lecture hall	71.04	5.45	82.07	62.26
	Pharmacy	65.00	3.19	81.60	55.40
	Corridor	66.81	1.62	81.20	59.11
	Reception	68.27	2.71	83.77	60.70
	Auditorium	62.37	5.39	74.96	56.31
	OP	64.06	5.92	79.50	56.10
10.30–13.00	Lecture hall	66.46	1.52	76.37	63.71
	Pharmacy	70.31	2.05	89.36	58.06
	Corridor	68.54	1.20	83.80	60.54
	Reception	71.87	1.19	90.49	60.66
	Auditorium	67.20	2.66	82.76	59.66
	OP	68.90	1.56	82.54	59.57
13.30–16.00	Lecture hall	68.81	1.62	81.21	63.11
	Pharmacy	72.51	2.35	91.20	60.23
	Corridor	68.03	1.45	82.34	61.16
	Reception	71.24	2.32	87.89	62.41
	Auditorium	67.34	2.54	80.29	61.26
	OP	68.33	1.42	82.47	60.34
16.30–19.00	Lecture hall	63.61	6.36	77.50	59.29
	Pharmacy	72.14	2.13	90.71	60.17
	Corridor	67.66	0.48	82.66	60.66
	Reception	70.84	3.01	87.16	62.47
	Auditorium	66.80	1.50	77.33	61.49
	OP	69.51	3.34	85.29	59.36

It can also be observed in Table 3 that morning was the quietest time and the afternoon, quite noisy. The L_{eq} measurements for each ward and their confidence intervals show that ICU and general ward tended to be the quietest because general ward was separated with partitions for individual patients (Table 4). The mean values of noise in general ward, ICU and single room turned out to be the same ($p > 0.05$) and the means were different in all the other wards ($p < 0.05$) It can also be observed in Table 4 that morning was the quietest time and the afternoon, quite noisy.

Table 3: Average L_{eq} dB(A) Over the Day in the Hospital

Time	L_{eq}	*SD*	*95% CI lower*	*95% CI Upper*
Morning	65.73	2.75	63.98	64.47
Forenoon	67.57	2.24	66.14	68.99
Afternoon	68.03	2.37	66.53	69.54
Evening	66.02	3.15	64.75	68.75

The noise levels were similar to the values measured in the post anesthesia care unit of a hospital in Lyon, France (67.1dB(A)) (Allaouchiche, 2002), and certainly greater than in countries like Taiwan (52.6 to 64.6dB(A)) (Yar-YuanPai 2007, Juang, 2010), Tanzania (45.7 to 62.dB(A)) (Moshi 2010, Philomini 2011) Athens, Greece() (Tsiou c 1998, Tsara v., 2008) and Slovok Republic (35 to 60 dB (A))(Sobotova.pdf)

The strategies should focus on the co-ordination and limitation of activities by nurses and visiting patients' relatives and conversation only in a low voice. It should be noted that the physical status of wards, proper site location for hospitals, municipal planning, traffic, equipment selection and maintenance are also necessary for the reduction of hospital noise.

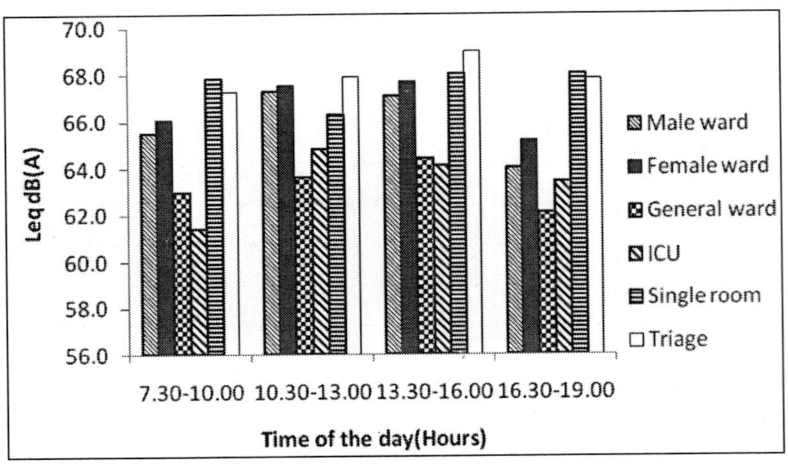

Fig. 1: Noise Levels Inside the Wards

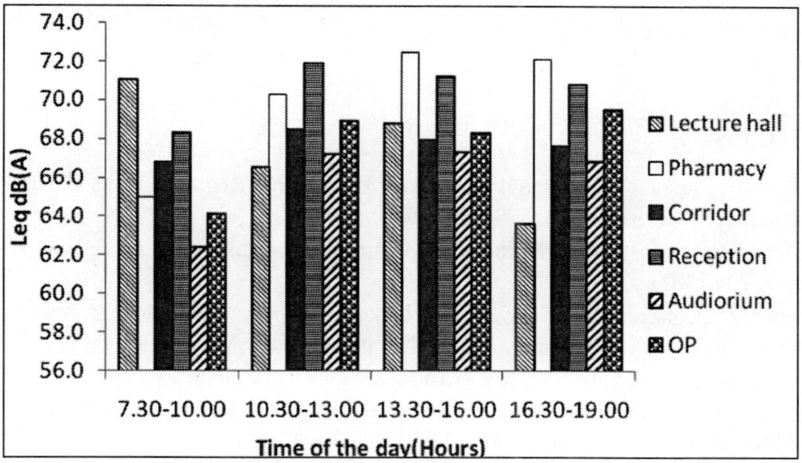

Fig. 2: Noise Levels Outside the Wards

Table 4: The L_{eq} dB(A) Values in the Hospital at Various Places and their Confidence Intervals

Places	L_{eq}	*SE*	*95% CI lower*	*95% CI Upper*
Male ward	68.51	0.32	67.88	69.14
Female ward	66.68	0.15	66.38	67.00
General ward	64.24	0.27	63.71	64.78
ICU	63.74	0.28	63.18	64.30
Single room	64.92	0.12	64.68	65.15
Triage	69.82	0.19	69.44	70.2
Lecture hall	66.95	0.07	66.8	67.1
Pharmacy	71.94	0.24	71.45	72.42
Corridor	69.79	0.15	69.5	70.09
Reception	69.16	0.18	68.81	69.51
Auditorium	64.42	0.13	64.16	64.68
OP	66.63	0.16	66.01	66.65

CONCLUSION

This study shows that there is noise pollution problem on wards of the chosen hospital in Thiruchirapalli city which can cause interference with the doctor-patient relationship and patient care services. The measured noise levels were found to be higher than WHO acceptable limit on hospital buildings. It is observed in this study that conversations of patients and nurses, movement of hospital equipment, overcrowding of patients' relatives vehicular noise are important contributors to noise levels on wards. Therefore, noise monitoring for an optimal hospital environment and the study of the influence of noise, on the quality care and the performance of staff are further needed.

REFERENCES

[1] Anna A. Moshi *et al.*, Noise Pollution on Wards in Iringa Regional Hospital, Tanzania. World Applied Sciences Journal 11 (5): 599–603, 2010.

[2] B. Allaouchiche *et al.*, Noise in the postanaesthesia care unit, British Journal of Anaesthesia 88(3): 369–73 (2002).

[3] Freedman, N.S., Kotzer, N. and Schwab, R.J., Patient perception of sleep quality and etiology of sleep disruption in the intensive care unit. Am. J. Respir. Crit. Care Med., 159(4), 1155–11622, 1999.

[4] Hospitals suffer from serious pollution—The Hindu, Dec. 10,2010.

[5] Ising, H. and B. Kruppa, Health effects caused by Noise: Evidence in the Literature from the past 25 years. Noise Health, 22: 5–13, 2004.

[6] D.F. Juang; 1C.H. Lee; T. Yang; M.C. Chang. Noise pollution and its effects on medical care workers and patients Environ. Sci. Tech., 7 (4), 705–716, Autumn 2010.

[7] Kulwa M. Philimoni *et al.*, Noise Pollution on Wards in Bunda District Hospital in Lake Victoria Zone, Tanzania. International Journal Of Environmental Sciences, Vol. 1, No 5, 2011.

[8] Sabotova *et al.*, The acoustical environment and health risk in hospitals http://isee.ehc.ro/past_celadna/session3/11_sobotova.pdf.

[9] Tsara V, Nena E, Serasli E, Vasileiadis V, Matamis D, Christaki P. Noise levels in Greek hospitals. Noise Health 10:110–2,2008

[10] Tsiou C., Eftymiatos D. *et al.*, Noise sources and levels in the Evgenidion Hospital intensive care unit. Intensive Care Med. Aug; 24(8): 845–7, 1998.

[11] WHO guidelines for Community Noise, World Health Organisation (WHO), 2010. <http://whqlibdoc.who.int/hq/2010>

[12] Yar-Yuan Pai. International Journal of Occupation Safety Ergon; Vol13: 83–90, 2007.

Comparison of Spectra of Flow Induced Turbulent Pressure Fluctuations with Sea State Noise

Sreejith S. Pillai, Jineesh George and D.D. Ebenezer

Naval Physical and Oceanographic Lab., Thrikkakara, Kochi, Kerala

ABSTRACT: *Flow noise can be a major performance limiting factor in towed array and flank array sonar systems. This noise is generated due to fluid flow over structures that are used to protect hydrophones. In this paper, the turbulent boundary layer pressure spectrum model for flow over flat plates proposed by Smol'yakov et al. and the spectrum model for flow over long circular cylinders proposed by Chase are discussed. The pressure spectra for flow over flat plate and cylinder at different flow speeds are compared with the ambient noise in the sea corresponding to Sea State 6. The turbulent wall pressure is seen to be significantly higher than the sea state noise levels in the passive sonar operating frequency range up to a few kilo Hertz.*

Keywords: Flow Noise, Sonar Hydrophone, Towed Array.

INTRODUCTION

High Reynolds number fluid flow over structures causes the formation of turbulent boundary layer that is characterized by random pressure fluctuations. These pressure fluctuations excite the structure causing it to vibrate and radiate acoustic and non-acoustic waves in to the fluid. The response of the sonar hydrophones to these undesirable waves is "flow noise" and it degrades the performance of the sonar. The pressure fluctuations are known as pseudo-sound because hydrophones respond to this pressure but the pressure propagates at a speed that is a fraction of the flow speed and much less than the speed of sound. In this paper, pseudo-sound is compared with sea-state noise because the latter is often used during estimation of active and passive sonar ranges.

Flow noise in flank and towed array sonar systems is of immense practical interest. Therefore, the turbulent wall pressure spectrum due to fluid flow over flat plates [1–6] and long, slender cylinders of circular cross section [7–9] has been studied by several authors. Flow over flat plates has been investigated for a few decades and numerous measurements of the spectra, carried out in various flow conditions and comprehensive reviews of such results are reported. Point pressure spectra measured by several investigators under various conditions collapse to a single curve in a frequency band when the scaling law appropriate for that band is used [3]. Despite its practical importance in the area of towed arrays, much less experimental data is found in the literature on the turbulent pressure due to flow over cylinders.

Point pressure measurements are used to compute wavenumber-frequency spectra. Knowledge of the latter is essential because it is used, in theoretical models, to estimate the pressure field acting on hydrophones and the flow noise. The structure, over which the flow occurs, acts as a filter and some regions of the wavenumber-frequency spectra are, therefore, more important than others. Special efforts have been made to measure the spectra in the important regions. Due to the lack of extensive data, some authors have used data obtained from plate measurements and physical arguments to obtain wavenumber-frequency spectra for flow over cylinders.

In this paper, point-pressure spectra obtained by integrating wavenumber-frequency spectra are compared with Sea State noise. In the case of plates, the integrated result presented in Ref. 3 is used. However, for cylinders, the integration is done by the authors. The comparison is done to get a quantitative understanding of the performance-limiting flow noise and to compare the noises due to flow over plates and cylinders.

FLOW OVER FLAT PLATE

Enormous amount of experimental data, covering a range of Reynolds numbers, on the noise generated by turbulent flow over flat plates is available in literature [1–6]. The general practice to deduce the wavenumber-frequency spectrum is to measure two point wall pressure cross spectra and taking spatial Fourier transform [5]. In another method, the spectra are measured directly using an array of sensors [2, 6].

A reasonably accurate expression for the wavenumber-frequency spectra of fluctuating pressure is given by Smol'yakov and Tkachenko [4]:

$$P(K,\omega) = A\frac{P(\omega)}{2\pi\, m_0}(\frac{U_C}{\omega})^2 h(\omega)[F(K,\omega) - \Delta F(K,\omega)] \tag{1a}$$

where $K = k_1^2 + k_2^2$ is the wavenumber, k_1 and k_2 are components in the convective and the transverse directions respectively, ω is angular frequency,

$$P(\omega) = \frac{\tau_w^2 \delta *}{U}\frac{5.1}{1+0.44(\omega\delta */U)^{7/3}}, \tag{1b}$$

$$\Delta F(K,\omega) = \frac{1}{n}\{1 + A^2 + \frac{n}{m_1}[(m_1 - k_1\frac{U_C}{\omega})^2 + (k_2\frac{U_C}{\omega})^2 - m_1^2]\}^{-3/2}, \tag{1c}$$

$$F(K,\omega) = [A^2 + (1 - k_1 \frac{U_C}{\omega})^2 + (k_2 \frac{U_C}{m_0\omega})^2]^{-3/2}, \tag{1d}$$

$n = 1.005$, $m_0 = 6.45$, $m_1 = (1 + A^2)/(5n - 4 + A^2)$, A is a generalized decay rate, U is free stream velocity, U_C is convective velocity, τ_w is wall shear stress, and δ^* is the displacement thickness of the boundary layer. The function $h(\omega)$ is approximately one. The wavenumber-frequency spectrum, obtained using Eq. (1), is shown in Fig. 1 at various frequencies for U corresponding to 12 knots.

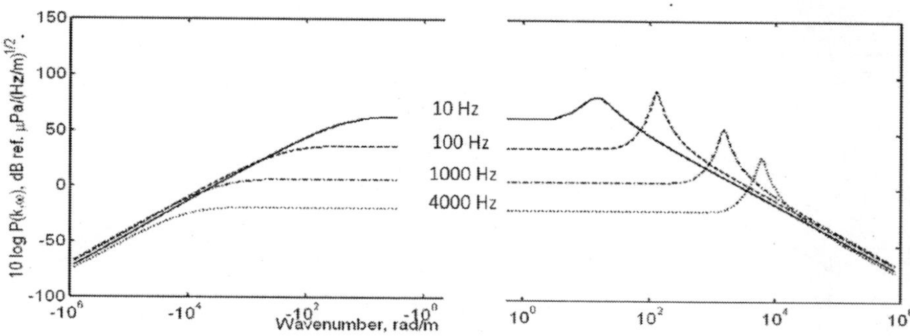

Fig. 1. Wavenumber – Frequency spectrum for water flow over a flat plate.
Flow speed is 12 knots.

It is seen that at all frequencies the energy of the pressure spectrum is concentrated near the convective wavenumber $k_C = \omega / U_C$. This concentration of spectral power is called the convective ridge because it looks like a ridge in a contour plot. The physics is that the turbulent flow pattern retains its spatial structure as it moves over the flat plate [9].

3. Axial Flow Over Long Circular Cylinder

An appropriate expression for axi-symmetric turbulent pressure spectrum due to axial flow over a long circular cylinder is presented by Carpenter and Kewely [7] based on the work by Chase [8]. This expression for cylindrical geometry is derived from the expression for flat plate wall pressure spectrum. To convert the plate spectrum to that for a long slender cylinder, Chase assumes that the important length scale becomes the cylinder radius a when $\Delta/a >> 1$, where Δ is boundary layer thickness, and is Δ when $\Delta/a \to 0$. For practical systems such as long towed arrays, Δ/a is at least 10. With this assumption, the expression for wavenumber-frequency spectrum on a cylinder is defined as

$$P_0(k_1,\omega) = C\rho^2 v_*^3 a^2 \{(k_1 a)^2 + 1/12\}\left[(\omega a - U_C k_1 a)^2 / h^2 v_*^2 + (k_1 a)^2 + b^{-2}\right]^{-2.5} \quad (2)$$

where $C = 0.063$, $h = 3.7$, $b = 1.08$ and U_C is convective speed which is equal to $0.68U$. The subscript '0' signifies that only the axi-symmetric components are considered. This expression was used by Knight [9] for analytically estimating the flow noise in towed arrays. $P_0(k_1,\omega)$ for a typical cylinder of 0.05 m radius at a flow speed of 12 knots is shown in Fig. 2. It is to be noted that, the phenomenon of convective ridge is present in the case of flow over cylinder as well.

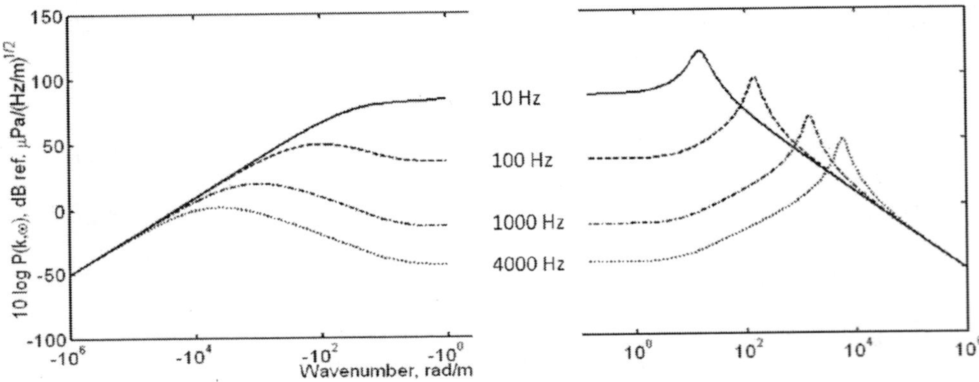

Fig. 2. Wavenumber – Frequency spectrum for axial water flow over a cylinder of 0.05 m diameter. Flow speed is 12 knots.

4. Comparison of Wall Pressure Spectra with Sea State Noise

In sonar systems, the frequency spectrum of various types of noise is of interest and the ambient noise levels in the sea [10] serve as an important reference level. The point-pressure spectrum of turbulent pressure fluctuations during flow over flat plate [3] and cylinder are compared with the Sea State 6 noise level that corresponds to rough sea under heavy storm, and is shown in Fig. 3. The flat plate results shown are also in agreement with dimensional spectra for various flow speeds deduced from the experimental observations by Abraham and Keith [2]. The pressure spectrum for the cylinder case is obtained by integrating the turbulent pressure wavenumber-frequency spectrum over wavenumber.

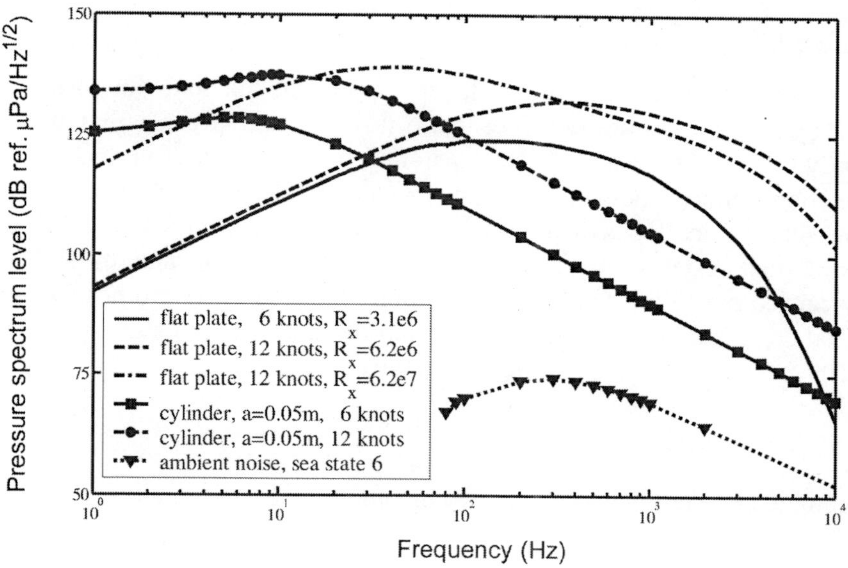

Fig. 3. Comparison of wall pressure spectra with sea state noise

The pressure levels are found to increase with flow speed for both flat plate and cylinder. However, for flat plate, there is no appreciable increase in pressure levels, as the flow speed is increased, for frequencies below 100 Hz at low Reynolds numbers (R_x). The peak in the spectra is found to occur at higher frequencies as the flow speed is increased. At a fixed flow speed, the pressure level increases considerably at low frequencies and decreases at higher frequencies as the Reynolds number is increased. The pattern of the flat plate spectrum for a large Reynolds number, corresponding to a larger value of boundary layer thickness, is comparable with the spectra for cylinder at lower frequencies. For a cylinder there is a constant increase of nearly 13 dB, with doubling the speed from 6 knots to 12 knots, for frequencies above 10 Hz. For frequencies below 10 Hz, the difference is less.

It is seen that the flow induced turbulent pressure fluctuations are significantly higher than the ambient sea state noise levels for frequencies up to 10 kHz. This shows the importance of filtering out the flow induced pressures through proper design of the protecting structure and sensors in sonar systems.

CONCLUSION

The comparison of turbulent boundary layer pressure fluctuations due to flow over flat plate and cylinder with ambient sea state noise levels reveals that the former is significantly higher in the passive sonar operating frequency bands up to a few kiloHertz. Therefore, flow induced noise is a major consideration in the design of sonar systems where the sensors are located close to the flow. Towed arrays and flank arrays are examples of such systems. Design of sensors and sensor enclosures of such systems is generally carried out to filter out a major portion of the flow induced noise in such a way that the total noise level due to flow is less than the ambient sea state noise. However, there exists further scope for optimization in this regard.

ACKNOWLEDGMENT

Encouragement and facilities provided by Director, NPOL are gratefully acknowledged.

REFERENCES

[1] W.K. Blake, Mechanics of Flow-Induced Sound and Vibration, Academic, New York, 1986.

[2] B.M. Abraham and W.L. Keith, "Direct Measurements of Turbulent Boundary Layer Wall Pressure Wavenumber-Frequency Spectra," J. Fluid Engg. 120 (1998), pp. 29–39.

[3] A. V. Smol'yakov, "Calculation of the Spectra of Pseudosound Wall-Pressure Fluctuations in Turbulent Boundary Layers," Acoust. Phys. 46 (3) (2000) pp. 342–347.

[4] A.V. Smol'yakov and V.M. Tkachenko, "Model of a field of pseudosonic turbulent wall pressures and experimental data," Sov. Phys. Acoust. 37(6) Nov.– Dec. 1991, pp.627–631.

[5] R.L. Panton and G. Robert., "The Wavenumber-Phase Velocity Representation for the Turbulent Wall-Pressure Spectrum," ASME J. Fluid Engg. 116 (1994) pp. 477–483.

[6] E. Manoha, "Wall Pressure Wavenumber-Frequency Spectrum Beneath a Turbulent Boundary Layer Measured with Transducers Calibrated with an Acoustical Method," Proceedings of the ASME Symposium on Flow noise Modeling, Measurement, and Control, Noise Control and Acoust. 11 (1991) pp. 21–35.

[7] A.L. Carpenter and D.J. Kewely, "Investigation of Low wavenumber Turbulent Boundary Layer Pressure Fluctuations on Long Flexible Cylinders", 8th Australasian Fluid Mechanics Conference, Univ. of New Castle, NSW, Australia, 1983, pp. 9A.1–9A.4

[8] D.M. Chase, "Further Modelling of Turbulent Wall Pressure on a Cylinder and its Scaling with Diameter", Tech. Memo 21, Chase Inc., December, 1981, AD A113820.

[9] A. Knight, "Flow Noise Calculations for Extended Hydrophones in Fluid and Solid-filled Towed Arrays", J. Acoust. Soc. Am., 100 (1), 1996, pp. 245–251.

[10] L.E. Kinsler, A.R. Frey, A.B. Coppens and J.V. Sanders, Fundamentals of Acoustics, 4th Edition, John Wiley & Sons, Inc, Singapore, 2000, p. 451.

Optimization of Secondary Sources for Active Noise Control in a Car like Rigid Cavity

T. Ramachandran, M.C. Lenin Babu and S. Muruga Poopathy

Department of Mechanical Engineering,
PSNA College of Engineering and Technology, Dindigul
E-mail: lenin_babu@yahoo.com

ABSTRACT: *In this paper genetic algorithm has been used for optimal placement of secondary sources in a three-dimensional rectangular rigid cavity for active noise control system. In this genetic algorithm a multivariable binary coding scheme along with a random initial population are used. Numerical examples are presented to illustrate the effect of location on sound pressure level. Result shows that the locations obtained by genetic algorithm are very effective than arbitrary locations.*

INTRODUCTION

The internal sound field in the enclosed cavity is significantly affected by the acoustic modal characteristics of the cavity, Porous material absorption efficiency is mostly apparent at high frequencies. Passive control has the disadvantage of minimal low frequency absorption; it can be bulky and heavy when used for low frequencies. To attenuate this low frequency noise, active noise control (ANC) system can be used in enclosed spaces without passive techniques. It has the specific advantage of reduction in size, volume and cost when compared with passive control. Nelson *et al.* (1987) examined the active noise control methods in harmonic enclosed sound fields. Further, the theory was verified by both numerical simulation (Bullmore *et al.* (1987)) and experimentally (Elliott *et al.* (1987)). They considered a rectangular enclosure with rigid walls and shown that minimizing the sum of the squared pressures at several points could lead to significant reduction in acoustic potential energy and sound pressure levels (SPL). Parkins *et al.* (2000) investigated active control system of three-dimensional enclosures based on acoustic energy density method. In their study, minimization of energy have been carried out by sensing both the potential and kinetic energy densities, while the most popular control systems of the past have relied on the potential energy density alone. Energy density control method shown to be more effectively than pressure based measurements especially for off-resonance control. Akl and Baz (2006) investigated the effectiveness of structural acoustics of the cavity/plate system, which is lined with smart foam.

In this work optimal location of secondary sources have been examined to enhance the reduction of the sound pressure level in a three dimensional rigid cavity as shown in Figure 1.

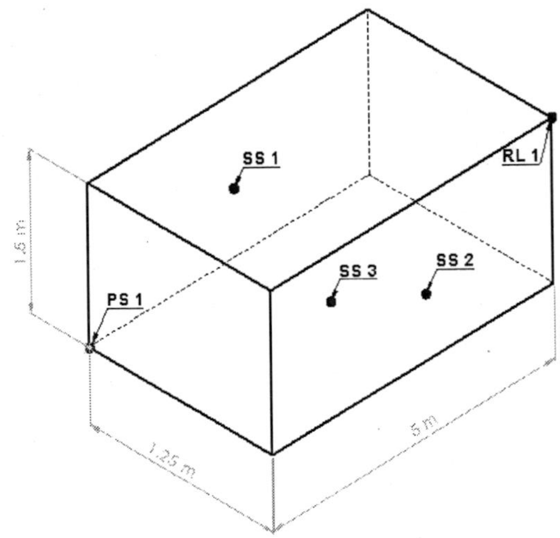

Fig. 1: Geometry of the Car Like Rigid Cavity

FORMULATION

The Helmholtz equation, for a cavity of volume V, enclosed by a rigid surface for harmonic variation in acoustic pressure p is given by

$$\nabla^2 p + \left(\frac{\omega}{c}\right)^2 p = 0 \qquad \ldots (1)$$

where ω is the circular frequency and 'c' the speed of sound. The weak integral form of the cavity equation can be found in Sgård *et al.* (2005) and is given as

$$\int_{\Omega_a}\left[\frac{1}{\rho_a}\nabla p_a . \nabla \delta p_a - \frac{\omega^2}{p_a c^2} p_a \delta p_a\right] d\Omega - \frac{1}{\rho_a}\int_{\Gamma_a}\frac{\partial p_a}{\partial n}\delta p_a d\Gamma = 0, \quad \forall \delta p. \qquad \ldots (2)$$

In this equation p_a and ρ_a representing pressure and density of the acoustic domain. In order to discretize the acoustic domain a 20-noded quadratic finite element is used which has only pressure as a degree-of-freedom (DOF) at each node. Figure 2 shows the reference 20-noded element and node numbering. The weak form of equation 2 discretized with this 20-noded quadratic finite element subject to boundary conditions can be in matrix form as,

$$\left([H]_a - \omega^2 [Q]_a\right)\{p_a\} = \{F_a\}; \qquad \ldots (3)$$

where $[H]_a$ and $[Q]_a$ represent the kinetic and compression energy matrices for the cavity, $\{p_a\}$ represents the global nodal acoustic pressures for cavity, while $\{F_a\}$ denotes the acoustic loading vector for the cavity. Above can be written in the form of

$$[A]\{Z\}=F; \quad [A]\{Z\}=F; \quad [A]=[H]-\omega^2[Q]_a; \qquad \ldots (4)$$

The amplitude produced by primary primary $\{Z_{ps}\}$/secondary $\{B_{ss}\}$ source distributions inside the cavity is given by,

$$\{Z_{ps}\}=[A]^{-1}\{F_{ps}\}; \quad \{B_{ss}\}=[A]^{-1}\{F_{ss}\}; \quad \{q_s\}=-\left[\{B_{ss}\}^H\{B_{ss}\}\right]^{-1}\{Z_{ps}\} \qquad \ldots (5)$$

Where $\{F_{ps}\}$ and $\{F_{ss}\}$ denote the loading vectors due to primary sources alone and secondary sources alone. The secondary source strengths for the ANC system are computed using quadratic minimization of acoustic potential energy as shown in Equation 5, suggested by Nelson *et al.* (1987). The optimized source strength due to primary and M secondary sources can be calculated for each exciting frequency as

$$Z_n = Z_{ps} + \sum_{m=1}^{M} B_{ssm} q_{ssm} \qquad \ldots (6)$$

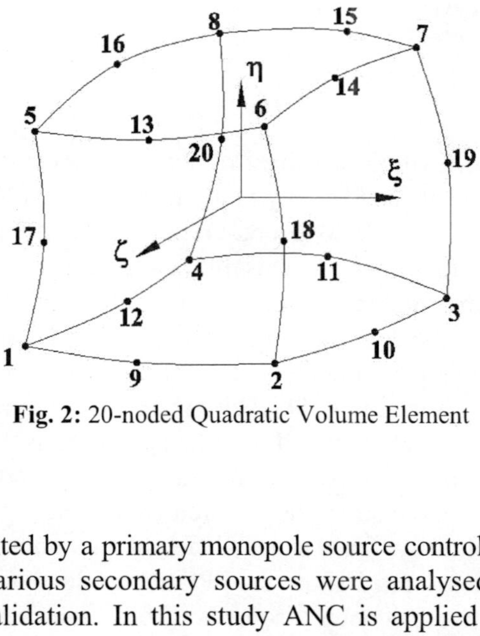

Fig. 2: 20-noded Quadratic Volume Element

Validation

The SPL of a cavity excited by a primary monopole source controlled by various secondary source controlled by various secondary sources were analysed by Raghu (2005) has considered for ANC validation. In this study ANC is applied to a rectangular cavity having dimensions of 4 m × 5 m × 3 m and the SPLs are calculated with two secondary sources (SS). It is assumed that a monopole primary source is located with coordinates (3.33 m, 0.833 m and 0 m) and a frequency range of excitation for primary source is 30–90 Hz and the source strength of the primary source is taken as 0.1 m^3/s. Two secondary sources are placed arbitrarily within the enclosure with coordinates (1.33 m, 1.67 m and 3 m) and

(1.33 m, 0 m and 1 m) respectively. Figure 3 shows the sound SPL at coordinates (0 m, 3.33 m and 1 m). From Figure 3, it is clear that with ANC the SPLs are considerably reduced at resonant frequencies. The agreement is good for the entire frequency range where the analysis performed.

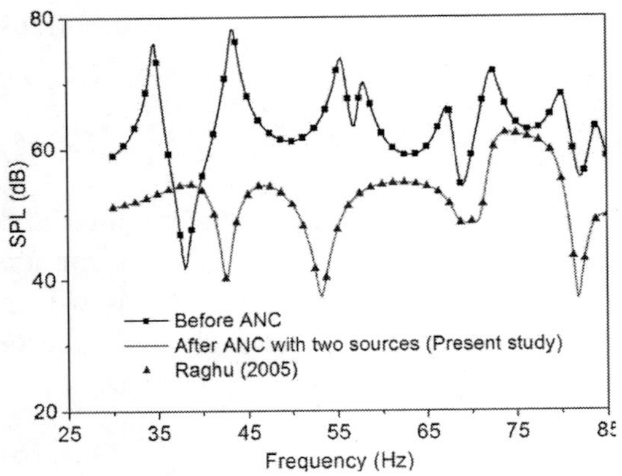

Fig. 3: Sound Pressure Level Comparison with Ragu

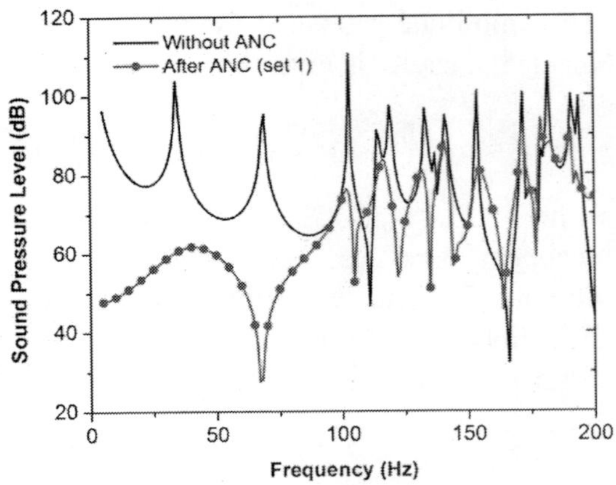

Fig. 4: Influence of ANC on SPL Based on Set 1 Locations

CAR LIKE CAVITY

The structure under investigation is a three dimensional rectangular car like rigid cavity as shown in Figure 1. The acoustic cavity is modeled using FEM with pressure as a field variable. Cavity is modelled with 20-noded brick finite element model with pressure as field variable as shown in Figure 2. This finite element model is implemented using MATLAB. The programs implement the equations derived by Nelson *et al.* (1987) in section 2. In order to find the converged mesh size calculated eigenvalues from the finite element model are compared with the natural frequencies of the cavity obtained by analytical method. It is assumed that the cavity is excited by a monopole primary source, and the frequency range of excitation for primary source is 5–200 Hz and the source strength of the primary source is taken as 0.1 m^3/s. Three secondary sources are placed inside the cavity and its source strengths are predicted using equation 5. Three secondary sources are placed arbitrarily for various different positions within the enclosure as shown in Table 1. Figures 4 and 5 show the SPL at coordinates (1.25 m × 1 m × 5 m) for set 1 and set 4 locations of secondary sources. From Figures 4 and 5, one can see that ANC is very effective below 100 Hz. In order to find the optimized secondary sources locations genetic algorithm has been used. The number of genes in each chromosome is assumed to be three to represent the magnitude of location of three secondary sources and their orientations with the Primary source. Here, the length of chromosome is 8 in which first four genes indicate the magnitude of location of secondary sources and the initial population size is assumed as 20. The main objective is minimization of the sound pressure level at receivers location by optimal location of secondary sources. The fitness function is established by mapping the objective function through mapping, scaling and normalization. Figure 6 shows the SPL at receivers location due to optimized secondary sources locations. It is clear that optimized secondary sources locations leads to 20–25 dB reduction in SPL from 100–200 HZ at receivers location.

Table 1: Sources and its Locations

Secondary Sources		*Location*	*Coordinate (m)*		
			X	*Y*	*Z*
Arbitrary Set 1	SS1	1285	0	0.75	2.5
	SS2	1292	1.25	0.75	2.5
	SS3	2524	0.72	1.5	4
Arbitrary Set 2	SS1	1153	0	0.65	3.25
	SS2	1160	1.25	0.65	3.25
	SS3	1979	0.72	1.18	4
Arbitrary Set 3	SS1	347	1.2	0.22	0
	SS2	486	1.25	0.22	3
	SS3	2543	0.36	1.5	4.5
Arbitrary Set 4	SS1	330	0.72	0.11	5
	SS2	425	1.25	0.22	1.75
	SS3	2521	0.45	1.5	4
Optimized	SS1	1971	0.72	1.18	3.5
	SS2	246	0.1	0.11	0.3
	SS3	806	0.1	1.5	5

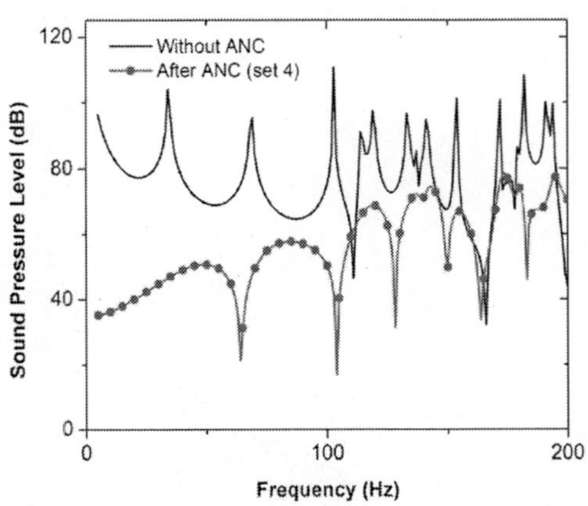

Fig. 5: Influence of ANC on SPL Based on Set 4 Locations

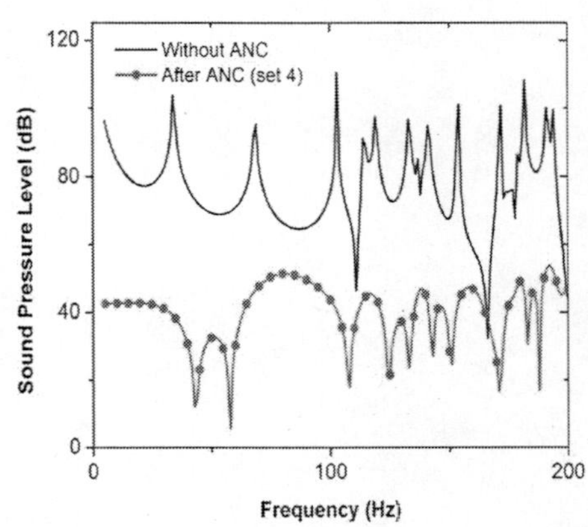

Fig. 6: Influence of ANC on SPL Based on Optimized Locations

CONCLUSION

In this work optimization of secondary sources for active noise control in a car like rigid cavity has been investigated. Sound pressure level at receives location has been obtained due to a point source excitation. Genetic algorithm has been used to obtain the optimum secondary sources locations in order to minimize the sound pressure level inside the cavity at receivers location. Results obtained from the model shows the reduction of mean sound pressure level up to 20–25 dB in 100–200 Hz band by changing the secondary sources positions from arbitrary to optimized locations.

REFERENCES

[1] P.A. Nelson, A.R.D. Curtis, S.J. Elliott and A.J. Bullmore, *The active minimization of harmonic enclosed sound fields, Part I: Theory*, Journal of Sound and Vibration, 117 (1987), 1–13.

[2] A.J. Bullmore, P.A. Nelson, A.R.D. Curtis and S.J. Elliott, *The active minimization of harmonic enclosed sound fields. Part II: A computer simulation,* Journal of Sound and Vibration, 117, (1987), 15–33.

[3] S.J. Elliott, A.R.D. Curtis, A.J. Bullmore and P.A. Nelson, *The active minimization of harmonic enclosed sound fields. Part III: Experimental verification,* Journal of Sound and Vibration, 117 (1987), 35–58.

[4] W. Parkins, S.D. Sommerfeldt and J. Tichy, *Narrowband and broadband active control in an enclosure using the acoustic energy density*, The Journal of the Acoustical Society of America, 108 (2000), 192–203.

[5] W. Akl, A. Baz, *Active vibration and noise control using smart foam*, Journal of Vibration and Control, 12 (2006), 1173–1203.

[6] F.C. Sgard, X. Olny, N. Atalla and F. Castel, *On the use of perforations to improve the sound absorption of porous material*, Applied Acoustics, 66(2005), 625–651.

[7] V. Raghu, *Optimization of secondary sources for active noise control*, M.S.Thesis, Indian Institute of Technology Madras, Chennai, 2005, pp. 10–55.

Using Directivity of Radiated Sound for Noise Control Applications

Gyanishankar Sharma, Abhijit Sarkar and N. Ganesan

Department of Mechanical Engineering, Indian Institute of Technology Madras, Chennai

ABSTRACT: *In this paper directivity based technique to control the noise radiated from vibrating plate in a baffle is presented. Finite element simulation is done in commercial software ANSYS to obtain the displacement response. The obtained displacement response is validated by deriving an analytical formula for response using modal summation technique. The surface velocity is calculated from displacement response which is used in a Rayleigh integral based calculation to arrive at sound pressure at a location and hence the directivity pattern. Control on directivity pattern is achieved by attaching discrete mass at different locations over the plate. Passive technique for noise and vibration control is known to have a poor performance for global control in low and medium frequency region. Hence control on the directivity could be a useful technique for local noise control in low and medium frequency vibration.*

INTRODUCTION

Noise control from vibro-acoustic sources has been a widely studied topic in the engineering community [1]. The passive control methods are broadly classified as (1) source control - wherein efforts are made to directly reduce the source of noise *viz.* vibration levels, flow ripples, etc. (2) transfer path control - wherein efforts are made to reduce the transmission of the generated noise to the receiver without affecting the source. Reactive mufflers are one of the best examples in this case wherein the sound generated at the source is reflected back. In this work, we attempt to investigate the possibility of using directivity as an additional technique for noise control.

In several applications there are instances where the objective of the noise control engineer is to minimize the audible noise at specific locations. For example, in automotive applications the target is to reduce the noise at the driver and passenger ear location. It is acceptable to achieve this target even at the cost of increasing sound levels at locations away from the passengers. Thus, we attempt to explore the possibility of redistributing the available acoustic energy in different directions such that the noise levels at the specific locations of interest are reduced at the cost of increased noise levels at other locations.

Chatillon [2] studied the influence of source directivity in noise levels inside an industrial hall through simulation and experiment. Noise level was observed to be changed by 4 dB as the source directivity changes. Zhang [3] observed different horizontal and vertical directivity pattern of railway noise through measurement. Though in the above works aim was not to control the directivity pattern, there are many literature in which the aim is to control directivity for different applications. Juhez *et al.* [4] developed an acoustic transducer system for long distance echo-ranging application. Directivity controlled piezo-electric transducer for sound reproduction was developed by Be'dard and Berry [5]. Seo and Kim [6] studied directional radiation pattern in structural acoustic coupled system. Three typical directivity patterns i.e. steered, focused and omnidirected are presented.

Voluminous literature is available on the study of the vibration response and acoustic radiation from mass loaded structures [7, 8, 9]. There was no literature found for acoustic directivity control from vibrating structures using passive techniques such as lumped mass addition.

DESCRIPTION OF THE SYSTEM

A simply-supported square thin plate is harmonically excited at the center. The structure is placed on an infinite baffle (refer Figure 1(a)). Table 1 shows the parameters for the structural and acoustic medium. The applicability of this problem lies in the fact that many structures of practical importance can be approximated as a rectangular plate to an extent. The objective is to determine the baseline directivity of this vibro-acoustic system. Further, the change in directivity pattern due to lumped mass loading is to be ascertained.

Table 1: Parameters Used for Model

Size of plate	1x1x0.01 m^3	Poisson's ratio (ν)	0.3
Density of plate (ρ)	7840 kg/m^3	Density of medium (ρ_0)	1.2 kg/m^3
Young's modulus (E)	210×10^9 N/m^2	Speed of sound in medium (c)	340 m/sec

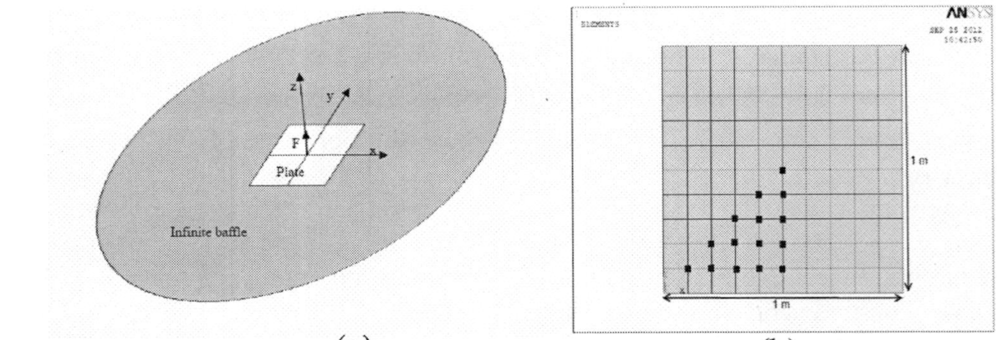

(a) (b)

Fig. 1: (a) Schematic of a SS Plate on a Baffle with Harmonic Excitation at the Center (b) Different Locations of Mass Considered Over the Plate

METHODOLOGY

To demonstrate the applicability of the above concept, we study the directivity patterns of a simply supported plate on a baffle with the objective of being able to control the directivity through simple means such as mass addition. Accordingly, an APDL script is written to develop the vibration model for Finite element Analysis in a commercial software ANSYS. The vibration model can be easily modified for the addition of masses at the desired location. A harmonic analysis is performed with a point load at the center to obtain the displacement response. The displacement response obtained is validated from analytical formula derived using modal summation technique. Surface velocity can be calculated once the displacement response is known which is then used for a Rayleigh integral based calculation [10] to arrive at the sound pressure at any point of interest. The Rayleigh integral calculation is implemented through a numerical quadrature program developed in MATLAB. Sound pressure is obtained at different locations over the plate and plotted on hemispherical surface to obtain the directivity pattern. Above procedure is repeated for different locations of mass.

Sample Directivity Pattern for Few Modes of Simply Supported Plate

Directivity pattern for first mode $\left(\sin\left(\frac{\pi x}{l}\right)\sin\left(\frac{\pi y}{l}\right)\right)$ and second mode $\left(\sin\left(\frac{2\pi x}{l}\right)\sin\left(\frac{2\pi y}{l}\right)\right)$ as a velocity profile is shown in the figure 2. It can be seen that the dominant radiation is normal to the plate surface for first mode whereas, dominant radiation for second mode is at some angle to the normal of the plate. It is also well known that harmonic response of plate can be obtained as sum of modes of baseline plate thus, it is inferred that directivity pattern can be altered by altering the modal contribution i.e. by attaching mass(es).

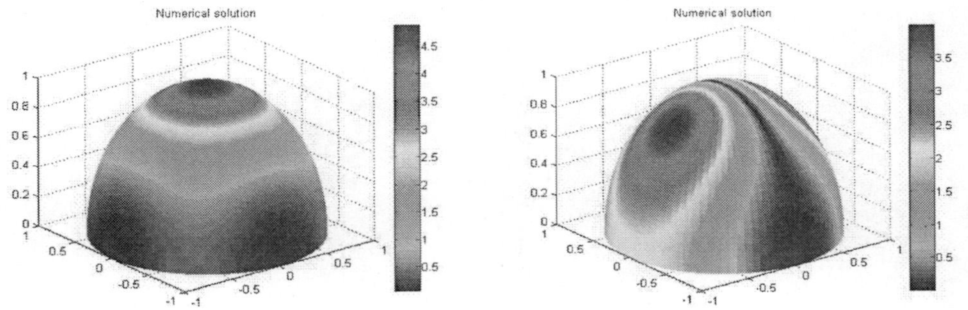

(a) First mode (b) Second mode

Fig. 2: Directivity Pattern at 500 Hz for (a) First Mode (b) Second Mode of Simply Supported Plate

NATURAL FREQUENCY ANALYSIS

Analytical formula for natural frequency of baseline plate is available in [11]. Analytical formula for mass loaded plate is derived using modal summation technique. A finite element simulation is done in ANSYS for validation and a match is obtained as shown in Table 2.

DIRECTIVITY PATTERN

The various locations of mass over the plate are considered as shown in figure 1(b). Directivity pattern for all these locations of mass over the plate is analyzed. The top view of hemispherical directivity pattern of baseline plate and mass loaded plate (with added masses = 20 % of plate mass) at 100 Hz excitation frequency and ka = 3 (where k is acoustic wave number and a is the distance between source and receiver) is shown in the figure below.

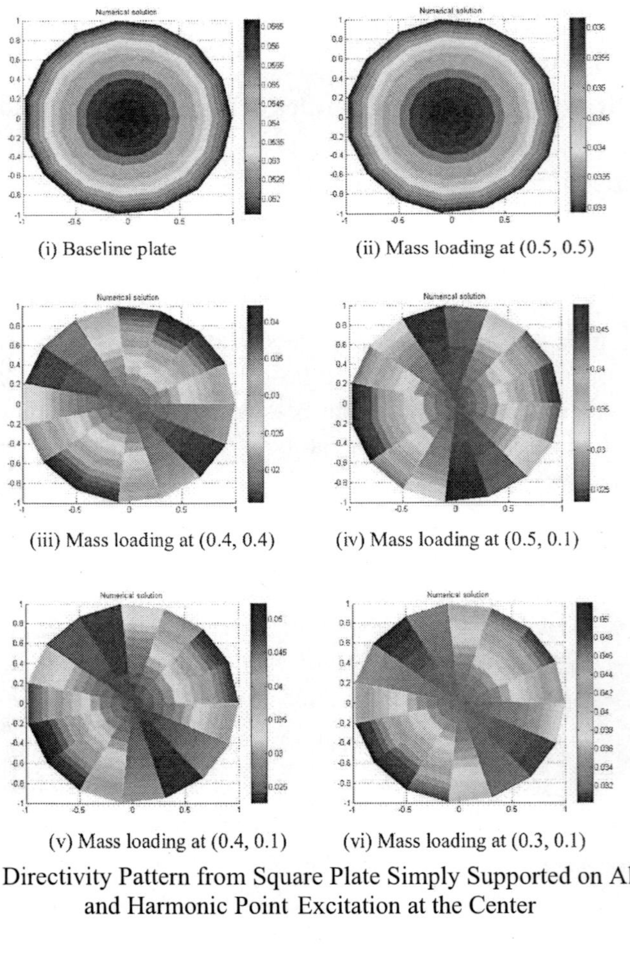

(i) Baseline plate (ii) Mass loading at (0.5, 0.5)

(iii) Mass loading at (0.4, 0.4) (iv) Mass loading at (0.5, 0.1)

(v) Mass loading at (0.4, 0.1) (vi) Mass loading at (0.3, 0.1)

Fig. 2: Directivity Pattern from Square Plate Simply Supported on All Side and Harmonic Point Excitation at the Center

Table 2: Natural Frequency of Baseline Mass Loaded Plate

	Base line plate (Hz)		Mass loaded plate(Hz) (with added mass = 20% of plate mass)	
Mode	Analytical	Ansys	Analytical	Ansys
(1,1)	49.20	49.20	46.58	46.48
(2,1)	123.00	123.00	94.31	91.56
(1,2)	123.00	123.00	122.97	122.97
(2,2)	196.81	196.77	162.37	157.91
(3,1)	246.01	245.98	227.50	224.25
(1,3)	246.01	245.98	245.90	245.90

DISCUSSION

On analyzing the above results of directivity pattern, it is seen that the directivity pattern of baseline plate can be altered by attaching additional mass. The location and magnitude of mass addition plays an important role in controlling the directivity pattern so as to reduce noise at different planes over the plate.

In the following we present a representative analysis of directivity pattern of baseline plate and mass loaded plate (0.4, 0.4). RMS value of displacement response of the mass-loaded plate calculated to be 1.74 times that of the baseline plate. The total sound energy radiated (using the formula $E\int_A \frac{p^2}{2\rho_0 c} dA$) for the baseline plate and mass loaded plate is computed. It is observed that the total radiated sound energy is decreased by a factor of 3.57 due to mass addition. Thus, the sound pressure is decreased by a factor of 1.88. But, when the ratio of sound pressure from baseline plate to sound pressure from mass loaded is plotted on hemispherical surface over the plate, it is seen that sound pressure is decreased at all points. The factor with which the sound pressure decreases, varies between 1.4 and 3.2 over different locations (refer Figure 3). Thus, the redistribution of acoustic energy is happening, resulting in higher sound pressure reduction at some locations. The result is interesting as we noted that the vibration increases. However, the observed decrease in acoustic radiation is attributed to phase cancellations between regions of positive and negative vibration response.

CONCLUSION

It is well known that for low and medium frequency vibration passive noise control measures such as point mass attachment etc. are not efficient to control noise in global sense. Thus, directivity of radiated noise can be employed to achieve the noise control locally. In this work we have demonstrated the applicability of the concept on a simple vibro-acoustic problem. As future work similar demonstrations need to be carried on for more complicated structures.

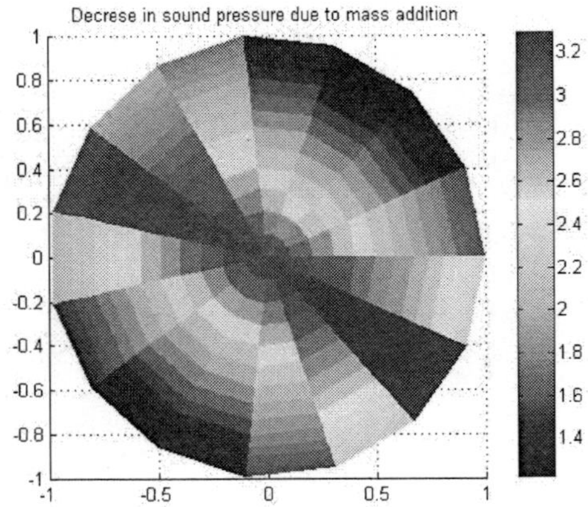

Fig. 3: Ratio of Sound Pressure from Baseline Plate to Mass Loaded (0.4, 0.4) Plate Plotted over Hemispherical Surface over the Plate

REFERENCES

[1] D.A. Bies and C.H. Hansen Engineering Noise Control: Theory and practice, Spon press, 2003.

[2] J. Chatillon (2007) Influence of source directivity on noise levels in industrial halls: Simulation and experiments, Applied Acoustics, 68, 682–698.

[3] X. Zhang (2010) The directivity of railway noise at different speeds, Journal of Sound and Vibration, 329, 5273– 5288.

[4] J.A. Gallego-Juhez, G. Rodnguez, J.L.S. Emeteno, P.T. Sanz and J.C. Lhzaro (1993) An acoustic transducer system for long-distance ranging applications in air, Sensors and Actuators A, 37–38, 397–402. [5] M. Be'dard and A. Berry (2008) Development of a directivity-controlled piezoelectric transducer for sound reproduction, Journal of Sound and Vibration 311, 1271–1285.

[6] H.S. Seo and Y.H. Kim (2005) Directional radiation pattern in structural-acoustic coupled system, J. Acoust. Soc. Am. 118(1), 92–103.

[7] A.E. Ekimov and A.V. Lebedev (1996) Experimental Study of Local Mass Influence on Sound Radiation From a Thin Limited Cylindrical Shell, Applied Acoustics, 48(1), 47–57.

[8] S. Li and X. Li (2008) The effects of distributed masses on acoustic radiation behaviour of plates, Applied Acoustics, 69, 272–279.

[9] J.S. Wu and S.S. Luo (1997) Use of analytical numerical combined method in the free vibration analysis of a rectangular plate with any number of point masses and translational springs, Journal of Sound and Vibration, 200(2), 179–194.

[10] F.J. Fahy and P. Gardonio Sound and Structural Vibration: Radiation, Transmission and Response, Academic press, 2007.

[11] A.W. Leissa Vibration of Plates, NASA SP-160, 1969.

Section–VIII

Acoustic Emission Basedcondition Monitoring

A Novel Approach of Substituting Boundary Conditions to Get Four Pole Parameters of Acoustic Element

Bhushan Singh[1] and M.L. Munjal[2]

[1]Ashok Leyland Technical Centre, Chennai

[1,2]Department of Mechanical Engineering, Indian Institute of Science, Bangalore

E-mail: munjal@mecheng.iisc.ernet.in

ABSTRACT: *There are analytical methods available in the literature for analysis of various muffler configurations. However, these are tedious, involve mathematical modeling, and result in lengthy and complex mathematical expressions on substituting the boundary conditions. Therefore, a semi-analytical 'Matrizant approach' is employed here for solving the basic governing equations (a set of linear first order differential equations), and an algebraic scheme has been worked out to incorporate the boundary conditions systematically in MATLAB. The TL predictions are shown to tally with those of the traditional transfer matrix based muffler program (TMMP).*

INTRODUCTION

Over time, muffler design has become more and more complex for enhanced performance [1]. It would be costly, iterative and very time consuming to experimentally evaluate the performance of all the mufflers at design stage. In the present work, a semi-analytical 'Matrizant approach' is employed for performance evaluation of muffler configurations having non-overlapping perforations. The system of basic governing differential equations are arranged in canonical differential matrix form and integrated. The boundary conditions are then substituted systematically using the MATLAB computer program. This eliminates the tedious work of handling lengthy expressions and reduces risk of oversights and human error. Besides, analysis of complex mufflers can be done easily without neglecting any higher order terms, and this eliminates the risk of any instability in the computed results.

BASIS OF MATRIZANT APPROACH

Generally speaking, the solution of a linear differential equation is exponential. Extension of this concept to a system of linear differential equations, whose solution is also exponential, subject to the complication that matrix multiplication is non-commutative, but can be solved easily by employing Matrizant approach using MATLAB software.

On integrating a linear ordinary differential equation of the form,

$$\frac{dy}{dz} = a(z) \times y$$

we get

$$\ln(y)\Big|_{z_0}^{z_l} = \int_{z_0}^{z_l} a(z)dz, \text{ or } y(z_0) = y(z_l) \times e^{-\int_{z_0}^{z_l} a(z)dz} \qquad \dots (1)$$

This gives a direct relation between the values of dependent variable (y) for the independent variable (z) values $z = z_0$ and $z = z_l$.

This concept can readily be extended to a system of linear differential equations arranged in the canonical matrix form,

$$\{\hat{V}\}' = [H]\,\{\hat{V}\} \qquad \dots (2)$$

where {V} is a state vector of order 2n for n interacting perforated ducts.

Making use of the Matrizant approach, the matrix Eq. (2) may be integrated between $z = z_0$ and $z = z_l$ as follows [2]:

$$\int_{\hat{V}_0}^{\hat{V}_l} \frac{1}{\hat{V}}\, d\hat{V} = \int_{z=z_0}^{z=z_l} [H]\, dz, \text{ or } \{\hat{V}_o\} = e^{-\int_{z_0}^{z_l}[H(z)]dz} \{\hat{V}_l\} \qquad \dots (3,4)$$

In case matrix [H] is independent of z, we get

$$\{\hat{V}_0\} = \left[e^{-Hl}\right] \{\hat{V}_l\} \text{ where } l = z_l \text{-} z_o \qquad \dots (5)$$

Using theory of linear algebra, we can calculate matrix exponential as,

$$\left[e^{Hl}\right] = [\Phi]\left[e^{\Lambda l}\right][\Phi]^{-1}, \text{ or } \left[e^{-Hl}\right] = [\Phi]\left[e^{-\lambda l}\right][\Phi]^{-1}, \qquad \dots (6)$$

where,

[Φ] is the modal matrix of [H],

[$e^{-\lambda l}$] is a diagonal matrix containing elements $\left[e^{-\lambda_1 l}, e^{-\lambda_2 l} \,.....\, e^{-\lambda_{2n} l}\right]$,

$\lambda_1, \lambda_2, \lambda_3, \ldots, \lambda_{2n}$ are eigen values of matrix [H], and n is the number of interacting perforated ducts.

The eigen-values and eigen-vectors of the matrix [H] can be computed by using a suitable algorithm. As [H] is a matrix consisting of elements independent of variable z, the eigen calculation and matrix inversions can be done easily on a digital computer. Thus, working in the frequency domain (time dependence $e^{j\omega t}$), and arranging basic governing differential equations in canonical differential matrix form, one can get the transfer matrix relation between the state vectors for different values of *z*.

GENERALIZED TRANSFER MATRIX FOR N-DUCT PERFORATED ELEMENT

An n-duct perforated element is characterized by (n–1) perforated ducts within a bigger duct or chamber as shown in Figure 1.

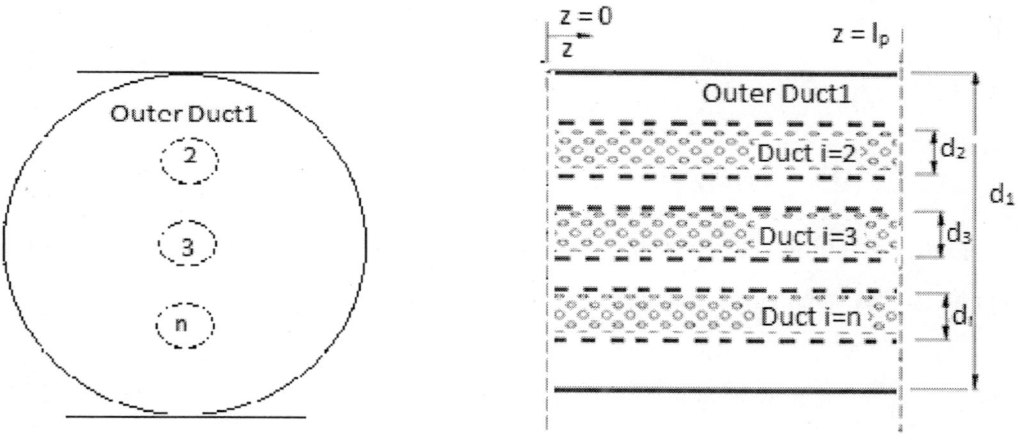

Fig. 1: A Perforated Element with n Interacting Ducts

Here we get 2n equations, n for continuity and n for momentum balance.

p_j, u_j, v_j are acoustic variables relating to the jth duct;

ς_j, d_j are perforate impedance and diameter of the jth duct.

As shown in Figure 1, j = 1 represents the outer chamber, hence d_1 is chamber diameter and there is no porosity. It interacts with the n–1 perforated ducts on the inside.

Let p, ρ, u, U, a, ζ denote pressure, density, particle velocity, mean flow velocity, sound speed and perforate impedance respectively. Assuming harmonic time dependence ($e^{j\omega t}$) and isentropicity, the continuity equations are as follows:

For the jth duct (j = 2,..., n):

$$M_j \frac{dp_j}{dz} + \rho_0 a_0 \frac{du_j}{dz} = \frac{4}{d_j \varsigma_j} p_1 - \left(jk_0 + \frac{4}{d_j \varsigma_j} \right) p_j \; ; \qquad \ldots (7)$$

For the 1st duct:

$$M_1 \frac{dp_1}{dz} + \rho_0 a_0 \frac{du_1}{dz} = -\left(jk_0 + \sum_{j=2}^{j=n} \frac{4d_j}{\varsigma_j (d_1^2 - \sum_{j=2}^{j=n} d_j^2)} \right) p_1 + \sum_{j=2}^{j=n} \frac{4d_j p_j}{\varsigma_j (d_1^2 - \sum_{j=2}^{j=n} d_j^2)} ; \quad \ldots (8)$$

and the momentum equations are:

$$\frac{dp_j}{dz} + M_j \rho_0 a_0 \frac{du_j}{dz} = -jk_0 \rho_o a_0 u_j , \ j = 1,2,..,n . \quad \ldots (9)$$

Eqns. (7) to (9) may be rearranged in canonical differential matrix form as

$$\begin{bmatrix} \frac{dp_1}{dz} & \rho_0 a_0 \frac{du_1}{dz} & \frac{dp_2}{dz} & \rho_0 a_0 \frac{du_2}{dz} & \cdots & \frac{dp_n}{dz} & \rho_0 a_0 \frac{du_n}{dz} \end{bmatrix}_{1\times 2n}^T =$$

$$[A]_{2n\times 2n} [p_1 \quad \rho_0 a_0 u_1 \quad p_2 \quad \rho_0 a_0 u_2 \quad \cdots \quad p_n \quad \rho_0 a_0 u_n]_{1\times 2n}^T \quad \ldots (10)$$

The matrix equation (10) is in the form, $\{\hat{V}\}' = [H] \{\hat{V}\}$. Therefore, making use of Eqs. (5) and (6), we can get the transfer matrix relation [3]

$$[p_1(0) \quad v_1(0) \quad p_2(0) \quad v_2(0) \quad \cdots \quad p_n(0) \quad v_n(0)]_{1\times 2n}^T =$$

$$[T]_{2n\times 2n} [p_1(l) \quad v_1(l) \quad p_2(l) \quad v_2(l) \quad \cdots \quad p_n(l) \quad v_n(l)]_{1\times 2n}^T \quad \ldots (11)$$

Eqn. (14) is a generalized transfer matrix relation for n-interacting perforated ducts across a section of length l_2, shown in Figure 2. This will be reduced to the required 2x2 matrix by substituting the boundary conditions systematically as shown below.

APPLICATION OF BOUNDARY CONDITIONS

The proposed novel method has been illustrated here for the cross-flow closed-end extended-perforation muffler shown below in Figure 2. This configuration has non-

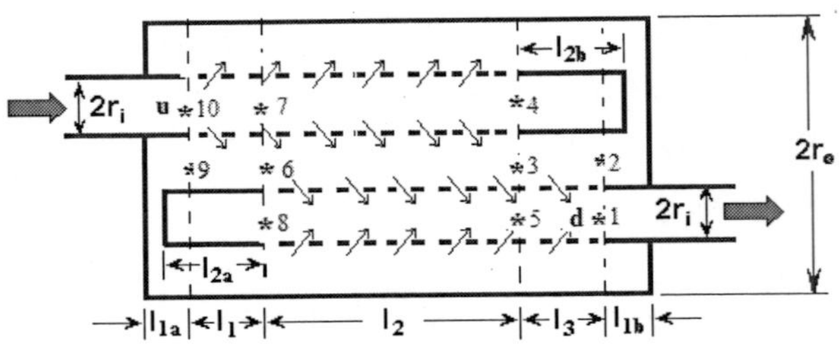

Fig. 2: Cross-Flow Closed-End Extended-Perforation Muffler

overlapping perforations. With this variation, the analysis becomes substantially different from the configurations having overlapped perforation, with different sets of equations and boundary conditions.

Using the generalized transfer matrix for an n-duct perforated element discussed above in section 3, we get the following matrix connecting state vectors at each section marked by dotted lines in Figure 2:

$$[p_9 \quad v_9 \quad p_{10} \quad v_{10}]^T = [A]_{4\times4}[p_6 \quad v_6 \quad p_7 \quad v_7]^T \qquad \ldots (12)$$

$$[p_6 \quad v_6 \quad p_7 \quad v_7 \quad p_8 \quad v_8]^T = [B]_{6\times6}[p_3 \quad v_3 \quad p_4 \quad v_4 \quad p_5 \quad v_5]^T \qquad \ldots (13)$$

$$[p_3 \quad v_3 \quad p_5 \quad v_5]^T = [C]_{4\times4}[p_2 \quad v_2 \quad p_1 \quad v_1]^T \qquad \ldots (14)$$

Boundary conditions, at the end of perforated tubes i.e., at points 8 and 4 are given by [1],

$$v_8 = \frac{-j\tan(k_0 l_{2a})}{Y_8} p_8 \text{ and } v_4 = \frac{j\tan(k_0 l_{2b})}{Y_4} p_4 \qquad \ldots (15, 16)$$

Similarly, boundary conditions, at the ends of bigger duct, i.e., at points 2 and 9 are given by,

$$v_2 = \frac{j\tan(k_0 l_{1b})}{Y_2} p_2 \text{ and } v_9 = \frac{-j\tan(k_0 l_{1a})}{Y_2} p_9 \qquad \ldots (17, 18)$$

Using boundary conditions (15) and (16), state variables v_8 and v_4 are eliminated from Eq. (13) and we get,

$$[D]_{6\times5}[p_6 \quad v_6 \quad p_7 \quad v_7 \quad p_8]^T_{5\times1} = [E]_{6\times5}[p_3 \quad v_3 \quad p_4 \quad p_5 \quad v_5]^T_{5\times1} \qquad \ldots (19)$$

which can be rewritten as,

$$[p_6 \quad v_6 \quad p_7 \quad v_7 \quad p_8 \quad p_4]^T_{6\times1} = [H]_{6\times4}[p_3 \quad v_3 \quad p_5 \quad v_5]^T_{4\times1} \qquad \ldots (20)$$

Taking the first 4 rows (4 equations) of Eq. (20), we get

$$[p_6 \quad v_6 \quad p_7 \quad v_7]^T_{4\times1} = [J]_{4\times4}[p_3 \quad v_3 \quad p_5 \quad v_5]^T_{4\times1} \qquad \ldots (21)$$

Using Eqs. (12) and (14), Eq. (21) can be written as,

$$[p_9 \quad v_9 \quad p_{10} \quad v_{10}]^T_{4\times1} = [K]_{4\times3}[p_2 \quad v_2 \quad p_1 \quad v_1]^T_{3\times1} \qquad \ldots (22)$$

where,

$$[K]_{4\times4} = [A]_{4\times4}[J]_{4\times4}[C]_{4\times4} \qquad \ldots (23)$$

Using Boundary condition (17) and (18), state variables v_2 and v_9 are eliminated from Eq. (23) and we get,

$$[L]_{4\times3}\begin{bmatrix}p_9 & p_{10} & v_{10}\end{bmatrix}^T_{3\times1} = [N]_{4\times3}\begin{bmatrix}p_2 & p_1 & v_1\end{bmatrix}^T_{3\times1} \quad \text{... (24)}$$

To retain only the downstream variables on the right hand side, Eq. (24) can be written as,

$$[O]_{4\times4}\begin{bmatrix}p_9 & p_{10} & v_{10} & p_2\end{bmatrix}^T_{4\times1} = [P]_{4\times2}\begin{bmatrix}p_1 & v_1\end{bmatrix}^T_{2\times1} \quad \text{... (25)}$$

or

$$\begin{bmatrix}p_9 & p_{10} & v_{10} & p_2\end{bmatrix}^T_{4\times1} = [Q]_{4\times2}\begin{bmatrix}p_1 & v_1\end{bmatrix}^T_{2\times1}, \quad [Q] = [O]^{-1} \quad \text{... (26)}$$

Finally, the transfer matrix relation between the upstream point 'u' and the downstream point 'd' can be obtained as,

$$\begin{bmatrix}p_{10} \\ v_{10}\end{bmatrix} = [T]_{2\times2}\begin{bmatrix}p_1 \\ v_1\end{bmatrix} \quad \text{... (27)}$$

Elements of the matrices in Eqs. (19)–(27) can easily be generated by compact statements in MATLAB [5].

RESULTS AND CONCLUSIONS

Figure 3 shows the TL spectra obtained by means of the semi-analytical 'Matrizant approach' presented here and Munjal's classical transfer matrix based muffler program,

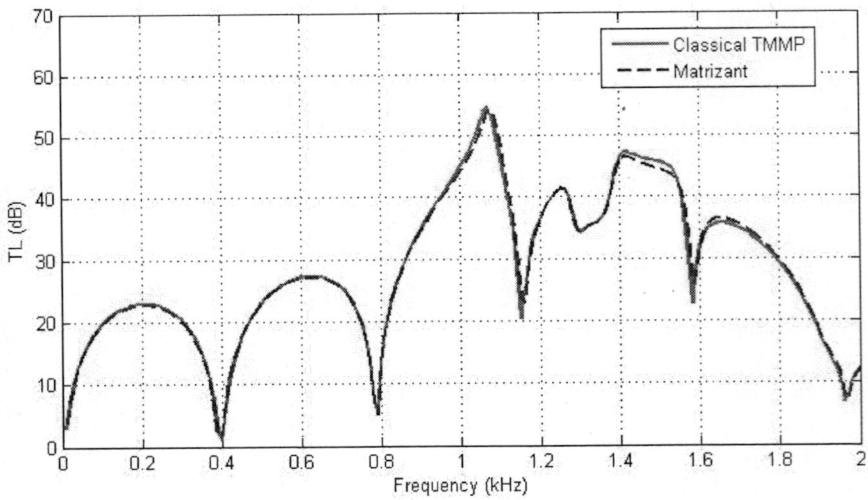

Fig. 3: Transmission Loss Spectra for Cross-Flow Closed-End Extended Perforation Muffler of Figure 2

TMMP. The following data have been used for plotting the curves in Figure 3: $\rho_0 = 1.225$ kg/m^3, $a_0 = 340$ m/s.

Dimensions of the muffler are (see Figure 2): $l_1 = 0.03$m, $l_2 = 0.3$m, $l_3 = 0.035$m, $l_{1a} = 0.02$m, $l_{1b} = 0.05$ m, $l_{2a} = 0.03$m, $l_{2b} = 0.035$m, $r_i = 0.02$m, $r_o = 0.1$m. Figure 3 shows that the TL spectrum obtained by employing the present approach matches very well with that obtained by Munjal's classical TMMP [4,5] thereby validating the Matrizant analysis described in section 4 above.

The matrizant approach along with the algebraic scheme for substituting boundary conditions, thus represents an easy alternative for analyzing the performance of the mufflers, especially for perforated configurations which are very tedious to analyze, as one has to go through long decoupling approach and other complex algebraic manipulations in the classical approach [1,4].

ACKNOWLEDGEMENT

The authors wish to acknowledge financial support of DST for the Facility for Research in Technical Acoustics (FRITA).

REFERENCES

[1] M.L. Munjal, 1987. Acoustics of Ducts and Mufflers, John Wiley, New York.

[2] E. Dokumaci, Matrizant approach to acoustic analysis of perforated multiple pipe mufflers carrying mean flow. *Journal of Sound and Vibration,* 191(4): 505–518, 1996.

[3] Bhushan Singh, M.L. Munjal, Flow Acoustic Analysis of Commercial Automotive Mufflers-Matrizant Approach. *Journal of Acoustical Society of India* (to appear).

[4] M.L. Munjal, Analysis of an Extended-tube Three-pass Perforated Element Muffler by Means of Transfer Matrices, Invited paper at the *International Congress On Sound and Vibration, (ICSV14)*, Cairns, Australia, July 2007.

[5] Bhushan Singh, Flow Acoustic Analysis of Commercial Automotive Mufflers-Matrizant Approach, ME Dissertation, Indian Institute of Science, July 2011.

Section–IX

Acoustics Sensor Signal Processing

Automatic Gear Teeth Defect Localization Using Acoustic and Vibration Signal in Time Domain

Dibya Prakash Jena and S.N. Panigrahi
School of Mechanical Science, Indian Institute of Technology Bhubaneswar
E-mail: psatyan@iitbbs.ac.in

ABSTRACT: *In present work a customized gear mesh test setup is prepared. The acoustic and vibration signals are captured by introducing seeded defects in gear teeth. The bursts have been observed in time domain signals with presence of defects and are focused in analysis. A signal processing scheme is proposed for automatic defect identification and localization of the fault in time domain. Towards this, the proposed scheme initially denoises the signals to improve the sharpness of the burst. The Discrete Wavelet Transform (DWT) has been shown suitable in enhancing the peak of the signal burst. The time frequency information of denoised signals are computed by implementing Complex Morlet Wavelet Transform (CMWT). Furthermore, the instantaneous power spectrum is computed by integrating coefficients in time domain. A multi scale based wavelet peak detection technique is used to extract the peaks of the spectrum in time domain. The proposed technique has been evaluated and validated with captured acoustic and vibration signals. Experimental investigations reveal that the proposed technique can be implemented in real-time condition monitoring and fault diagnosis of gear systems.*

Keywords: Condition Monitoring, Complex Morlet Wavelet Transform (CMWT), Gear Fault.

INTRODUCTION

Gear system is a prime component of rotating machines, power transmissions, automobiles etc. Fault diagnosis and condition monitoring of such system is highly desired for predictive maintenance purpose. The feature of the amplitude modulated–frequency modulated gearbox fault signal usually cannot be directly extracted. Li and Liang have implemented the time-frequency analysis to diagnose gear box fault [1]. In recent years, Li *et al.* investigated mechanical fault diagnosis based on redundant second generation wavelet packet transform, neighborhood rough set and support vector machine. Wang *et al.* have investigated rotary system fault using dual-tree complex wavelet transform [3]. Moreover, one can find many literatures on implementation of DWT, CWT and DCT in non-stationary signal processing [4]. Rafiee *et al.* reported the importance on mother wavelet in wavelet transform in fault diagnosis [5].

From literature, one can find many reports on implementation of various techniques of signal processing but there is no single robust method to localize the signal burst (acoustic and vibration) in time domain to corroborate the defect in gear or bearing. In present work, a signal processing method is proposed based upon Continuous Morlet Wavelet Transform (CMWT) of the denoised signal followed by peak detection of the instantaneous power spectrum generated from scalogram.

PROPOSED TECHNIQUE

It is quite natural, that some time defect signature is carried by wider signal burst depends upon type of signal and system. In such cases, the signals $x(t)$ is decomposed using discrete wavelet transform (DWT). In order to achieve a finer time frequency resolution, the decomposed detail component $D_n(t)$ processed through a continuous wavelet transform (CWT) to produce the two dimensional map of the CWT coefficients. The generated two dimensional maps of the CWT coefficients, $W_\psi D_n(a,b)$, as defined below, correspond to maximum energy zone which helps in identifying the signal bursts due to defects in the piston-bore. Mathematically, the continuous wavelet transform of $x(t)$ is defined as its integral transform with a family of wavelet functions, $\psi_{a,b}(t)$, described as [6,7]:

$$W_\psi D_n(a,b) = CWT(a,b) = \frac{1}{\sqrt{a}} \int_{-\infty}^{+\infty} D_n(t) \dot{\psi}\left(\frac{t-b}{a}\right) dt \text{, where } a>0, b \in \Re, \qquad \dots (1)$$

where, $\dot{\psi}$ is the complex conjugate of the mother wavelet ψ. In present case, the complex Morlet wavelet is used as mother wavelet, i.e., $\psi(t) = (1/\sqrt{f_b \pi})\, e^{-t^2/f_b}\, e^{2\pi f_c t}$. The main advantage of CWT is its ability to provide information simultaneously in time and scale with adaptive windows. Next, the instantaneous power spectrum of the CWT coefficient spectrums is generated. A wavelet based multi-level peak detection technique is investigated to locate the signal burst in time domain. Finally, a simulated signal is generated having sharp peaks at the derived burst location and overlapped with signal $x(t)$ to provide sufficient information in time domain.

EXPERIMENT

Experiments are performed on gear mesh test setup, with driver gear having seeded defect in one tooth as shown in Figures 1(a) and 1(b), respectively. One Behringer® ECM-8000 condenser microphone is mounted near the setup and a PCB® make internal excitation based piezoelectric (ICP) accelerometer with sensitivity of 50mV/g is mounted on bearing bracket to measure the generated acoustic emission and vibration. The acoustic and vibration signals are captured with the help of a NI SCXI-1530® DAQ system through a customized LabVIEW® based application. Experiments are conducted on a gear mesh test setup (the pinion, having 15 teeth, is meshed with driven gear of 30 teeth, and the driver shaft connected to a single phase 50Hz AC motor Crompton® make of power rating of 0.5 HP through a belt drive). The driver shaft operating frequency is maintained at 457 RPM. The faulty driven gear (gear defect type-1) is mounted on the driver shaft and the corresponding acoustic and vibration signals are captured and the signal sample of one second duration are shown in Figures 2(c) and 3(c), respectively.

Fig. 1: Experiment; (a) Gear Defect Simulation Test Setup; (b) Gear Defect Type-1

RESULTS AND DISCUSSION

At first the acquired acoustic signal is investigated with proposed algorithm. It is worth noticing that the signal burst is having prominent width which needs to be enhanced. In order to achieve narrower peak, the acoustic signal $x(t)$ (see Figure 2(c)) is decomposed up to level 3 using discrete wavelet transform (DWT). Experimental investigation reveals that the coefficient (01), i.e. detail from the first level of decomposition $D_1(t)$ is suitable for further diagnosis. It is important to observe that the decomposition should not lose the desired signal attributes due to its down sampling properties. Next, the denoised signal $D_1(t)$ is investigated with Complex Morlet Wavelet Transform (CMWT). The generated scalogram is shown in Figure 2(a). Then, the instantaneous power spectrum is generated from scalogram using a time marginal integration and is shown in Figure 2(b). It is worth noticing that the instantaneous power spectrum is having sharper peak which corroborates well with signal burst due to defect in time domain. Next, a multi-scale wavelet peak detection algorithm is implemented to extract the peak location, as shown in Figure 2(b). A synthetic signal is generated with sharper peak on time domain using extracted peak location and is overlapped with raw acoustic signal. Now, the defect localization is prominently visible without any ambiguity.

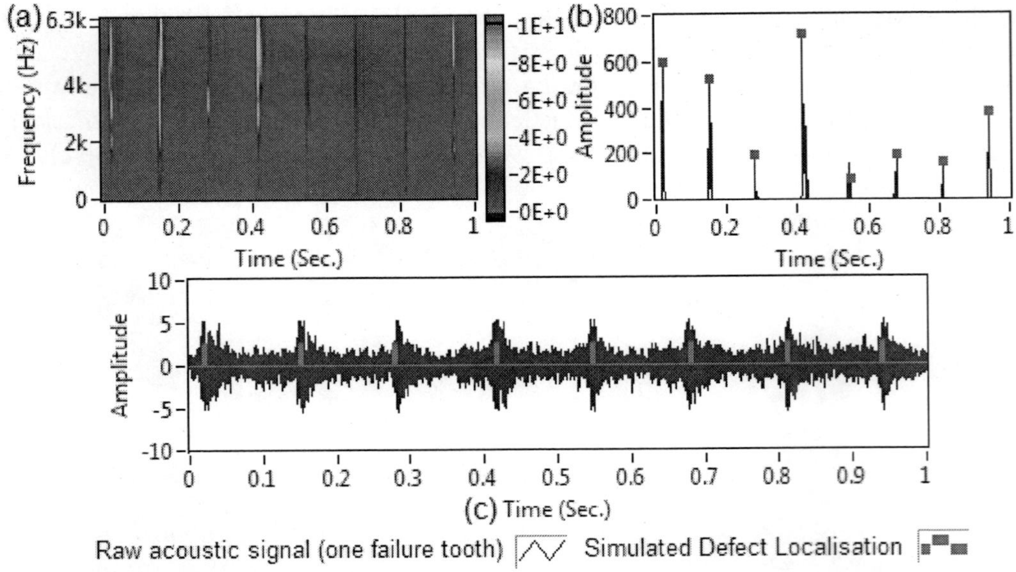

Fig. 2: Defect Localization on Acoustic Signal Using Proposed Technique; (a) Scalogram from CMWT, (b) Instantaneous Power Spectrum from Scalogram, and (c) Simulation of Detected Peak Overlapped with Raw Signal

Subsequently, the acquired vibration signal is investigated with proposed algorithm. It is worth noticing that the signal bursts are sharper compared to acoustic signal, as shown in Figure 3(c). In such cases un-necessary signal processing is not encouraged and can be investigated as it is. In line with proposed method, the vibration signal $x(t)$ is investigated with Complex Morlet Wavelet Transform (CMWT). The generated scalogram is shown in Figure 3(a). Then, the instantaneous power spectrum is generated from scalogram using a time marginal integration and is shown in Figure 3(b). It is worth noticing that the instantaneous power spectrum is having sharper peak which corroborates well with signal burst due to defect in time domain. Next, a multi-scale wavelet peak detection algorithm is implemented to extract the peak location, as shown in Figure 3(b). A synthetic signal is generated with sharper peak on time domain using extracted peak location and is overlapped with raw acoustic signal. Now, the defect localization is prominently visible without any ambiguity. The precise signal burst localization corresponding to defect in time domain is achieved which is not reported by other investigators so far.

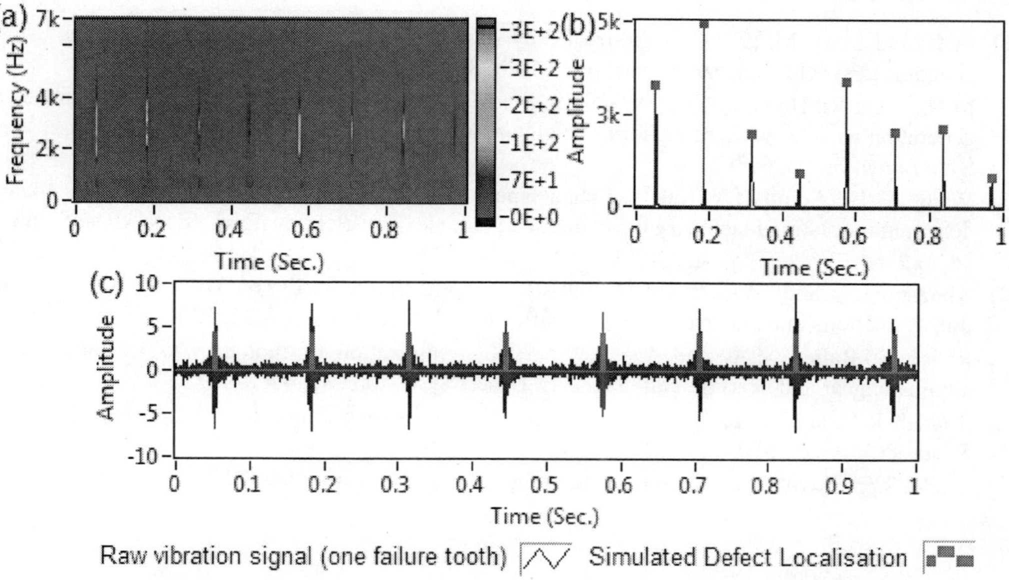

Fig. 3: Defect Localization on Vibration Signal Using Proposed Technique; (a) Scalogram from CMWT, (b) Instantaneous Power Spectrum from Scalogram, and (c) Simulation of Detected Peak Overlapped with Raw Signal

The x-axis (time domain) has been modified with a scale factor of 2742, i.e. (RPM/60) × 360° to convert the time axis into degrees of angular displacement of driver gear. It is worth noticing that the signal bursts are periodic in nature and having 360° periodicity. From experimental investigation, it is observed that acoustic signal inherent with more noise comparative to vibration signal. However, this proposed algorithm highly hinges on the signal and signal burst quality. On the demand of the situation, necessary denoising mechanism needs to be integrated to opt for the proposed method.

CONCLUSION

The experimental analysis presented here established a robust signal processing technique to localize the gear fault from acoustic and vibration signal together. The Continuous Morlet Wavelet Transform (CMWT) provides sufficient time-frequency resolution to analyze non-stationary acoustic and vibration signals. The instantaneous power spectrum generated from scalogram also corroborates well with signal burst, moreover, the peak detection technique localize the signal burst corresponding to defect precisely. Summarizing, the proposed method is suitable and reliable in localizing gear teeth fault in time domain using acoustic or vibration signals.

REFERENCES

[1] Li C. and Liang M. (2012). Time-frequency signal analysis for gearbox fault diagnosis using a generalized synchrosqueezing transform. Mech. Syst. Signal Pr. 26, 205–17.

[2] Li N., Zhou R., Hu Q. and Liu X. (2012). Mechanical fault diagnosis based on redundant second generation wavelet packet transform, neighborhood rough set and support vector machine. Mech. Syst. Signal Pr. 28, 608–21.

[3] Wang Y., He Z. and Zi Y. (2010). Enhancement of signal denoising and multiple fault signatures detecting in rotating machinery using dual-tree complex wavelet transform. Mech. Syst. Signal Pr. 24, 119–37.

[4] Khorrami H. and Moavenian M. (2010). A comparative study of DWT, CWT and DCT transformations in ECG arrhythmias classification. Expert Syst. Appl. 37, 5751–57.

[5] Rafiee J., Rafiee M.A. and Tse P.W. (2010). Application of mother wavelet functions for automatic gear and bearing fault diagnosis. Expert Syst. Appl. 37, 4568–79.

[6] Randall R.B. and Antoni J. (2011). Rolling element bearing diagnostics-A tutorial. Mech. Syst. Signal Pr. 25, 485–520.

[7] Mallat S., A wavelet tour of signal processing, 1st ed., Academic Press, 2009.

Acoustic Emission Technique for Condition Assessment and Monitoring Structural Integrity of Pressure Vessels and Components during Hydrotesting

T. Jayakumar, C.K. Mukhopadhyay and T.K. Haneef

Metallurgy and Materials Group, Indira Gandhi Centre for Atomic Research, Kalpakkam, Tamil Nadu

E-mail: tjk@igcar.gov.in

ABSTRACT: *Acoustic emission technique (AET) has been used for condition assessment of pressure vessels with artificial defects in the laboratory and for monitoring structural integrity of components. The laboratory studies are helpful for getting better insight into the deformation and fracture aspects in the vessels, towards better utilisation of AET for structural integrity of components.*

INTRODUCTION

Acoustic emission technique (AET), one of the advanced non destructive evaluation (NDE) techniques and possesses unique capabilities for on-line monitoring applications, can be employed for condition assessment and monitoring structural integrity of pressure vessels and components. AET is based on the generation of transient elastic waves by rapid release of energy from localized sources in materials and used for detecting dynamic changes taking place in materials. An AE sensor coupled to a sample undergoing dynamic changes detects a part of the elastic energy that is emitted in the form of elastic waves and gives information about the nature of changes taking place in the sample.

Hydro and pneumatic testing are employed for qualification and in-service assessment of components operating under pressure. While this testing can indicate that there are no through and through leaks, presence of harmful defects which would have grown during the hydrotest, would not be detected. On-line AE monitoring during hydrotesting of components could enable detection of deformation and/or damage to the components under pressure and also detection of growing defects, if any, during the hydrotest. This, in turn, could be used for condition assessment and for monitoring structural integrity of the components during the hydrotest. Working on these lines, AET has been applied for condition assessment of pressure vessels with artificial defects in laboratory and for monitoring structural integrity of components during hydrotesting, the details of which are presented in this paper.

CONDITION ASSESSMENT OF PRESSURE VESSELS WITH ARTIFICIAL DEFECTS

The cylindrical vessels fabricated from AISI type 304 stainless steel pipe of 250 NB and schedule 10 were used. Both the ends of the pipes were closed with sheets of the same material. At centre of one end of the vessels, closure sheets used were of reduced thickness in a circular region of 30 mm dia to represent loss of wall thickness, in order to induce stress concentration and initiate failure at these regions during hydrotesting. The AE generated was recorded using a 150 kHz resonant sensor. Variation of cumulative AE count vs. time for vessel 1 (vessel with reduced thickness 1.8 mm) is shown in Figure 1. The AE is generated from the beginning of the pressurization and increased with time. During hydrotesting, both ends of the vessels increase in diameter, giving rise to the generation of AE. It is known that the generation and motion of dislocations are the main activity giving rise to AE during the deformation processes in austenitic stainless steels [1–3]. The AE generated is also attributed to localized deformation at the defect region. Increase in AE with increasing pressure has been attributed to the initiation and growth of crack in the defect region at higher pressures, which gives rise to AE continuously till failure. The AE activity showed three distinct stages. As compared to first and second stages, the third stage is characterized by higher rate of AE and attributed to crack initiation and growth at the defect region with increasing pressure.

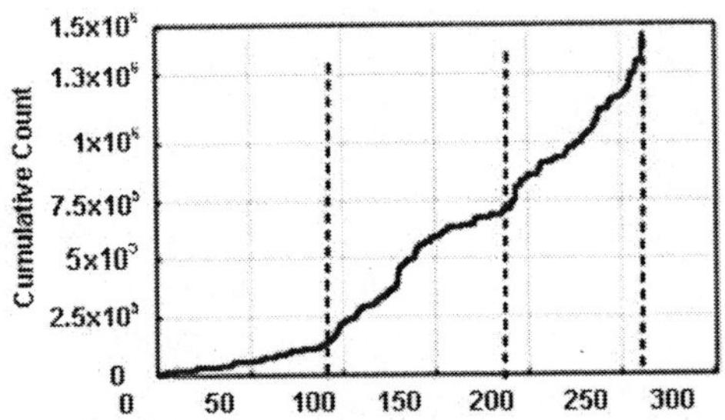

Fig. 1: Cumulative Count vs. Time for Vessel 1

STRUCTURAL INTEGRITY MONITORING OF COMPONENTS DURING HYDROTESTING

AET has been applied for monitoring structural integrity of several components during hydrotesting which include mounded LPG storage vessel; H_2S storage vessels and exchange unit towers of a Heavy Water Plant; and rocket motor casings.

Mounded LPG Storage Vessel

AE monitoring during hydrotesting of a mounded LPG storage vessel was carried out jointly by IGCAR, Kalpakkam and NML, Jamshedpur. During maintenance activities, the vessel got exposed to cryogenic nitrogen which was let in inadvertently while preparing the vessel for recommissioning after a planned statutory outage. The nitrogen supply was stopped immediately. Internal examination revealed the presence of cracks in the shell plate adjacent to the bottom nozzle through which liquid/gas nitrogen was introduced in the vessel. The damaged portion was confined to an area of 740 mm × 1317 mm around the nozzle. The rest of the vessel including all weldments was free from any damage. The vessel was brought back to normal working condition by cutting out the defective portion and by inserting and welding a new plate of same specification. The new weld joints were examined by NDT. Hydrotest of the vessel was recommended. It was also recommended to carry out AE monitoring during hydrotesting to confirm structural integrity of the repaired region. Towards this, AE monitoring alongwith hydrotest of the vessel was carried out. Figure 2 shows the vessel which is closed by hemispherical ends at both the sides. AE monitoring of the weld region (shown by arrow in Figure 2) during hydrotesting up to 18.7 kg/cm^2 with periodic holds at different pressures was done. A repress surising cycle was carried out immediately. Figures 3a–3b show the locations of the AE sources for

Fig. 2: Mounded LPG Storage Vessel

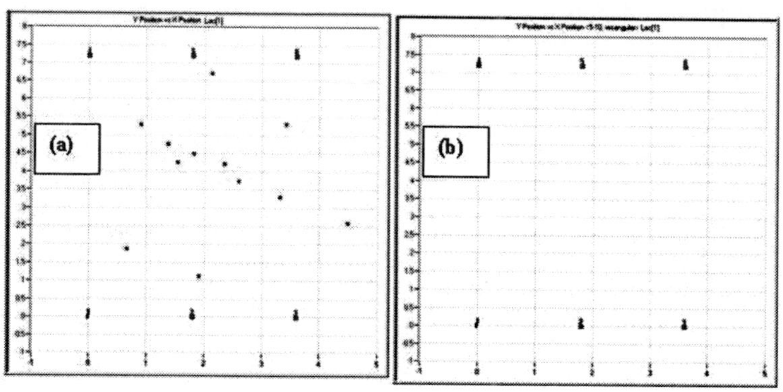

Fig. 3: Location of AE Sources in the Rectangular Region During (a) Pressurization and (b) Repressurization between 12 and 18.7 kg/cm^2

pressurization and repressuriation cycles respectively from 12–18.7 kg/cm^2. During pressurization, the AE is mainly generated at central region of the rectangle due to expansion of the vessel and noise by support structure and also due to microyielding of the vessel material. During repressurising cycle, absence of AE confirmed the absence of growing discontinuity at any locations monitored by AE.

H_2S Storage Vessels and Exchange Unit Towers of a Heavy Water Plant

Point type defect indications were reported during ultrasonic inspection of H_2S storage tanks (Figure 4a) and exchange tower units (Figure 4b) in a heavy water plant. These were thought to be due to loss of wall thickness by localized corrosion. Similar indications found in the virgin plates indicated that these defect indications are manufacturing defects and not service induced corrosion defects. It was decided to carry out AE monitoring during hydrotesting of the vessels in order to ensure that no defect indication is growing during the hydrotesting.

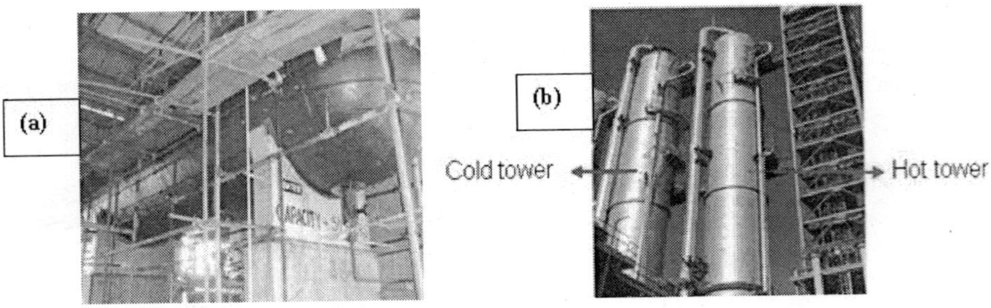

Fig. 4: View of (a) Storage Tank (b) Exchange Towers

The storage tanks are of cylindrical geometry and closed by hemispherical ends at both sides. The hydro test of the tanks was carried out to a maximum pressure of 42.5 kg/cm^2, with holds at different pressures. A reloading cycle from 35 kg/cm^2 to 42.5 kg/cm^2 was carried out immediately. Figure 5 shows the locations of AE sources on the tank for pressure between 20 and 42.5 kg/cm^2 and repressurisation from 35 to 42.5 kg/cm^2 respectively. AE events are generated only during the pressurisation. During repressurisation, AE was absent and this observation confirmed that there is no growing discontinuity at any locations during the hydro testing. The exchange unit is arranged in a 3-stage cascade for handling large quantities of process water and H_2S gas. It consists of three pairs of cold and hot towers operating at 30°C and 120°C respectively. The tower units are of cylindrical geometry with overall height of 51 m and 53 m for the cold and hot towers respectively. The towers are closed by hemispherical ends on top and bottom sides (Figure 4(b)). AE monitoring of the towers showed that AE events are generated only during pressurisation. No AE was observed during repressurization and this confirmed the absence of growing discontinuity in the towers monitored by AE. Ultrasonic testing carried out after hydrotesting for the tanks and the towers did not indicate detectable growth of the defects found earlier.

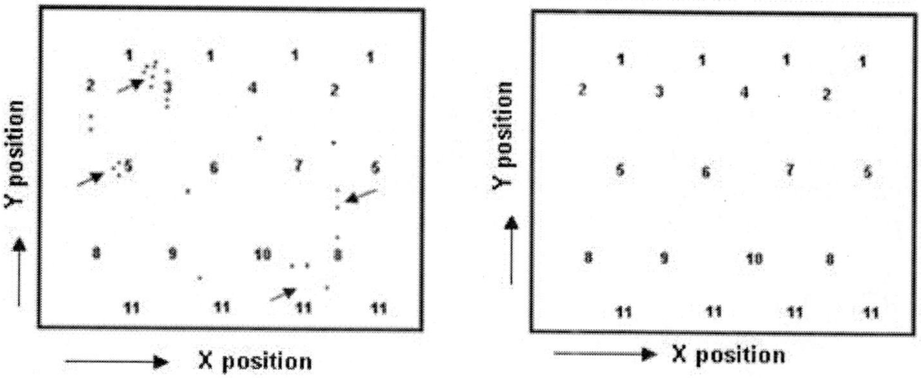

Fig. 5: Location of AE Sources During (a) Pressurization from 20 to 42.5 kg/cm^2 and (b) Repressurization from 35 to 42.5 kg/cm^2

AE Monitoring during Hydro-Testing of a Rocket Motor Casing

AE monitoring during hydrotesting of a rocket motor casing was carried out. A motor casing assembly was subjected to hydro-testing for structural integrity assessment. AE monitoring was carried out for early detection of deformation and nucleation and growth of defects, if any. Twelve sensors were used in triangular configuration to cover the full vessel. The hydro test was carried out to a maximum pressure of 52.5 kg/cm^2, with periodic holds at different pressures. A reloading cycle from 40 kg/cm^2 to 48 kg/cm^2 was also carried out immediately. It was observed that AE signals were generated only during the pressure rise. With increase in pressure, AE signals were generated in newer areas. The AE generated was attributed to local micro-plastic deformation of the material. AE was also generated from specific regions corresponding to joints by mechanical pins, only in a specific pressure range. This pressure range is found to be the pressure required for closure of fine gap between the mating holes and the pins due to expanding of the vessel during hydrotesting, and the consequent rubbing action. During repressurising, AE was absent and this confirmed the absence of growing discontinuity in the vessel during the hydrotesting, thus ensuring the integrity of the casing.

CONCLUSION

Acoustic emission technique has been effectively used for detection of deformation and/or damage and growth of cracks in vessels during hydrotesting in the laboratory. AET has also been applied for structural integrity monitoring of components. Acoustic emission generated during the hydrotesting of the components indicated that the AE signals generated were only due to structural noise and microyielding of the material. The absence of any acoustic emission during the repressurizing cycle confirmed the structural integrity of the components.

ACKNOWLEDGMENTS

Authors are thankful to Mr. S.C. Chetal, Director, IGCAR, Kalpakkam for encouragement and support. The authors thank Dr. B.P.C. Rao, Head, NDED, IGCAR for many useful discussions. The authors are also thankful to NML, Jamshedpur, for participating in the campaign of AE monitoring during hydrotesting of the mounded LPG storage vessel.

REFERENCES

[1] Heiple, C.R. and Carpenter, S.H., *J. Acoustic Emission,* 6, 177–204, 1987.

[2] Raj, Baldev and Jayakumar, T., "Acoustic Emission: Current Practices and Future Directions", ASTM STP 1077, 1990, pp. 218–41.

[3] Mukhopadhyay, C.K., Ray, K.K., Jayakumar, T. and Raj, Baldev, *Mater. Sci. & Engg.* A 255, 98–106, 1998.

Author Index